Mobile Multimedia/Image Processing, Security, and Applications 2009

Sos S. Agaian
Sabah A. Jassim
Editors

14–15 April 2009
Orlando, Florida, United States

Sponsored and Published by
SPIE

Volume 7351

Proceedings of SPIE, 0277-786X, v. 7351

SPIE is an international society advancing an interdisciplinary approach to the science and application of light.

The papers included in this volume were part of the technical conference cited on the cover and title page. Papers were selected and subject to review by the editors and conference program committee. Some conference presentations may not be available for publication. The papers published in these proceedings reflect the work and thoughts of the authors and are published herein as submitted. The publisher is not responsible for the validity of the information or for any outcomes resulting from reliance thereon.

Please use the following format to cite material from this book:
Author(s), "Title of Paper," in *Mobile Multimedia/Image Processing, Security, and Applications 2009*, edited by Sos S. Agaian, Sabah A. Jassim, Proceedings of SPIE Vol. 7351 (SPIE, Bellingham, WA, 2009) Article CID Number.

ISSN 0277-786X
ISBN 9780819476173

Published by
SPIE
P.O. Box 10, Bellingham, Washington 98227-0010 USA
Telephone +1 360 676 3290 (Pacific Time) · Fax +1 360 647 1445
SPIE.org

Printed in the United States of America.

Publication of record for individual papers is online in the SPIE Digital Library.

SPIE
Digital Library
SPIEDigitalLibrary.org

Paper Numbering: Proceedings of SPIE follow an e-First publication model, with papers published first online and then in print and on CD-ROM. Papers are published as they are submitted and meet publication criteria. A unique, consistent, permanent citation identifier (CID) number is assigned to each article at the time of the first publication. Utilization of CIDs allows articles to be fully citable as soon they are published online, and connects the same identifier to all online, print, and electronic versions of the publication. SPIE uses a six-digit CID article numbering system in which:

- The first four digits correspond to the SPIE volume number.
- The last two digits indicate publication order within the volume using a Base 36 numbering system employing both numerals and letters. These two-number sets start with 00, 01, 02, 03, 04, 05, 06, 07, 08, 09, 0A, 0B … 0Z, followed by 10-1Z, 20-2Z, etc.

The CID number appears on each page of the manuscript. The complete citation is used on the first page, and an abbreviated version on subsequent pages. Numbers in the index correspond to the last two digits of the six-digit CID number.

Contents

SESSION 7 POSTER SESSION

Conference Committee

Symposium Chair

Ray O. Johnson, Lockheed Martin Corporation (United States)

Symposium Cochair

Michael T. Eismann, Air Force Research Laboratory (United States)

Conference Chairs

Sos S. Agaian, The University of Texas at San Antonio (United States)
Sabah A. Jassim, University of Buckingham (United Kingdom)

Program Committee

David Akopian, The University of Texas at San Antonio (United States)
Salim Alsharif, University of South Alabama (United States)
Cesar Bandera, BanDeMar Networks (United States)
Chang Wen Chen, Florida Institute of Technology (United States)
Reiner Creutzburg, Fachhochschule Brandenburg (Germany)
Martin Dietze, Consultant (Germany)
Yingzi Du, Indiana University-Purdue University Indianapolis (United States)
Frederic Dufaux, École Polytechnique Fédérale de Lausanne (Switzerland)
Touradj Ebrahimi, École Polytechnique Fédérale de Lausanne (Switzerland)
Erlan H. Feria, College of Staten Island/CUNY (United States)
Phalguni Gupta, Indian Institute of Technology Kanpur (India)
Yo-Ping Huang, National Taipei University of Technology (Taiwan)
Jacques Koreman, Norges Teknisk-Naturvitenskapelige Universitet (Norway)
Maryline Maknavicius, Institut National des Télécommunications (France)
Alessandro Neri, University degli Studi di Roma Tre (Italy)
Gilbert L. Peterson, Air Force Institute of Technology (United States)
Salil Prabhakar, DigitalPersona, Inc. (United States)
Sonia Salicetti, GET/INT (France)
Harin Sellahewa, University of Buckingham (United Kingdom)
Xiyu Shi, University of Surrey (United Kingdom)
Yuri Shukuryan, National Academy of Sciences of Armenia (Armenia)
Gregory B. White, The University of Texas at San Antonio (United States)

Session Chairs

1. Image Enhancement/Restoration Techniques
   Sabah A. Jassim, University of Buckingham (United Kingdom)

2. Networking
   Sabah A. Jassim, University of Buckingham (United Kingdom)

3. Biometrics: Templates and Their Protection I
   Harin Sellahewa, Gray Cancer Institute (United Kingdom)

4. Security of Digital Media and Steganography
   Salim Alsharif, University of South Alabama (United States)

5. Image Quality/Evaluation Measures
   Sos S. Agaian, The University of Texas at San Antonio (United States)

6. Biometrics: Templates and Their Protection II
   Sabah A. Jassim, University of Buckingham (United Kingdom)

Introduction

While rapid technological advances provide new and exciting opportunities for wide ranging applications, they also present the research community with numerous challenges that are exacerbated by growing security concerns. Among the serious security challenges generated are those associated with modern multimedia systems transmitted and exchanged over wireless networks and pervasive computing environments. The main motivation for defence and security research activities is associated with the rapid growth in mass deployment of programmable mobile devices equipped with low-cost, high-resolution digital cameras, sensors, the rise of cybercrime, and identity theft. The most significant challenges in this respect include: efficient and secure processing of image/video, processing suitable for implementation on mobile devices that are constrained in their memory capacities, computational powers, and developing innovative solutions that facilitate the convergence of different wireless technologies (e.g. WiFi and WiMAX).

This year's conference was characterised by high quality research manuscripts presented as full papers or posters which together made significant contributions to meeting some of the challenges listed above. The dominating theme was the development of simple and efficient proactive security solutions for protecting computing infrastructures and sensitive information systems while preserving the privacy of the citizens. Several papers propose novel, secure, and efficient image/video encryption and steganography schemes for mobile devices/environments. Feature detection and extraction in images are dealt with in a number of papers that also propose approaches to improving the accuracy of such schemes. Various aspects of identification schemes are tackled with emphasis on the effect of image quality measures and adaptive human recognition schemes, faces, and irises, as well as mechanisms to protect biometric data. The conference also included number of invited presentations that fit the main issues and concerns raised in the symposium.

The various presentations encouraged quality questions and interactions among the researchers attending the conference.

Sabah Jassim
Sos Agaian

The Design of Wavelets for Image Enhancement and Target Detection

Stephen DelMarco[1], Sos Agaian[2]
[1]BAE Systems, 6 New England Executive Park, Burlington, MA, USA 01803
[2]The University of Texas at San Antonio, One UTSA Circle, San Antonio, TX, USA 78249

ABSTRACT

Detecting dim targets in infrared imagery remains a challenging task. Several techniques exist for detecting bright, high contrast targets such as CFAR detectors, edge detection, and spatial thresholding. However, these approaches often fail for detection of targets with low contrast relative to background clutter. In this paper we exploit the transient capture capability and directional filtering aspect of wavelets to develop a wavelet based image enhancement method. We develop an image representation, using wavelet filtered imagery, which facilitates dim target detection. We further process the wavelet-enhanced imagery using the Michelson visibility operator to perform nonlinear contrast enhancement prior to target detection. We discuss the design of optimal wavelets for use in the image representation. We investigate the effect of wavelet choice on target detection performance, and design wavelets to optimize measures of visual information on the enhanced imagery. We present numerical results demonstrating the effectiveness of the approach for detection of dim targets in real infrared imagery. We compare target detection performance to performance obtained using standard techniques such as edge detection. We also compare performance to target detection performed on imagery enhanced by optimizing visual information measures in the spatial domain. We investigate the stability of the optimal wavelets and detection performance variation, across perspective changes, image frame sample (for frames extracted from infrared video sequences), and image scene content types. We show that the wavelet-based approach can usually detect the targets with fewer false-alarm regions than possible with standard approaches.

Keywords: Wavelet, Target Detection, Infrared Imagery, Image Enhancement, Visibility

1. INTRODUCTION

Despite decades of research, automatic detection of targets in infrared imagery remains a difficult problem. Dim target detection is particularly challenging because standard techniques such as spatial thresholding, CFAR detection, and edge detection can fail due to the lack of contrast between target and background. To detect targets in clutter, there must be a set of characteristics that can be exploited to separate targets from clutter. Separation can be based on characteristics such as scale, shape, texture, pixel value dynamic range, pixel value statistical distribution, spatial frequency, brightness, and contrast differences.

In this paper we develop a wavelet-based approach to target detection. Wavelets have demonstrated some effectiveness for target detection, see for example the survey [1]. Traditionally, there are four primary applications of wavelet-based methods to target detection: These are

1) Wavelets as edge detectors;
2) Using wavelets to separate targets from clutter based on scale differences;
3) Using wavelets as approximate matched filters;

[1] stephen.delmarco@baesystems.com
[2] sos.agaian@utsa.edu

Mobile Multimedia/Image Processing, Security, and Applications 2009, edited by Sos S. Agaian, Sabah A. Jassim,
Proc. of SPIE Vol. 7351, 735103 · © 2009 SPIE · CCC code: 0277-786X/09/$18 · doi: 10.1117/12.816135

4) Capturing target dynamic range differences using wavelet filters.

Using wavelets as edge detectors assumes that target edges differ in some way from clutter edges. For example, edges from natural clutter may be more diffuse whereas edges from man-made objects such as vehicles may be harder, sharper, and more distinct. As edge detectors, wavelets may be designed to capture these edge differences [2],[3]. Targets may also differ from clutter by characteristic scales [4]. For scale separation, *a priori* knowledge of target or clutter characteristic scales may be exploited. Wavelet coefficients containing significant energy at clutter scales (or non-target scales) may be filtered out. Wavelets can be also designed to function as approximate matched filters. For such usage, wavelet filters are designed to produce a large response when matched against a target region [5]. Lastly, the low and highpass filters from the wavelet decomposition can be used to detect target regions of low or high pixel value dynamic range. Dim targets occur in low dynamic range regions; regions of high dynamic range can be rejected.

In this paper, we focus on using wavelets to capture some aspect of target edges, either internal to the target, or between target and background. The goal of this paper is two-fold. First to investigate the feasibility of using wavelets for fast detection of dim targets, and second to present the design of wavelets to optimize target detection performance. The approach we present is based on the single-level, undecimated wavelet transform. Previous work in [6] has indicated that such an approach has shown some utility for target detection. The approach is reasonably computationally efficient and parallelizable, with relatively low memory requirements thus making it potentially suitable for real-time automatic target detection (ATD) applications. Also, the undecimated wavelet transform is a tight frame [7], thus preserving image energy which makes it reasonable to believe that significant image features are not destroyed.

There are several other wavelet transform types that may have superior feature capturing capability, such as curvlets [8], directional directional wavelets [9], [10], steerable pyramids [11], dual-tree complex wavelet transform [12], and quaternionic wavelet transform [13]. However, these approaches are significantly more computationally expensive than the approach described in this paper and may be prohibitive for realtime application.

The rest of this paper is organized as follows. Section 2 presents an overview of wavelet design. Section 3 reviews the Michelson Visibility operator. Section 4 presents the wavelet filtered image representation. Section 5 presents the target detection algorithm. Section 6 presents numerical target detection results, and Section 7 presents conclusions.

2. WAVELET DESIGN OVERVIEW

The lattice decomposition [14] can be used to characterize the class of 2-band FIR wavelet filterbanks in the following way. This class of wavelet filterbanks is a subset of the class of two-channel, real, FIR perfect reconstruction filterbanks [14], which can be characterized by the polyphase components matrix (PCM). The PCM can be factored according to the lattice decomposition. Let $E(z)$ denote the PCM. Then the lattice decomposition consists of the factorization

$$E(z) = \alpha R_m \Lambda(z) R_{m-1} \Lambda(z) \cdots \Lambda(z) R_0 I \tag{1}$$

where α is a constant and

$$R_i = \begin{pmatrix} \cos(\theta_i) & \sin(\theta_i) \\ -\sin(\theta_i) & \cos(\theta_i) \end{pmatrix}, \qquad \Lambda(z) = \begin{pmatrix} 1 & 0 \\ 0 & z^{-1} \end{pmatrix}, \qquad I = \begin{pmatrix} 1 & 0 \\ 0 & \pm 1 \end{pmatrix}$$

The parameters θ_i, called lattice angles, therefore govern the design of this general class of filterbanks. The class of wavelet filterbanks results from imposition of the d.c. constraint on the lattice angles:

$$\sum \theta_i = -\frac{\pi}{4},$$

where we chose $\alpha = 1$ and $I(2,2) = -1$.

For a small number of stages in (1), the expression may be expanded, and a closed form solution of the wavelet filter taps may be obtained in closed form. Wavelets may be designed by numerically sweeping over the lattice angles and evaluating the closed form expression.

The lattice decomposition [14] has been used before for wavelet design, for example in [12] for complex dual-tree wavelet design and in [15] for radar signal detection.

3. MICHELSON VISIBILITY OPERATOR

The Michelson visibility operator [16],[17] has been used to quantify the strength of interference fringes, as depicted in Figure 1.

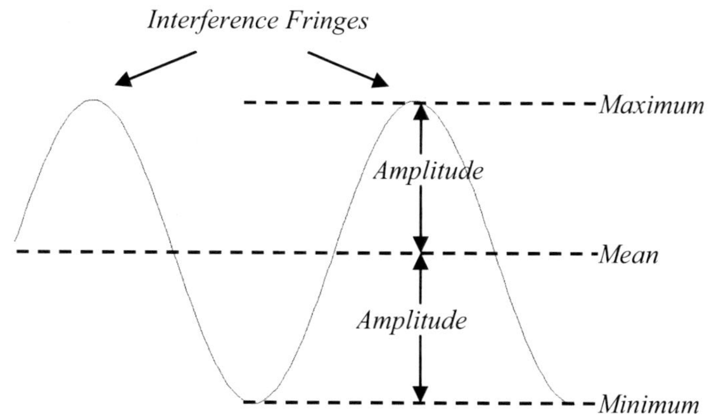

Figure 1. Depiction of Michelson Visibility Operator

The visibility image is created from an input image by extending the Michelson visibility operator to two dimensions. Let I denote an image containing M rows and N columns, and let $I(m,n)$ denote the pixel value at row and column coordinates , respectively (m,n) with $0 \le m \le M-1$, $0 \le n \le N-1$. To create the visibility image I_V, a window is placed about each pixel in the input image. Within this window, the maximum pixel value I_{MAX} and the minimum pixel value I_{MIN} are calculated. The center pixel is assigned the visibility value

$$I_V(m,n) = \frac{I_{MAX} - I_{MIN}}{I_{MAX} + I_{MIN}} \tag{2}$$

There are a number of extensions that can be made to the visibility operator in (2). These include a second-derivative-like visibility operator, and various forms of local visibility operators:

$$I_V(m,n) = \frac{I_{MAX} - 2I(m,n) + I_{MIN}}{I_{MAX} + 2I(m,n) + I_{MIN}}, \qquad I_V(m,n) = \sqrt{\left(\frac{I(m,n) - I_{MIN}}{I_{MAX} + I_{MIN}}\right)\left(\frac{I_{MAX} - I(m,n)}{I_{MAX} + I_{MIN}}\right)},$$

$$I_V(m,n) = \sqrt{\left(\frac{I(m,n) - I_{MIN}}{I(m,n) + I_{MIN}}\right)\left(\frac{I_{MAX} - I(m,n)}{I(m,n) + I_{MIN}}\right)}, \qquad I_V(m,n) = \sqrt{\left(\frac{I(m,n) - I_{MIN}}{I(m,n) + I_{MIN}}\right)\left(\frac{I_{MAX} - I(m,n)}{I(m,n) + I_{MAX}}\right)}.$$

These have been used for target detection with varying degrees of success.

4. WAVELET FILTERED IMAGE REPRESENTATION

The single-level 2D wavelet transform decomposes an image into four subbands which capture directional features. A vertical band captures vertical features, a horizontal band captures horizontal features, the LL band performs lowpass filtering in both horizontal and vertical directions, and the HH band, which performs highpass filtering in both directions captures diagonal features.

The wavelet filtered image representation I_{WF} that we use herein consists of performing the single-level, un-decimated wavelet transform on the image to generate overcomplete subbands, followed by combining the highpass subbands using a weighted sum.

$$I_{WF} = \alpha_{VER} I_{VER} + \alpha_{HOR} I_{HOR} + \alpha_{HH} I_{HH}. \quad (3)$$

Typically we take $\alpha_{HOR} = \alpha_{VER} = 1.0$ and $\alpha_{HH} = 0.0$.

The representation may be viewed as an image fusion approach where multiple subbands are fused via a weighted sum. Alternatively, the representation may be viewed as a single filtering operation where the wavelet subband filters are combined into a composite filter.

5. TARGET DETECTION ALGORITHM

The target detection algorithm consists of the following process (Figure 2). The wavelet filtering is applied to the input image to generate the representation in (3). Subsequently, the visibility image is generated. The Visibility image undergoes spatial thresholding to produce a set of candidate target pixels. The candidate target pixel locations are used as seeds in a region-growing approach to generate regions-of-interest containing candidate targets. A region growing operation, such as morphological dilation, can be used to reclaim additional target pixels. Metadata information, such as camera angles, range, etc. along with prior information, e.g., expected target sizes can be used to select dilation element size. Other *a priori* information such as road networks, maps, context information may be used for additional false-alarm suppression. Candidate target regions are dispatched to a recognition engine for ATR purposes.

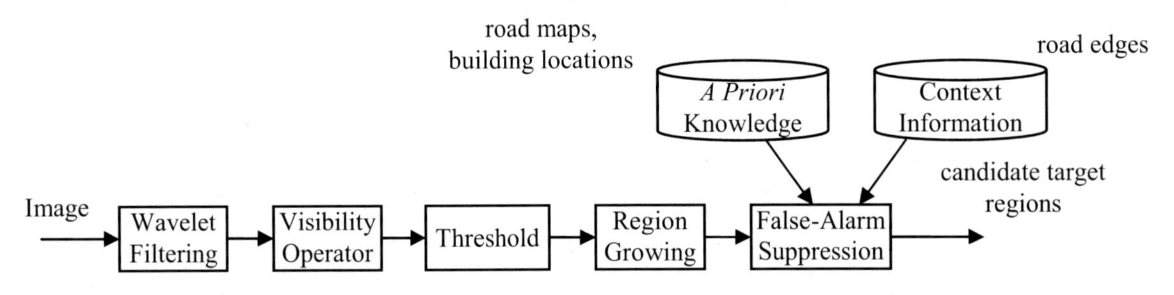

Figure 2: Target Detection Algorithm Process Flow

The detection algorithm is fast; the wavelet filtered image is a four-by-four mask filtering operation and the Visibility image generation generally uses a three-by-three window. Furthermore, the algorithm is highly parallelizable, thereby offering potentially significant runtime reductions.

6. NUMERICAL RESULTS

In this section we present numerical results demonstrating the performance of wavelet-based detection of dim targets in infrared imagery. We use imagery from the VIVID public release data set. Figure 3 contains a set of sample frames used to generate numerical results.

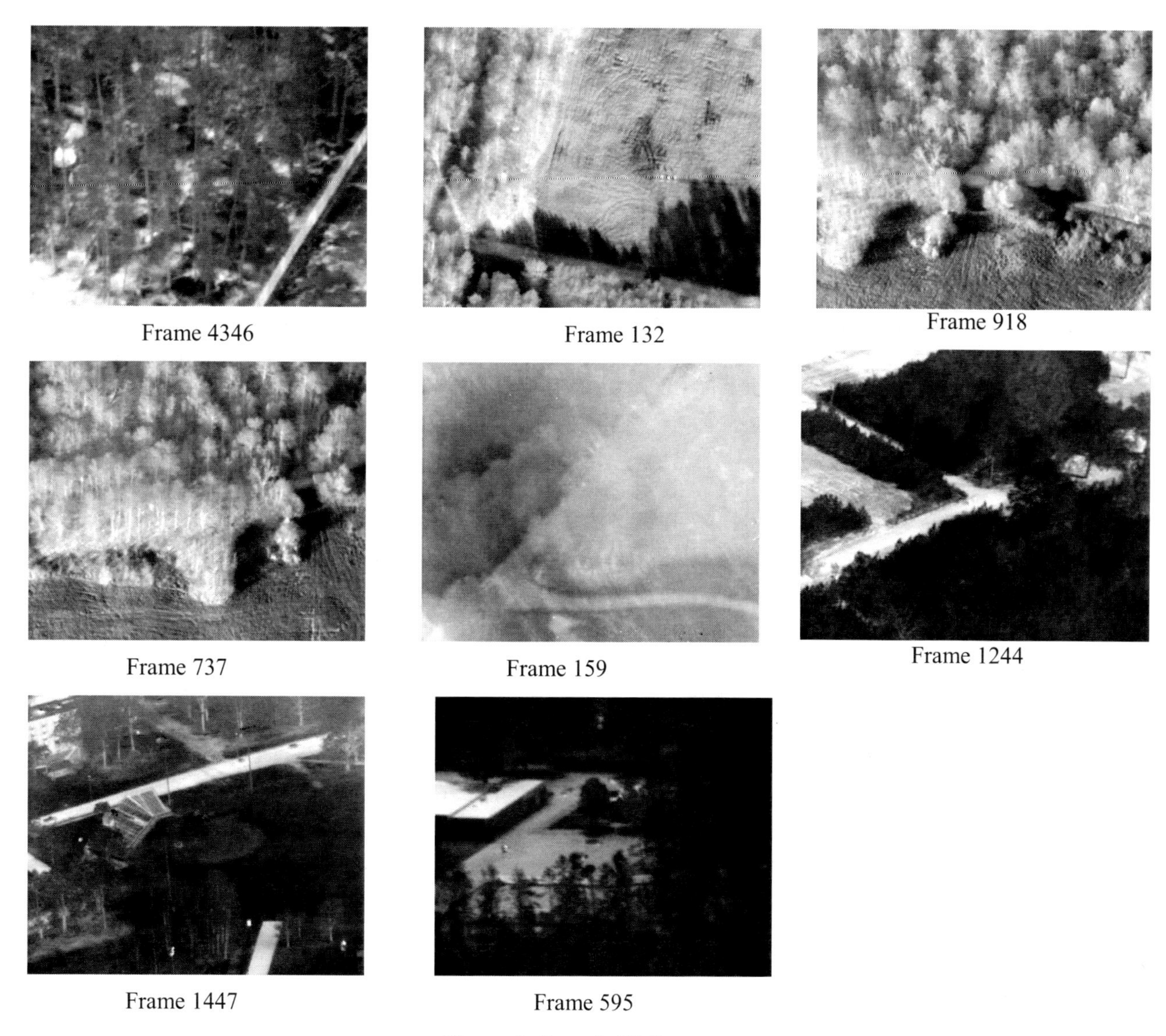

Figure 3: Sample IR Imagery

6.1 Examples of Wavelet Filtered Imagery

In this section we illustrate the effect of different wavelets on the filtered image representation. Figure 4 contains the wavelet filtered image representation of frame 4346 for nine different wavelets.

Figure 4 shows how different wavelets provide different visual representations of the image; from embossed-like visualizations to etched-like representations.

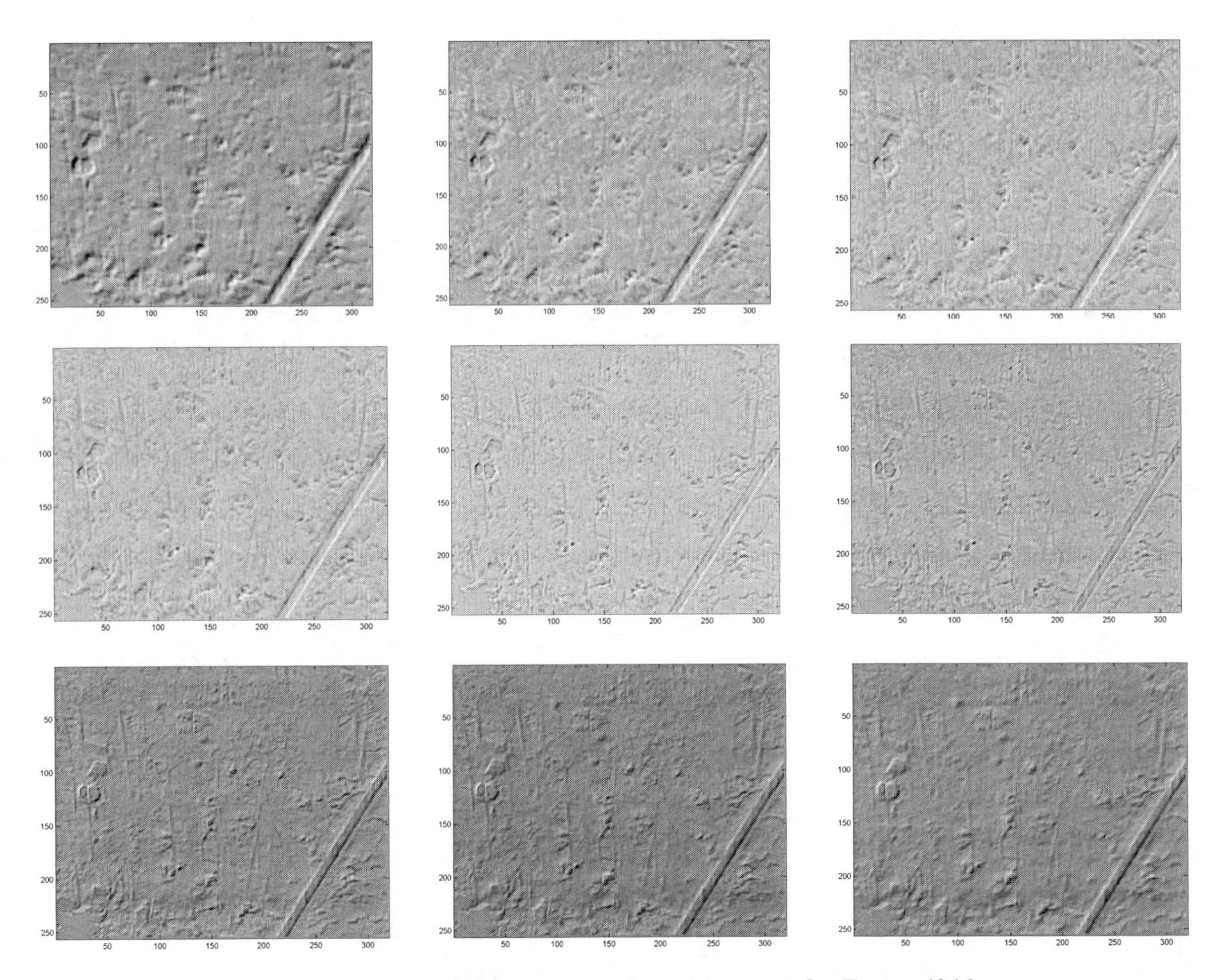

Figure 4: Examples of Wavelet Filtered Imagery for Frame 4346

6.2 Examples of Target Detection

In this section we present two examples of target detection using the detection algorithm. Figure 5 contains frame 737 and frame 4346. In each frame the highlighted area in the circle contains the target. For frame 737, the target consists of a vehicle traveling along a road and partially obscured by trees. For frame 4346, the target-of-interest consists of the dismount near the very bright vehicle. Figure 6 presents intermediate results of the detection algorithm for frame 737. The wavelet filtered image is generated (Figure 6(a)) upon which the Visibility image generation operates to create the Visibility image in Figure 6(b). The Visibility image is spatially thresholded producing the pixel survivors in Figure 6(c). A simple morphological dilation is performed to generate the candidate target detections as shown in Figure 6(d). The wavelet used to generate the detection results was chosen by sweeping over the single independent lattice angle, as discussed in Section 2, and choosing the value which provides minimum entropy over the wavelet filtered image.

The same processing is performed in Figure 7 for frame 4346. Note that Figure 6(d) only generates four false-alarm regions. Figure 7(d) shows that the dismount has been captured.

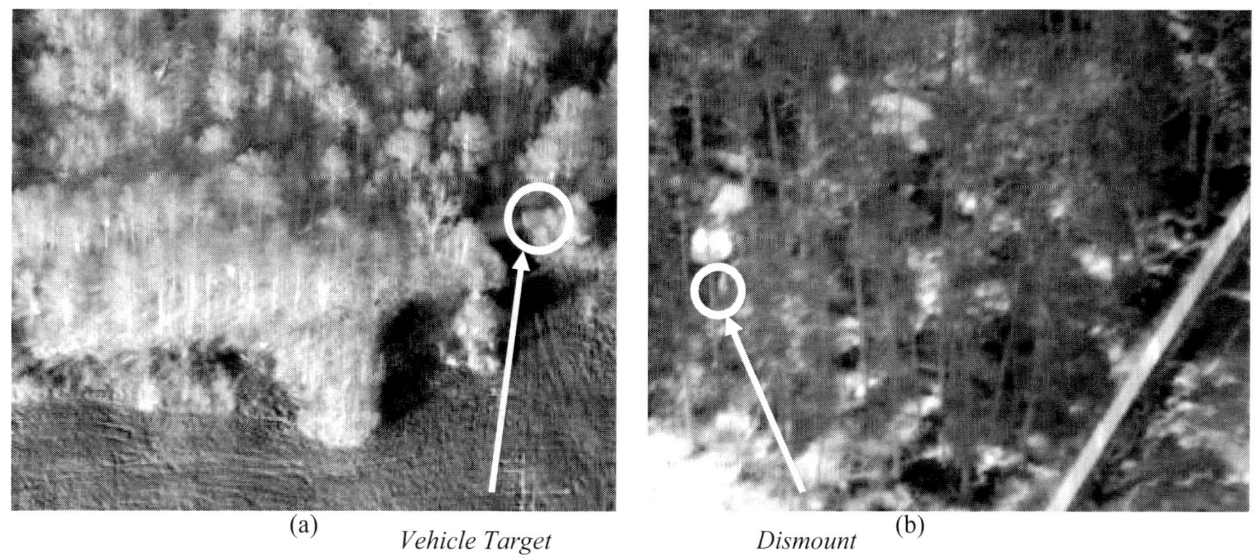

Figure 5: Targets in Image Fames; (a) Vehicle in Frame 737; (b) Dismount in Frame 4346

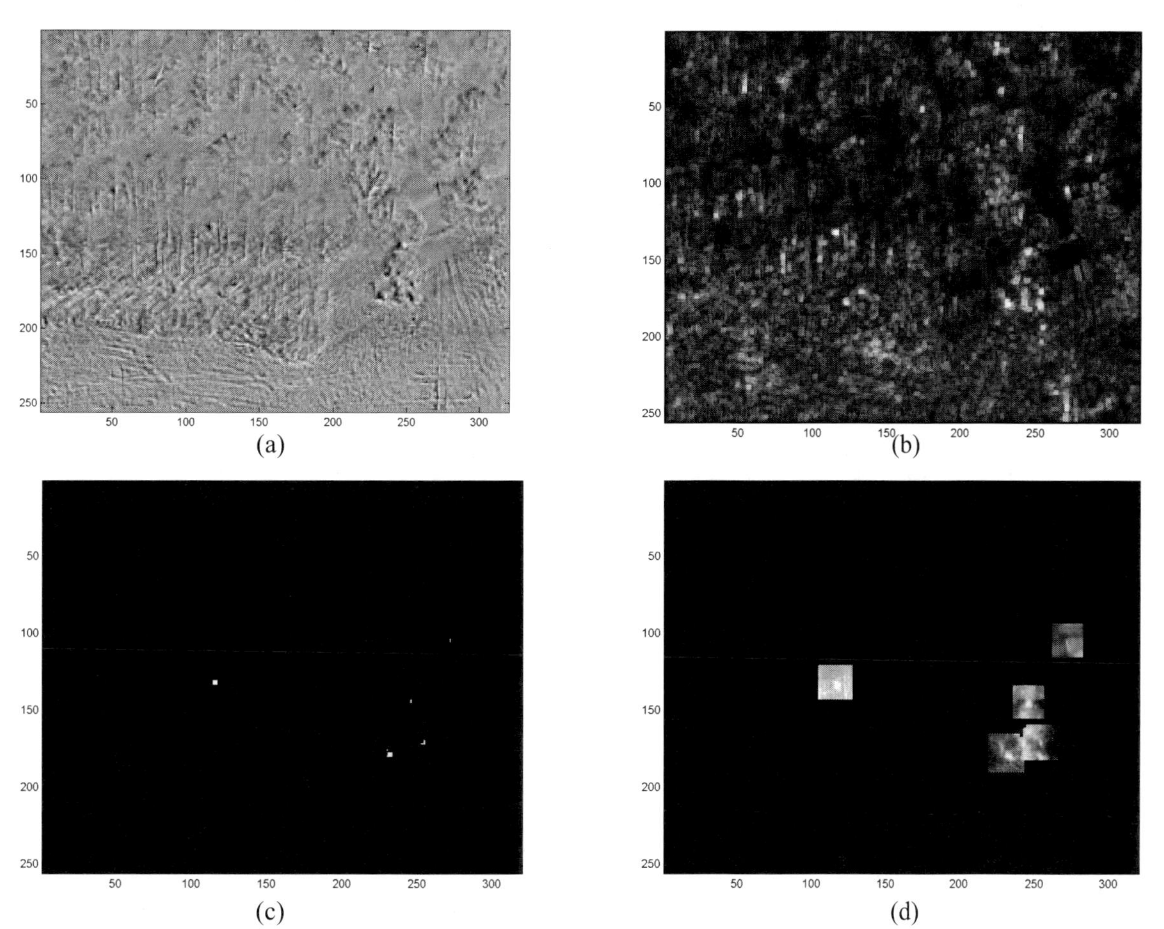

Figure 6: Target Detection for Frame 737; (a) Wavelet Filtered Image; (b) Visibility Image; (c) After Spatial Thresholding; (d) Region Grown

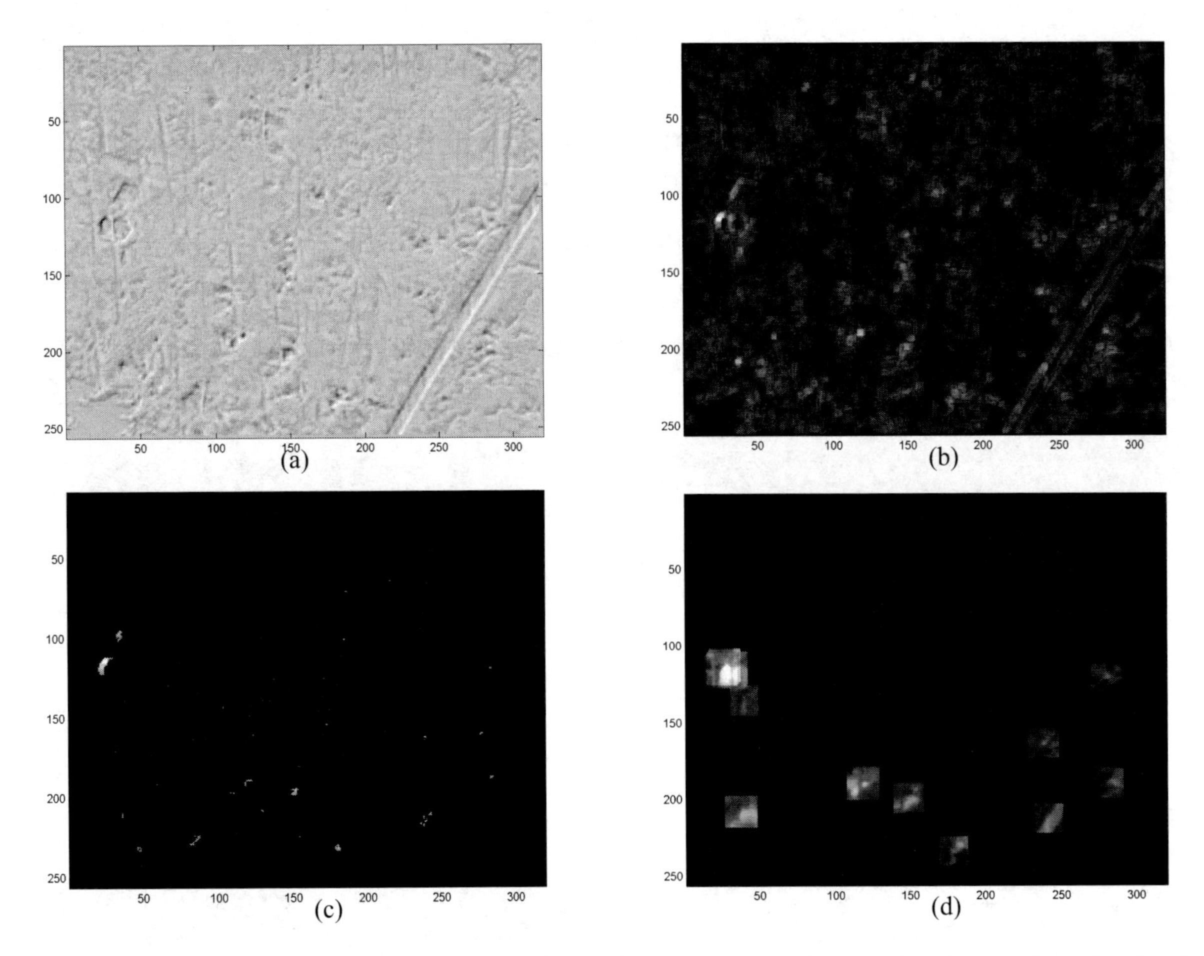

Figure 7: Target Detection for Frame 4346; (a) Wavelet Filtered Image; (b) Visibility Image; (c) After Spatial Thresholding; (d) Region Grown

6.3 Target Detection as a Function of Wavelet

In this section we generate optimal wavelets by optimizing several objective functions over the imagery. We present theoretical limits on example imagery of number of false-alarm regions by using optimal spatial thresholds. The optimal thresholds are determined interactively by varying the threshold until the last target pixel disappears. The optimal threshold, per image, is the largest spatial threshold which generated images containing target pixels. The region growing is performed, and the number of false-alarm regions is counted. Because there currently is no principled way to determine the optimum threshold *a priori*, we also present a sample case in which the spatial threshold is progressively reduced, to determine the rate of false-alarm region increase.

We apply the approach outlined in sections 4 and 5, and investigate use of different wavelet optimization criterion and baseline approaches, labeled as follows:

- *Visibility*: choose the wavelet which produces the minimum entropy of the visibility image;
- *Minimum Wavelet Entropy*: choose the wavelet which produces the minimum wavelet filtered image entropy;
- *Sobel Visibility*: calculate the maximum visibility of the sobel edge-detected image;
- *Minimum Visibility*: choose the wavelet which provides the maximum minimum visibility image value;
- *Mean Visibility*: choose the wavelet which provides the maximum mean visibility value of the wavelet filtered image;

- *Sobel*: standard sobel edge-detected image;
- *Image Visibility*: calculate the maximum image visibility value, for the visibility image calculated directly from the input image, without wavelet filtering.

We compare the wavelet-based approaches with an approach based on the standard sobel edge detector, and with a spatial domain approach in which the visibility images operates directly on the spatial image. Despite the fact that there are more advanced edge detectors, we use the sobel edge detector because the wavelet filtering has a similar level of computational complexity. More complex edge detectors, such as the Canny edge detector, may be used but these are usually more computationally expensive. Table 1 contains the number of false-alarm regions plus a single target region as a function of each approach, for each sample frame. The region count was calculated as the number of connected components, using a connected components algorithm.

	Visibility	Minimum Wavelet Entropy	Sobel Visibility	Minimum Visibility	Mean Visibility	Sobel	Image Visibility
Frame 4346	11	21	26	28	37	46	125
Frame 132	11	6	98	6	6	14	248
Frame 918	1	1	9	1	1	1	84
Frame 737	54	6	21	13	177	138	92
Frame 159	43	31	159	30	35	34	31
Frame 1244	53	12	146	15	138	109	too many
Frame 1447	127	80	490	70	71	150	152
Frame 595	37	60	376	52	62	111	219

Table 1: Number of False-Alarm Regions Plus One Target Across Frame for Different Detection Approaches

Table 1 indicates that the Visibility, Minimum Wavelet Entropy, and the Minimum Visibility approaches provide the best performance as measured by lowest number of false-alarm regions. Figure 8 demonstrates the effect of reducing the spatial threshold value on the false-alarm region rate for frame 737.

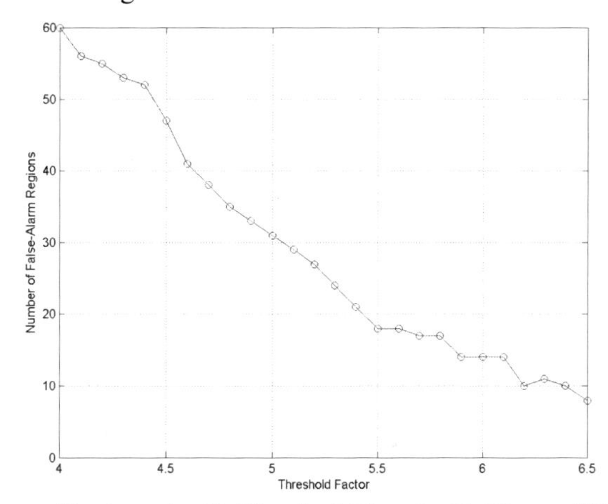

Figure 8: Number of False-Alarm Regions for Pullbacked Threshold (Frame 737, Minimum Entropy Wavelet) Showing Approximately Linear Increase

Figure 8 demonstrates that the false-alarm regions increase approximately linearly with reduced spatial threshold.

6.4 Stability of Optimal Wavelet Across Image Realization

In this section we examine the stability of the optimal wavelet choice across image frame realization. Figure 9 contains several consecutive frames. Table 2 contains number of false-alarm regions for each frame realization for three of the detection approaches.

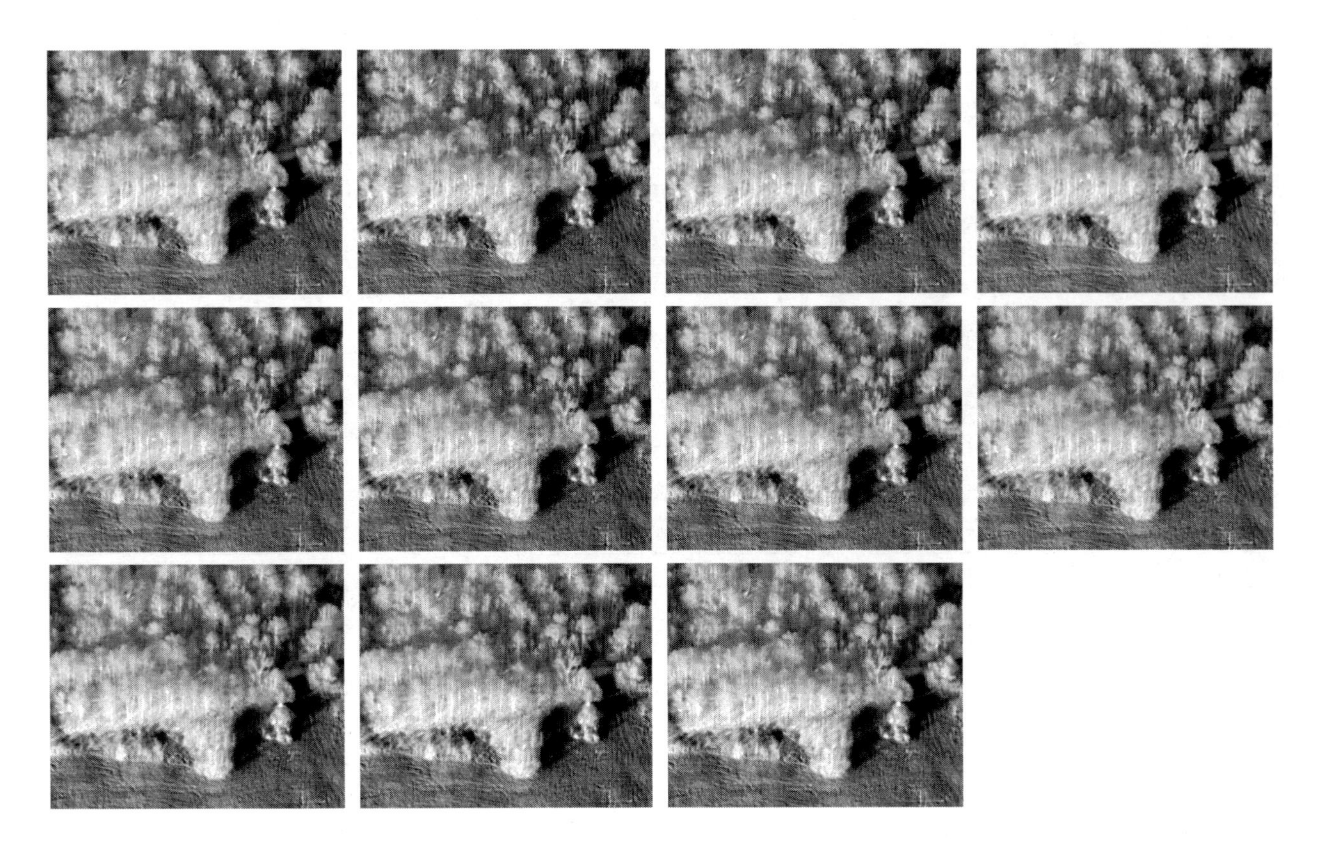

Figure 9: Consecutive Frames: Frame 737 Through 747

	Minimum Wavelet Entropy	Sobel Visibility	Sobel
Frame 738	28	232	90
Frame 739	19	115	101
Frame 740	99	203	64
Frame 741	70	16	68
Frame 742	14	258	101
Frame 743	19	359	79
Frame 744	9	42	77
Frame 745	14	305	127
Frame 746	24	904	98
Frame 747	3	314	147

Table 2: Number of False-Alarm Regions Plus One Target Region For Different Image Realizations

Table 2 indicates that the number of false-alarm regions does vary with image realization. There is sensitivity to frame realization that must be stabilized. However, the minimum entropy approach provides the best performance in Table 2.

6.5 Stability of Optimal Wavelet Across Perspective Change

In this section we examine the stability of the optimal wavelet choice across perspective change. Figure 10 contains example frames containing the same basic scene, but from different perspectives.

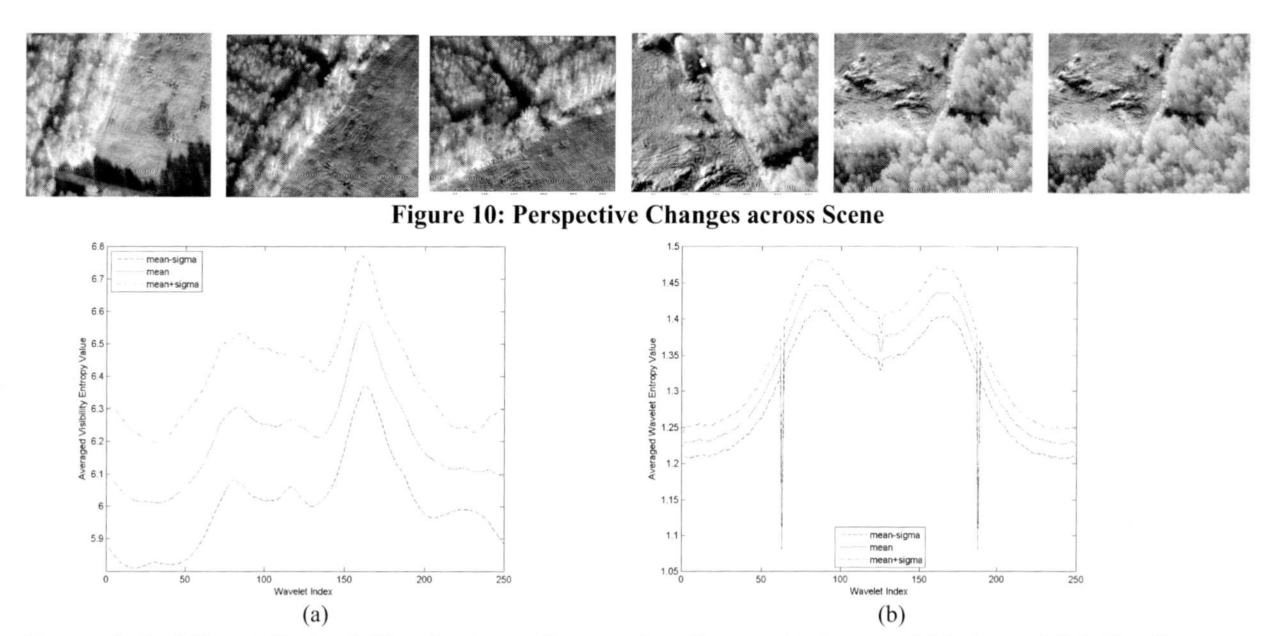

Figure 10: Perspective Changes across Scene

(a) (b)

Figure 11: Stability of Optimal Wavelet Across Perspective Change: (a) Averaged Minimum Visibility Entropy; (b) Averaged Minimum Wavelet Entropy

Figure 11 contains the mean Minimum Visibility Entropy and mean Minimum Wavelet Entropy averaged over the different perspective cases. The one standard deviation lines are also shown. Figure 11 indicates that the optimal wavelet, which displays the lowest entropy, is stable across perspective change.

6.6 Target Detection as a Function of Wavelet Using Global CFAR Approach

In this section we use a CFAR criterion for target detection and wavelet optimization. For each test frame, we perform the wavelet filtering and generate the visibility image. For each frame, we interactively determine the *a priori* target location within the visibility image. We then calculate a target CFAR threshold as the ratio of the maximum visibility

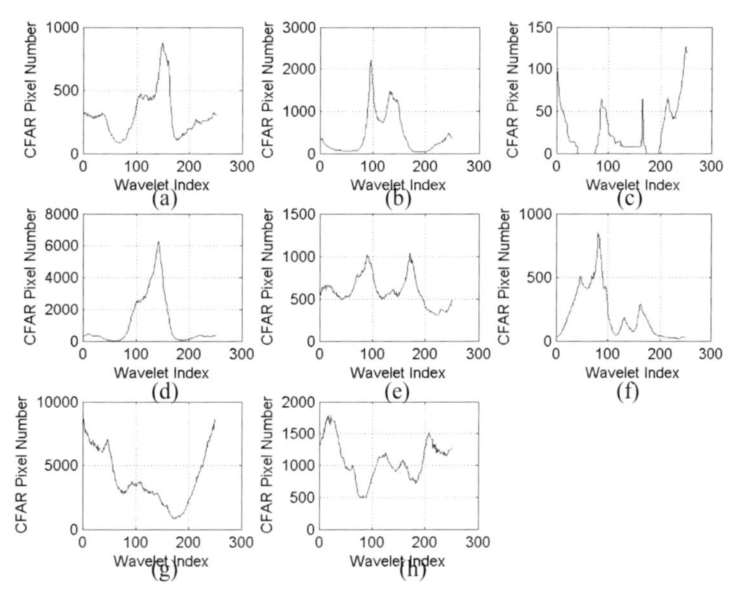

Figure 12: Optimal Wavelet for CFAR Optimality Criterion: (a) Frame 4346; (b) Frame 132; (c) Frame 918; (d) Frame 737; (e) Frame 159; (f) Frame 1244; (g) Frame 1447; (h) Frame 595

value in a small window containing the target, to the mean visibility value over the entire image. For each pixel outside of the target window, we calculate a CFAR value at that pixel as the ratio of the pixel visibility value to the mean image visibility. We count the number of pixels values that surpass the CFAR threshold, most of which correspond to false alarm pixels. The optimal wavelet is chosen to minimize the number of survivor pixels. The number of survivor pixels for each frame as a function of wavelet index are shown in Figure 12. Figure 12 indicates a large variability across wavelet type. Also, the optimal wavelet appears to be scene-content dependent.

7. CONCLUSIONS

This paper investigated the feasibility of using wavelets for detection of dim targets in infrared imagery. An approach was presented using fusion of un-decimated wavelet filters, together with further enhancement using the Michelson Visibility operator. Examples were presented showing that the approach can produce candidate target detections with manageable false-alarm region counts. However, some undesirable sensitivity to image frame was displayed. Also the optimal wavelet choice was seen to be dependent on scene content and wavelet optimization criterion, thus necessitating use of some form of dynamic selection algorithm for optimal wavelet choice. The approach can be generalized in several ways, through use of M – band wavelets, more stages in the lattice decomposition, more general paraunitary filterbanks, multi-wavelets, etc.. Also, other metrics could be used for defining the optimal wavelet. These are subjects of future work.

REFERENCES

1. E. Abdelkawy, and D. McGaughy, "Wavelet-Based Image Target Detection Methods", *Proc. SPIE,* Vol. 5094, 2003.
2. L. Li, and Y. Tang, "Wavelet-Hough Transform with Applications in Edge and Target Detection", *Int. Journ. Wavelets, Multiresolution, and Inform Proc.,* Vol. 4, no. 3, pp. 567-587, 2006.
3. F. Song and S. Jutamulia, "New Wavelet Transforms for Noise-Insensitive Edge Detection", *Opt. Eng.,* Vol. 41, no. 1, pp. 50-54, Jan. 2002.
4. D. J. Gregoris, S. K. W. Yu, and S. Tritchew, "Wavelet Transform-Based Filtering for the Enhancement of Dim Targets in FLIR images", *Proc. SPIE,* Vol. 2242, 1994.
5. D. Casasent, "Detection Filters and Algorithm Fusion for ATR", *IEEE Trans. Image Processing*, Vol. 6, no. 1, pp. 114-125, Jan. 1997.
6. S. Agaian, "Visual Morphology", *Proc. SPIE,* Vol. 3646, 1999.
7. M. Unser, "Texture Classification and Segmentation Using Wavelet Frames", *IEEE Trans. Image Processing*, Vol. 4, no. 11, pp. 1549-1560, Nov. 1995.
8. J.-L., Starck, F. Murtagh, E.J. Candes, D. L. Donoho, "Gray and Color Image Contrast Enhancement by the Curvlet Transform", *IEEE Trans. Image Processing*, Vol. 12, no. 6, pp. 706-717, Jun. 2003
9. V. Velisavljevic, P. L. Dragotti, and M. Vetterli, "Driectional Wavelet Transforms and Frames", *Proc. ICIP,* Vol. 3, pp. 589-592, 2002.
10. R. Zuidwijk, "Directional and Time-Scale Wavelet Analysis", *SIAM J. Math. Anal.,* Vol. 31, No. 2, pp. 416-430, 2000.
11. E. P. Simoncelli, W. T. Freeman, E. H. Adelson, and D. J. Heeger, "Shiftable Multiscale Transforms", *IEEE Trans. Inform. Theory*, Vol. 38, no. 2, pp. 587-607, Mar. 1992.
12. N. Kingsbury, "A Dual-Tree Complex Wavelet Transform with Improved Orthogonality and Symmetry Properties", *Proc. ICIP,* Vol. 2, pp. 375-378, 2000.
13. W. L. Chan, H. Choi, and R. G. Baraniuk, "Coherent Image Processing Using Quaternion Wavelets", *Proc. SPIE,* Vol. 5914, 2005.
14. P. P. Vaidyanathan, *Multirate Systems and Filter Banks*, New Jersey: Prentice-Hall, 1993.
15. S. Del Marco, "Use of the Wavelet Transform for Improved CFAR Detection in CW Radar Seekers", *Proc. SPIE,* Vol. 2491, 1995.
16. J. Strong, and G. A. Vanasse, "Interferometric Spectroscopy in the Far Infrared", *J. Opt. Soc. Amer..,* Vol. 49, no. 9, Sept. 1959.
17. E. F. Erickson, and R.. M. Brown, "Calculation of Fringe Visibility in a Laser-Illuminated Interferometer", *J. Opt. Soc. Amer..,* Vol. 57, no. 3, Mar. 1967.

Multi-view video segmentation and tracking for video surveillance

Gelareh Mohammadi, Frederic Dufaux*, Thien Ha Minh, Touradj Ebrahimi

Multimedia Signal Processing Group
Ecole Polytechnique Fédérale de Lausanne (EPFL)
CH-1015 Lausanne, Switzerland

ABSTRACT

Tracking moving objects is a critical step for smart video surveillance systems. Despite the complexity increase, multiple camera systems exhibit the undoubted advantages of covering wide areas and handling the occurrence of occlusions by exploiting the different viewpoints. The technical problems in multiple camera systems are several: installation, calibration, objects matching, switching, data fusion, and occlusion handling. In this paper, we address the issue of tracking moving objects in an environment covered by multiple un-calibrated cameras with overlapping fields of view, typical of most surveillance setups. Our main objective is to create a framework that can be used to integrate object-tracking information from multiple video sources. Basically, the proposed technique consists of the following steps. We first perform a single-view tracking algorithm on each camera view, and then apply a consistent object labeling algorithm on all views. In the next step, we verify objects in each view separately for inconsistencies. Correspondent objects are extracted through a Homography transform from one view to the other and vice versa. Having found the correspondent objects of different views, we partition each object into homogeneous regions. In the last step, we apply the Homography transform to find the region map of first view in the second view and vice versa. For each region (in the main frame and mapped frame) a set of descriptors are extracted to find the best match between two views based on region descriptors similarity. This method is able to deal with multiple objects. Track management issues such as occlusion, appearance and disappearance of objects are resolved using information from all views. This method is capable of tracking rigid and deformable objects and this versatility lets it to be suitable for different application scenarios.
Keywords: Multi-view, Object Tracking, Video Surveillance, Homography Transform

1. INTRODUCTION

Tracking moving objects is a key problem in computer vision and image processing. It is important in a wide variety of applications, like three-dimension (3D) broadcasting, virtual reality, special effects, image composition, human computer interaction (HCI), video surveillance, human motion analysis and traffic monitoring. Automatically monitoring people in crowded environments such as metro stations, city markets, or public parks, has nowadays become feasible for many reasons. First, from the accuracy's point of view, human operators are likely to fail in monitoring crowded and cluttered environments through tens of cameras. Automatic techniques have reached a degree of maturity to be employed at least as a first automatic step to alert human operators, reducing their effort and the sources of distraction. Second, from the economical point of view, the cost of mounting cameras and developing automatic solutions has declined in comparison to the cost of hiring human operators to watch them. Despite of the complexity increase, multiple camera systems exhibit the undoubted advantages of covering wide areas and enhancing the management of occlusions by exploiting the different viewpoints. Single camera tracking is limited in scope of its applications. While suited for certain applications like local environments, even simple surveillance applications demand the use of multiple cameras for two reasons. Firstly, it is not possible for one camera to provide adequate coverage of the environment because of limited field of view (FOV). Secondly, it is desirable to have multiple cameras observing critical areas, to provide robustness against occlusion.

Multiple-cameras provide us with more complete history of an object's actions in an environment. To take advantage of additional cameras, it is necessary to establish correspondence between different views. Thus, we see a parallel between

* frederic.dufaux@epfl.ch

Mobile Multimedia/Image Processing, Security, and Applications 2009, edited by Sos S. Agaian, Sabah A. Jassim,
Proc. of SPIE Vol. 7351, 735104 · © 2009 SPIE · CCC code: 0277-786X/09/$18 · doi: 10.1117/12.820786

the traditional tracking problem in a single camera and that in multiple cameras: tracking in a single camera is essentially a correspondence problem from frame to frame over time. Tracking in multiple cameras, on the other hand, is a correspondence problem between tracks of objects seen from different viewpoints at the same time instant. However, the automatic merging of the knowledge extracted from single cameras is still a challenging task.

Multi view tracking has the obvious advantage over single view tracking because of its wide coverage range. When a scene is viewed from different viewpoints there are often regions which are occluded in some views but visible in other views. A visual tracking system must be able to track objects which are partially or even fully occluded.

The technical problems in multiple camera systems are several and they have been summarized in [1] into six classes: installation, calibration, object matching, switching, data fusion, and occlusion handling.

In this paper, we address the issue of tracking moving objects in an environment covered by multiple un-calibrated cameras with overlapping fields of view, typical of most surveillance setups. Our main objective is to create a framework that can be used to integrate object-tracking information from multiple video sources and resolve the all mentioned technical drawbacks.

1.1. Related work

There are numerous single-camera detection and tracking algorithms, all of which face the same difficulties of tracking 3D objects using only 2D information. These algorithms are challenged by occluding and partially-occluding objects, as well as appearance changes. Some researchers have developed multi-camera detection and tracking algorithms in order to overcome these limitations. Haritaoglu et. al. [2] have developed a system which employs a combination of shape analysis and tracking to locate people and their parts (head, hands, feet, torso etc.) and tracks them using appearance models. In [3], they incorporate stereo information into their system. Kettnaker and Zabih [4] have developed a system for counting the number of people in a multi-camera environment where the cameras have a non-overlapping field of view. By combining visual appearance matching with mutual content constraints between cameras, their system tries to identify which observations from different cameras show the same person. Cai and Aggarwal [5] extend a single-camera tracking system by switching between cameras, trying to always track any given person from the best possible camera - e.g. a camera in which the person is un-occluded. All these systems use background subtraction techniques in order to separate out the foreground and identify objects, and would fail for cluttered scenes with more densely located objects and significant occlusion.

Approaches to multi-camera tracking can be generally classified into three categories: geometry-based, color-based, and hybrid approaches. The first class can be further subdivided into calibrated and un-calibrated approaches. A particularly interesting paper of calibrated approach is reported in [11] in which homography is exploited to solve occlusions. Single camera processing is based on particle filter and on probabilistic tracking based on appearance to detect occlusions. A very relevant example of the un-calibrated approaches is the work of Khan and Shah [15]. Their approach is based on the computation of the so called “Edges of Field of View”, i.e. the lines delimiting the field of view of each camera and, thus, defining the overlapped regions. Through a learning procedure in which a single track moves from one view to another, an automatic procedure computes these edges that are then exploited to keep consistent labels on the objects when they pass from one camera to the adjacent one.

With color-based approaches, the matching algorithm essentially uses of the color of the tracks, In [12] a color space invariant to illumination changes and histogram based information at low (texture) and mid-level are exploited to solve occlusions and match tracks with a modified version of the mean shift algorithm.

Hybrid approaches mix information about the geometry and the calibration with those provided by the visual appearance. These methods use probabilistic information fusion or Bayesian Belief Networks (BBN) [13].

In this paper we propose a new method which can be classified as a hybrid approach. In this technique, first we recover the homography relation between camera views. Homography mapping allows us to find the corresponding regions between different views by use of region descriptors. It is then possible to track objects in 2D/3D simultaneously across multiple viewpoints. In this paper, we consider a surveillance network of two cameras. However the same approach can be generalized to a network of more than two cameras.

2. METHODOLOGY

2.1. General overview of the algorithm

We first present an overview of the algorithm. More precisely this approach consists of a number of blocks. The basic step is to run a single-view tracking algorithm on each of the camera views. It is followed by a consistent object labeling algorithm on all views. Next step is to verify objects of each view separately: objects that are not consistent over time will not be considered in multi-view tracking step.

Correspondent objects are extracted through a homography transform from one view to the other and vice versa. Having found the corresponding objects in different views, each object is partitioned into homogeneous regions. In the last step, a homography transform is applied to find the region map of first view in the second view and vice versa. For each region (in the main frame and mapped frame) a set of descriptors are extracted and the best match of regions between two views is found based on region descriptors similarity. The block diagram of whole system is shown in Fig.1.

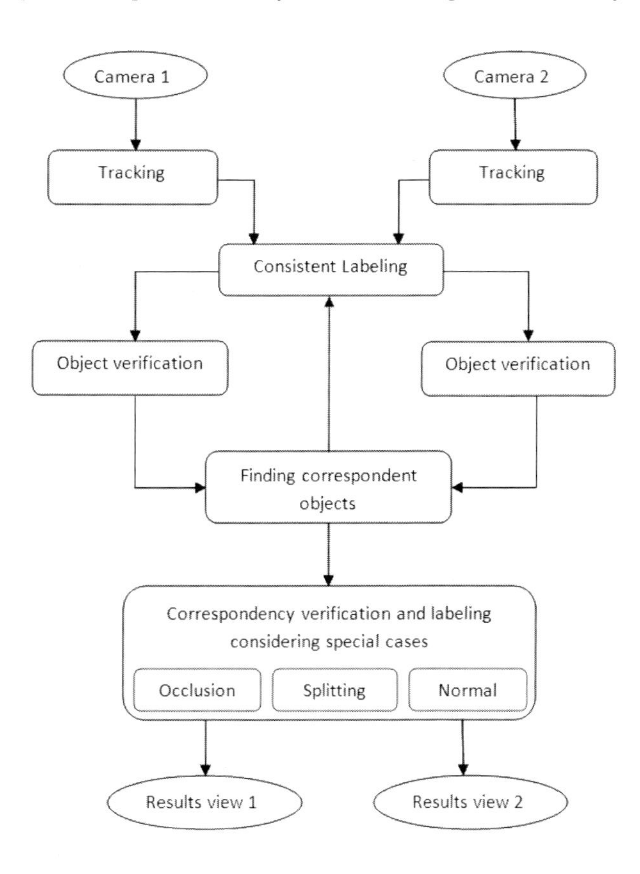

Fig. 1: Block Diagram of Multi-View Object Tracking

2.2. Single-view tracking

Many methods have been developed for object tracking in single view. In this paper, we used a multilevel object-region tracking algorithm [18]. A distinctive feature of the proposed algorithm is that this method operates on region descriptors instead of region themselves. This means that instead of projecting the entire region into the next frame, only region descriptors need to be processed. Therefore, there is no need for computationally expensive models. A brief description of this method is as follow:

1. Foreground object extraction: this decision is taken by thresholding the frame difference between the current frame and the frame representing the background.

2. Object Partition: each object is processed separately and is decomposed into a set of non-overlapping regions to produce the region partition. This step takes into account the spatio-temporal properties of the pixels in the computed object partition and extracts homogeneous regions.
3. Region descriptors: for each region a set of features are extracted as region descriptors. The feature space, used here, is composed of spatial and temporal features. Spatial features are color and a measure of local texture based on variance. The temporal features are the displacement vectors from optical flow computed via block matching. Then each region is represented by a region descriptor.
4. Region Tracking: the first step of tracking regions is the projection of the information of the current frame n into the next frame n+1. Regions of frame n and frame n+1 with most similarity considered as the correspondent objects and receive same labels. (Fig.2)
5. Object Tracking: after finding the corresponding regions between two successive frames, through a top-down and a bottom-up interaction with the region partition step, objects of current frame are validated and are given same labels as previous frame.

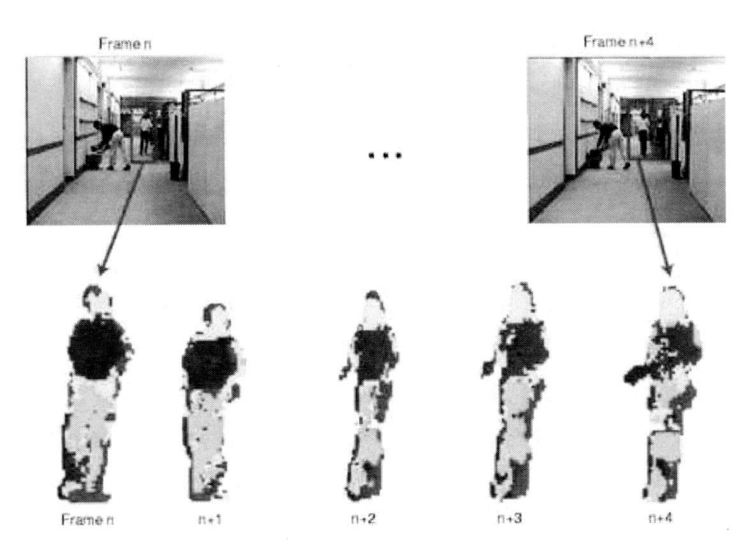

Fig. 2: Object-region extraction and tracking

This algorithm is capable of dealing with multiple simultaneous objects. Track management issues such as appearance and disappearance of objects, splitting and partial occlusion are resolved through interaction between regions and objects. Defining the tracking based on the parts of objects, identified by region segmentation, has led to a flexible technique that exploits the nature of the video object tracking. Fig.3 shows general block diagram of this method.

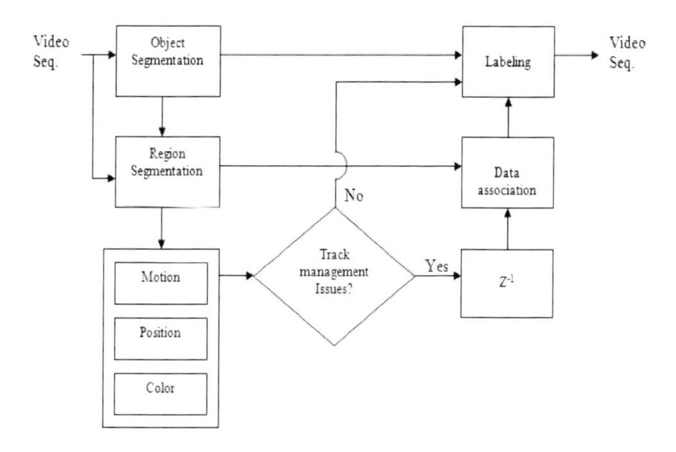

Fig. 3: General block diagram of single view tracking algorithm.

2.3. Consistent object labeling

In a multi-view object tracking scenario, it is essential to establish correspondence between different views of the same object, seen from different cameras, to recover complete information about the object. That is, all views of the same object should be given the same label, as illustrated in Fig. 4. Objects in common are given same labels and other objects receive distinctive labels so that there is no confusion between labels of all views.

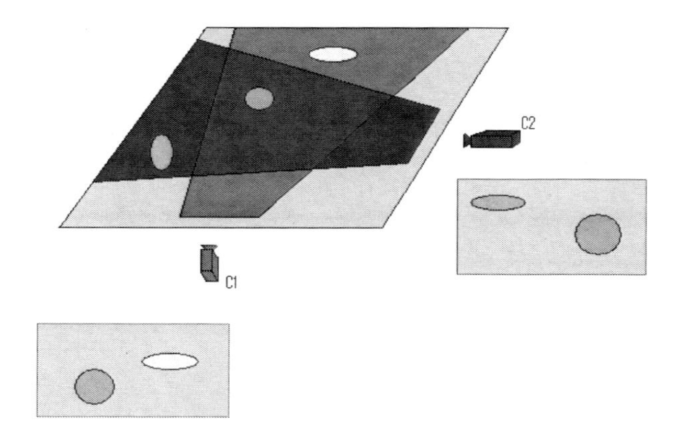

Fig. 4: Two cameras and their footprints are shown. The projections of boundaries of the footprint are also shown in the image that will be observed in two cameras.

2.4. Object consistency verification

Noise or segmentation methods deficiency may erroneously create some objects, called non-semantic objects, in some frames which interfere with the whole tracking system. In order to avoid the effect of such objects, the stability of each object is verified. Namely, this step considers the number of appearance in successive frames and tracking status from one frame to another. After initializing the single-view tracking procedure, a set of tracked objects $O_i^{C_p}(n)$ for each view in frame n are extracted where C_p denotes the camera p . Each object i is characterized by its regions $R_{i,j}(n)$ and the relation between objects and regions of each frame in two successive frames can be expressed as:

$$\forall O_i(n) \quad i=1,\dots,N_O^n \ \exists R_{i,j}(n) \quad j=1,\dots,N_{R_i}^n \ , \tag{1}$$

$$\forall O_i(n+1) \quad i=1,\dots,N_O^{n+1} \ \exists \widetilde{R}_{i,j}(n+1) \quad j=1,\dots,N_{R_i}^{n+1} \ , \tag{2}$$

In which N_O^n is number of video objects in frame n, and $N_{R_i}^n$ is number of regions for object i (Fig. 5-a). Based on the employed single-view algorithm, after tracking process, corresponding regions are described as follows:

$$\forall O_i(n), O_k(n+1)\ i=1,\ldots,N_O^n, k=1,\ldots,N_O^{n+1}\ \exists\ CRD_i = \{(R_{i,j}(n), \widetilde{R}_{k,h}(n+1))|\ R_{i,j}(n) \leftrightarrow \widetilde{R}_{k,h}(n+1)\}\ ,\quad (3)$$

In which CRD_i is a set of corresponding regions in object i from frame n and object k from frame $n+1$ (Figure 5-b). The stability factor for each object is defined as:

$$SF_i = |CRD_i| / N_{R_i}^n\ . \quad (4)$$

If $SF_i < T_1$ then object i will not be considered in the next step of multi-view object correspondence. By applying this method, objects which appear in one frame and disappear in the next frame are not considered as stable.

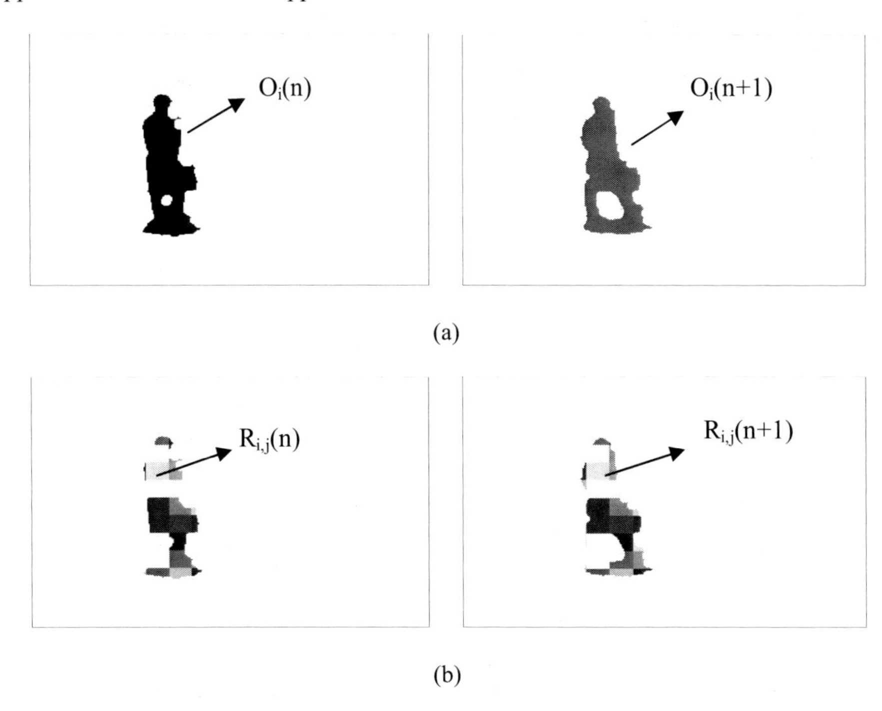

Fig. 5: (a) left image is object i in frame n and right image is object i in frame $n+1$. (b) Different regions of object in two successive frames (regions with same color are correspondent).

2.5. Objects Correspondence

Before we can jointly track objects between each camera view, it is necessary to recover some camera calibration information. We assume that the camera views are widely separated and moving objects are constrained to move along a dominant ground plane.

2.5.1. Homography alignment

A homography mapping defines a planar mapping between two overlapping camera views:

$$x' = \frac{h_{11}x + h_{12}y + h_{13}}{h_{31}x + h_{32}y} \qquad y' = \frac{h_{21}x + h_{22}y + h_{23}}{h_{31}x + h_{32}y}\ , \quad (5)$$

where (x, y) and (x', y') are image coordinates for the first and second views, respectively. Hence, each image point correspondence between two camera viewpoints results in two equations in terms of the homography coefficients. Given at least four correspondence points, those coefficients can be estimated. For this purpose we employed Singular Value Decomposition (SVD) for computing the homography [17] using a set of known landmarks as illustrated in Fig. 6.

With the homography matrix, we can find the transform of objects from the first view to the second one as well as from the second view to the first one.

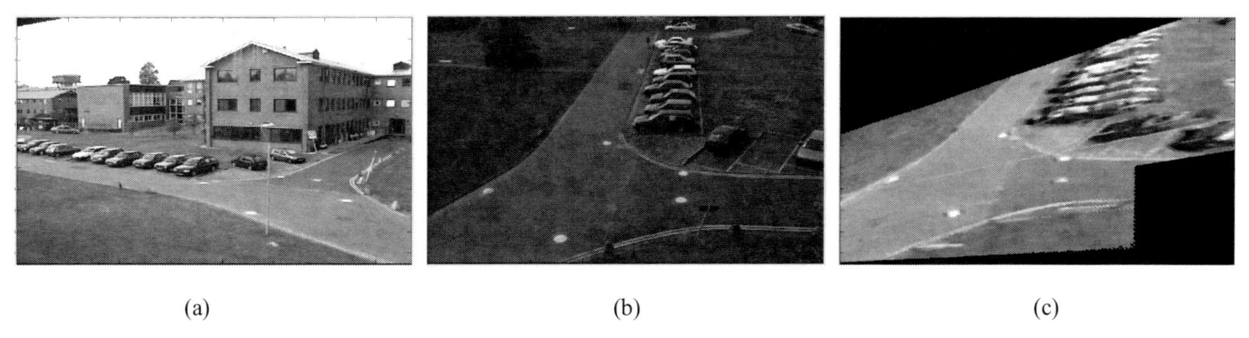

(a) (b) (c)

Fig. 6: (a) First view, (b) Second view, (c) Homography transform of the first view to the second.

2.5.2. Transfer error

The transfer error is the sum of the projection error in each camera view for a corresponding pair. It indicates the size of the error between correspondent objects and their expected projection according to the transfer function. The transfer error associated with a corresponding pair is defined as:

$$TE = (x' - Hx)^2 + (x - H^{-1}x')^2 \ , \tag{6}$$

where x and x' are projective image coordinates in the first and second camera views, respectively.

If $TE < T_2$ then x and x' are considered as a potential match. By considering all possible pairs of objects, we create a list of potential matches, which associates each label of the first view with its corresponding label or labels of second view.

2.6. Correspondence verification

In this step, the same objects receive the same labels in both views. Moreover occlusion or splitting of objects is handled. Two cases may occur. First if each object is just corresponding to one object of the other view straightforwardly it is assigned the same label and it will be tracked later with this new label. Conversely, if an object i in the first view matches with more than one object in the second view, it means that an occlusion has not been handled correctly with the single view information.

To address this problem, we use region level analysis. A set of features like gravity center, histogram and texture, are extracted for each region referred to as region descriptors. It can be expressed as follow:

$$\Phi_{i,j}(n) = (\phi_j^1(n), \phi_j^2(n), \ldots, \phi_j^n(n))^T \ , \tag{7}$$

where $\Phi_{i,j}(n)$ is the set of features of region j of object i in frame n. The homography transform of each region of object i from first view, $R_{i,j}^{C_1}(n)$ to the second view $\widetilde{R}_{i,j}^{C_1}(n)$ and also the homography transform of each region of object k from second view $R_{k,h}^{C_2}(n)$ to the first view $\widetilde{R}_{k,h}^{C_2}(n)$ are extracted as follow:

$$\widetilde{R}_{i,j}^{C_1}(n)=H\,R_{i,j}^{C_1}(n) \tag{8}$$

$$\widetilde{R}_{k,h}^{C_2}(n)=H^{-1}\,R_{k,h}^{C_2}(n)\quad , \tag{9}$$

in which H is the Homography matrix. The distance between regions of two views is calculated as follow:

$$D_{(i,j),(k,h)}^{C_1\rightarrow C_2}=\left|\widetilde{R}_{i,j}^{C_1}-R_{h,k}^{C_2}\right| \tag{10}$$

$$D_{(h,k),(i,j)}^{C_2\rightarrow C_1}=\left|R_{i,j}^{C_1}-\widetilde{R}_{h,k}^{C_2}\right| \tag{11}$$

So the total distance between each pair of regions is:

$$D_{(i,j),(k,h)}=D_{(i,j),(k,h)}^{C_1\rightarrow C_2}+D_{(k,h),(i,j)}^{C_2\rightarrow C_1} \tag{12}$$

Pairs of regions whose distances are above a given threshold, T_3, are discarded from the optimization process that follows. The best set of corresponding regions is obtained through applying the minimum mean square error (MMSE) method as follows:

$$MR=\left\{\left(R_{i,j}^{C_1},R_{h,k}^{C_2}\right)\right\}=\arg\min_{MR} E\{(R_{i,j}^{C_1}-R_{h,k}^{C_2})^2\} \tag{13}$$

in which MR is the set of corresponding regions. Furthermore, by utilizing top-down and bottom-up method, we assign object labels to each region. If some regions are not assigned to any object label during this procedure, they are given the label of the closest region.

3. RESULTS AND INTERPRETATION

The performance of the method described in this paper is evaluated using the first two PETS2001 data sets. We tried to modify the information of first camera based on second camera information. As explained, this helps us to handle the occlusion problem.

Here, the results of the algorithm for two sequences are shown. Fig. 7 displays a sequence which suffers from occlusions. In this Figure, top row represents frames of the first view and bottom row represents frames of the second view. Each column shows the same time instant from two views. In this sequence, from the first frame to frame #33 two objects are occluded in both views and these objects are presented by label #2 in fig.7-a, but as soon as these two objects are seen individually in the second view, the algorithm is able to correctly label them in the first view. In the Fig.7-b, (frame #41) it can be observed that occluded objects of first view are represented by two different labels base on the second view object splitting. A similar problem also occurs in frame #61 and the algorithm can follow the object in both views correctly.

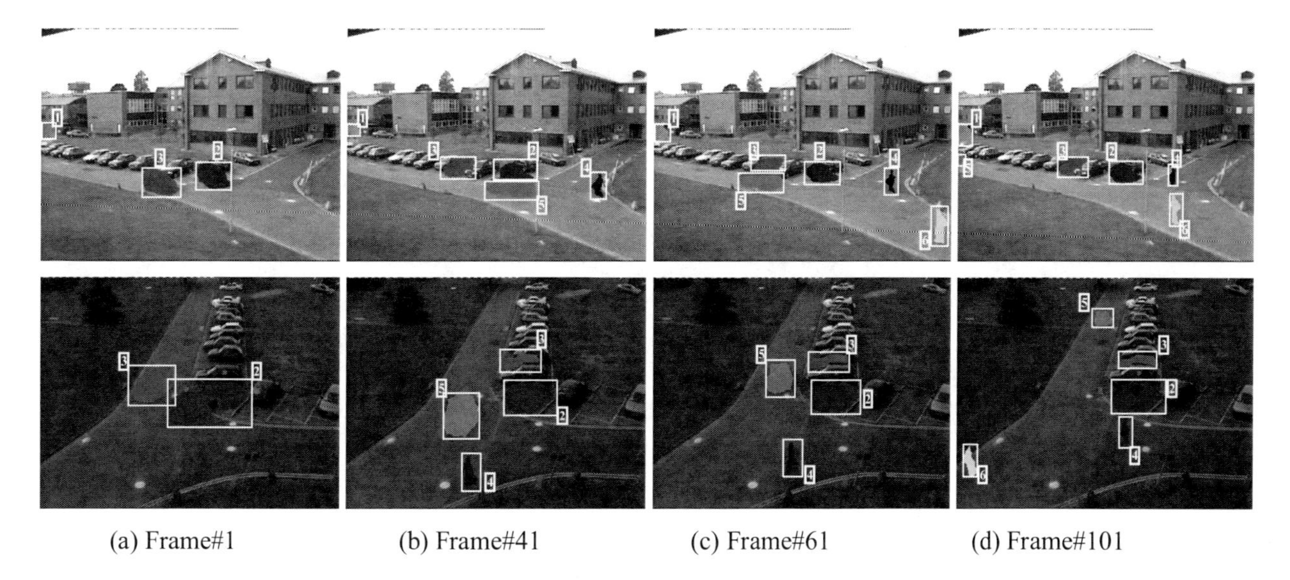

(a) Frame#1 (b) Frame#41 (c) Frame#61 (d) Frame#101

Fig. 7: Top row is the first view and bottom row is the second view at the same time.

The second sequence is showed in Fig. 8. A pedestrian is crossing the junction and in the first view, it has been occluded by a car. Based on the second view information, it is possible to track both pedestrian and car in both views correctly.

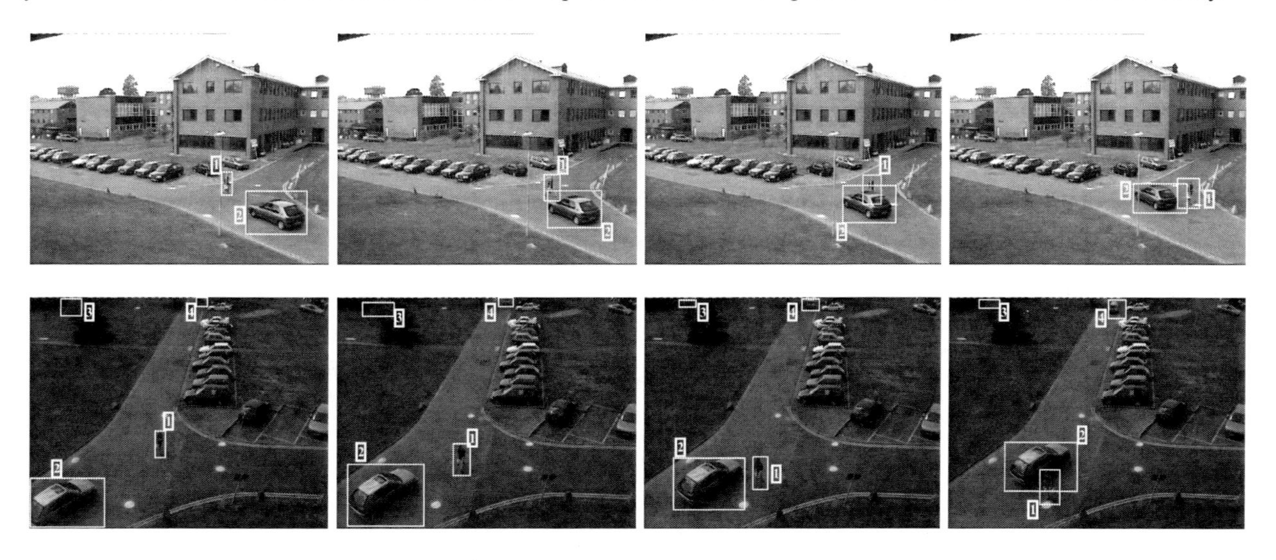

Fig. 8: Top row is the first view and bottom row is the second view at the same time.

To conclude this section, it is interesting to enumerate the advantages and disadvantages of the proposed algorithm. The proposed method based on Homography transform has the following advantages:

- It is fast enough to be used for online scenario on a standard PC (around 1 sec for each pair of 384×288 frames in MATLAB)
- It can handle the occlusion problem if objects are visible at least in one view
- It is possible to extend the method for more than two views
- Using the region descriptors to track multi-view objects is a novelty in multi-view object tracking

And the disadvantages are:

- If a group of objects enters the scene and then they get separated while they are occluded with another object/objects in the scene, the results will not be reliable
- Final results are rather dependent on single-view algorithm and if there is a tremendous error in single-tracking of both views, the algorithm might fail.

Region descriptors and homography transform provide an estimation of regions from one view to another, which helps to handle the occlusion and splitting issues and to refine the individual objects in a scenario with occlusion.

4. CONCLUSION AND FUTURE WORK

This paper presents an automatic multi-view object tracking algorithm based on interactions between object regions in one view and homography transform of object regions of other views. This method is able to deal with multiple simultaneous objects. Track management issues such as occlusion, splitting, appearance and disappearance of objects are resolved using information of other views.

This method is capable of tracking rigid and deformable objects and this versatility makes it suitable for different scenarios. All the component of this algorithm, included the single view tracking algorithm, can be run in real-time applications. We are currently investigating the method for more than two views to fuse the information of different views to get the most efficiency.

ACKNOWLEDGMENT

This work was partially supported by European Network of Excellence VISNET II (http://www.visnet-noe.org) IST Contract 1-038398, funded under the European Commission IST 6th Framework Program.

REFERENCES

[1] W. Hu, T. Tan, L. Wang, and S. Maybank, "A Survey on Visual Surveillance of Object Motion and Behaviours", IEEE Trans. on Systems, Man, and Cybernetics – Part C, 34(3), p. 334-352, 2004.

[2] I. Haritaoglu, D. Harwood, and L. Davis, "W4: Who, When, Where, What: A Real Time System for Detecting and Tracking People", 3rd Int. Conf. on Automatic Face and Gesture, 1998.

[3] I. Haritaoglu, D. Harwood, and L. Davis, "W4: A Real-time System for Detecting and Tracking People in 2 1/2D", 5th European Conf. on Computer Vision, 1998.

[4] V. Kettnaker, R. Zabin, "Counting People from Multiple Cameras", Proc. of IEEE ICMCS, p.267-271, 1998.

[5] Q. Cai, and J.K. Aggareal, "Automatic Tracking of Human Motion in Indoor Scenes across Multiple Synchronized Video Streams", Proc. Of 6th Int. Conf. on Computer Vision, p 356-362, 1998.

[6] J. Li, C.S. Chua, and Y.K. Ho, "Colour Based Multiple People Tracking", Proc. Of IEEE Int. Conf. on Control, Automation, Robotics and Vision, 1, p.309-314, 2002.

[7] J. Black, T.J. Ellis, and P. Rosin, "Multi-View Image Surveillance and Tracking", Proc. IEEE Workshop Motion and Video Computing 2002.

[8] N. Nguyen, H. Bui, S. Venkatesh, and G. West, "Multiple Camera Coordination in a surveillance system", ACTA Automatic Sinica, Vol 23(3), p408-422, 2003.

[9] J. Black, T. Ellis, 'Multi Camera Image Tracking", Image and Vision Computing (24), No. 11, p 1256-1267, 2006.

[10] S. Iwase and H. Saito, "Tracking Soccer Players based on Homography among Multiple Views", Proc. SPIE2003, v. 5150, p283-292, 2003.

[11] Z. Yue, S. Zhou, and R. Chellappa, "Robust two-camera Tracking using Homography", Proc. IEEE Intl Conf. on Acoustics, Speech, and Signal Processing, vol. 3, p 1-4, 2004.

[12] R. Cucchiara, A. Prati, R. Vezzani, "Posture Classification in a Multi-camera Indoor Environment", Proc. IEEE Intl Conf. on Image Processing(ICIP), vol. 1, p. 725-728, 2005.

[13] S. Calderara, R. Vezzani, A. Prati, and R. Cucchiara, "Entry Edge of Field of View for Multi-camera Tracking in Distributed Video Surveillance", Proc. IEEE Intl Conf. on Advanced Video and Signal-based Surveillance (AVSS'05), p. 93-98, 2005.

[14] I. Paek, C. Park, M. Ki, K. Park, and J. Paik, "Multiple_view object tracking using Metadata", Proc. Int. Conf. ICWAPR, vol. 1, no. 1, p 12-17,2007.
[15] S. Khan, and M. Shan, "Consistent Labeling of Tracked Objects in Multiple Cameras with Overlapping fields of View", IEEE Trans. On PAMI, 25(10), p. 1355-1360, 2003.
[16] A. Mittal, and L. Davis, "Unified Multi-camera detection and Tracking using Region_Matching", Proc. Of IEEE Worckshop on Multi-Object Tracking, p. 3-10, 2004.
[17] R. Hartley, and A. Zisserman, "Multiple View Geometry in Computer Vision", Cambridge University Press, 1998.
[18] A. Cavallaro, O. Steiger, and T. Ebrahimi, "Tracking Video Objects in Cluttereds Background", IEEE Trans. on Circuit and Systems for Video Technology, 2005.
[19] A. Yilmaz, and M. Shan, "Object Tracking: A Survey", ACM Computing Surveys (CSUR), vol. 38, 2006

A modified image restoration algorithm for multiframe degraded images

Geng Zexun, Zhao Zhenlei, Song xiang

Zhengzhou Institute of Surveying and Mapping, 66 Longhaizhong Rd., Zhengzhou, China, 450052

Abstract: A modified multiframe image restoration algorithm of degraded images is described in this paper, which is based on the method developed by V. Katkovnik. The projection gradient algorithm based on anisotropic LPA-ICI filtering, being proposed by V. Katkovnik, could only restore images contaminated by Gaussian noisy, also it was too complicated and time consuming. By improving Katkovnik's cost function and applying constraints on image intensity value in iteration, we reached a new multiframe recursive iteration restoration scheme in frequency domain. This method is suitable for reconstruction of images degenerated by both Gaussian noise and Poissonian noise, as well by mixed noise. Experimental results demonstrated that this method works efficiently, and could well restore images blurred heavily by multi-noise.

Key words: Multiframe image restoration, Cost function, Projection gradient minimization, Recursive iteration;

1 Introduction

Images acquired by any optical system are fundamentally limited in resolution by diffraction and corrupted by measurement noise. Aberrations intrinsic to the optical system and imaging medium result in further degradation and distortions of the observed images[1]. The loss caused by the imaging degradation is very heavy, thus, in order to enhance the visual quality and resolution of the degraded images, it is necessary to restore the images using the technique of digital images processing.

Usually, image restoration is called image deconcolution. Traditional deconvolution techniques always suppose that the point spread function (PSF) is exactly known. In many practical situations, however, it is impossible to obtain the accurate PSF, such as astronomical imaging and remote sensing. As a result, image deconcolution becomes blind deconvolution. A lots of blind deconvolution methods, such as iterative blind decovolution[6], simulated annealing algorithm[7] and Non-negativity and support constraints recursive inverse filtering algorithm[8], are developed.

Image blind deconvolution is an ill-conditioned inverse process, the quality of restoration image depends largely on the constraints posed on degraded image and plenty measurements observed in imaging. Christou et al [3] have argued that the use of multiframe observations could be an additional deconvolution constraint: the ratio of unknown variables to measured quantities being reduced from 2:1 for a single image frame to (L+1): L for L image frame observations. The simultaneous analysis of multiple observations implicitly accounts for correlations that may exist among variables as well as between variables and the data. Consequently, multiframe blind deconvolution should result in systematically lower error bounds and more reliable solution than when individual frame is blind deconvolved separately or when multiple frames are merged into an averaged "shift-and-added" image (i.e., an image generated by averaging the image frames

Mobile Multimedia/Image Processing, Security, and Applications 2009, edited by Sos S. Agaian, Sabah A. Jassim, Proc. of SPIE Vol. 7351, 735105 · © 2009 SPIE · CCC code: 0277-786X/09/$18 · doi: 10.1117/12.817831 after appropriate pixel shifts) and then deconvolved [5].

Generally, image blind deconvolution [10] restores image by minimizing a certain cost function. In 2006, V. Katkovnik et al [2] proposed a projection gradient algorithm based on anisotropic LPA-ICI（Local Polynomial Approximation- Intersection of Confidence Intervals） filters, which can efficiently restore images polluted by white Gaussian noise by recursively iterate in frequency domain and LPA-ICI filters in spatial domain alternately. This algorithm, however, is too complicated to implement and time consuming because of LPA-ICI filters. Moreover, it can only deal with the Gaussian noise.

To simplify Katkovnik's algorithm and enable it to deal with other type noise, we improved it's cost function, and presented a modified recursive projection gradient algorithm for blind restoration of multiframe images. The proposed method is easier to implement, and is capable of reconstructing image from observed date polluted by Gaussian noise or Poissonian noise, or by both of them. Also, the iteration procedure becomes more rapidly. Experiments results show that this method works well.

The rest of this paper is organized as follows. Section II describes the proposed algorithm with a modified cost function and a procedure to minimize this cost function by projection gradient algorithm in frequency domain, then reconstructs images using multiple recursion iterations. The principle of this new algorithm is more simply to implement, the calculation becomes more rapidly, and it is capable to reconstruct degradation images polluted by Gaussian noise or Poissonian noise or both of them, results of experiments demonstrate that this method works efficiently.

2. Modified Restoration algorithm for Multiframe Degraded

2.1 Image degradation model

The original image is $y(x), x \in R^2 = \{(x_1, x_2), x_1 = 1,2,...,n_1, x_2 = 1,2,...,n_2\}$, $y \in R^1$, the observed image is $z(x)$. Suppose that the image formation is linear and space invariant, Gaussian and signal dependent Poisson noise sources are present, then the observed image can be described as [1]:

$$z(x) = \{y(x) * v(x)\} \circ n_P(x) + n_G(x) \tag{1}$$

Where, $*$ is the convolution operator, $v(x)$ is PSF, $\circ$ denotes a pixel by pixel operation, $n_G(x)$ is the Gaussian random noise with the variance of σ_G^2, $n_P(x)$ is the stochastic Poisson noise, with $\sigma_P^2 \approx y(x) * v(x)$. When there are both Gaussian and Poisson noise in the images which are not photon-limited, a nonstationary but additive weighted-Gaussian noise model with variance $w(x)$ is a very good approximation [12]:

$$w(x) = \sigma_{n(x)}^2(x) = \sigma_G^2 + \sigma_P^2 \tag{2}$$

$$\sigma_G^2 = \frac{\pi}{2}\left(\langle z(x)\rangle_{\le 0}\right)^2, \quad \sigma_P^2 = \max[z(x), 0] \tag{3}$$

The first term in (3) σ_G^2 is the Gaussian detection-electronic readout noise, which can be estimated using the average over all negative pixels in the image. The second term in (3) σ_P^2 is the Poisson photonic noise [12], this term is derived from the fact that the variance equals the mean and the mode for a Poisson distribution. Using this noise model, the operation $\circ$ in (1) can be replaced by the simple addition. According to the convolution theorem, equation (1) could be expressed as:

$$Z(f) = Y(f)V(f) + N(f) \tag{4}$$

Where, $f = \{(f_1, f_2), f_i = 2\pi k_i / n_i, k_i = 1, 2, ..., n_i, i = 1, 2\}$. $Z(f)$, $Y(f)$, $V(f)$, $N(f)$ represents the Fourier transformation corresponding to the variables respectively.

Suppose the noise of each image is independent, so for the multi-frame observed images of the same object:

$$Z_j(f) = Y(f)V_j(f) + N_j(f), j = 1, \cdots, L \tag{5}$$

2.2 Improved Cost Function

The purpose of image restoration is to reconstruct the original image $y(x)$ from the noisy observations $\{z_j(x)\}_{j=1}^{L}$. V. Katkovnik supposed the noise is Gaussian white noise, and proposed the cost function (6):

$$J = \sum_{j=1}^{L} \frac{1}{\sigma_j^2} \sum_f |Z_j - YV_j|^2 + \lambda_2 \sum_f |Y|^2 + \lambda_1 \sum_{i,j=1}^{L} d_{ij} \sum_f |Z_i V_j - Z_j V_i|^2 + \lambda_3 \sum_{j=1}^{L} \sum_f |V_j|^2$$

$$d_{ij} = \frac{n_1 n_2}{\sigma_i^2 \sum_f |V_j|^2 + \sigma_j^2 \sum_f |V_i|^2} \tag{6}$$

The cost function is too complex, and when dealing with the non-Gaussian noise, σ_j^2 is difficult to be estimated with good accuracy.

This article simplifies the cost function (6), and gives a new cost function:

$$J = \sum_{j=1}^{L} \sum_f |Z_j - YV_j|^2 + \sum_{i,j=1}^{L} \sum_f |Z_i V_j - Z_j V_i|^2 \tag{7}$$

2.3 Recursive Projection Gradient Algorithm

The estimate of the image and the PSFs are solutions of the following problem:

$$(\hat{y}, \hat{v}) = \arg \min_{y \in Q_y, v_j \in Q_{v_j}} J \tag{8}$$

Where the admissible sets Q_y for y and Q_{v_j} for v_j are defined as:

$$Q_y = \{y : \mathrm{T} \geq \mathrm{y} \geq 0\}, \quad T = \rho * \max(y(:)) \tag{9}$$

$$Q_{v_j} = \{v_j : \sum_{v_j} v_j(x) = 1, v_j(x) \geq 0, x \in D_{\sup}\} \tag{10}$$

The $\rho > 0$ in (9) is an adjusting parameter. Through adjusting ρ, we can enhance the visual quality of the restoration and restrain the amplification of the noise. The $D_{\sup}$ in (10) is the support region of PSF.

In order to minimize the equation (8), let $\partial_{Y^*} J = 0$, $\partial_{V_j^*} J = 0, j = 1, ..., L$, then:

$$\partial_{Y^*} J = -\sum_{j=1}^{L} (Z_j - V_j Y) V_j^* \tag{11}$$

$$\partial_{V_j^*} J = -(Z_j - V_j Y) Y^* + \sum_{i,j=1,\, i \neq j}^{L} (Z_i V_j - Z_j V_i) Z_i^* \tag{12}$$

The superscript asterisk (*) in (11) (12) represents complex conjunction. Then use the

Recursive Projection Gradient Algorithm to calculate (8). Firstly, the values $Y^{(k)}$ and $V_j^{(k)}$ are calculated:

$$Y^{(k)} = Y^{(k-1)} - \alpha \partial_{Y^*} J(Y^{(k-1)}, V^{(k-1)}) \tag{13}$$

$$V_j^{(k)} = V_j^{(k-1)} - \beta \partial_{V_j^*} J(Y^{(k)}, V^{(k-1)}),\quad k = 1, \cdots \tag{14}$$

Where, $\alpha > 0,\ \beta > 0$ are step-size parameters. The ill-conditioning of the inverse problem means that the criterion J has different scale behavior for different frequency. In order to get stable iterative results for all frequencies, step parameter α and β should be choose the smaller ones, but at the same time resulting into the slower convergence speed [2].

In order to improve the convergence speed of (13)-(14), we take the Hessian matrix $H_{V^*V^T}$ and $H_{Y^*Y^T}$ as scaling factors of the step sizes:

$$Y^{(k)} = Y^{(k-1)} - \alpha \frac{1}{H_{Y^*Y}} \partial_{Y^*} J(Y^{(k)}, V(k)) \tag{15}$$

$$H_{Y^*Y} = \partial_Y \partial_{Y^*} J = \sum_j |V_j|^2 + \lambda \tag{16}$$

$$V_j^{(k)} = V_j^{(k-1)} - \beta \frac{1}{H_{V_j^*V_j}} \partial_{V_j^*} J(Y^{(k)}, V(k)) \tag{17}$$

$$H_{V_j^*V_j} = (\partial_{V_j} \partial_{V_j^*} J)_{i,j} = |Y|^2 + \sum_i |Z_i|^2 \tag{18}$$

The λ in (16) is a regularization parameter, which is to prevent denominator from equating to zero. This paper chose the machine's minimizing value as the value of λ. Substitution (16) into (15), Substitution (18) into (17), obtains the following final formulas for the iterations:

$$Y^{(k)} = (1-\alpha) Y^{(k-1)} + \alpha \frac{\sum_j Z_j V_j^{*(k-1)}}{\sum_j |V_j^{(k-1)}|^2 + \lambda}$$

$$y^{(k)} = P_{Q_y}\left\{ FT^{-1}\{Y^{(k)}\} \right\} \tag{19}$$

$$V_j^{(k)} = (1-\beta) V_j^{(k-1)} + \beta \frac{Z_j Y^{*(k-1)} + Z_j \sum_i V_i^{(k-1)} Z_i^*}{|Y^{(k)}|^2 + \sum_i |Z_i|^2}$$

$$v_j^{(k)} = P_{Q_{V_j}}\left\{ FT^{-1}\{V_j^{(k)}\} \right\} \tag{20}$$

2.4 Steps of Proposed Algorithm

The detailed computation processes of the algorithm can be described as follows:

1) Initialization: give the Gaussian PSF to $v_j^{(0)}$, give the average of the observed images to $y^{(0)}$, $y^{(0)} = \sum_{j=1}^{L} z_j(x)/L$;

2) Imaging evaluation and projection: calculating $Y^{(k)}$ and $y^{(k)}$ using equation (19), and calculating $Y^{(k)} = F\{y^{(k)}\}$;

3) PSF evaluation and projection: calculating $V_j^{(k)}$ and $v_j^{(k)}$ with equation (20), to accelerate the convergence speed, calculating equation (20) for M times repeatedly, this internal iterations imbedded in the main recursive algorithm. This paper chose $M = 7$. Calculating $V_j^{(k)} = FT\{v_j^{(k)}\}$;

4) Increasing k and repeating the iterative process (2)-(3) K times.

3. Experiment results and analysis

The test images are the image of letters with size 128×128 and the image of ISS (International Space Station) with size 256×256, the software is MATLAB 7.0. Separately using the 19×19 out of focus PSF and 19×19 motion PSF to degrade the images, and adding different noises into the images. Define the support region of PSF as 20×20, parameter $\alpha = 0.6$, $\beta = 0.9$, $\rho = 0.8$. The iterative time in each experiment is 50.

3.1 Experiment of blurred images with Gaussian noise

Adding Gaussian white noise to degraded images blurred by the 19×19 out of focus PSF and 19×19 motion PSF, making the BSNR (Blurred Signal to Noise) of the images $z_j (j = 1, 2)$ equal to 40, where BSNR [2] is calculated by equation (21):

$$BSNR = 10\log_{10}\left(\frac{1}{n_1 n_2 \sigma_j^2} \| (y * v_j)(x) - \frac{1}{n_1 n_2} \sum_{\bar{x}} (y * v_j)(x) \|_2^2\right) \tag{21}$$

The experiment results are showed in figure (1) (2) (3). For the letters image, the calculating progress of the algorithm proposed by V. Katkovnik costs 90s, while the process of the algorithm proposed in this paper only lasts 46s; and for the ISS image, the consumed time is separately 230s and 110s.

In figure 1: (a) is the original letters image; (b) is the degraded letters image blurred by the 19 ×19 out of focus PSF and the adding noise, whose BSNR is 40; (c) is the degraded letters image blurred by the motion PSF and the adding noise, whose BSNR is 40; (d) is the original ISS image; (e) is the degraded ISS image blurred by the 19×19 out of focus PSF and the adding noise, whose BSNR is 40; (f) is the degraded ISS image blurred by the motion PSF and the adding noise, whose BSNR is 40.From the figure (a)-(f), we can see the image is degraded very heavily.

In figure 2: (a) is, when BSNR is 40, the restored letters image using the method proposed by V. Katkovnik; (b) is, when BSNR is 40, the restored letters image using the method proposed in this article; (c) is, when BSNR is 40, for the letters image, the SNR graph of the arithmetic proposed by V. Katkovnik and this article; (d) is, when BSNR is 40, the restored ISS image using the method proposed by V. Katkovnik; (e) is, when BSNR is 40, the restored ISS image using the method proposed in this article; (f) is, when BSNR is 40, for the ISS image, the SNR graph of the arithmetic proposed by V. Katkovnik and this article. From figure (c) and (f), we can get the conclusion that the method presented in this article has a higher SNR than that of V. Katkovnik; and from pictures (a) (b) (d) (e), we can see that both the methods reconstruct the image effectively, but the image (b) (d) (reconstructed with the method proposed by V. Katkovnik) still has some blurred part.

In figure 3: (a) is the local amplificatory picture of the original letters image;(b) is the local amplificatory picture of the restored letters image using the method proposed by V. Katkovnik; (c) is the local amplificatory picture of the restored letters image using the method proposed in this article; (d) is the local amplificatory picture of the original ISS image; (e) is the local amplificatory picture of the restored ISS image using the method proposed by V. Katkovnik; (f) is the local amplificatory picture of the restored ISS image using the method proposed in this article. Figure 3(b) and (e) show the method of V. Katkovnik's not only leaving some blurred part but smoothing some image details; figure 3(c) and (f) show the restored image by the method in this article is very close to the original one.

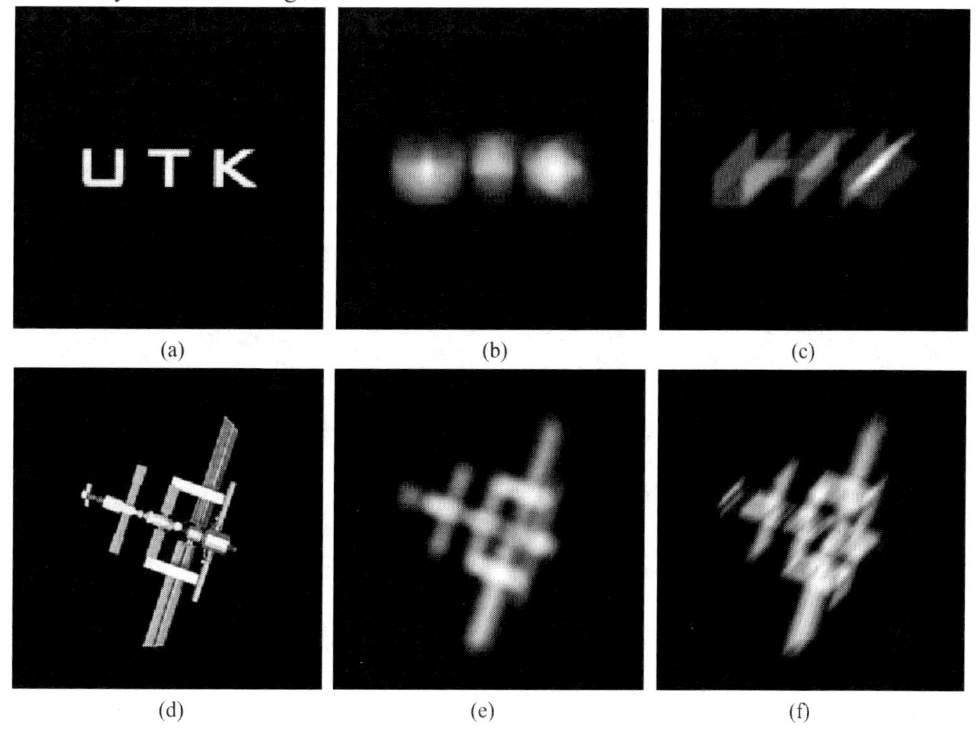

(a) (b) (c)

(d) (e) (f)

Figure 1: (a) the original letters image, (b) the degraded image of letters blurred by out of focus PSF and the adding noise, whose BSNR is 40, (c) the degraded image of letters blurred by motion PSF and the adding noise, whose BSNR is 40. (d) the original ISS image, (e) the degraded image of ISS blurred by out of focus PSF and the adding noise, whose BSNR is 40,(f) the degraded image of ISS blurred by motion PSF and the adding noise, whose BSNR is 40.

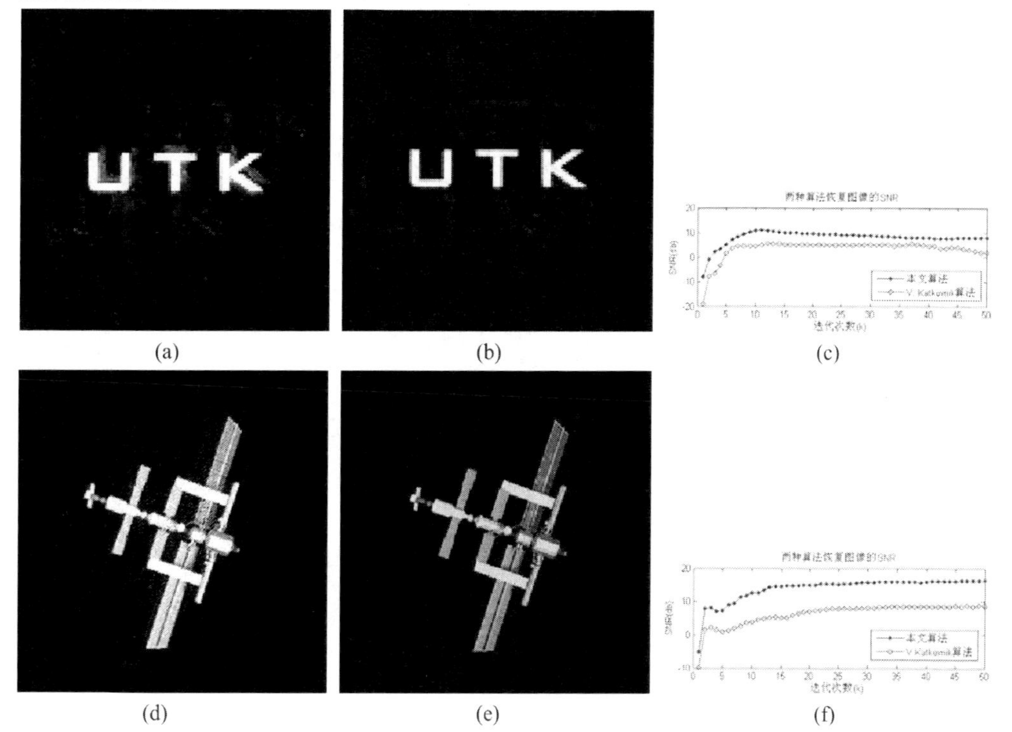

(a) (b) (c)

(d) (e) (f)

Figure 2: (a) when BSNR is 40, the restored letters image using the method proposed by V. Katkovnik, (b) when BSNR is 40, the restored letters image using the method proposed in this article, (c) when BSNR is 40, for the letters image, the SNR graph of the algorithms proposed by V. Katkovnik and this article; (d) when BSNR is 40, the restored ISS image using the method proposed by V. Katkovnik, (e) when BSNR is 40, the restored ISS image using the method proposed in this article, (f) when BSNR is 40, for the ISS image, the SNR graph of the algorithms proposed by V. Katkovnik and this article.

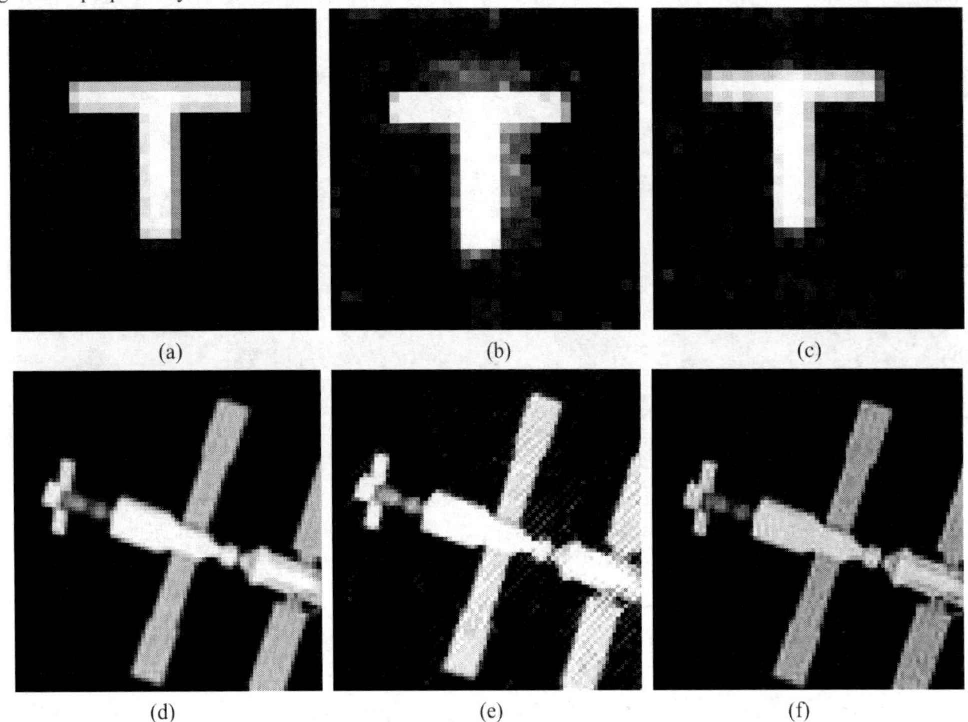

Figure 3: (a) the local amplificatory picture of the original letters image, (b) the local amplificatory picture of the restored letters image using the method proposed by V. Katkovnik, (c) local amplificatory picture of the restored letters image using the method proposed in this article; (d) the local amplificatory picture of the original ISS image, (e) the local amplificatory picture of the restored ISS image using the method proposed by V. Katkovnik, (f) local amplificatory picture of the restored ISS image using the method proposed in this article.

3.2 Experiment of blurred images with Poisson noise

Add Poisson noise with a mean value being zero to the degraded images by 19×19 out of focus PSF and 19×19 motion PSF. In order to get the Poisson noise with a desired level, firstly, make the original image $y(x)$ (the gray region is [0, 1]) multiply a zoom parameter χ, the observed image z_j can be obtained by equation (22):

$$z_j \sim P(yy * v_j), \quad yy = \chi \cdot y \tag{22}$$

Thus, $E\{z_j\} = \sigma^2_{z_j} = \chi \cdot y * v_j$, the larger χ corresponds to a higher blurred SNR.

In this article $\chi = 255$, figure 4 (a), (b), (d), (e) are the degraded images, (c) and (f) are the restored results. From the figures, we can see the profile of the original images is reconstructed, though there still remains some noise.

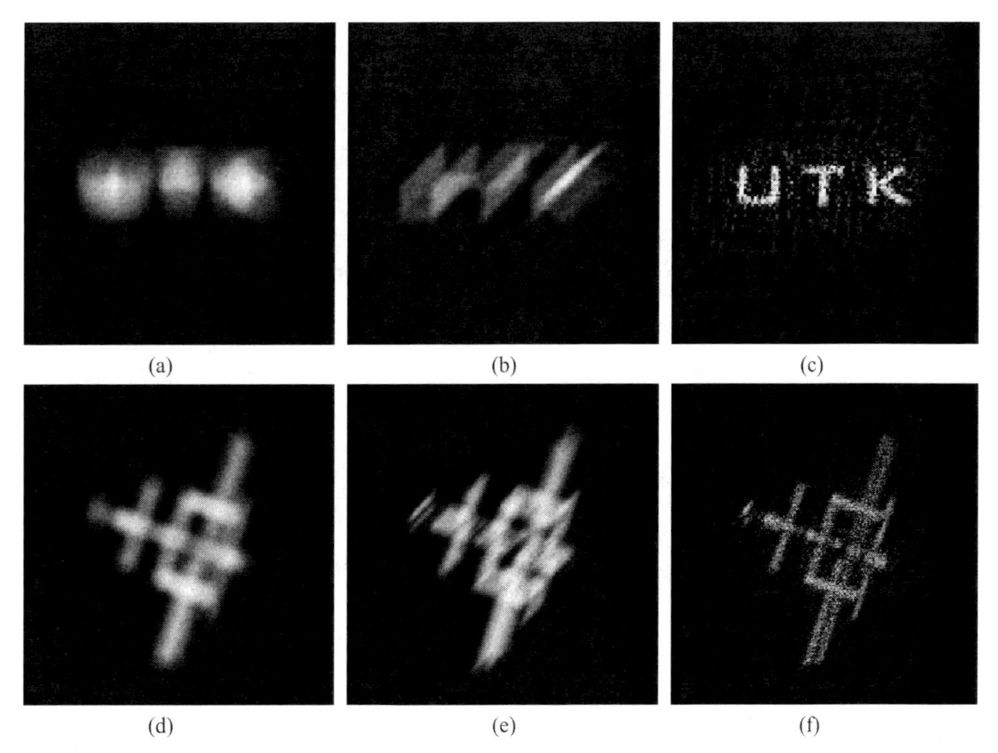

(a) (b) (c)

(d) (e) (f)

Figure 4: (a) the degraded letters image by 19×19 out of focus PSF, and adding Poisson noise, (b) the degraded letters image by 19×19 motion PSF, and adding Poisson noise, (c) the restored letters image using the method proposed in this article; (d) the degraded ISS image by 19×19 out of focus PSF, and adding Poisson noise, (e) the degraded ISS image by 19×19 motion PSF, and adding Poisson noise, (f) the restored ISS image using the method proposed in this article.

3.3 Experiment of blurred images with mixed noise

Firstly, adding the Poisson noise (with a mean value of zero, $\chi = 500$) to the degraded images by 19×19 out of focus PSF and 19×19 motion PSF respectively, then adding the Gaussian white noise (with $\sigma^2 = 0.015$), getting the degraded images showed in figure 5 (a), (b), (d), (e). Figure (c) and (f) are the restored images. From the results, we can see that even with the mixed noise it is still able to get the reasonable restored results.

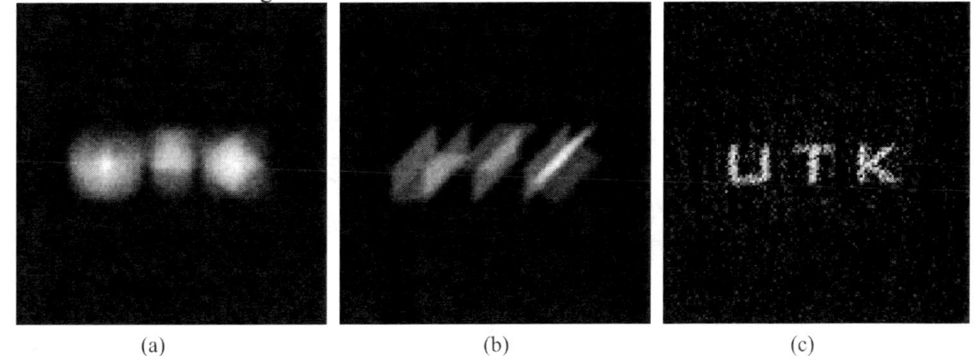

(a) (b) (c)

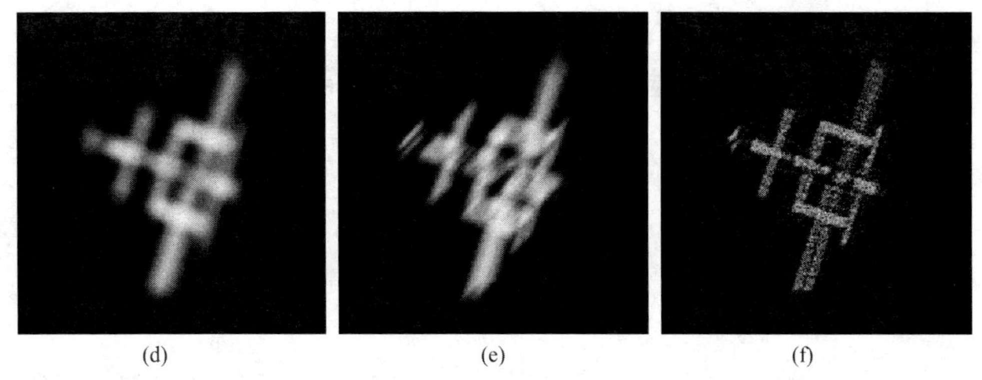

(d) (e) (f)

Figure 5: (a) the degraded letters image by 19×19 out of focus PSF, and adding mixed noise, (b) the degraded letters image by 19×19 motion PSF, and adding mixed noise, (c) the restored letters image using the method proposed in this article; (d) the degraded ISS image by 19×19 out of focus PSF, and adding mixed noise, (e) the degraded ISS image by 19×19 motion PSF, and adding mixed noise, (f) the restored ISS image using the method proposed in this article.

4 Conclusions

In this article, we make some improvement to the cost function in the method proposed by V. Katkovnik, present a new cost function, then do the minimizing process using the projection gradient algorithm, finally, get a new recursive projection gradient algorithm of blind deconvolution of multi-frame images. The principle of this algorithm is simply and the algorithm can be easily implemented. The stimulant experiment results prove that the computation time of this method is attractive, and we can get effective results even when dealing with the images degraded by different noise and various PSF.

ACKNOWLEDGEMENTS

This work was supported by the National Natural Science Foundation of China (NSFC, No. 60778051) and the National High Technology Research and Development of China (863, No. 2006AA12Z110, No. 2008AA7034180A). For further information contact Geng Zexun by email at zxgeng@126.com.

Reference

[1] Erik F. Y. Hom, F. Marchis, Timothy K. Lee, S. Haase, D. A. Agard, John W. Sedat. "AIDA: an adaptive image deconvolution algorithm with application to multi-frame and three-dimensional data." J. Opt. Soc. Am. A. Vol. 24, No. 6, June 2007: 1580~1600.

[2] V. Katkovnik, D. Paliy, K. Egiazarian, and J. Astola. "Frequency domain blind deconvolution in multiframe imaging using anisotropic spatially-adaptive denoising." 2006. http://www.cs.tut.fi/~lasip/.

[3] J. C. Christou, A. Roorda, and D. R. Williams, "Deconvolution of adaptive optics retinal images." J. Opt.Soc. Am. A 21, 2004: 1393~1401.

[4] V. Katkovnik, K. Egiazarian, J. Astola. "A spatially adaptive nonparametric regression image deblurring." IEEE Trans. on Image Processing, Vol. 14, Issue 10, 2005: 1469~1478.

[5] H.R. Ingleby, D.R. McGaughey, "Experimental results of parallel multiframe blind deconvolution using wavelength diversity." Proc. SPIE 5578, 2004: 8~14.

[6] G. Ayers, J. C. Dainty, "Iterative blind deconvolution method and its applications." Opt. Lett. 13, 1988: 547~549.

[7] B. C. MacCallum, "Blind deconvolution by simulated annealing." Opt. Commun. Vol. 75, No. 2, 1990: 101~105.

[8] D. Kundur, D. Hatzinakos. "A novel recursive filtering method for blind image restoration." IASTED Int Conf on Signal and Image Proc. Nov. 1995: 428~431.

[9] Gonzalez R C, Woods R E. Digital Image Processing 2nd Edition. Beijing: Publishing House of Electronics Industry. 2002.

[10] Zou Mouyan. Deconvolution and signal recovery. Beijing: National Defense Industry Press, 2001.

[11] A. Foi, S. Alenius, M. Trimeche, etc. "A spatially adaptive poissonian image deblurring." 2005.

[12] L. M. Mugnier, T. Fusco, and J.-M. Conan, "MISTRAL: a myopic edge-preserving image restoration method, with application to astronomical adaptive-optics-corrected longexposure images," J. Opt. Soc. Am. A 21, 2004: 1841~1854.

Compensating Image Degradation due to atmospheric Turbulence in anisoplanatic Conditions

Claudia S. Huebner*,
Dept. of Signatorics, FGAN-FOM, Gutleuthausstrasse 1, 76275 Ettlingen, Germany

ABSTRACT

In imaging applications the prevalent effects of atmospheric turbulence comprise image dancing and image blurring. Suggestions from the field of image processing to compensate for these turbulence effects and restore degraded imagery include Motion-Compensated Averaging (MCA) for image sequences. In isoplanatic conditions, such an averaged image can be considered as a non-distorted image that has been blurred by an unknown Point Spread Function (PSF) of the same size as the pixel motions due to the turbulence and a blind deconvolution algorithm can be employed for the final image restoration. However, when imaging over a long horizontal path close to the ground, conditions are likely to be anisoplanatic and image dancing will effect local image displacements between consecutive frames rather than global shifts only. Therefore, in this paper, a locally operating variant of the MCA-procedure is proposed, utilizing *Block Matching* (BM) in order to identify and re-arrange uniformly displaced image parts. For the final restoration a multi-stage blind deconvolution algorithm is used and the corresponding deconvolution results are presented and evaluated.

Keywords: Image restoration, atmospheric turbulence, motion compensated averaging, block matching, blind deconvolution

1. INTRODUCTION

The most pronounced image degrading effects of atmospheric turbulence comprise image dancing and image blurring. These image-degradation effects arise from random inhomogeneities in the temperature distribution of the atmosphere, producing small but significant fluctuations in the index of refraction. Light waves propagating through the atmosphere will sustain cumulative phase distortions as they pass through these turbulence-induced fluctuations, illustrated in Fig. 1. When imaging over horizontal paths, as opposed to vertical imaging, the degree of image degradation is particularly severe. As a consequence, image resolution is generally limited by atmospheric turbulence rather than by design and quality of the optical system being used.

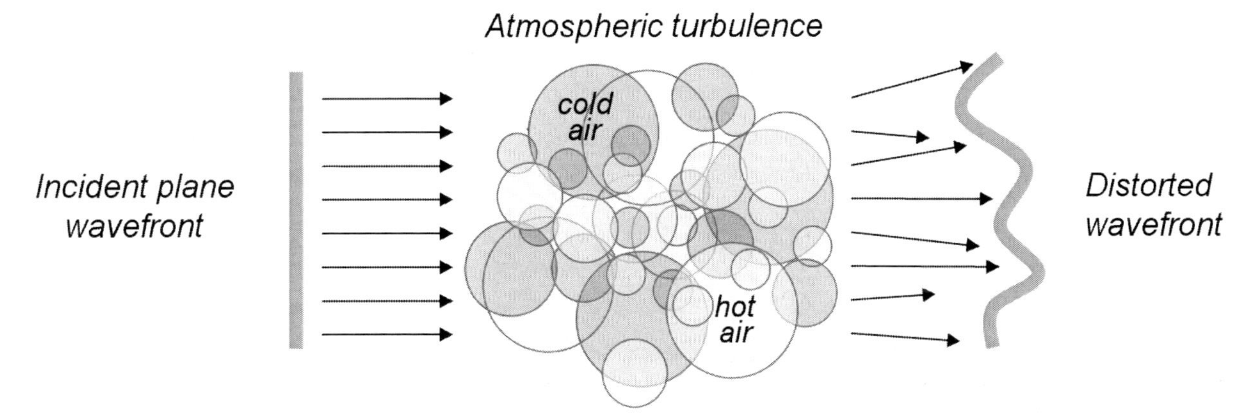

Fig. 1. Schematic, illustrating the propagation of a plane wave of light through a turbulent atmosphere consisting of turbulence cells of hot and cold air.

Suggestions from the field of image processing to compensate for these turbulence effects and restore degraded image sequences include *Motion-Compensated Averaging* (MCA). In isoplanatic conditions, such an averaged image can be considered as a non-distorted image that has been blurred by an unknown *Point Spread Function* (PSF) of the same size

*huebner@fom.fgan.de; phone 49 7243 992-252; fax 49 7243 992-299; www.fom.fgan.de

Mobile Multimedia/Image Processing, Security, and Applications 2009, edited by Sos S. Agaian, Sabah A. Jassim,
Proc. of SPIE Vol. 7351, 735106 · © 2009 SPIE · CCC code: 0277-786X/09/$18 · doi: 10.1117/12.818560

as the pixel motions due to the turbulence and a blind deconvolution algorithm can be employed for the final image restoration. However, when imaging over a long horizontal path close to the ground, conditions are likely to be anisoplanatic and image dancing will effect local image displacements between consecutive frames rather than global shifts only. The inevitable conclusion to be drawn when faced with anisoplanatic conditions, as described more closely in the following section, is the necessity to look for a locally operating algorithm. As the MCA in conjunction with blind deconvolution has proven reasonably successful in the past, especially if applied to data acquired in isoplanatic conditions. Therefore, in this paper, the MCA-procedure, as was described before in [1], is now enhanced by utilizing a *Block Matching* (BM) technique in order to identify and re-arrange uniformly displaced image parts. The algorithm is described in detail and implementation methods are discussed in Section 2. For the final restoration a multi-stage blind deconvolution algorithm is employed in the following section and the corresponding deconvolution results are presented and evaluated in Section 5, followed by a summary and conclusion.

1.1 Anisoplanatism

The spatial frequency content of an image is determined mainly by phase distortions in the received wave fronts due to turbulence close to the optics while volume turbulence over a distance transforms the phase deformations into amplitude distributions, commonly known as scintillation. When imaging extended sources, i. e. objects with a large angular extent, through a turbulent medium, light emanating from different areas of the object will traverse different parts of the turbulence before reaching the optics, entailing a spatial and temporal decorrelation of the wave front. The maximum angle within which incident light waves can be assumed to have passed through almost identical regions of atmospheric perturbations is generally referred to as *isoplanatic angle*. For image acquisition this means the best case is, of course, for the isoplanatic angle to be larger than the *Field Of View* (FOV) of the optics employed since then the whole image has been exposed to the same kind of degradation. Accordingly, the worst case occurs if the isoplanatic angle is smaller than the *Instantaneous Field Of View* (IFOV) of the individual detector elements, meaning *total anisoplanatism*, where the degradation for each pixel has undergone a completely different degradation process. *Local isoplanatism* is a mixture of both, meaning parts but not all of the image will have been degraded uniformly. Assuming that the structure parameter of the refractive index fluctuations C_n^2 is constant along the path L in the case of horizontal propagation, the Fried parameter r_0 and the corresponding isoplanatic angle θ_0 can be expressed for a given wavelength λ as [2]:

$$r_0 = 3.02 \cdot \left(\frac{2\pi}{\lambda}\right)^{-6/5} L^{-3/5} \left(C_n^2\right)^{-3/5} \qquad \text{(Eq. 1)}$$

$$\theta_0 = 0.95 \cdot \left(\frac{2\pi}{\lambda}\right)^{-6/5} L^{-8/5} \left(C_n^2\right)^{-3/5} \qquad \text{(Eq. 2)}$$

1.2 Data and Equipment specifics

The test sequence used in this work was acquired by FGAN-FOM during a field campaign of TG11 (NATO RTG SET) on the Naval Air Weapons Station China Lake, located on the western edge of the Mojave Desert, CA, USA. It should be noted that all information regarding the acquisition of the test sequence (specs, C_n^2, etc.) was taken from [3].

Table 1. Equipment specifications of high-speed camera RETICON MD4256C and optics.

RETICON MD4256C by EG&G High-speed camera with Si-Detector	
Max. responsivity	λ = 800 nm
Bits per pixel	8 bpp
Framerate	Max. 1000 fps
Image size	256 × 256 pixels
Eff. pixel size	14 µm × 16 µm (H × W)
Detector pitch	16 µm × 16 µm

OPTICS	
Aperture	D = 125 mm
Focal length	f = 1250 mm
F-number	10
FOV	0.19° (3.28 mrad)
IFOV	11.2 µrad × 12.8 µrad (H × W)
Angular resolution	dφ = 1.22 λ/D → dφ=7.8 µrad (λ=800 nm)

Time and date of the acquisition was at 1.18 p. m. on July 18th, 2001. Sequence length is 4096 frames with an image resolution of 256 × 256 pixels, recorded at a frame rate of 300 fps. The propagation path between reference target and camera was parallel to the ground at a height of about 1.5 m and with a path length of 1.3 km. The equipment utilized in the acquisition of the data was a RETICON MD4256C high-speed camera with Si-Detector. Its specifications and those of the optics employed are listed in Table 1.

Fig. 2. (a) Reference target on location, object distance to camera: 1300 m; (b) Sample frame from the test sequence.

Fig. 2 (a) shows the pattern chart that was used for the sequence, giving an impression of the location. The complete measurements of the chart can be viewed once more in detail in Fig. 3 (b). Fig. 2 (b) indicates on a sample frame the number of pixels corresponding to the width of the chart, yielding a resolution of approx. 70 pixels per meter, i. e. 1 pixel in the image corresponds to about 1.4 cm in the real scene.

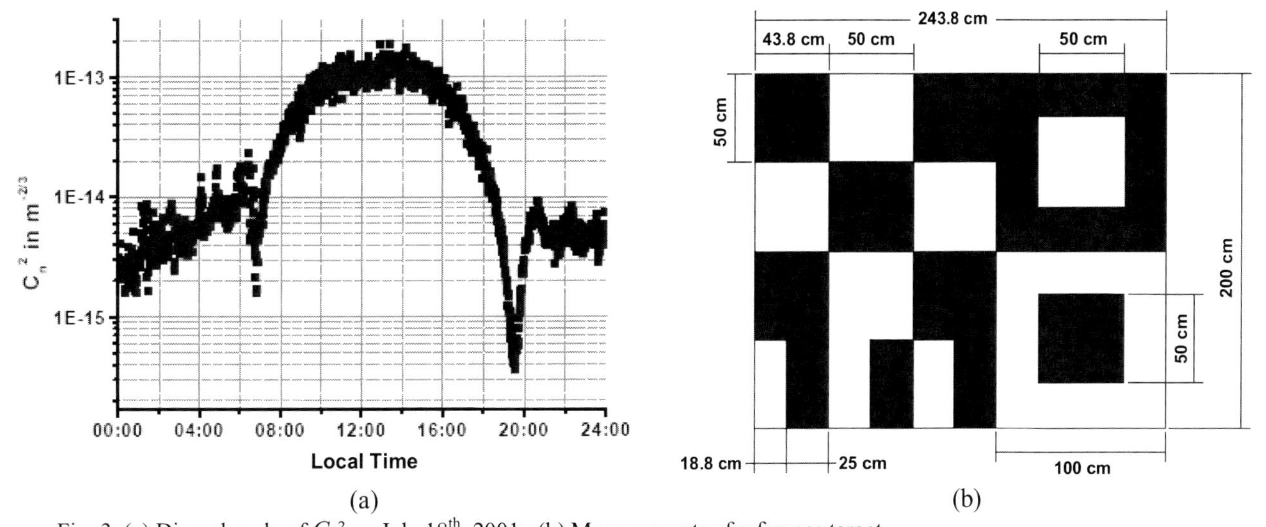

Fig. 3. (a) Diurnal cycle of C_n^2 on July 18th, 2001; (b) Measurements of reference target.

Measurements of C_n^2, the structure parameter of the refractive index fluctuations, were taken with a Boundary Layer Scintillometer BLS900 of Scintec AG. The complete diurnal cycle of C_n^2 for the day of acquisition can be viewed in Fig. 3 (a). As the sequence was recorded during the midday heat, with $C_n^2 = 1.3*10^{-13}\,\mathrm{m}^{-2/3}$ the turbulence at the time can be classified as *strong turbulence*. Accordingly, the exposure time of ≈3.3 ms (300 fps) qualifies for the *long exposure* case even though, in weaker turbulence conditions, it could be considered to be *short exposure.*

Exploiting the information given with regard to the test sequence, (Eq. 1) yields a coherence length $r_0 = 1.17$ cm for $\lambda = 800$ nm, and $r_0 = 0.75$ cm for $\lambda = 550$ nm. Accordingly, (Eq. 2) yields isoplanatic angles $\theta_0 = 2.84$ µrad for $\lambda = 800$ nm and $\theta_0 = 1.81$ µrad for $\lambda = 550$ nm, respectively. As the IFOV of the optical system is significantly larger than the calculated isoplanatic angle(s), conditions during image acquisition can be considered as totally anisoplanatic.

2. MOTION-COMPENSATED AVERAGING

Motion Compensated Averaging (MCA) is in essence the same as normal image integration, the main difference being, that before integrating the next frame of the input sequence it is shifted slightly within a given search space of a number of pixels in every direction such that the input frame better matches the running average. A sliding window of length W is used for calculating this running average or, alternatively, a temporal median as was proposed by [4]. With growing turbulence strength this window length ought to be increased, keeping in mind that the desired characteristics of averaging, like the mitigation of warping effects and noise reduction, need to be weighed against the undesired but unavoidable loss of fine detail which single frames might be able to provide. Since the turbulence is rather strong, a window length of 100 frames was chosen. Originally, greater lengths were tested, e. g. 256 frames) but the disadvantages (motion blur) outweighed the advantages as can be seen when comparing the various average images in Fig. 4 (top).

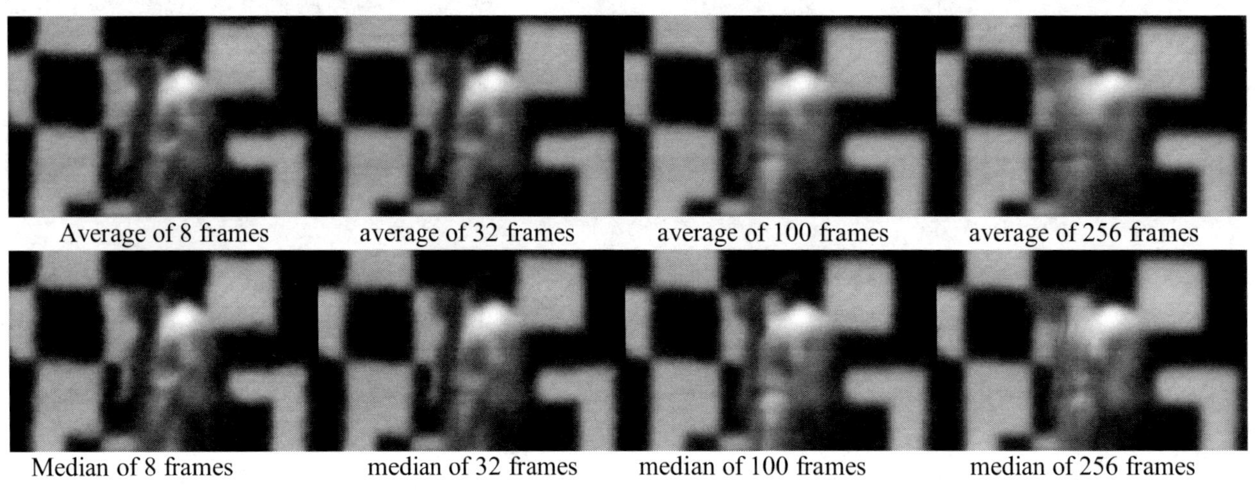

Fig. 4. Comparison of average (top) and median images (bottom) for increasing (from left to right) number of frames.

Obviously, increasing the number of averaged frames proves highly advantageous for obtaining a geometrically correct if not very detailed representation of static objects (background) in that the warped edges of the pattern appear considerably straighter than before. Unfortunately, it has the adverse effect when moving objects are concerned: motion blur is added to the atmospheric blur. The same holds true for the temporal median Fig. 4 (bottom): even though a few more details are discernible, due to the movement of the person on the left, artifacts are introduced when the motion is slow enough to be picked up into the median image. Furthermore, it takes almost twice as many frames as the normal average to achieve the same kind of geometric correctness of the background, so consequently the average was chosen above the temporal median to serve as reference image in the absence of a perfect image.

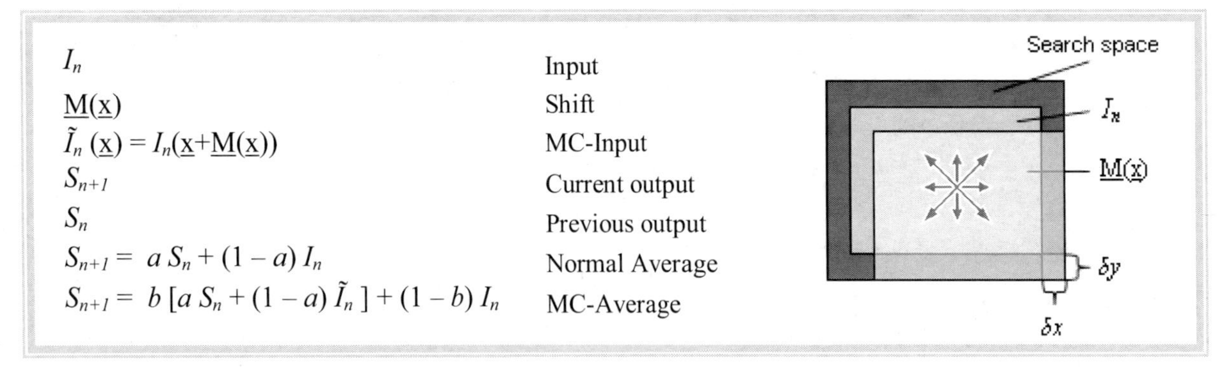

Fig. 5. Illustration of the basic MCA-algorithm.

Increasing the search space, automatically also increases the computational complexity by $(2*N+1)^2$ where N denotes the maximum pixel shift. Accordingly, the shift operation results in a number of $(2*N+1)^2$ shifted images from which the one, that matches the previous output image the best, needs to be determined. To keep calculating time down, the size of the search space was limited to no more than 3 pixels in every direction. Fig. 5 shows a schematic of the basic algorithm.

3. BLOCK-MATCHING

Block Matching (BM) is generally known as a standard technique for encoding motion in video sequences. Motion between consecutive frames is detected in a block-wise fashion by shifting each block from the previous frame to best match the corresponding block in the current frame over a pre-defined search space of a number of N pixels in every direction. In the case of effecting turbulence compensation, the block-shifting is applied to the next frame instead of the previous and it is done to best match the reference image, i. e. the running average, instead of the current frame as a means for correcting the effects of local image dancing where whole image parts are moving in different directions. The basic idea of rearranging the individual blocks of an image to match a given reference image is illustrated in Fig. 6.

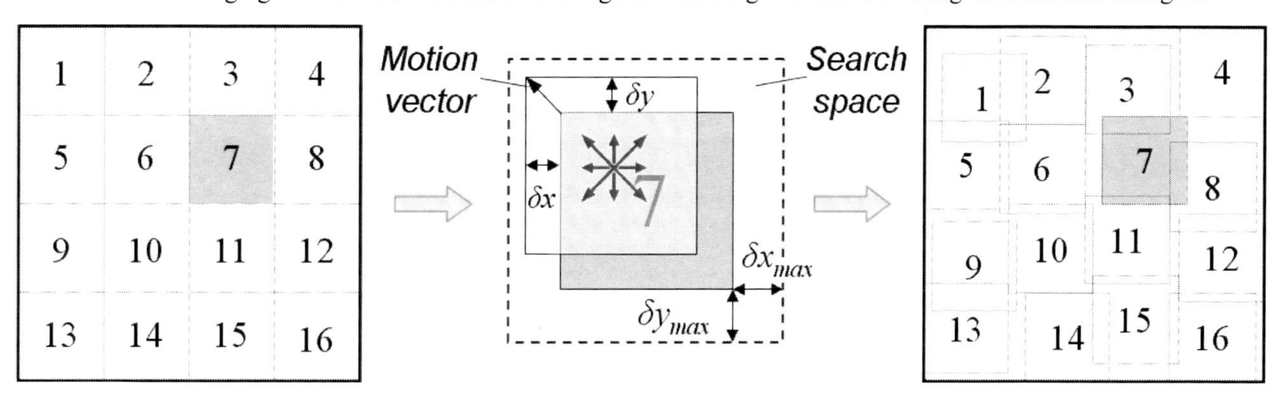

Fig. 6. Schematic, illustrating the block-wise operation of fitting the blocks from the next image frame to match the reference image.

3.1 The Algorithm

Let $I^{(n)}$ denote the n-th input image, $\underline{\mathbf{x}} = (x, y)$ the pixel coordinates, N the size of the search space and N_{max} the total number of shift directions such that $\underline{\mathrm{M}} \in \mathbb{Z} \times \mathbb{Z}$, $\underline{\mathrm{M}}(\underline{\mathbf{x}}) = (\underline{\delta x}, \delta y)$ denotes the vector containing all possible motions $\delta x, \delta y \in \{-N, -N\text{-}1, \ldots, -1, 0, +1, +2, \ldots, +N\}$. The N_{max} shift images $S_i^{(n)}$ for frame $I^{(n)}$ are then given by:

$$S_i^{(n)}(\underline{\mathbf{x}}) = I^{(n)}(\underline{\mathbf{x}} + \underline{\mathrm{M}}(\underline{\mathbf{x}})), \qquad (i = 1, \ldots, N_{max}) \tag{Eq. 3}$$

The n-th reference image R is given by the average of the last W unchanged frames:

$$R^{(n)}(\underline{\mathbf{x}}) = \frac{1}{W} \sum_{k=1}^{W} I^{(n-k)}(\underline{\mathbf{x}}) \tag{Eq. 4}$$

with W denoting the length of the sliding window. While using one of the methods discussed in the next section to determine the *best match*, for every pixel $\underline{\mathbf{x}}$ the corresponding index Idx $(\underline{\mathbf{x}})$ of the best matching block is found by block-wise implementation of:

$$\mathrm{Idx}(\underline{\mathbf{x}}) = \underset{i \in \{1, \ldots, Nmax\}}{\text{Best match}} \left(R^{(n)}(\underline{\mathbf{x}}), S_i^{(n)}(\underline{\mathbf{x}})\right) \tag{Eq. 5}$$

This index is now split up into a number of *mask images* $M_i^{(n)}$, one for every shift direction. $M_i^{(n)}$ will contain ones where Idx $(\underline{\mathbf{x}}) = i$ and zeroes otherwise such that $I_{mc}{}^{(n)}$, the motion compensated version of frame $I^{(n)}$, can be written as the sum of products of corresponding mask and shift images:

$$I_{mc}^{(n)} = \sum_{i=1}^{Nmax} M_i \cdot S_i^{(n)} \tag{Eq. 6}$$

The process of integrating the pieces taken from each shift image into the composite motion compensated image is in five steps exemplified in Fig. 7. The index image on the left cannot truly be considered to be a map of the turbulence insofar as consecutively evaluated pixel motions $\underline{\mathrm{M}}(\underline{\mathbf{x}})$ are not necessarily similar to each other although most of them are. In any case, regions of the same shade of grey are taken from the same shift image.

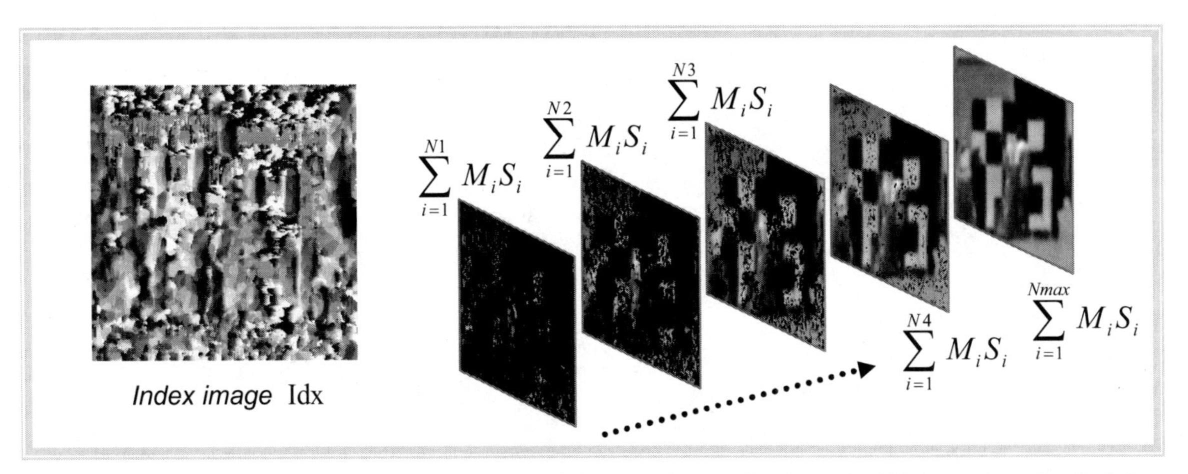

Fig. 7. Schematic, illustrating the integration process of the matching results for each shift image into the final frame. Regions of the same shade of grey in the index image (on the left) are taken from the same shift image.

3.2 "Best Match"-Criteria

There are a number of different approaches to making the decision of what constitutes the "best" or rather "closest" match in the given context. Simple error summation methods as the *Mean Squared Error* (MSE), *Root Mean Squared Error* (RMSE) or *Mean Absolute Error* (MAE) have the advantage of having relatively low computational complexity. Another common but more complex and therefore rather slow method is the *Normalized Cross-correlation*. The RMSE, for instance, also known as *Euclidean Distance*, is defined by:

$$D_E(i, j, \delta_i, \delta_j) = \sqrt{\sum_{m=-\delta x_i}^{+\delta x_i} \sum_{n=-\delta y_j}^{+\delta y_j} \left(g(t)_{i+m, j+n} - g(t+\Delta)_{i+\delta_i+m, j+\delta_j+n} \right)^2} \qquad \text{(Eq. 7)}$$

All of the methods mentioned above have been tested for their suitability in connection with the MCA-algorithm. However, there was little or no discernible improvement over using the simple MAE, so in the end calculating time was the deciding factor. When faced with typical sequence lengths of 1000 up to 20000 frames (or more), computing time becomes an issue even if real-time processing was not given top priority in this work. Converting the MAE to the terminology used in the previous section, it can be written as:

$$MAE(\underline{\mathbf{x}}) = \frac{1}{XY} \sum_{\underline{y} \in Nhb(\underline{\mathbf{x}})} \left| R^{(n)}(\underline{y}) - S_i^{(n)}(\underline{y}) \right| \qquad \text{(Eq. 8)}$$

where $\underline{y} \in Nhb(\underline{x})$ means that $\underline{y} = (x_0, y_0)$ must lie within a defined neighborhood around $\underline{x} = (x, y)$. In this case it is the pixel block surrounding $\underline{x}$ by a number of $(B\text{-}1)/2$ pixels in all directions if B denotes the block size. X and Y denote the image dimensions and therefore XY gives the number of pixels. Accordingly, we get the best match condition by minimizing the argument over all shift images:

$$\text{Best match}_{MAE}(\underline{\mathbf{x}}) = \min_{i=1,\dots,Nmax} \left(\frac{1}{XY} \sum_{\underline{y} \in Nhb(\underline{\mathbf{x}})} \left| R^{(n)}(\underline{y}) - S_i^{(n)}(\underline{y}) \right| \right) \qquad \text{(Eq. 9)}$$

Analogously follows for the best match-condition using RSME:

$$\text{Best match}_{RMSE}(\underline{\mathbf{x}}) = \min_{i=1,\dots,Nmax} \sqrt{\left(\frac{1}{XY} \sum_{\underline{y} \in Nhb(\underline{\mathbf{x}})} \left| R^{(n)}(\underline{y}) - S_i^{(n)}(\underline{y}) \right|^2 \right)} \qquad \text{(Eq. 10)}$$

3.3 Results of Motion Compensation

To avoid a blocky structure of the motion compensated image, the BM was implemented as a sliding neighborhood operation. The disadvantage is that large block sizes invariably slow the algorithm down. For a block size of $B = 7\times7$ pixels, for instance, processing takes an average of 7.3 sec per image when using column-wise sliding neighborhood operation. This reduces to 3.8 sec for a block size of $B = 5\times5$ pixels and to 1.6 sec for 3×3 pixel blocks. If the block size is supposed to be $B = 1\times1$, it takes only 0.26 sec per frame because simply that pixel from all shift images is chosen whose intensity value is closest to the one in the reference image. Since there is no real block matching involved, only a distance metric needs to be minimized, this method is referred to here as the "direct" motion compensation (MC) method. In Fig. 8 a comparison of the MC-results with the original frame is shown for direct method as well as for increasing block sizes B=2×2, B=4×4, B=8×8 and B=16×16, respectively. As the turbulence conditions during acquisition of the test sequence were so severe, there was little use in displaying any results for even larger block sizes. Block sizes of one pixel were strictly speaking already too large due to the small isoplanatic angle.

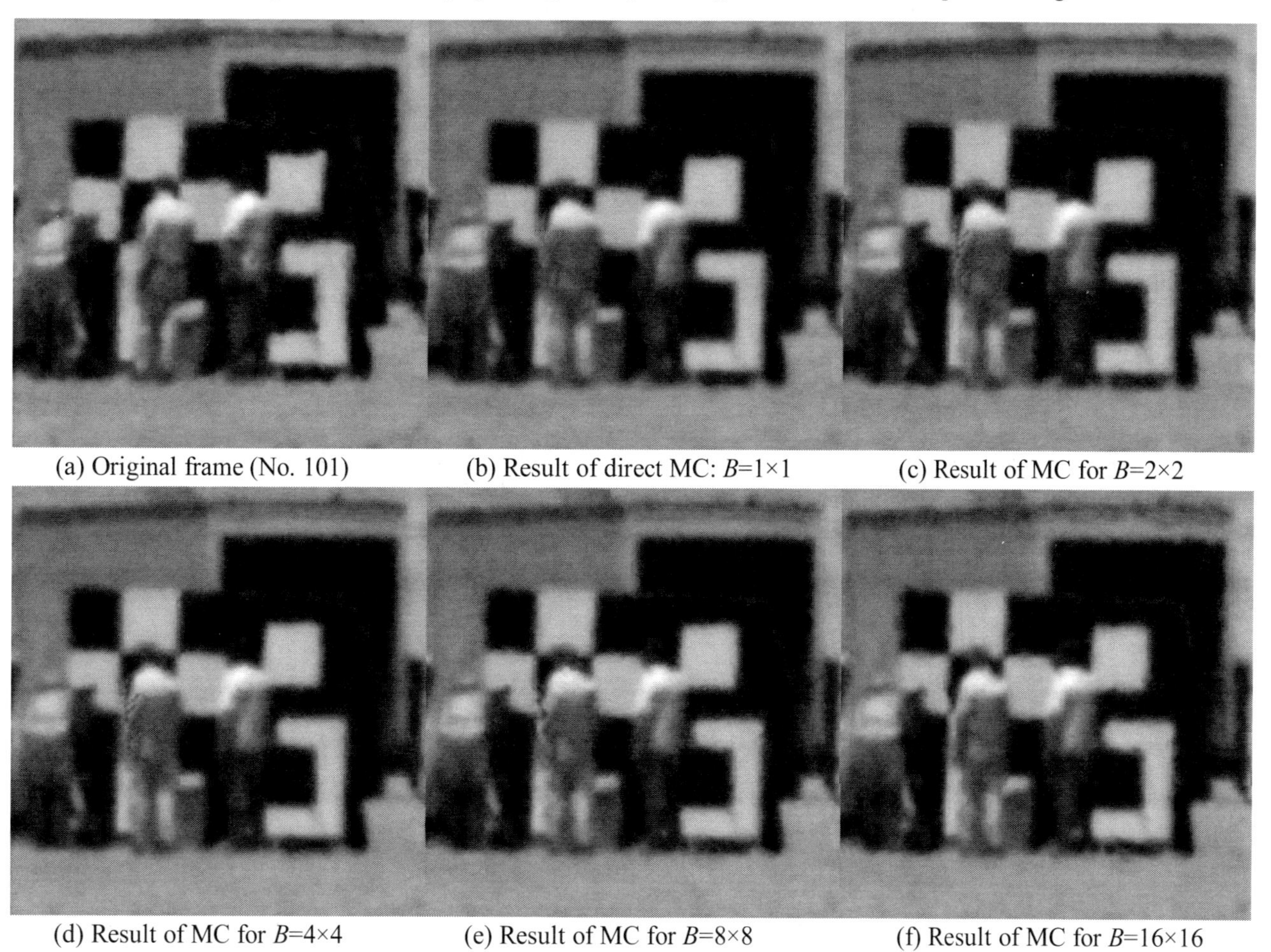

(a) Original frame (No. 101) (b) Result of direct MC: B=1×1 (c) Result of MC for B=2×2

(d) Result of MC for B=4×4 (e) Result of MC for B=8×8 (f) Result of MC for B=16×16

Fig. 8. Comparison of MC-results with original frame for increasing block sizes.

For purposes of clarity all of the sequence images were downsized from the original 256×256 pixels. The stabilizing effect of the block-wise MC for single frames is clearly visible in Fig. 8, especially for very small block sizes, even though it must be noted that the full stabilization effect can only be fully appreciated when seen in a video. Also visible is the unfortunate formation of artifacts when directed movement (as opposed to the undirected movement caused by the turbulence) is involved. This results, of course, from the use of the average as reference image. Unfortunately, increasing the search space above the block size is also prone to create artifacts and even yield erratic results.

4. MULTI-STAGE BLIND DECONVOLUTION

Essentially, a deconvolution describes the procedure of separating two convolved signals f and h. In the spatial domain the blind deconvolution problem takes the general form as illustrated with the image degradation process in Fig. 9:

$$g(x,y) = h(x,y) * f(x,y) + n(x,y) \qquad \text{(Eq. 11)}$$

where g denotes the blurred image, h the unknown blurring function, generally referred to as *Point Spread Function* (PSF), f the true image, * the convolution operator and n an equally unknown additive noise component. To simplify further steps, it is common practice to transfer the problem into the Fourier domain where the relatively complicated convolution-operation becomes a simple multiplication:

$$\mathrm{G}(u,v) = \mathrm{H}(u,v) \cdot \mathrm{F}(u,v) + \mathrm{N}(u,v) \qquad \text{(Eq. 12)}$$

with G, H, F and N denoting the Fourier transforms of g, h, f and n, respectively. Unfortunately, the problem is ill-posed: due to the noise the true image can never be recovered even if the exact blurring function is known. Many attempts at solving this deceptively simple equation for a wide variety of applications can be found throughout literature. An overview of the most popular of these blind image deconvolution algorithms is detailed in [5].

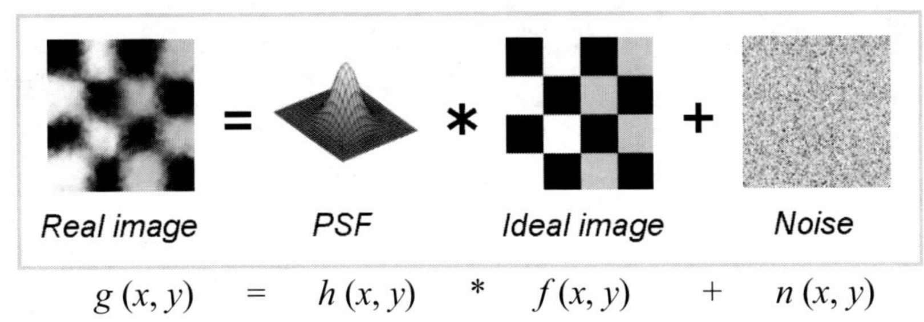

$$g(x,y) \quad = \quad h(x,y) \quad * \quad f(x,y) \quad + \quad n(x,y)$$

Fig. 9.Illustration of the image degradation process to visualize the blind deconvolution problem.

4.1 LRD – Lucy-Richardson Deconvolution

The *Lucy-Richardson Deconvolution* (LRD) algorithm was developed independently by [6] and [7] and is a nonlinear and basically non-blind method, meaning the PSF, or at least a very good estimate, must be a priori known. It has been derived from Bayesian probability theory where image data are considered to be random quantities that are assumed to have a certain likelihood of being produced from a family of other possible random quantities. The problem regarding the likelihood that the estimated true image, after convolution with the PSF, is in fact identical with the blurred input image, except for noise, is formulated as a so-called *likelihood function*, which is iteratively maximized. The solution of this maximization requires the convergence of [8]:

$$\hat{f}_{k+1}(x,y) = \hat{f}_k(x,y)\left[h(-x,-y) * \frac{g(x,y)}{h(x,y) * \hat{f}(x,y)}\right] \qquad \text{(Eq. 13)}$$

where k denotes the k-th iteration. It is the division by $\hat{f}$ that constitutes the algorithm's nonlinear nature. The image estimate is assumed to contain Poisson distributed noise which is appropriate for photon noise in the data whereas additive Gaussian noise, typical for sensor read-out, is ignored. In order to reduce noise amplification, which is a general problem of maximum likelihood methods, it is common practice to introduce a dampening threshold below which further iterations are (locally) suppressed. Otherwise high iteration numbers introduce artifacts to originally smooth image regions.

4.2 Iterative Blind Deconvolution

The *Iterative Blind Deconvolution* (IBD) algorithm, proposed by [9], is mainly a blind version of the LRD algorithm where the PSF needs not to be known, only its support. The IBD is a so-called *Expectation Maximization* (EM) algorithm which is an optimization strategy for estimating random quantities corrupted by noise. In the case of blind deconvolution this means that the likelihood function from the LRD algorithm is again maximized iteratively but with specified constraints until an estimate for the blurring PSF is retrieved from the data along with the estimate for the true image. The IBD algorithm is characterized by a computational complexity in the order of $O(N \log_2 N)$ per iteration where N is the total number of pixels in a single frame while normally more than one iteration is required for its convergence.

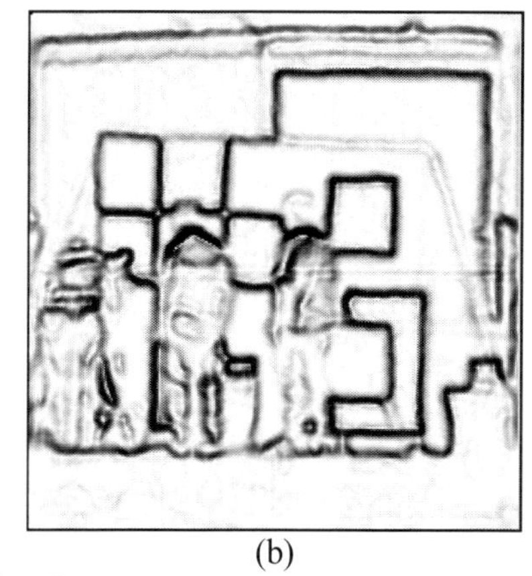

(a) (b)

Fig. 10. Weighting function for multi-stage deconvolution. (a) Sobel filtered average image as weight function (b) Result of Sobel filtering of deconvolution result.

4.3 Multi-stage IBD

The multi-stage variation of the IBD-algorithm used for the image restoration in this work is mainly a concatenation of several ordinary deconvolution steps with varying support size and number of iterations. The main difference is the introduction of a weighting function in form of the Sobel-filtered edge image (see Fig. 10 (a)) of the reference/average image and intensifying the deconvolution where there are edges in the image. The procedure can be broken down into three stages:

1. *Stage*: Deconvolution of the complete image I:

   ➔ $D_1 = \text{IBD}(I)$, PSF sizes: 7, 5, 3, number of iterations: 1, 3, 2

2. *Stage*: Further deconvolution of D_1

   ➔ $D_2 = \text{IBD}(D_1)$, PSF sizes: (7), 5, 3, number of iterations: (1), 3, 5

3. *Stage*: Weighted combination of D_1 and D_2

   ➔ $C_{orr} = E\,D_2 + (1 - E)\,D_1$

The Sobel-filtered edge image of the thus corrected average image is shown in Fig. 10 (b) by way of comparison. The choice of the parameters for PSF size and iteration number depends, of course, on the input data and in this case they had to be appraised by a fair amount of testing.

4.4 Results of Multi-stage IBD

(a) Average of 100 frames (b) Multi-stage deconvolution (c) Ordinary deconvolution

Fig. 11. Comparison of result of multi-stage deconvolution method (b) with the simple average (a) and the result of ordinary blind deconvolution (c). For better visibility, an additional adaptive contrast enhancement (CLAHE) has been used, the same on all images.

Deconvolution, as used on the example, took an average of 5.6 sec per frame and another 0.6 sec if some level of additional image contrast enhancement is employed, e. g. *Contrast Limited Adaptive Histogram Equalization* (CLAHE), as has been done to all of the images shown in Fig. 11. In Fig. 12 the respective cross-sections of the checkerboard pattern on the reference target, indicated on the right, from the corrected image, the average and the current frame are plotted against each other. For an ideal image the transitions would take the form of a perfect rectangle. The improvement of the correction at the transitions between black and white is significant – if slightly exaggerated, considering the little peaks at both transitions. By the human eye transitions with such peaks are perceived as areas of high contrast, giving the impression of heightened sharpness. Reducing the number of deconvolution iterations would reduce the peaks, though.

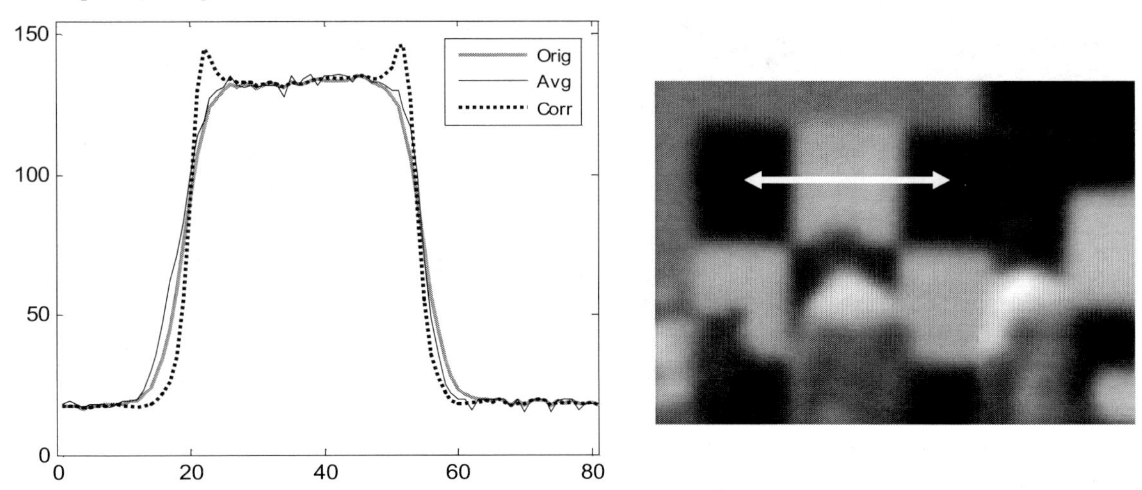

Fig. 12. Comparison of the respective image cross-sections, indicated on the right, of the images in Fig. 11: a single frame ("*Orig*", thick grey line), the average of 100 frames ("*Avg*", thin black line) and the multi-stage deconvolution result from the average image (*'Corr*", dotted black line).

5. RESULTS

(a) Correction of average (b) Correction of MCA for B=2×2 (c) Correction of MCA for B=16×16

Fig. 13. Comparison of the respective results of the multi-stage IBD algorithm applied to (a) the average of 100 frames, (b) the average of 100 motion-compensated frames for a block size of B=2×2 pixels and (c) [same as (b)] for B=16×16.

As demonstrated in Fig. 13, the respective results of the multi-stage IBD algorithm when applied (a) to the average of 100 frames, which is the same as in Fig. 11 only without the CLAHE, or (b) and (c) to the averages of 100 motion-compensated frames for block sizes B=2×2 pixels and B=16×16, respectively, exhibit hardly any differences. The few differences there are have been made visible by difference images in Fig. 14, where (a) shows the difference between the results for the normal and the MC-average with B=2×2 pixels while (b) shows the difference between the results from the MCAs with the two block sizes which is greatest where there is movement in the image (person in center).

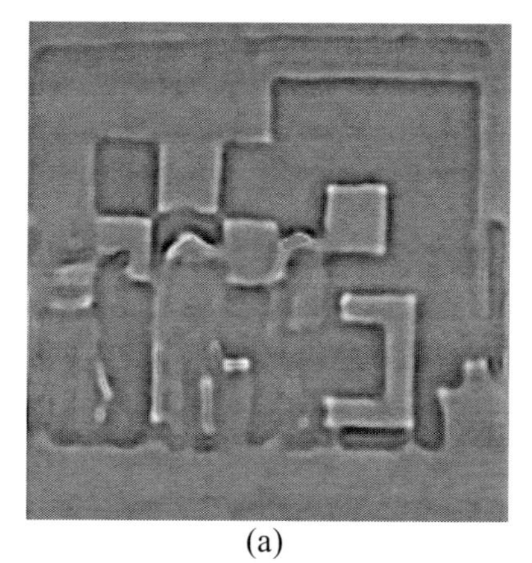

(a)

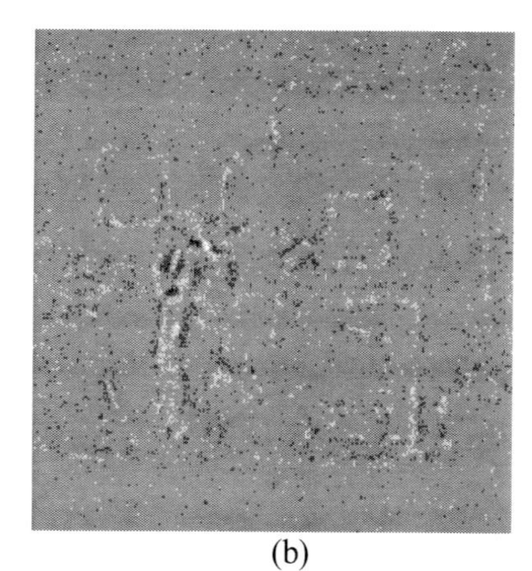

(b)

Fig. 14. (a) Difference image of the results from Fig. 13 (a) and (b); (b) Difference image of Fig. 13 (b) and (c).

6. CONCLUSION

In this work a locally operating, block-wise implementation of the MCA-algorithm has been given and the stabilizing properties of the block-wise motion compensation for single frames, described in sections 2 and 3, have been shown, even though it must be noted that the full stabilization effect can only be fully appreciated when seen in a full-video. Furthermore, for the final restoration a multi-stage procedure involving the IBD-algorithm has been proposed and proven to be successful on the basis of its restoration results. Concerning the results of the combined forces of block-wise MCA and multi-stage blind deconvolution, it seems as if there is no significant improvement over employing the global MCA that might justify the lengthened calculating time. Nevertheless, the augmented stability of a single frame suggests that the number of frames necessary for averaging can be reduced notably, thus potentially increasing the amount of detail to be recovered. Another part of future research will concern moving objects and an adaptation of the block matching principle to the task of identifying image regions that exhibit directed movement such that different correction methods may be applied.

It should be noted that all calculations for this work have been done using MATLAB and implementations in other programming languages may execute faster than mentioned here.

REFERENCES

[1] Huebner, C. S., Greco, M., "Blind Deconvolution Algorithms for the Restoration of atmospherically degraded Imagery: a comparative Analysis", Proc. SPIE 7108, 7108M (2008).

[2] Fried, D. L., "Anisoplanatism in adaptive optics", J. Opt. Soc. Am. 72 (1), pp. 52-61 (1982).

[3] Seiffer, D., "Messung der optischen Turbulenz und räumlichen Turbulenzeffekte", Report FGAN-FOM 2001/30 (2001).

[4] Gilles, J., "Restoration algorithms and system performance evaluation for active imagers", Proc. SPIE 6739, pp. 6739B (2007).

[5] D. Kundur, D. Hatzinakos, "Blind Image Deconvolution", IEE Signal Processing Magazine, 1053-5888/96, pp. 43-64, (1996)

[6] Richardson, W. H. ,"Bayesian-Based Iterative Method of Image Restoration", J. Opt. Soc. Am. 62 (1), pp. 55-60 (1972)

[7] Lucy, L., "An iterative technique for the rectification of observed distributions", Astron. J. 79, pp. 745 (1974).

[8] Gonzalez, R. C., Woods, R. E., "Digital Image Processing", 2nd ed., Prentice Hall, NJ (2002).

[9] Ayers, G. R., Dainty, J. C., "Iterative blind deconvolution method and its applications," Opt. Letters, vol. 13, no. 7, pp. 547–549 (1988).

Deghosting based on In-Loop Selective Filtering using Motion Vector Information for Low-Bit-Rate-Video Coding

Niranjan Dayanand Narvekar[a], Wei-Jung Chien[a], Nabil G. Sadaka[a],
Glen P. Abousleman[b], and Lina J. Karam[a]
[a]Department of Electrical Engineering, Arizona State University, Tempe, AZ 85287
[b]Video Compression & Exploitation Lab, General Dynamics C4 Systems, Scottsdale, AZ 85257

ABSTRACT

In this paper, a technique is presented to alleviate ghosting artifacts in the decoded video sequences for low-bit-rate video coding. Ghosting artifacts can be defined as the appearance of ghost like outlines of an object in a decoded video frame. Ghosting artifacts result from the use of a prediction loop in the video codec, which is typically used to increase the coding efficiency of the video sequence. They appear in the presence of significant frame-to-frame motion in the video sequence, and are typically visible for several frames until they eventually die out or an intra-frame refresh occurs. Ghosting artifacts are particularly annoying at low bit rates since the extreme loss of information tends to accentuate their appearance. To mitigate this effect, a procedure with selective in-loop filtering based on motion vector information is proposed. In the proposed scheme, the in-loop filter is applied only to the regions where there is motion. This is done so as not to affect the regions that are devoid of motion, since ghosting artifacts only occur in high-motion regions. It is shown that the proposed selective filtering method dramatically reduces ghosting artifacts in a wide variety of video sequences with pronounced frame-to-frame motion, without degrading the motionless regions.

Keywords: deghosting, ghosting artifact, in-loop filter, wavelet-based coding, motion vector information

1. INTRODUCTION

In recent years, the transmission of digital video over IP networks has become widespread because increased bandwidth is becoming readily available to a larger number of consumers. However, the scarcity of broadband network facilities in many of the rural areas necessitates the need for low-bit rate, error-resilient video coding techniques.

In [2], a region-of-interest (ROI) based ultra-low bit-rate video coding system based on a modified JPEG2000 image compression core [1] was proposed. This system can accommodate various bandwidth requirements with selectable compression parameters, such as transmission rate, frame rate, and GOP, and with the ability to select multiple ROIs. The ROI coding capability is very important in the case of ultra-low-bit rate video coding applications since it allows the user to specify regions within the video frame to be coded with higher fidelity as compared to the remainder of the frame. On the other hand, in numerous low bit-rate applications, it may be necessary to code the entire frame without specifying any ROI. In these situations, the obtained video quality may be significantly degraded due to various compression artifacts such as blocking, ringing, and ghosting, as a result of low-bit rate video coding.

In the present work, ghosting artifacts are targeted and a selective in-loop filtering method is proposed to alleviate the problem. Ghosting artifacts result from the use of a prediction loop in the video codec and appear as a blurred remnant trailing behind fast moving objects. These ghosting effects become even more prominent at low bit rates since there are not enough bits to represent the information. Since the ghosting artifacts are prominent only in the regions with considerable motion, the in-loop filter is applied only to the blocks where motion vectors are non-zero.

This paper is organized as follows. Section 2 describes the proposed video coding system along with the concept of in-loop selective filtering. Performance results are presented in Section 3. A conclusion is given in Section 4.

a {Niranjan.Narvekar, Wei-Jung.Chien, Nabil.Sadaka, Karam}@asu.edu; phone: 1-480-965-0493
b Glen.Abousleman@gdc4s.com; phone: 1-480-441-2193
This work was supported by General Dynamics C4 Systems.

Mobile Multimedia/Image Processing, Security, and Applications 2009, edited by Sos S. Agaian, Sabah A. Jassim,
Proc. of SPIE Vol. 7351, 735107 · © 2009 SPIE · CCC code: 0277-786X/09/$18 · doi: 10.1117/12.820315

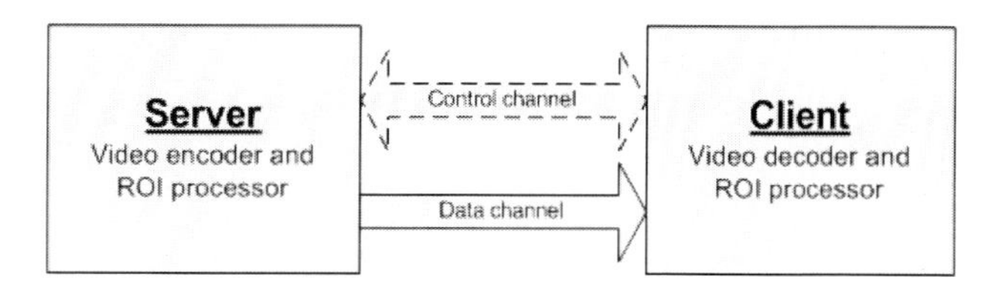

Figure 1. System overview.

2. SYSTEM OVERVIEW

The proposed system is based on a server-client architecture as shown in Figure 1. The system is capable of operating in either point-to-point (one-to-one) or multicast (one-to-many) configurations depending on the chosen network settings. There are two available communication modes between the client and the server, namely, full-control mode and server control mode. The full control mode involves communication via two disparate network channels. One of the channels is a control channel which is a guaranteed-delivery TCP/IP connection between the client and the server (shown as dotted arrow in Figure 1). This is designed for full-duplex communication systems that can support two-way transmission. The critical system information, such as video coding parameters and ROI codewords can be exchanged between the server and the client via this channel. The second channel in the full control mode is a data channel which is a guaranteed-throughput UDP/IP connection (shown as solid arrow in Figure 1). This is a one-way communication channel and is used to transmit time-critical coded video data from the server to the client achieving true real-time video transmission. The server control mode consists of only the UDP/IP data channel and is designed for one-way network communication systems. In such a scenario, there is a possibility of packet loss during UDP/IP data transmission and hence the critical video coding parameters and ROI codewords are transmitted with each video frame.

2.1 Server

A block diagram of the server (transmitter) is shown in Figure 2. It mainly consists of four components: a parameter processor, an ROI processor, a video encoder, and a packet transmitter. The role of the parameter processor is to continuously monitor for any update events from the graphical user interface (GUI), or from the client if the full-control mode is selected. If any update events occur then the parameter is stored into the parameter buffer resulting in a restart of another GOP coded sequence. This is essential because the parameter update could be a resize of the video resolution, for example. The update events are also monitored by the ROI processor to check for any ROI related update events. In order to synchronize the transmitter and receiver, the update parameters and ROIs are transmitted to reflect the status change. In the full-control mode, if the update event is from the server, the parameters and compressed ROIs would be transmitted. If the update event is from the client, only a confirmation message is transmitted. The aim is to minimize network traffic by exchanging only necessary events. In the server control mode, the parameters and the ROI codewords would be transmitted through the data channel along with each compressed video frame.

The video encoder is based on a modified JPEG2000 image compression core [1] to which a motion compensated predictive coding wrapper, with built-in zero-motion, full-search and hexagonal search algorithms, is applied as in [3]. The proposed system in this paper consists of an additional block, which is the in-loop selective filter (discussed in Section 2.3) that is applied after the motion compensation block. This gives rise to three additional predictive coding schemes namely, zero-motion with filtering, full-search with filtering and hexagonal search with filtering. The in-loop selective filter could be switched on or off as required.

The flow of data at the server (transmitter) side can be described as follows. First, the original frame is acquired from the source, which could be a file or a camera. It is then encoded using either intra or inter-frame coding, depending on its location within the GOP. A motion compensated prediction loop is implemented along with the in-loop selective filtering for inter-frame coding, which generates error frames. The discrete wavelet transform (DWT) is then applied on either the original frame or the error frame, and the transform coefficients of each sub-band are quantized using a family of uniform quantizers with refinement quantization steps determined by subband levels and ranges. Based upon the ROI wavelet mask that is stored in the ROI mask buffer, the bitplanes of quantized coefficients are either unchanged or shifted up according to the applied ROI method. The sub-band indices are then divided into codeblocks of fixed sizes and bitplane-coded independently using embedded block coding with optimum truncation (EBCOT) [4] in the order of most-significant to least-significant bitplanes. EBCOT is applied on each bitplane progressively in three coding passes to form an embedded bit stream with three dimensions of scalability (distortion, resolution, and spatial scalability). The

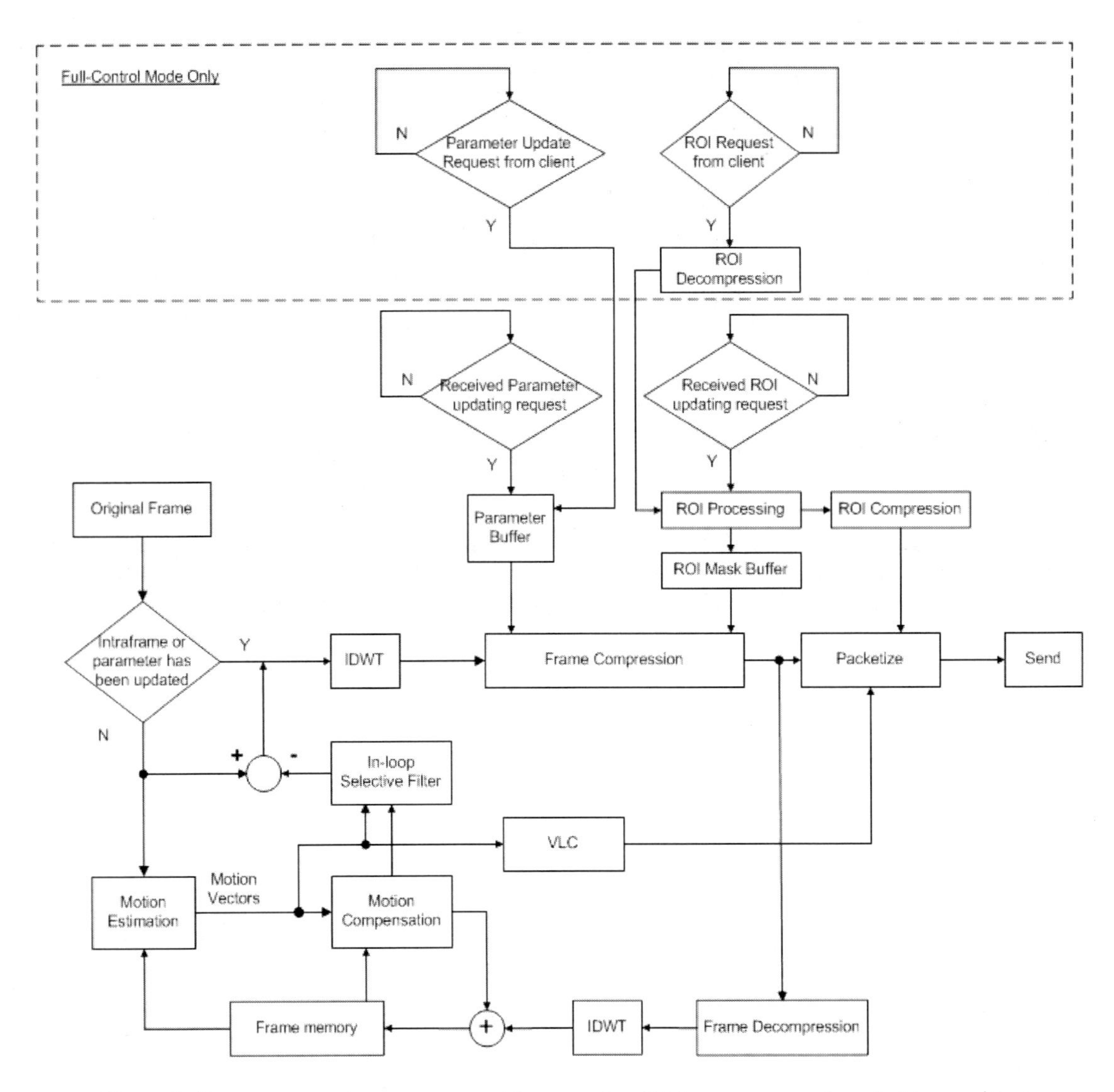

Figure 2. Server: parameter processor, video encoder, ROI processor and packet transmitter.

output symbols from each pass are entropy coded using a context-adaptive arithmetic encoder, and a set of truncation points of rate and distortion (R-D) pairs are calculated for each codeblock. After the bitplanes have been coded, the rate control algorithm is applied to determine the contribution of each to the final bit stream. The packet transmitter packetizes the final bit stream into video packets by adding the frame number, packet type, and packet number. The packets are then transmitted over the network.

2.2 Client

The client (receiver) also has four components as shown in Figure 3. If the system is in full-control mode, the parameter processor and the ROI processor would examine the update events; however, the update events cannot be applied directly to the client. Instead, the events are transmitted to the server and the client waits for the confirmation messages from the server before applying the update events. As for the video decoding, the procedure can be described as follows. First, a packet receiver is responsible for receiving video packets from the predefined channel and extracting the video codewords from the video packets and placing them in a video buffer. The buffer is designed to overcome network transmission problems such as jitter and out-of-order packets. Secondly, the parameter processor extracts all video parameters and the ROI processor decompresses all ROI codewords. Finally, the video decoder decodes the received bit stream into the quantized coefficients, which have shifted or non-shifted bitplanes. The bitplanes of the quantized coefficients are either shifted down or unchanged according to the ROI mask, which is stored in the ROI mask buffer.

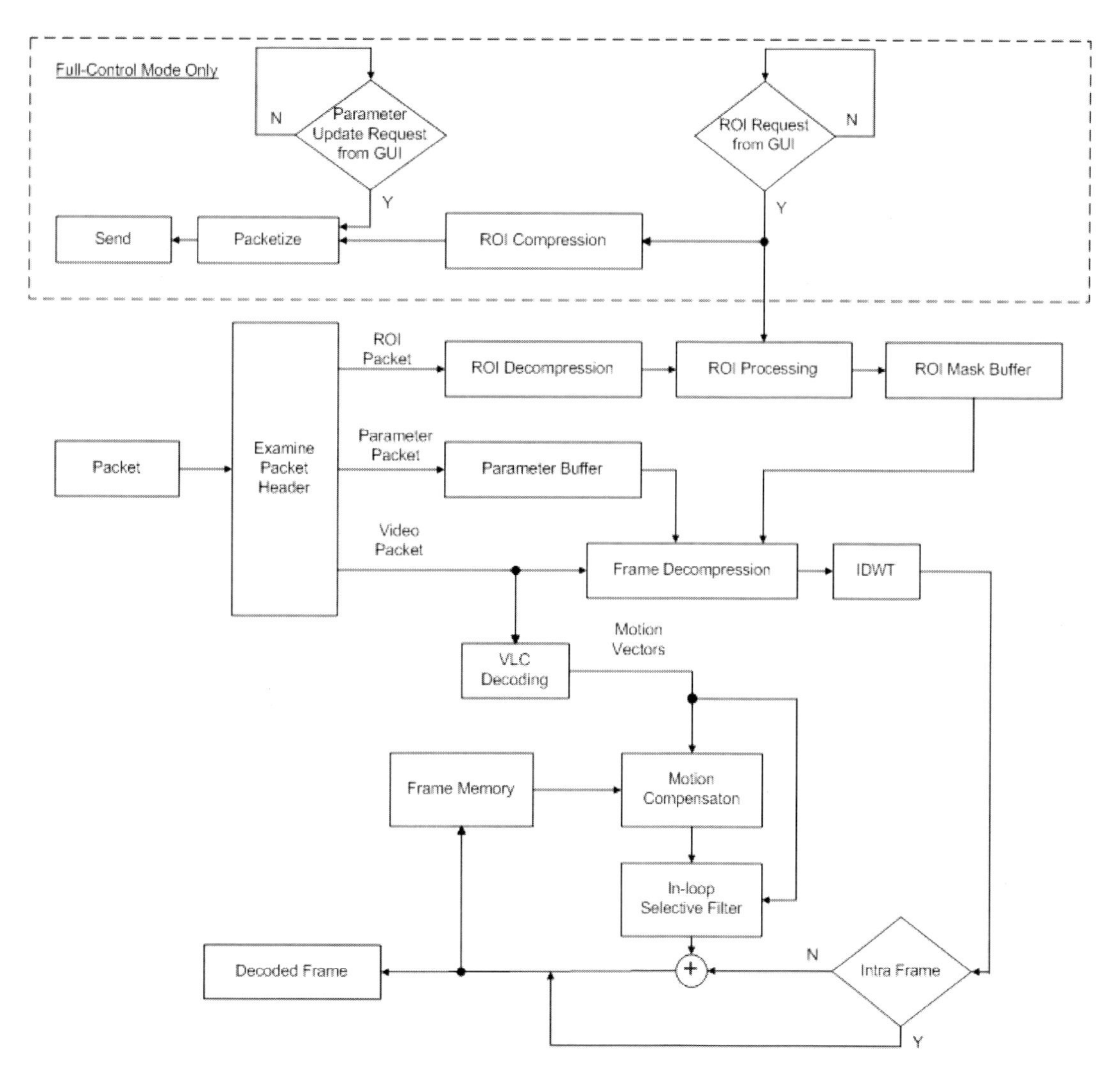

Figure 3. Client: parameter processor, packet processor, video decoder, and ROI processor.

After applying the inverse wavelet transform, the result is added to the selectively filtered motion compensated frame and the reconstructed frame is obtained.

2.3 In-loop Filter

The system architecture explained earlier is similar to the one proposed in [2] except for the addition of a selective in-loop filter. The role of the in-loop filter is to mitigate the ghosting effect. Ghosting occurs due to the predictive coding loop used in the described system which tends to introduce high frequency artifacts in the presence of quantization noise. It becomes very prominent when significant motion occurs from one frame to another, especially at low bit rates.

The proposed in-loop selective filtering scheme is summarized in Figure 4. As shown in Figure 4, while encoding an inter-frame, motion estimation is performed with respect to the reference frame and the blocks that contain motion within the considered frame, are identified. Before obtaining the residual frame, only those blocks that contain motion are filtered. The residual frame is then obtained by subtracting the newly formed, selectively filtered, motion-compensated frame from the original frame. The selective in-loop filtering is performed so as not to affect the regions that are devoid of motion, since ghosting artifacts mainly occur in high-motion regions. The application of selective filtering results in the smoothening of the high frequency artifacts and helps in significantly reducing the ghosting effects.

An important requirement of the in-loop filter is that the decoder should be able to mimic the operation at the encoder. To accomplish this, the filter coefficients that are used for the selective filtering are needed at the decoder

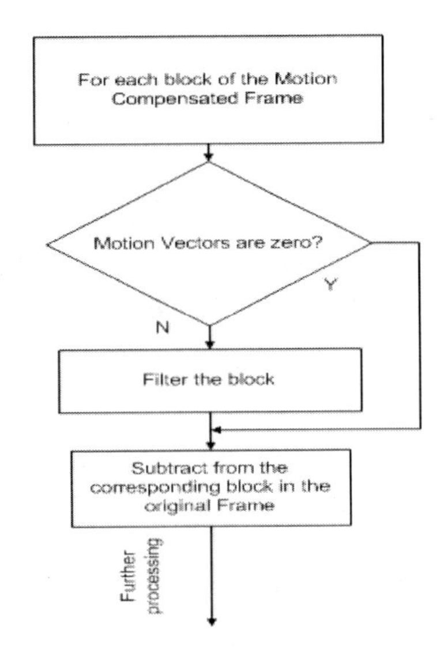

Figure 4. Flowchart summarizing the operation of the proposed in-loop selective filtering.

together with the motion vector information. In the proposed scheme, the same fixed filter coefficients are used at the encoder and the decoder; so, no additional information regarding the filter coefficients needs to be transmitted between then encoder and decoder. As indicated earlier, the proposed system uses three types of motion estimation techniques namely, zero motion, full search and hexagonal search. In the cases of full search and hexagonal search motion estimation, the motion vectors are transmitted from the encoder to the decoder and, hence, no additional information needs to be transmitted in order to be able to selectively filter the blocks with motion content. But, in the case of zero motion (DPCM) mode, motion estimation is performed at the encoder in order to check whether the considered block has motion or not. In this latter case, the motion vectors are not sent to the decoder (client), but instead a binary map (one bit per block) is generated in the raster scan order of the blocks and is sent to the decoder. In the generated and transmitted binary map, a bit value of 1 indicates that the corresponding block has a non-zero motion vector, whereas a bit value of 0 indicates that the block has a zero motion vector (no motion). This map is then utilized at the decoder to perform selective filtering for the blocks with motion.

In our current implementation, the in-loop selective filtering was performed by means of a lowpass separable Gaussian FIR filter with a standard deviation of 1.2. Results obtained with different filter sizes are reported in Section 3.

3. SIMULATION RESULTS

The performance of the proposed system is demonstrated by compressing the CIF (352 x 288) Foreman and Mother-Daughter video sequences. The sequences were compressed at 10 frames per second (fps) at a bit rate of 50 kilobits/second (kbps). Table 1 gives the average peak signal-to-noise ratio (PSNR) of the first 300 frames of the test sequences which are compressed without using the in-loop selective filter. Tables 2, 3 and 4 list the average PSNR of the first 300 frames of the test sequences in which the frames are compressed using the proposed system with in-loop selective filtering. The results in Table 2 are generated using an in-loop selective 3x3 Gaussian filter with a standard deviation 1.2. Similarly, the results in Tables 3 and 4 are generated using in-loop selective Gaussian filters having a standard deviation equal to 1.2 and of size 5x5 and 7x7, respectively. For these simulations, motion estimation was performed on 16x16 blocks with integer-pel accuracy.

From Tables 1 to 4, it can be seen that the proposed method results in PSNR gains of 0.2 to 0.5 dB. Also, it can be observed that the 7x7 Gaussian filter performs best when the DPCM (no motion estimation) mode is used, while the 3x3 Gaussian filter performs best when any of the motion estimation methods namely, full search and hexagonal search, are used. Although the PSNR gains are not very significant, the proposed method results in a significantly improved visual quality. This fact is illustrated in Figures 5, 6 and 7. Figure 5(a) shows the 139th frame of the Foreman sequence compressed using the DPCM mode without using the in-loop selective filtering. For comparison, Figure 5 (b) shows the 139th frame of the Foreman sequence using the proposed in-loop selective filtering. It can be clearly observed that, while the ghosting artifacts are very prominent in Figure 5(a), their effect is significantly reduced using the proposed method as shown in Figure 5(b). Figures 6(a) and 6(b) show, respectively, the 155th frame of the Mother-Daughter sequence compressed using the DPCM mode without and with the proposed in-loop selective filtering. Here again, it can be observed that the ghosting artifacts are significantly reduced in the areas where there is motion. Another observation worth noting is that the background information in Figures 6 (a) and 6 (b) remains unchanged. This is because of the selective filtering nature of the proposed method, wherein areas devoid of motion are not filtered. Similar observations can be made by comparing Figures 7(a) and 7(b), which show, respectively, Frame 158 of the Mother-Daughter sequence compressed using the full-search motion estimation mode without and with in-loop selective filtering.

Table 1. Average PSNR of the first 300 frames coded at a bit-rate of 50 kbps and a frame rate of 10fps without in-loop selective filtering.

Sequence	PSNR		
	No Search	Full Search	Hexagonal Search
Foreman	26.1679	26.3103	26.1562
Mother-Daughter	33.8615	34.0258	33.9891

Table 2. Average PSNR of the first 300 frames coded at a bit-rate of 50 kbps and a frame rate of 10fps with in-loop selective Gaussian filtering of size 3x3 and standard deviation 1.2.

Sequence	PSNR		
	No Search + Filtering	Full Search + Filtering	Hexagonal Search + Filtering
Foreman	26.5169	26.4287	26.2578
Mother-Daughter	33.967	34.1935	34.1501

Table 3. Average PSNR of the first 300 frames coded at a bit-rate of 50 kbps and a frame rate of 10fps with in-loop selective Gaussian filtering having of size 5x5 and standard deviation 1.2.

Sequence	PSNR		
	No Search + Filtering	Full Search + Filtering	Hexagonal Search + Filtering
Foreman	26.5451	26.3971	26.2506
Mother-Daughter	33.9701	34.1532	34.1274

Table 4. Average PSNR of the first 300 frames coded at a bit-rate of 50 kbps and a frame rate of 10fps without in-loop selective Gaussian filtering of size 7x7 and standard deviation 1.2.

Sequence	PSNR		
	No Search + Filtering	Full Search + Filtering	Hexagonal Search + Filtering
Foreman	26.5581	26.3872	26.2462
Mother-Daughter	33.99	34.1547	34.1202

4. CONCLUSION

In this paper, an in-loop selective filtering scheme based on motion vector information was proposed to alleviate ghosting artifacts. This scheme does not require transmission of any additional information from the encoder to the decoder except in the DPCM mode where a binary map corresponding to motion content information is transmitted.

Simulation results show that the proposed scheme improves the visual quality of the transmitted video by reducing the ghosting artifacts without degrading the motionless regions.

REFERENCES

[1] ISO/IEC 15444-1, "JPEG 2000 Part I Final Committee Draft," ISO/IEC, version 1.0, March 2000.

[2] Chien, W., Sadaka, N.G., Abousleman, G.P., Karam, L.J., "Region-of-Interest-Based Ultra-Low-Bit-Rate Video Coding," Proceedings of the SPIE, The International Society for Optical Engineering, v 6978, p 69780C-1-9, 2008

[3] Chien, W., Lam, T., Abousleman, G.P., Karam, L.J., "Automatic network-adaptive ultra-low-bit-rate video coding," Proceedings of the SPIE, The International Society for Optical Engineering, v. 6246, n. 1, p 624606-1-10, 2006.

[4] ISO/IEC 15444-2, "JPEG 2000 Part II Final Committee Draft," ISO/IEC, December 2000.

(a) (b)

Figure 5. CIF Foreman sequence: 139th frame at 10 fps compressed at 50kbps. (a) Frame compressed using the DPCM mode without filtering, PSNR: 24.8977 dB; (b) Frame compressed using the DPCM mode and in-loop selective filtering, PSNR: 25.7944 dB.

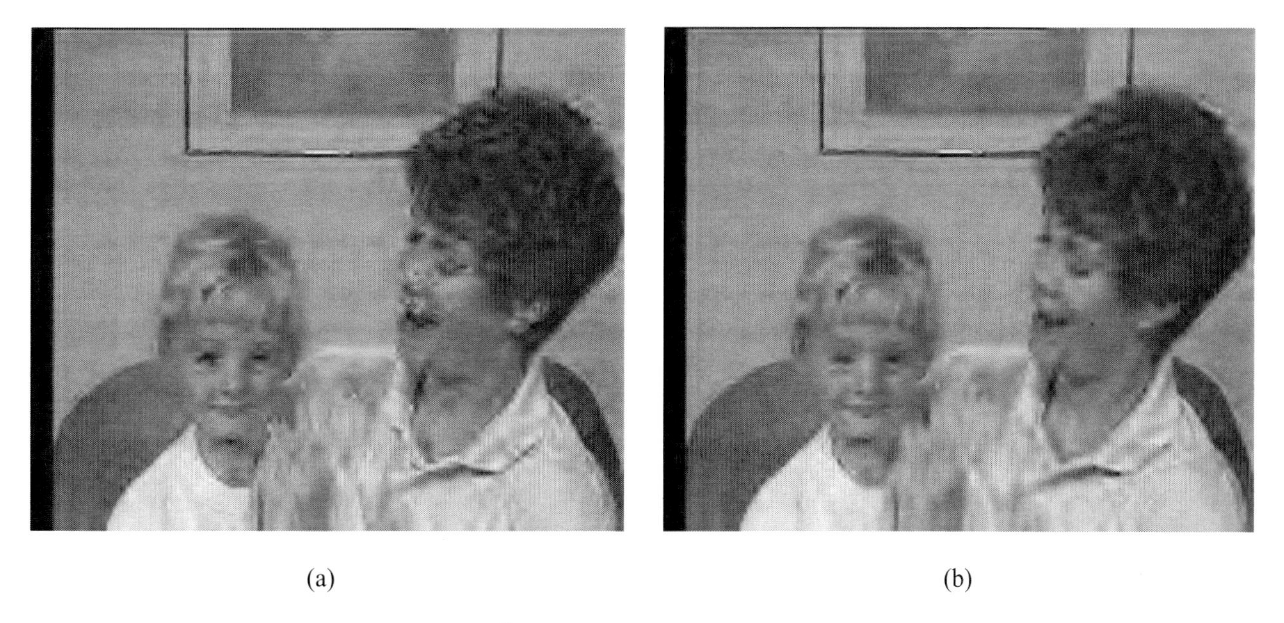

(a) (b)

Figure 6. CIF Mother-Daughter sequence: 155th frame at 10fps compressed at 50kbps. (a) Frame compressed using the DPCM mode without filtering, PSNR: 31.9590 dB; (b) Frame compressed using the DPCM mode with in-loop selective filtering, PSNR: 32.7021 dB.

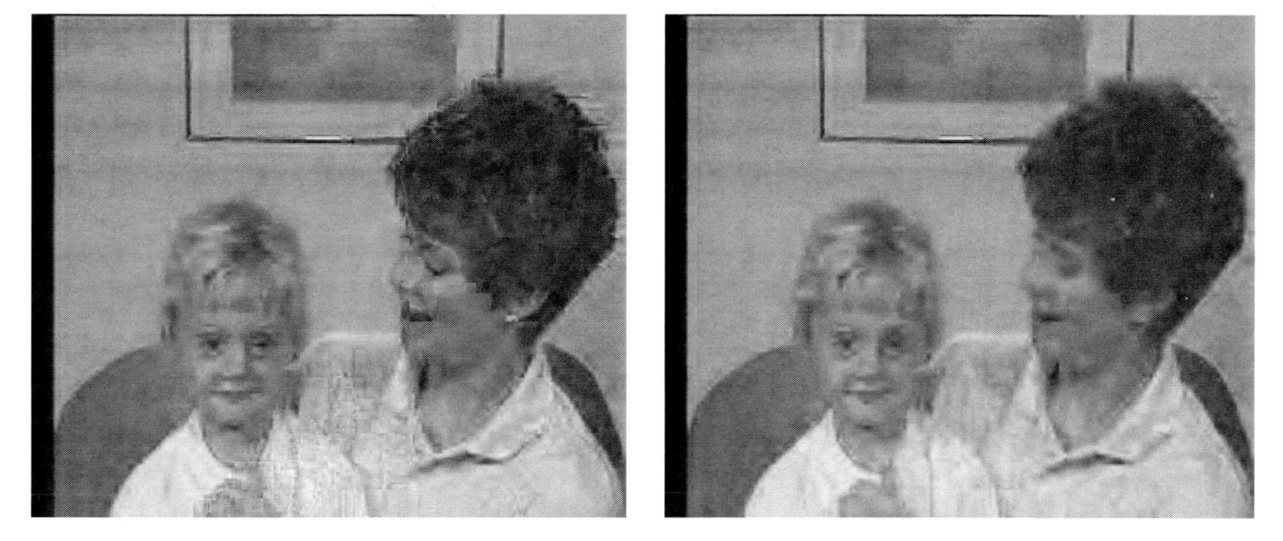

Figure 7. CIF Mother Daughter Sequence: 158th frame at 10fps compressed at 50kbps. (a) Frame compressed using full search motion estimation and motion compensation without filtering, PSNR: 31.3515 dB; (b) Frame compressed using full search motion estimation and motion compensation with in-loop selective filtering, PSNR: 32.4839 dB.

Integrity Monitoring In WLAN Positioning Systems

Sri Phani Yerubandi, Bhargav Kalgikar, Maheedhar Gunturu, David Akopian, Philip Chen
The University of Texas at San Antonio, San Antonio, TX, USA

ABSTRACT

Indoor Positioning Systems using WLANs have become very popular in recent years. These systems are spawning a new class of applications like activity recognition, surveillance, context aware computing and location based services. While Global Positioning System (GPS) is the natural choice for providing navigation in outdoor environment, the urban environment places a significant challenge for positioning using GPS. The GPS signals can be significantly attenuated, and often completely blocked, inside buildings or in urban canyons. As the performance of GPS in indoor environments is not satisfactory, indoor positioning systems based on location fingerprinting of WLANs is being suggested as a viable alternative. The Indoor WLAN Positioning Systems suffer from several phenomena. One of the problems is the continual availability of access points, which directly affects the positioning accuracy. Integrity monitoring of WLAN localization, which computes WLAN positioning with different sets of access points is proposed as a solution for this problem. The positioning accuracy will be adequate for the sets which do not contain faulty or the access points which are offline, while the sets with such access points will fail and they will report random and inaccurate results. The proposed method identifies proper sets and identifies the rogue access points using prediction trajectories. The combination of prediction and correct access point set selection provides a more accurate result. This paper discusses about integrity monitoring method for WLAN devices and followed by how it monitors and developing the application on mobile platforms.

Keywords: Indoor Positioning, Integrity Monitoring, WLAN Positioning.

1. INTRODUCTION

It is now a common idea that users can determine the client's position with Global Positioning System (GPS) [1] receivers at any location in the world. When the user is in the open view of sky, his GPS receiver is expected to operate with high accuracy. While GPS provides outstanding performance outdoors it is not designed for indoor and urban canyon areas where satellite signals are very weak. Indoor areas are covered by cellular signals and the position can be estimated using base stations as beacons and applying trilateration techniques. The accuracy of these approaches is very poor though. Another widely deployed indoor system is WLAN[2] also known as Wi-Fi. If we go through the history of WLAN, we can say that positioning and tracking are new services for WLAN. Nowadays, Wi-Fi is widely deployed in indoor areas like universities, hospitals, hotels, and airports. Positioning solutions using Wi-Fi infrastructure is considered as a competitive and promising technology as reviewed in the following. WLAN penetration is remarkable evidenced by double digit growth of WLAN semiconductor products and an expected growth of $4 billion by 2012 [3]. This growth is also reflected in wide deployment of WLAN applications in cellular phones which made it possible to use this network for positioning tasks as well. The newest popular entrant into the mobile phone industry, Apple iPhone makes use of this self localization feature to pinpoint position in indoor and urban areas[4]. There are various applications of indoor positioning systems. Some are listed below:

- Finding in-demand persons like doctors in hospital, or in-demand equipment like closest available printer.
- Positioning and guiding fire-fighters in a building to rescue hostages.
- Guided tours with a wireless device in museums, universities or exhibitions.
- Locating an illegal user in the secured network like the university network.
- Locating a friend on campus.

WLAN technology is a form of wireless Ethernet networking standardized by the IEEE 802.11 Working Group (WG). Since the formation of the WG in 1990, they produced a series of 802.11 standards[5]. Table-1 gives the information about the list of frequencies and data rates defined by the IEEE 802.11 WG for WLAN networks. The areas standardized by the IEEE 802.11 within the first and second layers of the OSI Seven Layer Model, which is referred to as the Physical and Data Link Layers. The Figure-1 shows the graphical representation of how the IEEE 802.11 standard varies with

Mobile Multimedia/Image Processing, Security, and Applications 2009, edited by Sos S. Agaian, Sabah A. Jassim,
Proc. of SPIE Vol. 7351, 735109 · © 2009 SPIE · CCC code: 0277-786X/09/$18 · doi: 10.1117/12.818338

OSI seven layers Model. The Data Link Layer is subdivided into the Logical Link Control (LLC) and Media Access Control (MAC) sub-layers. The network MAC sub-layer for 802.11 is based on the Carrier Sense Multiple Access with Collision Avoidance (CSMA/CA) channel access methods. It serves as a common interface for Physical Layer Protocols below it. The Physical Layer is responsible for the actual Radio Frequency (RF) transmission and defines the frequencies used. The Subsequent Physical standards (802.11a/b/g) based on better modulation techniques which are now in wide use. Lots of research and projects are conducted to compute the effective position of a user in positioning systems; for example, Google has recently launched a commercial positioning system called Google Lattitude[6] for tracking of a client. Depending on the obstructions and signal strengths, the accuracy achieved may be 6-7 feet in rooms and 100-200 meters outdoors. Nowadays, Ekahau[7], Newbury[8] and Bluesoft[9] companies are providing commercial versions of WLAN based positioning systems. The mentioned vendors are not too open and they just discuss how their products determine the user's position according to their own technologies and methods. We use the Ekahau positioning system as a reference for our research. An example of a commercially available Wi-Fi positioning system on handsets is the one offered by Skyhook Wireless[3]. Skyhook provides a reference database that contains information about every access point in a given place of interest and updates them on an ongoing and continuous process for data monitoring and analysis. When a mobile unit wants to find its position, it scans for the available Wi-Fi access points in its vicinity and sends this information to Skyhook servers, which compares the received data with the stored database information and relays the location information back to the mobile unit. The known approaches are based on theoretical foundations which encounters problems if WLAN access points are frequently inaccessible. The fingerprints collected during training phase will not match fingerprints of online phase. To overcome these problems, this paper presents a new approach of integrity monitoring to reduce tracking errors and robustness in such environments.

Table 1: WLAN Data rates of IEEE 802.11 standards

IEEE Standard	Year Released	Maximum Data Rate	ISM Frequency Band
802.11	1997	2 Mb/s	2.4 GHz & IR
802.11b	1999	11 Mb/s	2.4 GHz
802.11a	1999	54 Mb/s	5.0 GHz
802.11g	2003	54 Mb/s	2.4 GHz

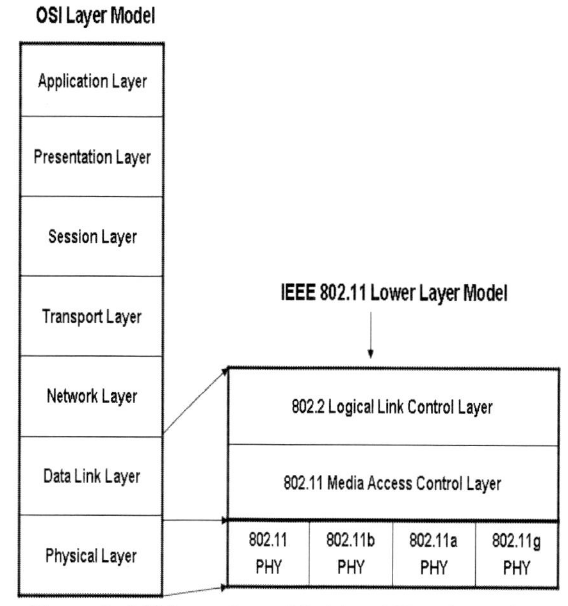

Figure-1: OSI Seven Layer Model and IEEE layered

2. LOCATION FINGERPRINTING

Our main goal of this paper is to decrease the location and tracking errors in indoor environments with integrity monitoring algorithm using existing WLAN structure. This technique takes the major advantage of one of the available outputs of a standard WiFi card, which is the Received Signal Strength Indicator (RSSI) from each AP. Given this consideration outputs of a standard WiFi card, this is the Received Signal Strength Indicator (RSSI) from each AP. Given this consideration, it is possible to get a list of RSSI coming from all the APs covering the area where the laptop/mobile is moving. The simplest approach for locating a WiFi device in a WLAN environment, using this available information, is to approximate its position by the position of the APs received at that location with the strongest signal strength, but its main drawback is its large estimation error. To reduce this error we developed an integrity monitoring algorithm. Fingerprinting mainly consists of signal power footprints or signatures that define a position in the environment. The footprints are the received signal strengths from different APs that cover the environment. Generally, the fingerprinting based positioning systems can be divided in two phases. The first step, called the training or calibration phase, is necessary to build this mapping between collected received signal strength and certain positions in the environment. This leads to a database that is used during the positioning phase. The location fingerprinting are collected by making a site survey of the RSSI from all available multiple access points. Some access points which are not available or offline will be ignored. The site survey is performed in the form of rectangular grid of points. Equally spaced points with a distance of 5 feet are marked in the zone and they are referred to as reference points. The distance between two closest grid points is called grid spacing and is usually represented in meters or feet. The RSSI values are measured with enough statistics to create a database. All the values of RSSI are stored in a database in the form of tables that are used to compare with online phase RSSI values which computes the user position by different methods. We use a commercially available system from NetStumbler which measures the signal strengths from available access points and which can be stored. Secondly, the online phase consists of retrieving the data related to the signal strengths from the NetStumbler software[10] and writing data into a SQL database. We compare the data obtained during the tracking phase to the reference data obtained during the survey phase using this conventional probabilistic approach to obtain the position of the Wi-Fi enabled device.

2.1 Probabilistic Approach

Now, let us explore about the conventional probabilistic approach. The Probabilistic method assumes that location fingerprinting is described by a conditional probability distribution, or likelihood function $P(L/F)$. This method deals with mean and standard deviation of the received signal strength values to compute the position. For example, assuming Gaussian measurement distribution, one can estimate mean and variance at each grid point using multiple observations. A statistical model is designed for all the points on the grid. Alternatively, different observation sets can be combined as a weighted sum of several equally weighted Gaussian kernel functions at locations L, given the fingerprint F. A conditional probability and Bayesian inference can be used to estimate location. Let us assume that mean, ρ_i and standard deviation σ_i is adequately described. Assuming that N access points are found at one location and at time window T, the fingerprint vector can be described as

$$F = (\rho_1, \rho_2, \rho_3 \ldots\ldots\ldots, \rho_n)^T \tag{1}$$

And the standard deviation is another vector that can be described as

$$D = (\sigma_1, \sigma_2, \ldots\ldots\ldots, \sigma_n)^T \tag{2}$$

At a particular fingerprint location on the map, n RSSI samples are taken from a single access point. Each sample is assumed to be Gaussian distributed with a mean of ρ and a measured standard deviation of σ. The resulting likelihood function is the weighted sum of n equally weighted Gaussian kernel functions at a particular location given by,

$$P(L \mid F) = \frac{1}{n}\sum_{i=1}^{n}\left[\frac{1}{\sqrt{2\pi}\sigma}\exp\left(-\frac{(s-\rho_i)^2}{2\sigma^2}\right)\right] \tag{3}$$

Taking into account multiple access points and assuming that they are independent of each other, we can estimate the conditional probability $P(F \mid L) = P(s_1 \mid L)P(s_2 \mid L)\cdots P(s_N \mid L)$. It is then possible to estimate a posteriori distribution of a particular location. Assuming that the a priori probability $P(L)$ of each location is known (initially could be equally likely), we can apply Bayes' rule to find the conditional probability of the location, L given the fingerprint F

$$P(L \mid F) = \frac{P(F \mid L)P(L)}{P(F)} = \frac{P(F \mid L)P(L)}{\sum_{k \in L} P(F \mid L_k)P(L_k)} \tag{4}$$

We compute the position of a user by matching the RSSI values of a respective location with other grid points, and we calculate probability using Gaussian kernel functions with other locations. According to Gaussian curve, the result shown must be having the highest probability of received signal strength at that location. Hence this method would select the location fingerprint that has the highest estimated posterior probability. By this method, we can locate the position of a user accurately. However, the signal strength fluctuations introduced many unexpected errors in the final trajectory. Reducing these tracking errors can be done by integrity monitoring algorithm. Many of these errors are caused due to obstructions for the signals and noise in the environment.

2.2 Problems of Error in Location Detection

Changes to the signal due to propagation indoors is difficult to predict because of the dense environment and propagation effects such as Reflection, Obstacles and Noise in the environment which causes the received signal to fluctuate around a mean value at a particular location. All these issues are causing the tracking errors and problems in location detection. There are so many cases for a received signal disappearing for a moment and appearing after some period of time. According to our research, we classify the following as the causes for RSSI disappearing:

- Weak signals;
- Strong signals, but disappear for a moment and;
- When Access points are overloaded or malfunctioned.

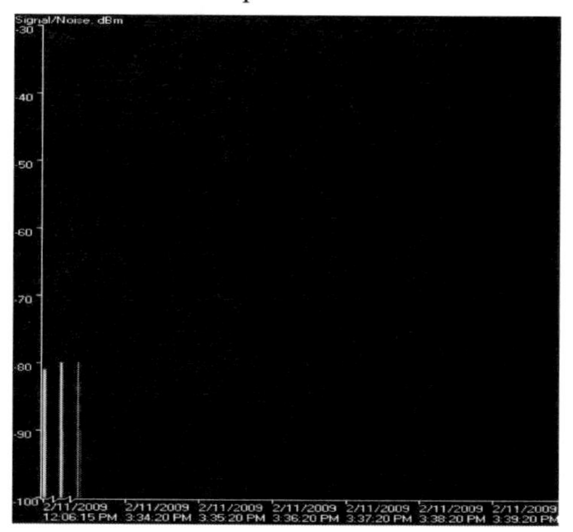

Figure-2(a): Sample for Weak signal represented in dbm.

Figure-2(b): Sample for strong signal, but disappeared for some instance.

We are using Proxim Gold WLAN card for measuring signal strengths with our laptop. We collected some samples for signal strengths of selected access points which are found in our survey environment very often. All these samples are measured in a certain time period for days long and are represented in dbm. Some of the measured RSSI values appear in our training phase fingerprinting may not appear in the online phase after some period of time i.e. after some months. These are called as weak signals. This problem raises due to signal strength values of one access point at a particular grid point is found with some variations in signal strength of the same access point at same grid point after some months. The sample Figure-2(a) shown above is taken after one month after our training phase of selected access point. In this case, we can observe that the signal appeared for some instance and it is no longer found after that. But, the same signal may appear one month back in your training phase. This may cause errors while we calculate the mean values for RSSI with probabilistic method and computing the position. We ignore this type of weak signals while comparing and computing the user location in the integrity monitoring algorithm. Secondly, the signal that appears to be strong in training phase will disappear and reappear for instance in online phase. These cause tracking errors when we are tracking the position, and the mean values will differ from the survey results. We can observe in the sample Figure-2(b) shown, that a signal of one access point appears to be strong for some seconds, suddenly skewed around the zero value, and appeared again after some minutes. In this case, the integrity monitoring algorithm automatically triggers from probabilistic algorithm when the signal gets skewed up for some minutes. Some access points are designed and programmed which gets turned off when the access point is overloaded i.e. when it is having more users rather than defined users. In this case, the access point gets turned off or restricts additional users to associate. This causes problems when we need to track the restricted user's location. In some cases, due to the hardware malfunction of access point gets malfunctioned or some rogue access points will cause the fluctuation and tracking errors in location estimation. To overcome these conflicts, we designed the integrity monitoring algorithm. However, a representative distribution of the underlying RSSI process is needed to gain more understanding of location fingerprinting. If the RSSI distribution can be identified and modeled, then analytical models of location fingerprinting and the indoor positioning system can be developed.

3. INTEGRITY MONITORING IN WLAN USING LAPTOPS

3.3.1.1 System Design

In this section, the systems software architecture is described. We developed an application for authenticating WiFi (802.11) enabled devices. The application is developed using VB.Net and VB Script, and must be installed on the server. The main purpose of this application is to track the client periodically at very short intervals and to validate the position of the client with the predefined boundaries defined by the user. We obtain the signal strengths from the Network Stumbler software and then import this data with the help of a VB script into a Microsoft SQL server database and then apply our positioning algorithm and determine the position of the Wi-Fi user. The System Model see Figure-3 is described below.

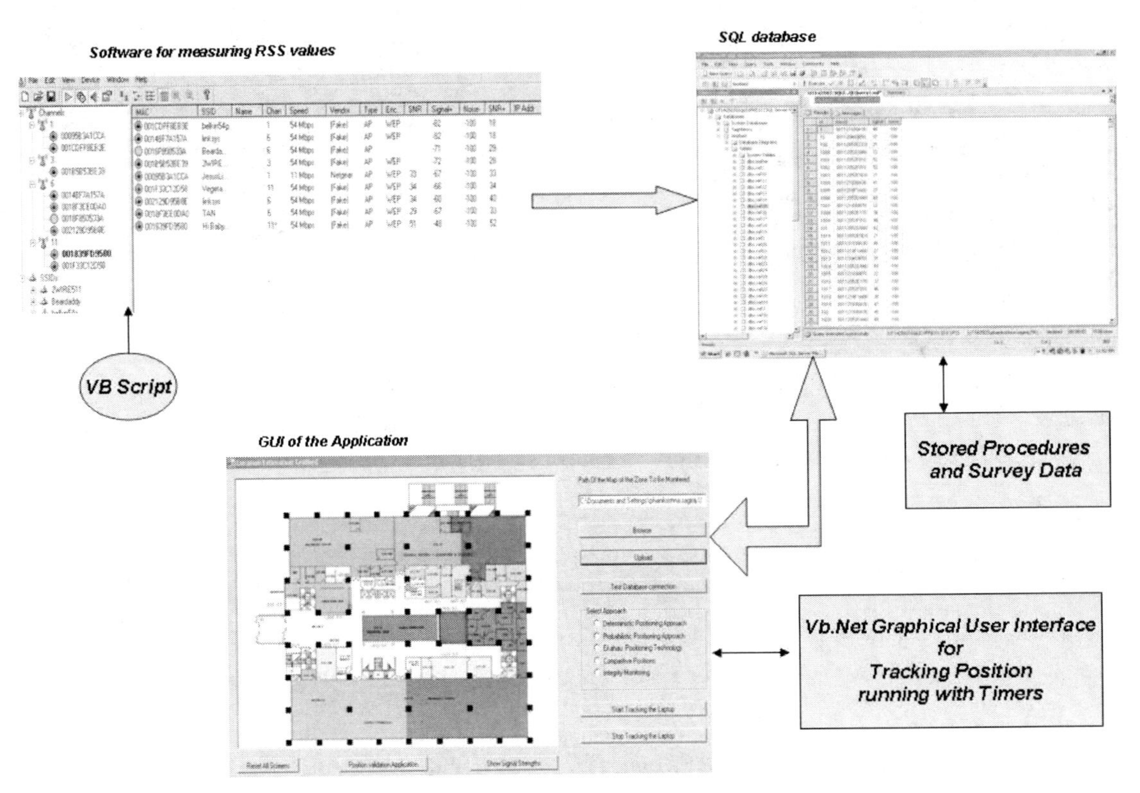

Figure- 3: System Model and Software Architecture.

An experimental setup was established in the lab BSE 2.210 on the second floor of the BSE building in the Department of Electrical Engineering, at The University of Texas At San Antonio. Access points mounted at different locations in the second floor of the BSE building were considered. The access points act as wireless signal transmitters and receivers. We used a Dell Inspiron 600m (OS: WINDOWS XP Professional) as the Wi-Fi device that is being tracked. An Intel® PRO/wireless 2915ABG wireless card was used as the network card. The NetworkStumbler software measures the RSS values from these access points. In the training phase, the reference data was collected for a period of seven minutes at each reference point at a rate of two samples per second. The data was collected in intervals of one minute (i.e. for every one minute of data collected we took an interval of one minute and then resumed collecting data). This is done to avoid the case of missing the radio signature of a hibernating access point which wakes up in due time. VBscript supported by NetworkStumbler, runs in the backend which writes all the reference data of defined output parameters to the SQL Database in the form of tables. The online phase consists of retrieving the data related to the signal strengths from the NetStumbler software and inserting it into buffer table in database. The algorithm for probabilistic approach is coded in the form of stored procedures in database. We compare probability of the data obtained during the tracking phase to the reference data obtained during the training phase using probabilistic approaches to obtain the position of the Wi-Fi enabled device. This gives the output of a grid point with highest probability, the best match of the RSSI value of the tracking phase to the reference data. The GUI which is coded in Vb.Net helps us to track the position of the client by uploading the map of the respective environment. This is a user-friendly application which shows the position of the client, and which runs with timers on the backend for tracking. When the program runs, all the timers are initiated and the RSSI values from the NetworkStumbler are written to buffer table of the database. The timer is run every 1.5 seconds, and so all the data which is exported from the NetworkStumbler software is imported into the database. The timer then initiates a stored procedure which collects the last thirteen snapshots of the data from the buffer table and inserts them into the trial data table, which is used for the location estimation. This is done to calculate the latest position update of the Wi-Fi enabled device using only the last few signal strengths of various access points. Now, the database computes the position of the user by a probabilistic approach and returns the value to the application, which shows the position of the user on the map graphically.

3.2 Integrity Monitoring Operation

The above shown environment consists of several access points. We selected only six of them, which are found very often, to compute the user position accurately. Assuming that all the access points are working perfectly (as measured during the finger printing phase), then we get the client position accurately and correctly without any tracking errors by a probabilistic approach. Now, consider a case where one of the access points is overloaded or malfunctioning or mislead by a rogue access point then we cannot compute the position accurately by a probabilistic approach as the signal measurements change with the reference data, which leads to a large error in position. There may be some jumps in the tracking due to this malfunctioning. To avoid this problem we use the integrity monitoring approach. This approach is automatically triggered when there is a fault found in the probabilistic approach. In this integrity monitoring approach, to compute the position of client, the following are the steps involved:

Step-1: We store the previously calculated position by a probabilistic approach, which is overwritten for iteration of the code, while tracking the client. This value is used to find any fault with the probabilistic approach. We compare this by assigning a threshold value i.e. seven feet, which implies that the user cannot move more than seven feet in 1.5 seconds. If the distance between current position and previous position is more than threshold value, then that implies that the error is in the position of the location, and triggers the integrity monitoring method. This technique is used to reduce the compilation time of the algorithm.
Step-2: Now, to compute the user position by the integrity monitoring approach, we first eliminate one of the six access points and then we calculate the position using the remaining access points. We compute the probability of the position using the IM algorithm by comparing the online RSSI values with stored measurements. Similarly, we eliminate a different AP each successive time and compute the position. Now, we achieve coverage of the client which is called 'Position cloud' as shown in Figure-4. To test this, we manually turned off one access point in the algorithm as it is difficult to make an access point

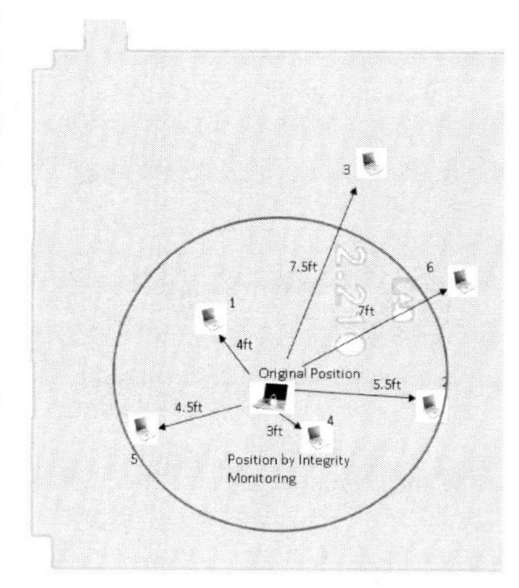

Figure-4: Computing Position Cloud.

Step-3: From the Position Cloud we can get the circle of the user's area. Now, we compare each position, or rounded-off average, of all the obtained positions from the cloud with the previous position to find the closest match and predicted trajectory of the user. All the grid points assigned are irrespective of 2-dimensional coordinates, which refers to the pixels in the GUI.
Step-4: After we get the final trajectory and final position from the algorithm by comparing with the distance from previous position, the obtained position is shown in the GUI, which is a better result than the position obtained by the probabilistic approach and reduces tracking errors. This position gets updated every 1.5 seconds once this integrity monitoring method is triggered.

4. IMPLEMENTATION OF WLAN POSITIONING & INTEGRITY MONITORING ON MOBILE PLATFORMS

4.1 Introduction

In this world, where technology is ever changing, more and more information is being made available at the tip of the fingers. Many mobile phone makers are increasingly investing highly in technologies that can lure the public into buying new phones by providing the cutting edge applications that help them in their day-to-day lives. One of the most prevalent is location-based applications which cater to the need of wide ranging fields of applications, such as medical emergencies for tracking doctors/patients with need, geographic location navigation, location-specific information to travelers, location sensitive billing etc [18]. The Global Positioning System (GPS) achieves this, with the aid of satellites orbiting the earth providing real time transmission of location information, and helps the user to accurately determine their location at any point in time. However, GPS systems can only be used outdoors due to their high signal degradation in indoor settings. One technology that is most prominent indoors, not just in workplaces but also in university research labs, is 802.11 backhand networks. A significant amount of research has already been performed on WLAN-based Real Time Location Systems (RTLS).

In WLAN positioning, the system either monitors propagation delays among the wireless nodes (access points and users) to triangulate and calculate relative positions, or it maintains the database of location fingerprints, which is used to identify the most likely match of incoming signal data with those in the preliminary survey saved in the database. The latest version is the one launched by researchers at the Fraunhofer Institute for Integrated Circuits IIS[19]. This localization software depends on the finger printing approach. There are many techniques for estimating the RTL; they can be classified into three broad categories, 1) those which depend on Received Signal Strength Indicator (RSSI), 2) those which are time-based, and 3) the combination of first two. The most common time-based approaches are Time of Arrival (TOA), Angle of Arrival (AOA), and Time Difference of Arrival (TDOA) [20] to name a few. The most effective method of them all is finger-printing, as it accounts for attenuation due to obstructions caused by reflections, refraction, and scattering of the electromagnetic waves along structures of the building, and thus tends to provide more accuracy in WLAN position estimation (Li *et al.*, 2006). This paper investigates the issue of deploying WLAN positioning software on mobile phone platforms. In this paper we focus on a modular probabilistic approach for inferring location that uses Bayesian networks for estimation by building Finger-Printing (FP) data based on the distribution of signal strengths, as obtained in a series of calibration measurements using the user's mobile phone. We further discuss a software approach for improving the accuracy of such systems by a mechanism known as Integrity Monitoring (IM) and additional match compensation. Our implementation is based on a stand-alone mobile phone approach i.e. the application runs without interacting with other external entities such as outside servers or databases. Any updates such as new location finger-print data or information updates can be manually configured by the software programmer or automatically configured using the OTA mechanism. The traditional, probabilistic approach to location estimation is based on an empirical model that describes the distribution of received signal power similar to that specified in [17], but we further refine the algorithm by focusing on the problem of reducing the ambiguity and increasing the accuracy around corners of a room or near the intersection of many closely spaced FP data, caused due to rogue AP or the absence of certain WLAN AP configurations.

4.2 IM based WPS System Overview

This system can be divided into two main modules. The first module deals with the collection of Wi-Fi AP scan data. This is achieved with the help of free to download WiFiFoFum software, developed by Aspecto software [22], which makes use of the inbuilt Wi-Fi capability of the mobile phone (Windows mobile 5 Pocket PC and Smartphone editions) to obtain the Wi-Fi scan results. This feature is based on .NET CF and provides information such as WEP, SSID, MAC, Signal Strength (RSSI), latitude and longitude information. The scan data can be stored in XML format or .txt file (for this project we used XML format due to its descriptive and ease of use). The second module interprets end process the WiFi AP scan data by parsing the values of the xml file obtained through WifiFoFum. This module is implemented using Java Platform, Micro Edition or Java ME, formerly known as Java 2 Micro Edition (J2ME), which has a dedicated set of technologies and specifications developed for small devices like mobile phones. This platform has many

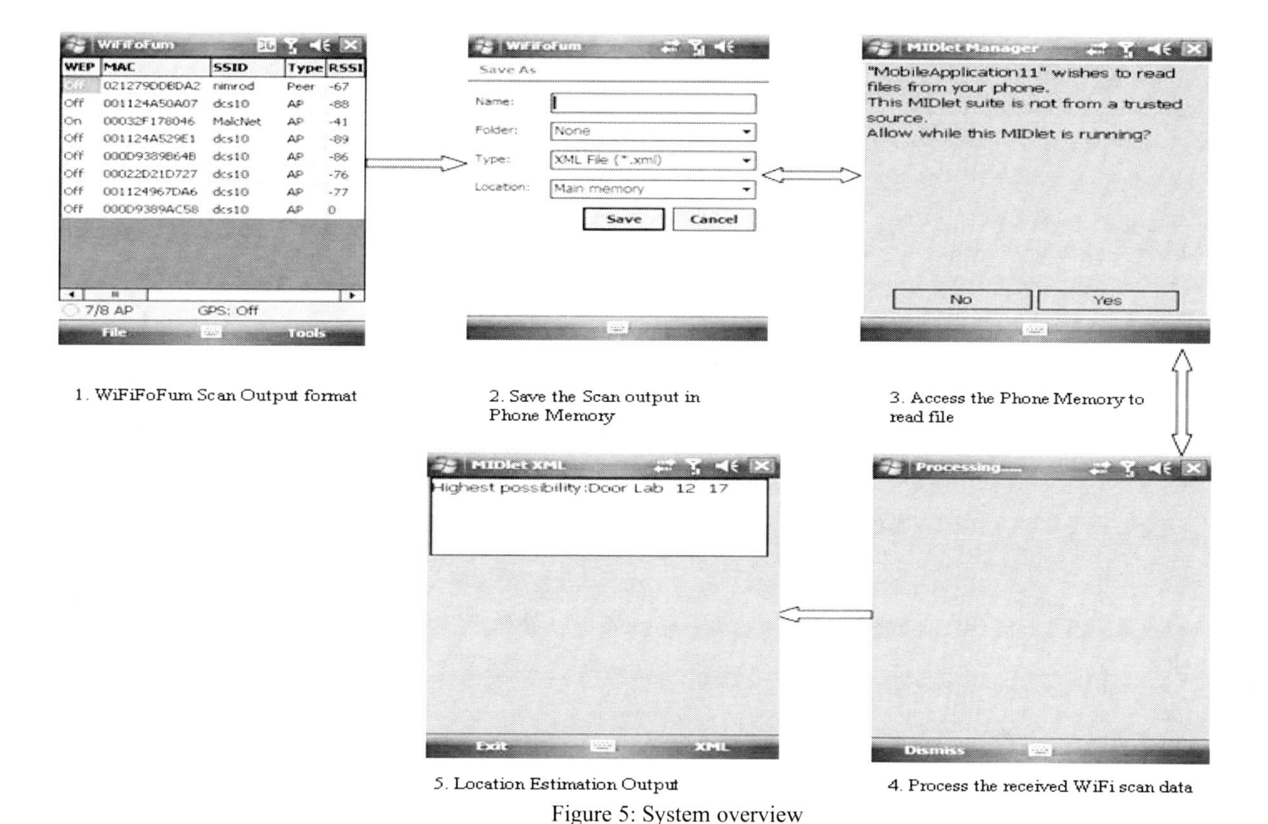

Figure 5: System overview

advantages, but the key factor for choosing this platform was due to its cross platform capabilities. When developing applications on mobile phones, the developer need not worry about the underlying platform as Java source code can be deployed to any of the mobile operating systems that have a Java Virtual Machine (JVM).Its pure object oriented approach and automatic allocation and de-allocation of memory (reducing the incidence of errors) also gives it an edge over other languages. Additionally, its worldwide acceptance by mobile phone manufacturers; according to[23] around 80% of 1.5 billion devices all over the world were powered by Java Me Technology in 2007. Other key advantages of using Java Me as a platform of choice could be found at [24].

4.3 Midlet Development

As mentioned earlier, this is standalone software that doesn't interact with any external entities such as servers to process the scan data obtained using WiFiFoFum. The principle advantage of such an approach is the speed of processing that removes the latency involved in interacting with external entity and on processing speed of the external entity for responding to client request. We also acknowledge the drawback of postulating large memory and processing power to execute such an application but considering the evolution, in terms of processing and memory capabilities, of mobile phones in the past few years it can easily be concluded that such requirements should not be a major constraint for development and moreover the speed gained by such an approach clearly outweighs the drawback involved.

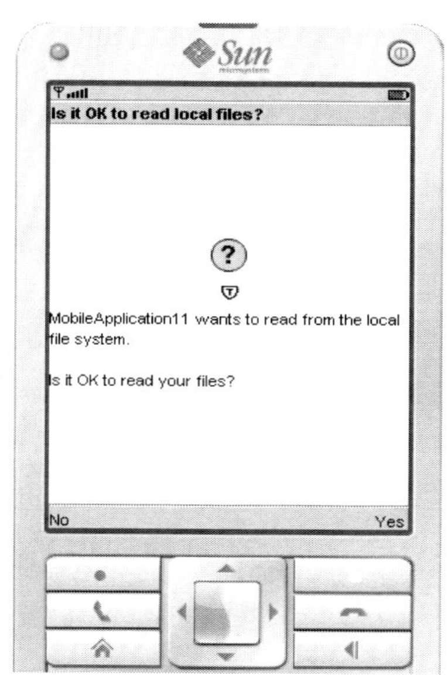

Figure-6: J2ME Wireless Emulator

This module is coded, compiled and run using the NetBeans IDE v6.5. Before running the MIDLet application on the actual MIDP device, we validate the software with the help of J2ME Wireless Tool Kit (J2MEWTK), a package of tools for building and testing MIDlets, as shown in Figure 5. As mentioned earlier, by using the probabilistic approach we predict/infer the location of the mobile handset by mapping the FP data and real time measured data using the below equation.

$$P(L|F) = \frac{1}{n}\sum_{i=1}^{n}\left[\frac{1}{\sqrt{2\pi}\sigma}\exp\left(-\frac{(s-\rho_i)^2}{2\sigma^2}\right)\right] \quad (5)$$

For our testing purposes we selected 5 sample locations around our Lab which has 70 x 40 feet area see Figure-7 and collected the finger print data. We maintain a database for each location which contains the MAC address, associated Mean RSSI $(\rho)$and variance $(\sigma)$of an access point (AP) computed over 20 iterations, at different intervals of time. We base our location prediction output using two parameters, the maximum number of AP matches with a particular location FP data and the least count of

$$Total = \sum[(s_i - \rho)^2 / 2\sigma^2] \quad (6)$$

Where si is the real time measured RSSI data for a given AP

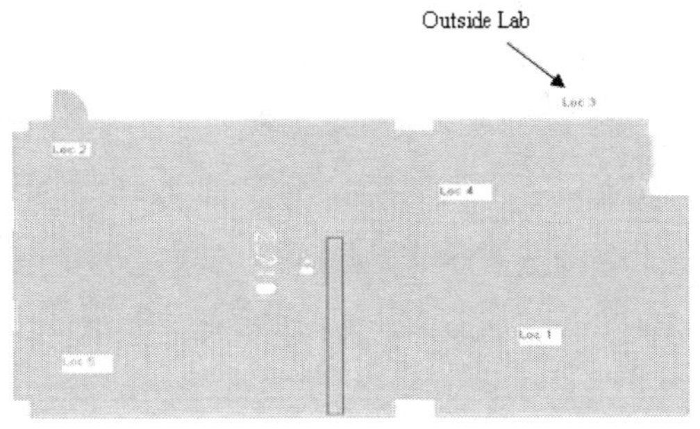

Figure 7: Floor plan of our Lab at UTSA

In this approach we do not apply probabilistic framework for AP's that are not found common between a particular location FP and real time measured scan data. Also, we predict output as *Unknown Location / No database info found* if there are less than 3 AP matches with any location FP data. Hence this approach inherently requires at least 3 AP matches between FP and real time measured scan data to estimate the location.

4.3.1 Code Implementation

This application [3] necessitates inclusion of following classes:
→*javax.microedition.lcdui.**, for MIDP GUI programming
→*javax.microedition.io.**, for file connection programming
→*org.kxml.**, KXML for xml file parsing
Upon initiation of the application, the first step is to open file connection using
SAMPLE CODE: *fconn = (FileConnection)Connector.open(url, Connector.READ);*
Next, each character of the XML file is read and appended, by opening an *InputStream* on our file connection, and storing it, in a local buffer as a string.
SAMPLE CODE: *InputStream is = fconn.openInputStream();*

```
........
xmlString.append( new String(char));
```

Based upon the command action, our application responds and notifies the MIDlet to start XML parsing or exit the application. The parsing of data is performed with the help of kxml 2.0 pull type parser [26], written for J2ME platform.

4.3.1.1 Integrity Monitoring:

- For the first iteration of parsing, the location estimate is computed without applying IM scheme i.e. we consider all the AP's & corresponding RSSI values, found in the scan result, to base our location estimate. IM scheme is applied if, lets say, at the end of the above cycle, we have a situation wherein, the maximum AP matches is associated with location FP 1 and the least *Total* with location FP 2.

- For implementation of IM approach, we maintain a global variable *GLB_IM_COUNT* that is initialized to an appropriate value that helps in iterating through all possible cases of 'skipping one AP from the list of measured AP scan output' and computing the location estimate (Probabilistic approach) for each iteration until all the AP's have been skipped at least once. At the end of each iteration, we add the X & Y location co ordinates of the predicted output locations to the closest matching position.

4.3.2.1 XML parsing:

This function operates on the xml string by reading the data as byte array using the *InputStreamReader* and then calls `XmlParser` to parse the whole XML String

Step 1: SAMPLE CODE*:* `byte[] xmlByteArray = xmlString.getBytes();`

```
…..
InputStreamReader Reader = new InputStreamReader(xmldata);
XmlParser parser = new XmlParser( xmlReader );
```

During the XmlParser's read operation, it returns *ParseEvent* object when it encounters an event type of the form *Xml.START_TAG, Xml.END_TAG, Xml.END_DOCUMENT* etc. Based on the event type we switch to the particular case

Step 2: SAMPLE CODE*: case: Xml.START_TAG*

```
if ("MAC".equals(event.getName()))
{
  //extract the MAC data
  cnt++;      //counter to incrementaly skip the AP during IM
  ............
  macdata = pe.getText();
  ............
}
```

Check if we need to skip this AP if IM is set based on *GLB_IM_COUNT* if so then continue with *parser.read()* operation, if not then go ahead with compare the real time AP scan data with all the location FP data values for that AP.

Step 3: SAMPLE CODE: *if(macdata.compareTo(fp_mac[j]) == 0) {.....}*

If there is a match found between a given FP & measured data, then extract the RSSI information and convert the corresponding value to an integer to perform arithmetic operation as per Equation (2).

Step 4: SAMPLE CODE: `if ("MaxRSSI".equals(event.getName()))`

```
{ ............
  temp1 = Integer.parseInt(tempdesc);
  .........
}
```

Step 5.a: Update and store the *Total* value, specific to particular location, after each operation is completed based on Equation (6). The above set of operations (`step 1`) is continued until *Xml.END _DOCUMENT* parse event is generated.

Step 6: When the *case: Xml.END _DOCUMENT* is reached, verify which location count has the maximum AP matches and least *Total* obtained using Equation (2).

Outcome possibility 1: Both the 'AP Maximum matches' and 'Least Total' parameters are associated with same location. If so, display that location as the position estimate. Exit the application

Outcome possibility 2: The parameters 'AP Maximum matches' and 'Least Total' are associated with two different location FP data sets. Then, initiate IM and set the *GLB_IM_COUNT* appropriately so as point 2 mentioned in 3.1.1.1 is satisfied.

Step 7: Re-Compute the location estimation for position coordinates that are found within the threshold radius (without skipping any real time measured AP scan data). Display the outcome at the end of step 6.

4.3.3.1 Additional Match compensation penalty:

While performing the tests around the corner/edges of or few feet away from a fixed FP data calibration point it was found that sometimes the location estimation is erroneous (within small range ~ 2 meters). This was successfully corrected using the below mentioned Equation (7):

$$\text{Compensation} = \frac{\text{Total range of variation of the signal}}{|(s_i - \rho)|} \qquad (7)$$

The output of the compensation parameter is added to the *Total* parameter value maintained by every location FP data set which doesn't have corresponding AP MAC match

Consider if there are two FP data sets, then

Step 5.b: SAMPLE CODE*: if(loc2 == -1) // check if MAC match is not found*

```
{
if(buf == 0)        // check if |(s_i − ρ)| of loc 1 is zero
        tmp = 15;   // if so (this indicates complete match of AP RSSI with FP
                    data) add 15 (the maximum, chosen based on trail &
                    error)to the Total of loc2 as a penalty.
```

}

If $|(s_i - \rho)|$ of loc 1 is non zero then compute the compensation output & then adds it to the *Total_2* corresponding to location 2. This way the possibility of predicting the outcome as location 2 is reduced variably based on amount of difference between the RSSI values of location 1FP data and real time measured AP scan output. In our calculation we used the total signal variation range of 60 (-30 dB to -90 dB),

Step 5.c: SAMPLE CODE: *tmp = (60 / $|(s_i - \rho)|$);*

total_2 = total_2 + tmp;

The above algorithm is easily extendable in scenario's with multiple location FP data, with a small variation in computing compensation parameter by finding the sum of compensation parameters computed by adding all the $|(s_i - \rho)|$ of matched AP FP data and finding the average and continuing with step 5.c

5. PERFORMANCE EVALUATION

The Position Estimation and Tracking of a WI-FI enabled device with integrity monitoring using standard Wireless networks (802.11) for enhanced accuracy is implemented and tested. This is tested and all the observations are measured by turning off one access point manually in the algorithm as it is difficult to make an access point faulty in our selected network. The error and working is calculated by comparing with the position by probabilistic approach. The test methodology was to collect fingerprints from specific locations on the map. There are 50 reference points in our lab collected having fingerprints in the database and all are located in a zigzag manner. Table-2 shows the information how the integrity monitoring computes the position. The measurements are taken when each access point is ignored and computed the position. It compares the obtained position with the previous position to find the closest match and shows the accuracy in feet of the obtained position compared with original position. All the measurements are approximated values.

Table-2: Variations and Success rate compared with probabilistic approach

Position of Client	Threshold Distance(7ft) Error	Probabilistic Position Accuracy(ft)	Integrity Monitoring Accuracy(ft)
Position-1	12	8.5	4.5
Position-2	11	9	3
Position -3	10	7.3	3.5
Position -4	7	7	6
Position -5	9	12	5
Position -6	10	6.5	4
Average	9.8	8.3	4.3

It also shows that if the distance between the previous position and the probabilistic position is greater than the threshold value, then it triggers the integrity monitoring approach and computes the new position. Note that the threshold distance error is irrespective to the error in feet, not the difference with the grid points. It gives the information about the success rate of integrity monitoring compared with the probabilistic approach. We reduced the tracking error when compared to probabilistic approach and achieved the accuracy of 4-5 feet with this method.

6. CONCLUSIONS

In this research an area localization algorithm is developed for the application of room localization. An experimental testbed for comparative analysis of various WLAN positioning systems is designed and an enhanced WLAN position computing approach has been implemented. Because of the average standard deviation of the location fingerprints, a designer could identify the minimum distance separation between present and previous positions that could achieve required performance. And we are also developing this method based on mobile platform for better for more robustness and to enhance the performance. This method provides a more robust means to avoid frequent variations of the client position. A systematic study was used to analyze the location fingerprint and discover its unique properties.

7. ACKNOWLEDGEMENTS

This work was in-part supported by National Instruments and National Science Foundation under Grant numbers SGER # 0833852 and CRI # 0551501 respectively without which this work would not have been possible. We also sincerely thank all those directly or indirectly involved in the successful completion of this work.

REFERENCES

[1] Understanding GPS: Principles and Applications, Kaplan E.D., Editor, Artech House.
[2] W. Stallings. Wireless Communications & Networks. Pearson Education, Inc. 2005.
[3] Ajit Deosthali, "worldwide WLAN semiconductor 2008–2012 forecast," Sep 2008 http://www.idc.com/getdoc.jsp?containerId=214316
[4] www.skyhookwireless.com
[5] Ken Masica, "Securing WLANs using 802.11i", Idaho Falls, February 2007, pp.1- 4.
[6] http://www.google.com/mobile/default/latitude.html
[7] Documentation from the website of Ekahau, http://www.ekahau.com. Accessed Feb 18, 2008.
[8] http://www.newburynetworks.com/
[9] http://www.directionsmag.com/companies/Bluesoft,_Inc./
[10] Network Stumbler at http://www.netstumbler.com/ as an software. Accessed Feb 18, 2008.
[11] Vinod Padmanabhan, "Area Localization Using WLAN", Master of Science Thesis, Stockholm, Sweden, 2006.
[12] D. Akopian and P. Chen, "Implementation of an intrusion detection using WLAN," Proc. SPIE conference, 2007.
[13] Kamol Kaemarungasi, "Design of Indoor positioning systems based on location fingerprinting", university of Colorado at Boulder, 1999.
[14] Alirez Nafareih and Jacek Ilow, "A Testbed for localizing Wireless LAN devices Using RSS", Dalhousie University, Proc. IEEE Conference, 2008, pp. 4-7.
[15] F.Barcelo and M.Ciurana, "Indoor location for safety applications using wireless networks", University of polytechnic at Catalunya.
[16] Eugene A. Gryazin, "Indoor Positioning And Navigation Using WLAN," Helsinki University of technology, Espoo, Finland, 2001, pp. 3-7.
[17] Teemu Roos, "A Probabilistic Approach to WLAN User Location Estimation," International Journal of Wireless Information Networks, Vol. 9, No. 3, July 2002 (q 2002)
[18] Ulf Rerrer and Odej Kao, "Suitability of Positioning Techniques for Location-based Services in wireless LANs," proceedings of the 2nd workshop on positioning, navigation and communication (wpnc'05) & 1st ultra-wideband expert talk (uet'05)
[19] Fraunhofer-Gesellschaft, "WLAN leads the way," http://www.physorg.com/news121445857.html
[20] Sinan Gezici, " A Survey on Wireless Position Estimation," Department of Electrical and Electronics Engineering, Bilkent University Bilkent, Ankara 06800, Turkey
[21] http://www.navizon.com
[22] http://www.aspecto-software.com/rw/applications/wififofum
[23] http://www.ovum.com
[24] http://developer.att.com
[25] Martyn Mallick, "Mobile and Wireless Design Essentials," Published by John Wiley and Sons, 2003, ISBN 0471214191, 9780471214199
[26] http://kxml.objectweb.org/software/downloads
[27] Soma Ghosh, http://www.ibm.com/developerworks/library/wi-parsexml

A New Approach for Non-Cooperative Iris Recognition

Craig Belcher, Yingzi Du*

Department of Electrical and Computer Engineering
Purdue School of Engineering and Technology
Indiana University-Purdue University Indianapolis
Indianapolis, IN 46202
Email {cbelche,yidu}@iupui.edu

***Abstract*—**Traditional iris recognition algorithms can work well for the frontal iris images. However, when the gaze of an eye changes with respect to the camera lens, many times the size, shape, and detail of iris patterns will change as well and cannot be matched to enrolled images using traditional methods. Additionally, the transformation of off-angle eyes to polar coordinates becomes much more challenging and noncooperative iris algorithms will require a different approach. In this paper, we propose a new approach for iris recognition. This new method does not require polar transformation, affine transformation or highly accurate segmentation to perform iris recognition. Our research results using the remote non-cooperative iris Image database show that the proposed method works well on frontal look images and off-angle images as well.

Index Terms—Video-Based Non-Cooperative Iris Recognition, Off-Angle Iris, Non-cooperative iris recognition

I. INTRODUCTION

Iris recognition in cooperative situations has been tested to be one of the most discriminating biometrics in use today with recognition rates as high as 99.99% [1, 2]. Several methods have been developed for cooperative iris recognition [2-12]. Within, the 2-D Gabor filter approach is the most popular used method in the commercialized systems [2]. All of these methods require the iris region to be converted to polar coordinates. However, transformation to log-polar/polar coordinates can be difficult for off-angle eyes in a non-cooperative situation without an affine transformation and can be severely affected by segmentation error. In Ref. [2], Daugman proposed using active contours based on the Fourier series expansion of pupil and limbic edge data instead of a circle model and reported a ten-fold increase in recognition performance, which is due to more accurate segmentation results. Conversely, it shows that the typical iris recognition approach is sensitive to the segmentation accuracy.

For non-frontal iris recognition, Daugman proposed using Fourier-based trigonometry to estimate the two spherical components of angle of gaze and used an affine transformation to "correct" the image and center the gaze [2]. This method is limited because "the affine transformation assumes the iris is planar, whereas in fact it has some curvature" [2]. The eye is a 3-D object and the

Mobile Multimedia/Image Processing, Security, and Applications 2009, edited by Sos S. Agaian, Sabah A. Jassim,
Proc. of SPIE Vol. 7351, 73510A · © 2009 SPIE · CCC code: 0277-786X/09/$18 · doi: 10.1117/12.818886

deformed iris patterns may present different correlations of iris patterns. It would be ill-posed to use a 2-D feature extraction model to apply for 3-D object recognition. It is true that some eyes have patterns that do not change very much when gaze changes and these eyes respond well to an affine transformation. However, our experimental results show that many iris patterns do change substantially with change of gaze and, therefore, require a different approach to iris recognition in a noncooperative environment. Schuckers *et al.* proposed two methods to calculate angle of gaze: using Daugman's integrodifferential operator and also an angular deformation calibration model [13]. It is assumed that an estimate of the degree of off-angle is available for the algorithms and subjects are required to place their heads on a chin rest. It is often difficult to accurately estimate the degree of off-angle.

In a non-cooperative situation, iris images can be blurred, severely occluded, poorly illuminated and/or severely dilated in addition to off-angle. As an iris changes gaze with respect to the camera lens, the size, shape, and relative centroids of the limbic and pupil regions change. Given these conditions and lower quality of iris image, segmentation error may not be avoidable. This requires that the recognition method should be tolerant of the segmentation error.

Recently, we proposed the Regional Scale Invariant Feature Transform (SIFT) approach [14] for non-cooperative iris recognition. Iris features are described without a polar or affine transformation and the feature point descriptors are scale and rotation invariant. However, the iris region consists of both noise and patterns and Regional SIFT describes the area around a feature point without extracting the underlying iris patterns.

In this paper, we propose a new method for Non-Cooperative Iris Recognition. In this new method, iris features are extracted using 2-D Gabor wavelets without a polar transformation or affine transformation including for off-angle irises, and the method is more tolerant to segmentation error and changes in iris dilation. In our experiments, we used the IUPUI Remote Iris Image Database and found that the proposed method is capable of reliably classifying iris images; iris images of different angles exhibit changes in iris features with respect to the camera; and using iris video sequences along with enrollment images of multiple angles and majority vote score fusion can result in zero False Accepts with reasonable False Rejects.

The remainder of this paper is organized in the following manner: Section II – Gabor Descriptor Based Non-Cooperative Iris Recognition; Section III – Experimental Results; and Section VI – Conclusion.

II. Gabor Descriptor Based Non-Cooperative Iris Recognition

A. Feature Point Selection

In this paper, feature points are used to locally describe the features of an iris as opposed to the global approaches mentioned previously. To match two images, it is much faster and more accurate to compare feature points that can reasonably be expected to represent a similar feature instead of comparing all feature points in one image to all feature points in another image. To facilitate this constraint, we divide the iris region into 720 sub-regions or bins where, no matter the amount of dilation or scale, there are 10 bins between the pupil and limbic boundary and 72 bins from 0 to 2π in the angular direction. Since the pupil and limbic boundaries are modeled as ellipses, the sizes of these sub-regions vary in the radial direction for each of the 72 angular bins. In this way, a normalized map of size 10 by 72 is formed where each bin can potentially have a feature point with a total of 720 feature points possible. This is a major difference from the previous Regional SIFT method in that the entire iris area can potentially have feature points and every bin size changes with dilation. In addition, to compensate for feature points that are on the boundaries of sub-regions, a second 10 by 72 normalized feature point map is created with a 5 degree angular offset.

DoG is used to find stable feature points within an iris image. Stability of feature points is important since it is necessary to compare the same feature in two images from the same point of reference. To find stable feature points, the first step is to apply a nominal Gaussian blur, Eq. 3-9, resulting in $I(x, y)$.

$$G(x, y) = \frac{1}{2\pi\sigma_n^2} e^{-(x^2+y^2)/2\sigma_n^2} \quad (1)$$

Here $\sigma_n = .5$. Then the nominally blurred image, $I(x, y)$, is progressively Gaussian blurred. The size of the Gaussian filter is always the closest odd number to 3σ. These parameters were selected empirically and are the same for all images. Then the four DoG images are created by subtracting each Gaussian image from the previous Gaussian image in scale:

$$D(x, y, s) = G(x, y, s+1) - G(x, y, s) \quad (s=0,1,2,3). \quad (2)$$

This is different than the DoG method proposed by Lowe [17] in that, only one potential feature point is found per scale within our defined sub-region and only two scales are used since the scale of useful iris images is not changing drastically due to a constant camera focal length. Lowe's method finds many points stable within scale space with many points possible in a very small region. The goal of our approach is to increase the opportunity to correctly match feature points within a similar relative position with respect to the pupil across multiple iris images, which is why we restrict each sub-region to only two feature points. Once potential feature points are identified and mapped to the feature point map, the 3-D quadratic method proposed by Brown and Lowe [16] is used to eliminate unstable feature points. Fig. 1 shows an example of stable feature points found for an iris.

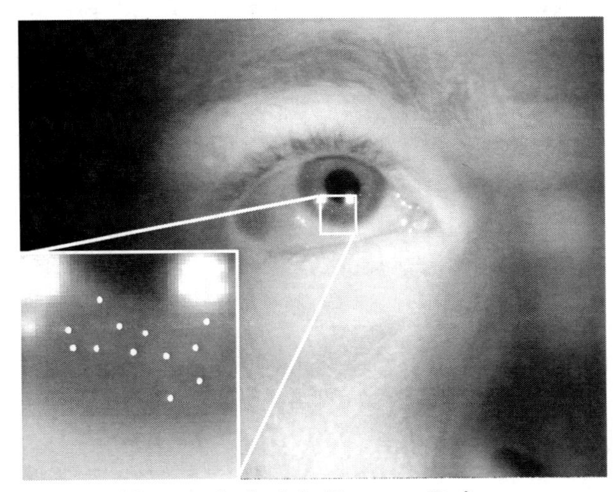

Figure 1. Stable Feature Points

B. Feature Description

In order to capture the iris features around a given feature point, a bank of 2-D Gabor filters are used with the angle and width of each filter changing based on the angle of a feature point with respect to the pupil centroid and the distance between the pupil and limbic boundary around the feature point, respectively. For each feature point, a feature description of length 64 is created based on the normalized and Gaussian weighted position of each point within a normalized window around a feature point (4 x-bins and 4 y-bins) and the magnitude and phase response (4 phase orientation bins). Since the normalized window can be rotated based on the relative orientation, some points will no longer be in the window. Therefore, only points within a window of N/2 are used for the feature descriptor. Fig. 2 shows the window of points around two feature points being rotated to match their respective angles in reference to the pupil center.

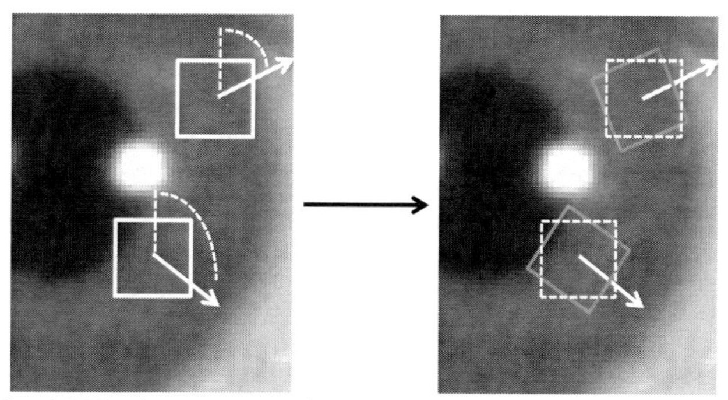

Figure 2. Original Window Rotated to Match Relative Orientation around Pupil

For each point in the normalized window around the feature point, the magnitude and phase response of the appropriate 2-D Gabor wavelet is calculated with the wavelet centered on the point being considered. The magnitude is then Gaussian weighted based on the relative spatial distance from the feature point so that points in the window closest to the feature point carry the most weight and points further away carry less. The weight of the Gaussian, *wn*, is calculated as:

$$wn = e^{-.5\left(\frac{(nx)^2}{2\sigma_x^2}+\frac{(ny)^2}{2\sigma_y^2}\right)} \quad (3)$$

where $\sigma_y = N/2$ and σ_x changes based on the dilation around the feature point. Finally, the weight of each point is calculated as

$$weight = wn \cdot mg \quad (4)$$

where *mg* is the magnitude response of the 2-D Gabor wavelet, and *weight* is added to one of 64 bins based on relative distance from the feature point and quantized phase response of the 2-D Gabor wavelet. The resulting 64 bin feature point descriptor is then normalized to a unit vector by dividing by the 2-norm of the descriptor:

$$descr_{norm} = \frac{descr}{\|descr\|_2}. \quad (5)$$

Since each descriptor is normalized, the relative difference in magnitude response from the 2-D Gabor filter remains the same for the same points around a feature point across iris images with different global illumination. And since phase is not affected by illumination, the same points in two iris images affect the same descriptor bins. Therefore, each feature point descriptor created has each of the 64 bins uniquely affected by the surrounding points based on distance from the feature point, and 2-D Gabor wavelet response magnitude and phase; and an accurate descriptor is formed based entirely from the annular iris data.

This approach differs from the SIFT approach in several ways. Most significant is the use of the 2-D Gabor filter to extract the iris features and describe each feature point as opposed to using the local gradient magnitude and orientation. In addition, the window around a feature point is specifically adjusted for each iris based on dilation so that the same iris at a different scale and with varying dilation can be correctly matched. And more subtly, only 64 bins are used in the descriptor because 128 bins were found to be more susceptible to errors from noise.

C. Region Based Matching

To match two iris images, the set of two 10 by 72 feature point maps are compared to find which overlapping sub-regions contain feature points and the Euclidean distance is found between each feature point descriptor. In other words, one of the feature point maps from image A is compared to a feature point map from image B and for each sub-region between the two feature maps that both contain a feature point, the Euclidean distance is calculated, normalized to fall between 0 and 1, and stored. The final distance score between two feature maps is the average of the distance scores between all overlapping feature points. Since there are two feature point maps for each iris image, four complete comparisons are made and the minimum average Euclidean distance is found to be the matching distance between two iris images. Recall that the two feature point maps for an iris image describe the same regions, but are offset by 5 degrees. This is done in order to accommodate feature points that fall on boundaries of sub-regions within a feature point map.

Segmentation of a non-ideal iris image can be difficult, so despite the success of our video-based non-cooperative iris image segmentation [17], it is necessary to make allowances for segmentation error when matching two feature point maps [18]. Also, rotation of an iris image can occur due to natural head movement and it is necessary to compensate for up to 10 degrees rotation. Therefore, each feature point in image A is compared to each feature point in the fifteen surrounding bins (two bins on either side and one bin above and below) in image B, and the minimum average distance score is stored for the two feature point maps compared. In this way, the proposed method is less sensitive to the segmentation error that is prone to occur in non-ideal iris images since feature points can occur anywhere within a bin and allowances are made to maximize the opportunity for the same two feature points in two images to be compared. Algorithms that sample the iris region and encode globally require more stringent segmentation results so as to correctly match each encoded point.

III. EXPERIMENTAL RESULTS

A. IUPUI Remote Iris Image Database

The IUPUI Remote Iris Image Database consists of 2 sessions, 31 subjects, 62 irises, 731 video sequences, and 205,538 video frames. Images were acquired at 10.3 feet from the camera to the subject using a MicroVista NIR camera with Fujinon zoom lens. The light sources are simple, off-the-shelf near-infrared Sony HVL-IRM Nightshot video lights and were placed about 2~3 feet from the subject. 6 videos were captured for each subject with multiple angles of the iris with respect to the camera lens: the first video was only frontal look; the second and third videos captured the iris while the subject read from posters 15 feet from the subject and 5 feet behind the camera, Fig. 3(a); the fourth and fifth videos recorded the subject as he/she searched the wall to count the number of occurrences of a certain symbol, Fig. 3(a); and the sixth video captured the iris while the subjects performed simple calculations of numbers posted on the ceiling, Fig. 3(b).

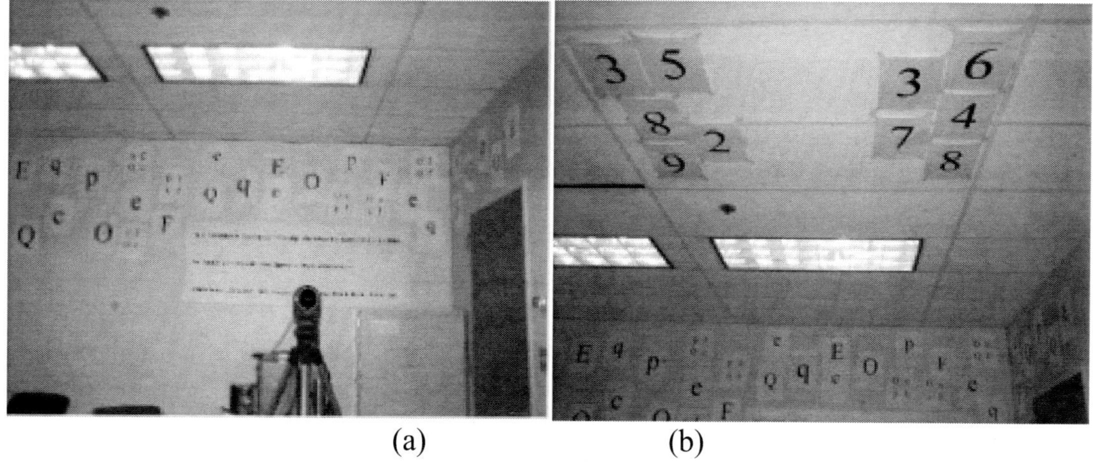

(a) (b)

Figure 3. Remote Iris Image Acquisition Station Set Up

In this way, subjects were able to move their heads freely while not being required to look directly at the camera, which simulates a remote, non-cooperative situation similar to capturing an iris while a subject looks at flight times in an airport. In addition, we did not restrict the subject for eliminating emotion or body movement, which allows for greatly varying conditions across all videos acquired including differences in angle with respect to camera, iris location within the frame, iris size, illumination, and dilation. Each video was acquired at 30 frames per second with 1280x1024 resolution. The average iris radius of the video images in the database is about 95 pixels.

B. Recognition Results and Comparison

1) Frontal Look Recognition Results and Comparison

Table I show that our results using the proposed method and the Regional SIFT method are comparable to the results achieved using traditional matching on the centered eyes from our non-cooperative database. The pupil and limbic boundaries were modeled as circles which is a simple and reasonable approximation of the pupil and limbic boundaries' geometries. We did not perform this same matching algorithm on the other classes since they are not frontal looking images and it would be difficult to reliably sample the iris pattern for off-angle images without some transformation such as Daugman proposed [2]. While this approach seems reasonable, we argue that due to the 3-D nature of iris patterns, it is more reasonable to encode iris patterns without a transformation and more accurately represent the patterns presented to the camera.

Table I: Cooperative Recognition Algorithms Compared to Non-Cooperative for Frontal Look Eyes

Algorithm	# Images	EER	GAR @ FAR .1%	GAR @ FAR .01%
2-D Gabor	618	0.0177	0.9310	0.8870
1-D Log Gabor	618	0.0291	0.9240	0.8990
Regional SIFT	618	0.0348	0.9180	0.8595
Proposed	618	0.0268	0.9210	0.8845

2) Multiple Angle Recognition Results and Comparison

In this experiment, 10 video frames for each iris for six classifications of angle with respect to the camera were selected: looking center, left, right, up-left, up-right, and up. This resulted in 60 images per iris with the exception of those iris videos that did not contain 10 segmented frames for one of the classifications mentioned. The total number of images used for this experiment was 3707 and included both left and right eyes from 31 subjects. For the classification of "up," the images are a mixture of irises looking up, up-left and up-right since all three classifications can be considered looking up.

Fig. 5 shows the respective ROC curves using the proposed method for the dataset. The results show that the Equal Error Rate (EER) and Genuine Accept Rate (GAR) = 1 – FRR (False Reject Rate) for each class of images are very reasonable with the exception of the "up" eyes. This is because of the variance in the "up" images from up-left to up-right which shows that images that are not in the same classification will not match as well due to the 3-D nature of iris patterns mentioned previously. The up-left to up-left and up-right to up-right comparisons do not perform as well as looking left, right, and center because of variations in the degree of off-angle when classifying up-left or up-right. For many methods, including the Regional SIFT method, center gaze would achieve better recognition accuracy than off-angle eyes. However, for the proposed method, our experiment results show that the left and right looking eyes could achieve higher accuracy than frontal look images (Fig. 5). This shows that the proposed method is working well in a non-frontal gaze situation.

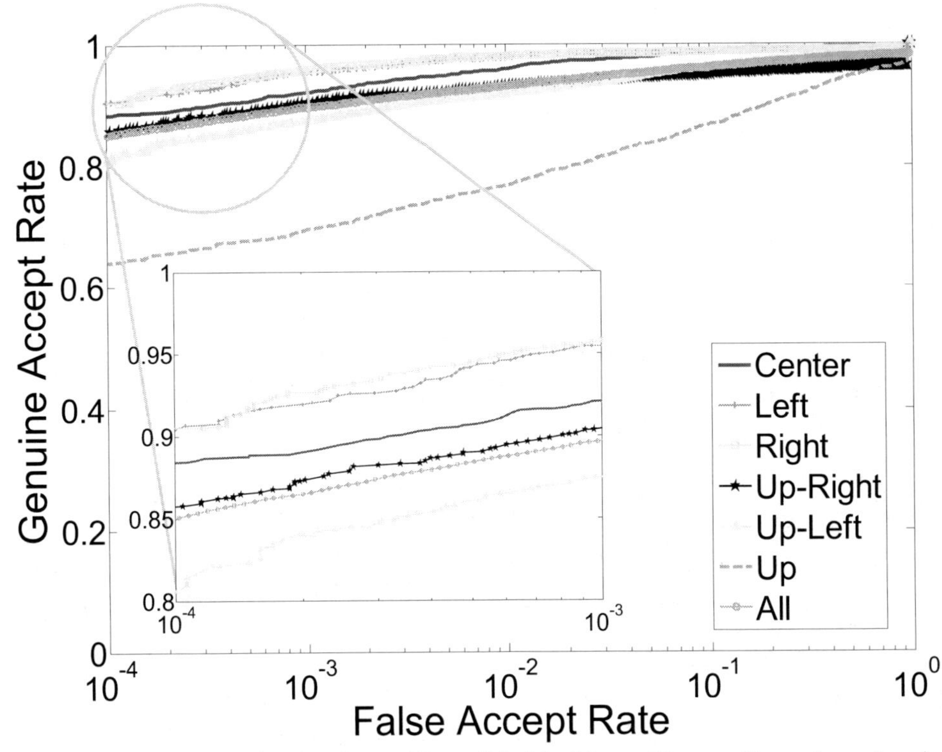

Figure 5. Recognition Results for Same Eyes Divided into Classes Based on Angle of Gaze

IV. CONCLUSION

In this paper, we proposed a new approach for Non-Cooperative Iris Recognition. The proposed method does not transform the iris to polar coordinates, is normalized for changes in dilation, and is tolerant of the segmentation errors that are likely to occur in a non-cooperative situation. Experimental results show that the proposed method is comparable to traditional methods on frontal eyes and performs well for the non-frontal iris images.

ACKNOWLEDGMENT

We would like to thank MicroVista for partial support of our camera equipment. This project is sponsored by the ONR Young Investigator Program (award number: N00014-07-1-0788) and National Institute of Justice (award number: 2007-DE-BX-K182).

REFERENCES

[1] F.H. Adler, Physiology of the Eye. St. Louis, MO: Mosby, 1965.

[2] J. Daugman, "New Methods in Iris Recognition," IEEE Transactions on Systems, Man, and Cybernetics - Part B: Cybernetics, Vol. 37, No. 5, October 2007.

[3] J. Daugman, "How Iris Recognition Works," IEEE Transaction on Circuits and Systems for Video Technology, Vol. 14, No. 1, pp. 21- 30, 2004.

[4] L. Ma, T. Tan, Y. Wang, and D. Zhang, "Personal Identification Based on Iris Texture Analysis," IEEE Transactions on Pattern Analysis and Machine Intelligence, 25(12), pp. 1519-1533, 2004.

[5] L. Ma, T. Tan, Y. Wang, and D. Zhang, "Efficient Iris Recognition by Characterizing Key Local Variations," IEEE Transactions on Image Processing, Vol. 13, pp. 739-750, 2004.

[6] Y. Chen, S.C. Dass, and A.K. Jain, "Localized Iris Image Quality Using 2-D Wavelets", IEEE International Conference on Biometrics, 2006.

[7] L. Masek, and P. Kovesi, MATLAB Source Code for a Biometric Identification System Based on Iris Patterns, The University of Western Australia, 2003.

[8] V.A. Pozdin, and Y. Du, "Performance Analysis and Parameter Optimization for Iris recognition Using Log-Gabor Wavelet," SPIE Electronic Imaging, Vol. 6491, 649112-1~11, 2007.

[9] Y. Du, R. W. Ives, D.M. Etter, and T.B. Welch, "Use of One-Dimensional Iris Signatures to Rank Iris Pattern Similarities," Optical Engineering, Vol. 45, No. 3, 037201-1~10, 2006.

[10] Z. Sun, Y. Wang, T. Tan, and J. Cui, "Improving Iris Recognition Accuracy Via Cascaded Classifiers," IEEE Transactions on Systems, Man, and Cybernetics – Part C: Applications and Reviews, Vol. 35, No. 3, August 2005.

[11] J. Thornton, M. Savvides, and B.V.K.V. Kumar, "A Bayesian Approach to Deformed Pattern Matching of Iris Images," IEEE Transactions on Pattern Analysis and Machine Intelligence, Vol. 29, No. 4, April 2007.

[12] W.W. Boles and B. Boashash, "A Human Identification Technique Using Images of the Iris and Wavelet Transform," IEEE Trans. Signal Processing, Vol. 46, No. 4, pp. 1185-1188, 1998.

[13] S.A.C. Schuckers, N.A. Schmid, A. Abhyankar, V. Dorairaj, C.K. Boyce, and L.A. Hornak, "On Techniques for Angle Compensation in Nonideal Iris Recognition," IEEE Transaction on Systems, Man, and Cybernetics-Part B: Cybernetics, Vol. 37, No. 5, pp. 1176-1190, 2007.

[14] C. Belcher, and Y. Du, "Region-based SIFT Approach to Iris Recognition," Optics and Lasers in Engineering, Vol. 47, Iss. 1, pp. 139-147, 2009.

[15] D.G. Lowe, "Object Recognition from Local Scale-Invariant Features," In Proceedings of the International Conference on Computer Vision, Vol. 2, pp. 1150–1157, Corfu, Greece, September 1999.

[16] M. Brown, and D.G. Lowe, "Invariant Features from Interest Point Groups", In British Machine Vision Conference, Cardiff, Wales, pp. 656-665, 2002.

[17] E. Arslanturk, "Video Based Non-Cooperative Iris Processing Method for Iris Surveillance System," MS Thesis, Indiana University-Purdue University of Indianapolis, 2008.

[18] Z. Zhou, "Iris Quality Filter and Segmentation Evaluation for Iris Recognition Systems," MS Thesis, Indiana University-Purdue University of Indianapolis, 2008.

An FPGA-based design of a modular approach for integral images in a real-time face detection system

Hau T. Ngo*[a], Ryan N. Rakvic[a], Randy Broussard[b], Robert W. Ives[a]
[a]Electrical and Computer Engineering Department, U. S. Naval Academy, Annapolis, MD 21402;
[b]System Engineering Department, U. S. Naval Academy, Annapolis, MD 21402;

ABSTRACT

The first step in a facial recognition system is to find and extract human faces in a static image or video frame. Most face detection methods are based on statistical models that can be trained and then used to classify faces. These methods are effective but the main drawback is speed because a massive number of sub-windows at different image scales are considered in the detection procedure. A robust face detection technique based on an encoded image known as an "integral image" has been proposed by Viola and Jones. The use of an integral image helps to reduce the number of operations to access a sub-image to a relatively small and fixed number. Additional speedup is achieved by incorporating a cascade of simple classifiers to quickly eliminate non-face sub-windows. Even with the reduced number of accesses to image data to extract features in Viola-Jones algorithm, the number of memory accesses is still too high to support real-time operations for high resolution images or video frames. The proposed hardware design in this research work employs a modular approach to represent the "integral image" for this memory-intensive application. An efficient memory manage strategy is also proposed to aggressively utilize embedded memory modules to reduce interaction with external memory chips. The proposed design is targeted for a low-cost FPGA prototype board for a cost-effective face detection/recognition system.

Keywords: Face detection, Viola-Jones detector, FPGA-based design, integral image, memory intensive applications, real-time image processing

1. INTRODUCTION

Object detection, and in particular, face detection is an important element of various computer vision areas, such as image retrieval, video surveillance, and human-computer interaction. The goal is to find an object of a pre-defined class in a static image or video frame. Sometimes this task can be accomplished by extracting certain image features such as edges, color regions, textures, contours, etc. and using heuristics to find configurations and/or combinations of those features specific to the object of interest. For more complex objects, such as human faces, it is hard to find these features and heuristics that will handle the huge variety of instances of the object class (e.g., faces may be slightly rotated in all three directions; some people wear glasses; some have moustaches or beards; often one half of the face is in the light and the other is shadow, etc.). For such objects, a statistical model (classifier) may be trained instead and then used to detect the objects.

Statistical model-based training takes multiple instances of the object class of interest, or "positive" samples, and multiple "negative" samples, i.e., images that do not contain objects of interest. Positive and negative samples together make a training set. During training, different features are extracted from the training samples and distinctive features that can be used to classify the object are selected. This information is "compressed" into the statistical model parameters. If the trained classifier does not detect an object or mistakenly detects the object, it is easy to make an adjustment by adding the corresponding positive or negative samples to the training set. To build a system capable of automatically labeling features on the face it is first necessary to localize the face in the image.

There have been many different approaches proposed for face detection [1-3]. For example, Rowley et al [1] propose a neural network approach to partition the image sub-windows into face or non-face regions. Schneiderman et al propose a statistical method for 3D object detection which can be used to detect faces very effectively [2]. The algorithm proposed in [2] can be used to detect human faces that are out-of-plane rotation which is very useful for real world applications. A method based on support vector machines is proposed by Osuna et al [3]. Another proposed method for face detection is based on sparse network of winnows (SNoW) functions to discriminate faces from background [4]. Most methods are effective but the main drawback is speed. A massive number of sub-windows within each image must be evaluated.

Mobile Multimedia/Image Processing, Security, and Applications 2009, edited by Sos S. Agaian, Sabah A. Jassim,
Proc. of SPIE Vol. 7351, 73510B · CCC code: 0277-786X/09/$18 · doi: 10.1117/12.820248

Therefore, it is a time consuming process. Viola and Jones have introduced a robust face detection system capable of detecting faces in real-time with both high detection rate and very low false positive rates [5]. The main advantage of this approach is the higher speed compared to other methods due to the representation of pixels as an integral image which requires a fixed small number of operations to evaluate features in a sub-window.

Even with the reduced number of memory accesses for feature computations in Viola-Jones algorithm, the number of memory accesses is still very high for embedded systems. Some have proposed hardware implementations for face detection systems on Field Programmable Gate Array (FPGA) chips, microcontrollers as well as custom VLSI chips [6-9]. VLSI design requires the longest design, debug and testing cycles which might not be very attractive for fast prototyping of the robust face detection algorithm. The approach using a microcontroller does not present significant speedup due to frequent accesses to external memory chips. In addition, there is a limit in the parallel operations for this approach. FPGA boards provide a low-cost platform to realize the face detection algorithm in a short design time with the flexibility of fine-tuning the design for more parallel operations as needed. In recent years, new generations of FPGAs with embedded DSP resources have provided an attractive solution for image and video processing applications. In this paper, we present an efficient, FPGA-based architecture design for the Viola-Jones face detector in video streams. The design utilizes a modular representation of the integral image to reduce external memory accesses.

2. FACE DETECTION ALGORITHM OVERVIEW

The face detection algorithm used in this work is one of the most efficient methods with high detection rate, low false detection rate and fast computational time. The method was first proposed by Viola and Jones [5]. The Viola-Jones detector consists of three parts. The first is an efficient method of encoding the image data known as an "integral image". This allows the sum of pixel responses within a given sub-rectangle of an image to be computed quickly and is the main contribution to the speedup of the Viola-Jones detector. The second element is the application of a boosting algorithm known as AdaBoost to select appropriate features that can form a template to model human face variation. The third part is a cascade of classifiers that further speeds up the search by quickly eliminating unlikely face regions. The speed of the algorithm is obtained by using a cascade of simple Haar-like features to progressively filter out non-face regions.

Using simple templates such as those shown in Fig. 1, the intensity ratios for sub-windows within the image can be calculated. For example, when template *b* is applied to the face, the value of this feature would be the sum of the pixel intensities in the white section over the sum of the intensities in the black section. Similarly, for more complex templates *c* and *d*, the value is the ratio of the sum of intensities in the white sections over those in the black. These sub-windows can be scaled to any size to find features over any sub-window within the image.

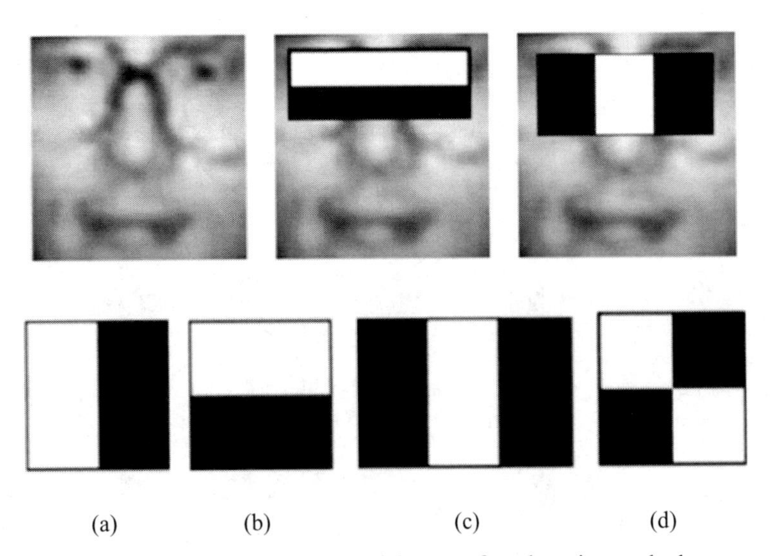

(a) (b) (c) (d)

Fig. 1. Haar features used for Viola Jones face detection method

2.1 Integral image

Since summing the pixel intensities many times over can be a slow process, the Viola-Jones detector uses the integral image representation to improve the speed. The integral image is constructed by replacing each image pixel with a value that corresponds to the pixel sum above and to the left of the pixel as shown in Fig. 2a.

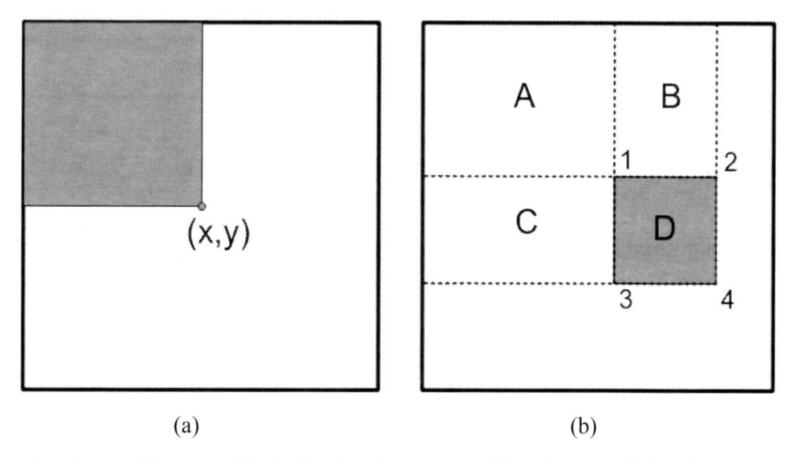

(a) (b)

Fig. 2. Construction of an integral image with pixel at (x, y) represented by the sum of the pixel above and the left (shaded region), (b) the pixel sum of a region D is computed with four array references (4+1-2-3)

Let *i(x, y)* be the intensity value of an image at points *x* and *y*. We will define the integral image to be:

$$ii(x,y) = \sum_{\substack{x' \le x \\ y' \le y}} i(x',y') \tag{1}$$

By using the following recurrences:

$$s(x,y) = s(x,y-1) + i(x,y) \tag{2}$$

$$ii(x,y) = ii(x-1,y) + s(x,y) \tag{3}$$

The integral image can be computed in a single pass over the image. With integral image, computing a sum of a rectangular region becomes a simple sequence of accessing four sums at the corner of the rectangle and performing the appropriate operations. For example in Fig. 2b, the sum of intensities at region *A* would be the value of the integral image at point 1. The sum of intensities in region *B* would be the value of 2 minus those at point 1. For region *D*, the sum would be 4 + 1 - (2 + 3).

The second part of the Viola-Jones algorithm is the selection of the image features for forming the template. This step is done during the training procedure. Given a training set containing labeled examples of faces (positive) and non-faces (negative), a complex and robust classifier is built by multiple weak classifiers using a procedure called boosting. The boosted classifier is built iteratively as a weighted sum of weak classifiers. On each iteration, a new weak classifier is trained and added to the sum. The weight of all the training samples is then updated, so that on the next iteration the roles of those samples that are misclassified by the already built classifiers are emphasized. The Viola-Jones detector assumes that out of the hundreds of thousands of possible features within a window, only a small number are necessary to form an effective strong classifier. Training the full feature set discussed in [5] is a difficult and time consuming task which might take days or weeks to complete. However, in this work, we focus only on the actual detection task and we assume that the training procedure has been done off-line and the feature set has been selected.

2.2 Cascade of filters

To increase the speed of the detector, it is best to remove as many non-face sub-windows from consideration as possible early on. The key point is that smaller, and therefore more efficient, boosted classifiers can be constructed to reject many of the negative sub-windows while detecting almost all positive instances. Simpler classifiers are used to reject the majority of sub-windows before more complex classifiers are called upon to achieve low false positive rates. Stages in the cascade are constructed by training classifiers using AdaBoost. Starting with a feature strong classifier, an effective face filter can be obtained by adjusting the strong classifier threshold to minimize false negatives. The initial AdaBoost threshold is designed to yield a low error rate on the training data. A lower threshold yields higher detection rates and higher false positive rates. Based on performance measured using a validation training set, the classifier can be adjusted to detect 100% of the faces with a higher false positive rate (about 50%). Although the detection performance of the classifier is far from acceptable as an object detection system, the classifier can significantly reduce the number of sub-windows that need further processing with very few operations.

The overall form of the detection process is that of a degenerate decision tree shown in Fig. 3. A positive result from the first classifier (T) triggers the evaluation of a second classifier which has also been adjusted to achieve very high detection rates. A positive result from the second classifier triggers a third classifier, and so on. A negative outcome (F) at any point leads to the immediate rejection of the sub-window.

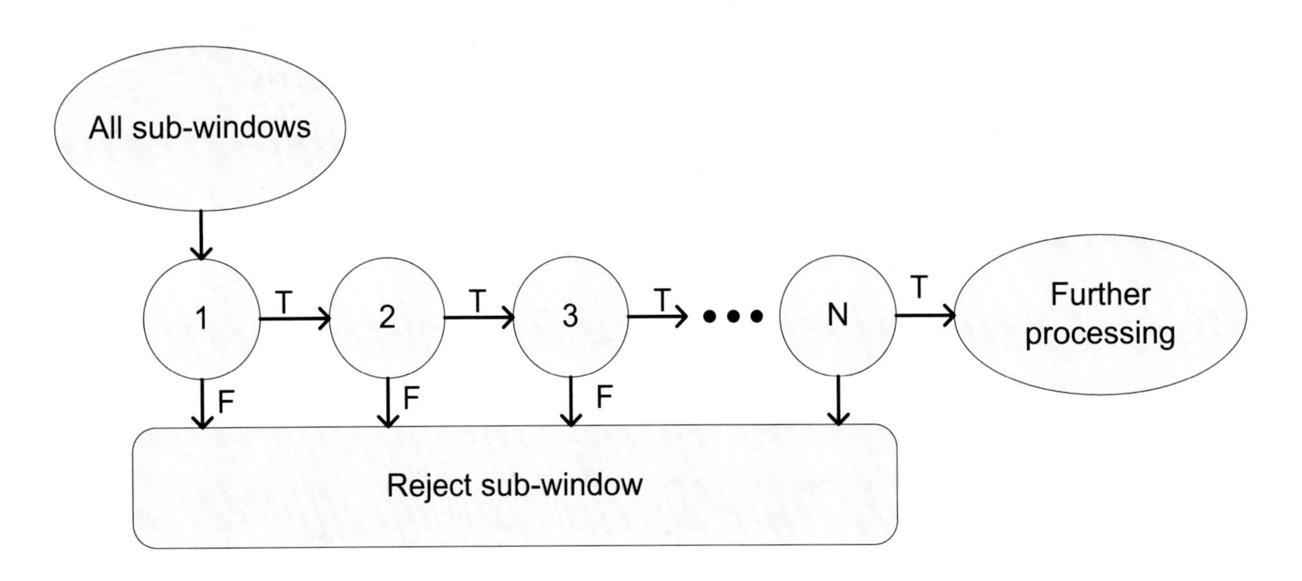

Fig. 3. Cascade of filters

3. FPGA-BASED DESIGN

FPGA-based systems have become an attractive solution for real-time image processing and computer vision problems because of their high density, high performance and the configurability to support specific applications. An FPGA offers a combination of the flexibility of a general purpose computer and the hardware-based high speed of Application Specific Integrated Circuits (ASICs). An architecture design for FPGA technology can fully exploit the data and I/O parallelism in any particular image processing application. Additionally, DSP-supported FPGAs, readily available in the market contain substantial number of technology-optimized and built-in components that are frequently used in DSP applications such as multipliers, Multiply-and-Accumulator (MAC) units, etc.

In this particular algorithm, the bottleneck of the system is the memory interface since this is a memory intensive application. In this work, we target the architecture design for a low cost FPGA board from Altera. The development board used for this work is the DE2 development and education board. The board consists of a Cyclone II FPGA with 105 embedded M4K RAM blocks, each of which can be configured as 512×8, 256×16 or 128×32 block. Other external RAM modules available are a 512 KB SRAM module which is configured as 256K×16 block, 8 MB SDRAM (4 banks of 2M×8 or 1M×16). In order to detect faces at different sizes, features at different scales are accessed and evaluated. With the modest amount of on-chip memory for this FPGA device, most of the sum values in the integral image would be stored in the off-chip memory which require significant more time to access. It is essential that the system is design to focus on limiting the number of off-chip memory accesses so that the it can support real time operations. Fig. 4 shows an overview of the proposed system. The face detection algorithm is realized in hardware design with two main processing modules: the integral image unit which computes the modular representation of the integral image, and a filter cascade which consists of multiple processing elements to perform addition, subtraction, comparison and multiplication operations. The main contribution of this research work is the modular representation of the integral image to limit accesses to external memory.

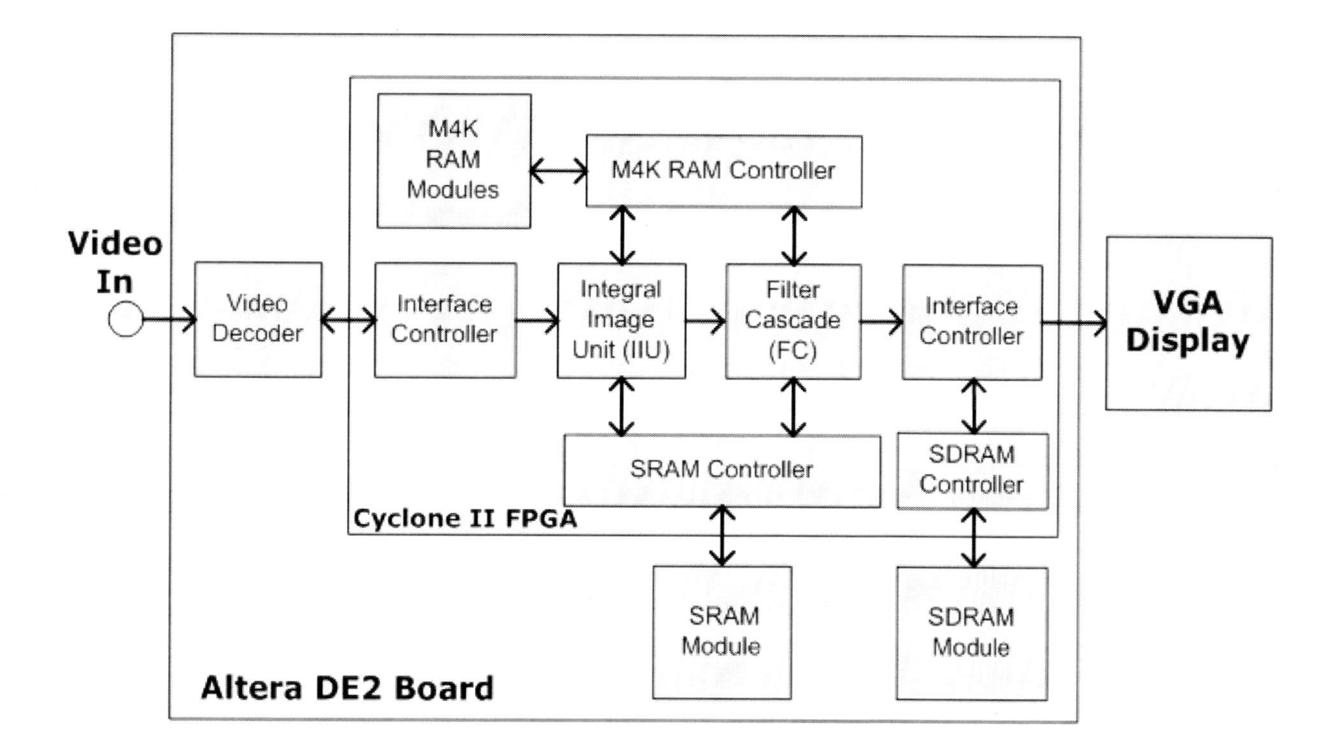

Fig. 4. Block diagram of the FPGA-based face detection system

3.1 Modular representation of the integral image

The conventional and straightforward implementation of the integral image would require each sum value of a pixel (x,y) to be represented with 32 bits. Since the SDRAM module requires longer latency to read/write a word in addition to refreshing times, it is used as an image buffer for displaying video sequences only. The embedded M4K RAM of the FPGA is too small to contain the whole integral image on-chip, so SRAM is used to store the integral image. Accessing the SRAM module is fast but problems occur due to its configuration as a 256K×16 block; therefore, accessing a 32-bit word (sum value) requires at least 2 read operations. For example, to compute a sum for a rectangular region such as D in Fig.2, eight external memory read operations (at least 16 cycles assuming each read operation takes 1 cycle) must be performed. More complex features would require more accesses to SDRAM since they might consist of more than one rectangles.

The main contribution in this work is to represent the integral image as a modular structure and to aggressively utilize embedded memory to reduce external memory interactions. The integral image is partitioned into smaller sub-modules such that any particular sum in the sub-module can be accessed in one cycle. The modular representation of the integral image is shown in Fig. 5 where an integral image is partitioned into M×N sub-modules. In order to represent any sum in a sub-module with a 16-bit word (for fast memory accessing due to memory configuration), the sub-module size is set to be 16×16 as shown in Fig. 5. The size of the sub-module is selected such that the sum values within the sub-module can be represented by 16 bits which can be accessed by one read operation to the external memory.

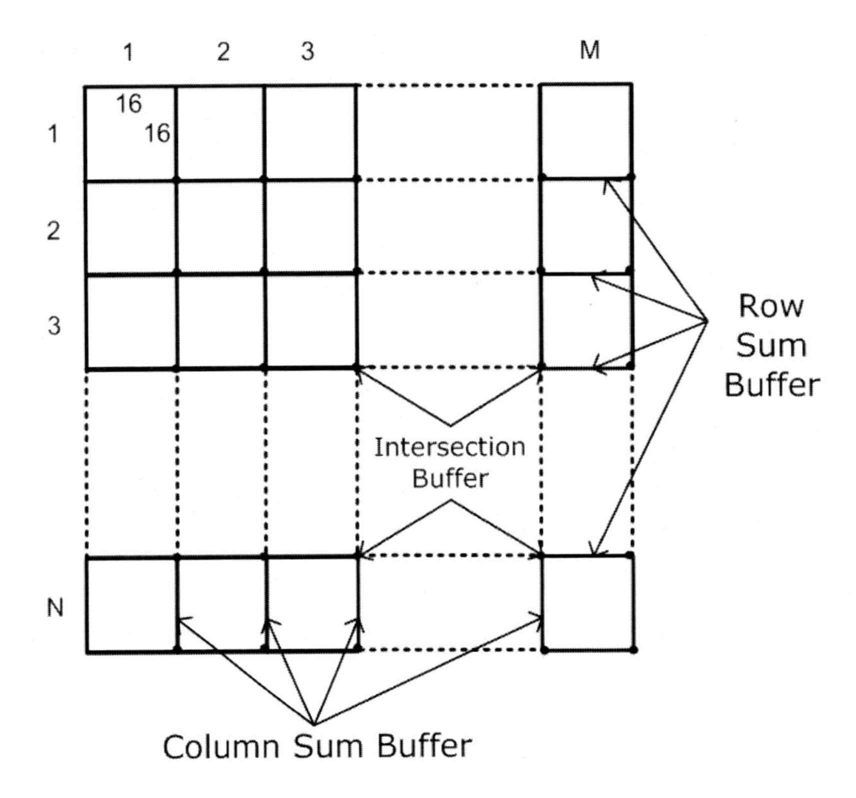

Fig. 5. Modular approach for computing integral image

For example, for a 320×240 video frame, 300 sub-modules are constructed to represent the integral image. The sums lie at the row or column lines are stored in specialized row and column buffers which are implemented using embedded RAM blocks. For this example, each row buffer is divided into $\lfloor 320/16 \rfloor$ segments each of which has 15 sum values. Each segment has 15 sums instead of 16 to avoid overlapping values with column buffer and intersection buffer. Similarly, each column buffer is divided into $\lfloor 240/16 \rfloor$ segments each of which also has 15 sum values. The intersection buffer has a total of $\lfloor 320/16 \rfloor \times \lfloor 240/16 \rfloor$ values. The sum values within each sub-module (those sums that are not row, column or intersection values) are stored in the SRAM module as before; however, with this approach, a value can be accessed with one read/write operation. To support quick computation of a sum within a region that spreads over several sub-modules, 20 column buffers and 15 row buffers are implemented with embedded M4K RAM blocks. If the sub-module size is less than 16, more memory blocks are needed for column and row buffers. Any sub-module size that is larger than 16 would require two memory accesses to the external SRAM for each value in an integral image. Each column buffer consists of 225 32-bit words and each row buffer consists of 300 32-bit words. In addition, the intersection points of rows and columns are also needed and are stored in the intersection buffer, which is also implemented with the embedded M4K RAM.

3.2 Filter cascade with module approach

The cascade of filters is useful since a simple classifier with small number of features can reject most sub-windows while maintaining almost all face sub-windows. Additional stages of filters are added to further refine the results until all or most non-face sub-windows are eliminated. In this design, we design a smaller version of the detector presented in [5]. The cascade of filters consists of three main stages. The first stage has one filter that has one feature; the second stage has one filter with five features and stage three has one filter with 20 features. As indicated in [5], the first classifier in the cascade is constructed using one or two features and it can reject about 50% of non-face while correctly identify close to 100% of faces. The classifier in the next stage of the cascade has five features and it detects almost 100% faces while rejecting 80% of non-faces. The next classifier has 20 features and it maintains a detection rate of almost 100% of faces and rejects 98% of non-faces. For applications that require more accuracy, more features can be incorporated into additional subsequent stages. In this work, we consider three stages of classifiers as shown in Fig. 6.

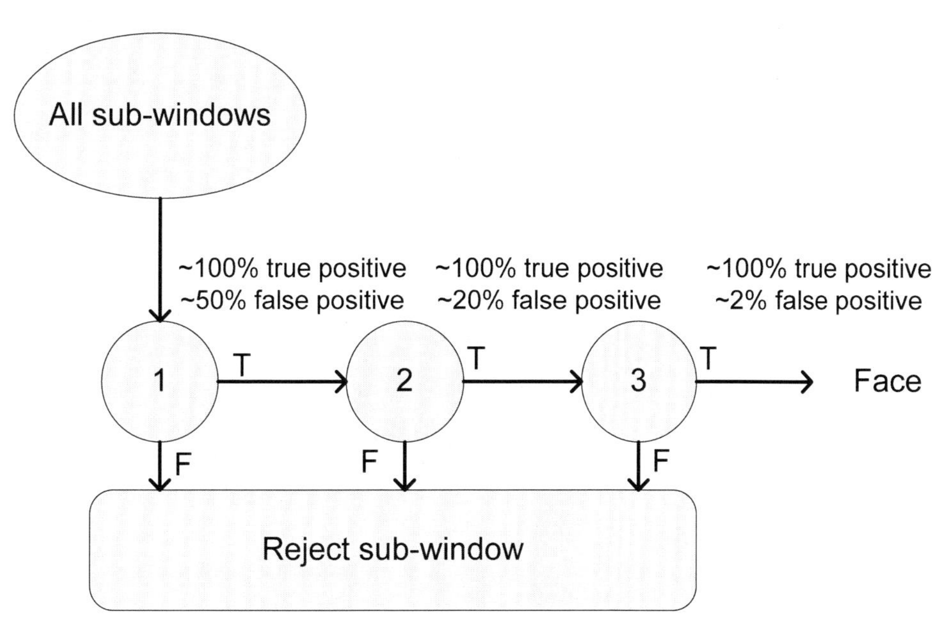

Fig. 6. Cascade of classifiers implemented in this work.

For each stage, one or more processing elements are designed to support parallel processing; however, more parallel processing require additional memory bandwidth. Therefore, parallel processing is not essential to the final design. The main contribution is the design of a modular architecture for calculation of the integral image which utilizes fast embedded M4K RAM to quickly access a sum in the integral image. Once the modular integral image is constructed and stored, reading a sum is then considered either a local or global operation. Local operations are evaluated if the entire feature rectangle is contained within one sub-module. If the rectangle spans over two or more sub-modules, global operations are needed. Local operations require accessing only sub-modules, while global operations require accessing sub-module, row, column and intersection buffers.

For example, to compute the sum of rectangle R1 in Fig. 7, local operations are performed. The operations for R1 are the same as the conventional approach which is computed by 4+1-2-3. In the local operation mode, one memory access is carried out for each corner point of the rectangle. Computing the sum for region R2 as shown in Fig. 7 requires global operation mode. For each corner point of the rectangle R2, one read from each of the RAM modules (specifically SRAM, row sum buffer, column sum buffer and intersection buffer) is needed. For example, to obtain the value at point 8 shown in Fig. 7, a read from SRAM is needed to obtain a sum value of point 8 in the sub-module. A read from row sum buffer n+1 (on-chip RAM) is needed for point 9 and a read from column buffer m+1 (on-chip RAM) for point 10. Point 11 is obtained from the intersection buffer (on-chip RAM). All four read operations can be carried out simultaneously since they are read from different memory modules. Then, point 8 for global operation can be computed by performing 8+9+10-11. Similarly, points 5, 6 and 7 are obtained for global operations with four memory reads, two additions and one subtraction for each point. The final step to obtain the sum of region R2 is computed the same way as in the conventional approach by performing 8+5-6-7.

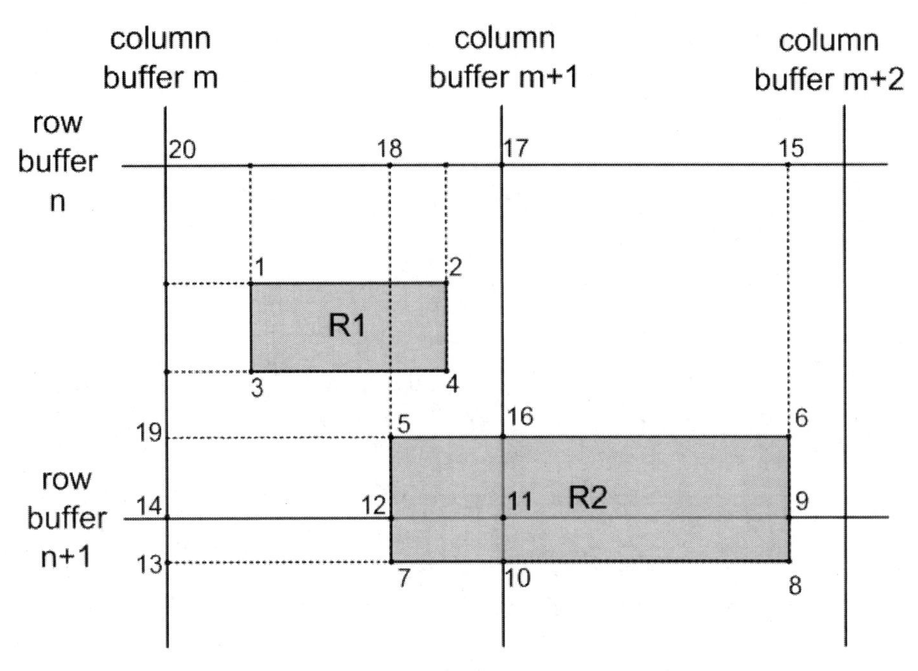

Fig. 7. Illustration of local and global operation modes to compute sums of rectangular regions in the proposed modular approach.

4. EVALUATION

The proposed modular architecture for the face detection system based on Viola-Jones algorithm is designed for Altera DE2 development and education FPGA board. As discussed earlier, the targeted FPGA has very limited amount of on-chip RAM blocks which can be used to store row, column and intersection buffers. Table 1 lists the required RAM blocks for some of the most common image sizes. In order to efficiently take advantage of the proposed modular approach for integral image representation, the number of on-chip RAM blocks must meet the requirement. For the Cyclone II FPGA available in the DE2 development board, there are 105 128×32 RAM blocks. Hence, the largest image size supported in this work would be 320×240. Some of the latest FPGA chips such as Stratix IV form Altera have much more embedded RAM bits that can be used to support larger image size such as 1025×768. Table 2 gives a summary of resource utilization and maximum speed achieved from the implementation for a 320×240 video stream. A three stage filter cascade with 26 features is considered. The design requires one 128×32 M4K RAM modules for storing the features, 74 128×32 M4K RAM modules for row and column sum buffers, three 128×32 M4K RAM modules for the intersection buffer and seven 128×32 M4K RAM modules for FIFO queues for the memory controlling units.

Table. 1 On-chip RAM blocks (128×32) required for different image sizes

Image size	Row buffer	Column buffer	Intesection buffer	Total RAM blocks required
160×120	9	10	1	20
320×240	36	38	3	77
640×480	141	150	10	301
800×600	220	235	15	470
1025×768	360	384	24	768

Table. 2 FPGA Implementation result for the proposed modular approach to represent integral images (320×240)

Description	Result
Device	Cyclone II 2C35 FPGA
Logic elements (LEs)	4232 (13%)
RAM bits	376832 (78%)
Max. clock freq.	235 MHz

For analysis, let's consider 320×240 image size. For this size, eleven sub-window scales are considered and a sub-window shift parameter Δ is set for 1 and 1.5 (for each scale level) in the simulations. The shifting parameter affects both speed and accuracy of the face detector. A larger value of Δ would result in faster performance with less accuracy. Two values for Δ are selected based on simulation results shown in [5]. Since all sub-windows are considered in the first stage of the cascade, simulations for a conventional design approach and a modular approach for first stage are performed. The results of total memory accesses and total time based on a 50 MHz clock (available from the board) are shown in Table 3. The modular approach reduces the total memory bits in the external RAM by half; therefore, the number of memory reads from external SRAM and the total access time are reduced about 50%. The first stage of the cascade is the main processing block since all sub-windows are considered. Stage 2 and 3 require only a small number of sub-windows since the majority of the sub-windows will be rejected after the first stage. The performance of the modular approach after the first stage simulation shows that this method is very suitable for real-time processing of at least 30 frames per second.

Table. 3 Simulations of the first stage in the filter cascade with 11 scales and different shifting parameters.

Method	Number of sub-windows	Number of memory accesses	Total access time (ms)
Conventional $(\Delta=1)$	94,441	1,511,064	30.22
Modular $(\Delta=1)$	94,441	755,532	15.11
Conventional $(\Delta=1.5)$	29,077	465,233	9.34
Modular $(\Delta=1.5)$	29,077	232,617	4.65

5. CONCLUSION

The design of an efficient modular architecture for detection faces in video streams has been presented in this paper. The main contribution is the modular approach to represent the integral image in the Viola-Jones algorithm to reduce memory bits and access time. The design was targeted for Altera's low cost DE2 development and education board. Utilization of M4K RAM embedded in the Cyclone II FPGA is managed efficiently to support the modular approach. Simulation results show a 50% reduction in total memory bits needed to store the integral image in the external SRAM, hence reducing the number of memory accesses and access time by about 50%. Simulation results also show that this approach is very suitable for real-time applications that require the minimum processing rate of 30 frames per second.

REFERENCES

[1] Rowley, H., Baluja, S. and Kanade T., "Neural network based face detection," IEEE Trans. on Pattern Analysis and Machine Intelligence 20(1), 23-38 (1998).

[2] Schneiderman, H., and Kanade, T., "A statistical method for 3d object detection applied to faces and cars," Proc. Computer Vision and Pattern Recognition Conference, 746-751 (2000).

[3] Osuna, E., Freund, R. and Girosi, F., "Training support vector machines: an application to face detection," Proc. Computer Vision and Pattern Recognition Conference, 130 (1997).

[4] Yang, M. H., Roth, D. and Ahuja, N., "A snow-based face detector," Proc. Advances in Neural Information Processing Systems, 855–861 (2000).

[5] Viola, P. and Jones M., "Robust real-time face detection," International Journal of Computer Vision, 57(2), 137-154 (2004).

[6] McReady, R., "Real time face detection on a configurable hardware system," Proc. Intl. Workshop on Field Programmable Logic and Applications, 157-162 (2000).

[7] Theocharides, T., Link, L., Vijaykrishnan, N., Irwin, M. J. and Wolf, W., "Embedded hardware face detection," Proc Intl. Conf. on VLSI Design, 133-138 (2004).

[8] Kianzad, V., Saha, S., Schelessman, J., Aggarwal, G., Bhattacharyya, S., Wolf, W. and Chellappa, R., "An Architectural level design methodology for embedded face detection," Proc. Intl. Conf. on Hardware/Software Codesign and System Synthesis, 136-141 (2005).

[9] Hori, Y., Shimizu, K., Nakamura, Y. and Kuroda, T., "A real time multi face detection technique using positive-negative lines-of-face template," Proc. Intl. Conf. on Pattern Recognition, 765-768 (2004).

Low-Cost Mobile Video-Based Iris Recognition for Small Databases

N. Luke Thomas, Yingzi Du, Sriharsha Muttineni, Shing Mang, Dylan Sran

Purdue School of Engineering and Technology at Indianapolis

Electrical and Computer Engineering

723 W. Michigan St. SL160, Indianapolis, IN 46202

Abstract:

This paper presents a low-cost method for providing biometric verification for applications that do not require large database sizes. Existing portable iris recognition systems are typically self-contained and expensive. For some applications, low cost is more important than extremely discerning matching ability. In these instances, the proposed system could be implemented at low cost, with adequate matching performance for verification. Additionally, the proposed system could be used in conjunction with any image based biometric identification system. A prototype system was developed and tested on a small database, with promising preliminary results.

Keywords: Low-cost biometrics, iris recognition, video-based identity verification

1. Introduction

Biometrics is the automatic identification of a person automatically using their physiological and/or behavioral traits, characteristics, or features [1-4]. Due to their convenience, accuracy, and difficulty in defrauding, biometrics are an important and desirable method for identifying people.

Iris recognition is one of the most accurate biometric measures, with reported false match rates of less than 1 in 200 billion [5]. Iris patterns are very stable and unique to each person [6-8]. For example, a person's left and right eyes have different iris patterns, and identical twins have different iris patterns. Iris recognition is non-intrusive for the user, and there are few social issues with exposing one's eyes, as compared to exposing one's face (such as with some religious groups) [1-3, 9, 10].

The goal of the proposed system is to use a low cost webcam for iris recognition. The proposed system could have several benefits over existing handheld iris matching devices—lower cost, lower operating expenses, simple expansion of the system, and simplified database management. It could be used in the medical field, the department of corrections, or other situations where cost is a concern. There are several challenges to develop such a system: communication bandwidth, image quality, computational power, database size, and flexibility of the system architecture.

Mobile Multimedia/Image Processing, Security, and Applications 2009, edited by Sos S. Agaian, Sabah A. Jassim, Proc. of SPIE Vol. 7351, 73510D · © 2009 SPIE · CCC code: 0277-786X/09/$18 · doi: 10.1117/12.818896

2. Review of Existing Iris Recognition Systems

Existing iris recognition systems can be categorized into four groups:

- Standalone Handheld Devices – These are typically small hand-held devices that provide stand-alone iris recognition. Many of these systems are all-inclusive systems which include image acquisition, processing, and database hardware in the same device. Examples include the HIDE system used by the Army from L1 systems [11], or systems run on cellular phones and PDA's [12].
- PC-Based Systems – These systems use a PC for their data processing, and typically use add-on or plug-in hardware for image acquisition. These systems are typically not portable. An example is the Panasonic Authenticam [13].
- Access Control Systems – These systems use stand alone hardware for image acquisition and processing, and typically require the user to actively participate with the acquisition and recognition process. Examples include LG's IrisAccess [14] and Panasonic's BM-ET200 [15].
- Surveillance Systems – These systems require less active participation from the user for acquisition and recognition as compared to standalone systems. In some cases, they may be able to acquire iris images without a user's knowledge. An example is Sarnoff's Iris on the Move system [16].

While each of these systems can produce excellent matching results when used in their intended operating environments, they can be expensive. Even the portable commercially available systems require the user to be habituated to the system and actively compliant with the image acquisition process. In addition, for standalone handheld devices, their matching capabilities are limited due to the computational limitations of handheld devices and storage limitations of their on-board databases. Since these systems have databases that are distributed throughout the company or departments, their database maintenance and control is much more complicated. Lastly, these types of devices are typically very expensive because of the need for high-speed portable computation devices. Some uses of biometric verification, however could require less stringent accuracy along with significantly reduced cost. For this project the goal is to design such a system.

3. Proposed System

A simple block diagram of the system's operation is shown below (Fig. 1). The proposed system uses low-cost decentralized image acquisition devices, such as a webcam, that acquires video images of user's eyes for matching. Then, using a quality filter and frame selection algorithm, some portion of the acquired images would be sent to a centralized iris recognition system for matching, with a final matching output using a score fusion algorithm to combine each of the individual iris images results. The matching result would be transmitted back to the hand-held device for display. This type of system is novel because it could be implemented using many existing iris recognition system for matching, or even other biometric systems. For example, this same type of system could be used for face recognition by replacing the iris recognition system with a face recognition system. This allows the system very broad application and implementation, since the approach could be used for any number of biometric or image-based recognition systems.

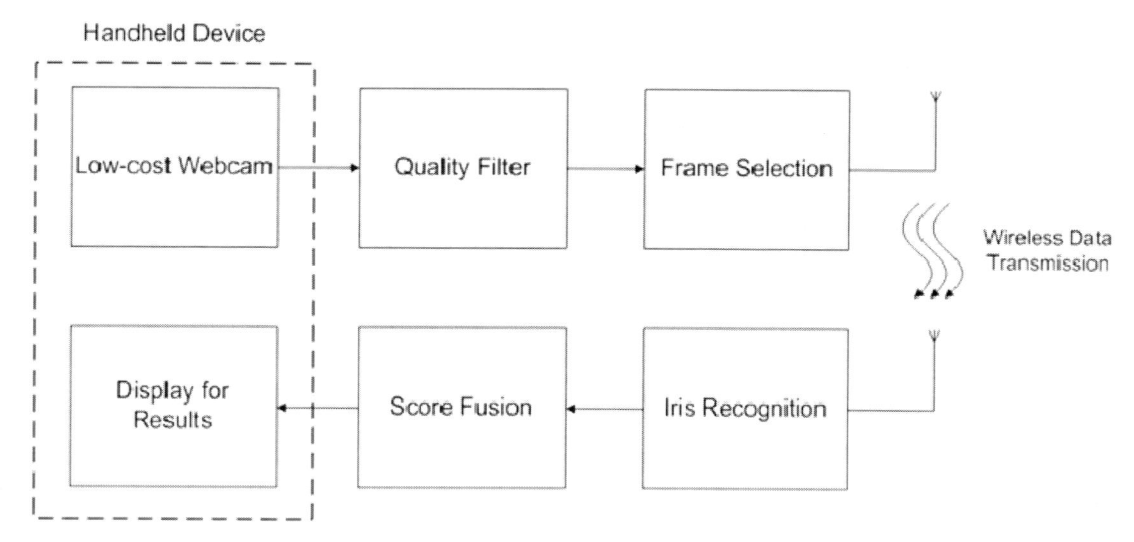

Figure 1. A block diagram of the systems operation.

The handheld device is composed of a low-cost 'webcam' type camera for acquiring iris images and a display to show the results to the user. Depending on the application, this hand-held device might have little or no additional computational ability. Instead, it may just transmit the iris video frames to a centralized server for further processing. This would make each hand-held device very inexpensive and easy to maintain. Also depending on the application, the hand-held device could include a card-reader or input device for the user to assert their identity to be matched.

The quality filter is a high-speed filter capable of evaluating each frame in the acquired video to ensure that it fulfills a minimum quality necessary for the iris recognition process. Specifically, it evaluates if there is an iris in the frame, if that iris is in focus, and if there is significant occlusion. Based on these criteria, bad-quality frames are removed from the video stream prior to further processing [17].

Similarly, the frame selection process chooses a certain number of frames based on their quality measure and other criteria to be sent to be recognized. This portion of the system can be tailored to the specific needs of the implementation or the current demand on the system—more frames can be sent if the matching results need to be more accurate, or less if processing speed is more important or if there are many active requests for matching.

The iris recognition system could be any system capable of matching iris images. For cost issues, it would be most cost-effective to utilize an existing system. Given the assumptions of compressed images and a small database size, the iris recognition system would not need to be particularly complex or expensive, as the system is intended for verification of a user's asserted identity and not broader identification.

Lastly, the score fusion system takes the individual matching results from the multiple images submitted to the iris recognition system and combines their results into a single result. This system uses both the iris recognition system's outcome, matching quality measure if it exists, and the outcome from the quality filter to intelligently combine the individual results into a single one.

The image acquisition device would not have to perform much computation or matching, so would not require an on-board database or intensive computational abilities. This would make the individual hand-held devices low cost, with little or no maintenance. The heavy computation and database management could be centralized, both to reduce costs and increase ease of use for the users and maintenance personnel. Since the system is not intended to be used in a continuous fashion, like in an airport or border security checkpoint; but to allow many separate users to access the system periodically, so it should be able to handle a large number of periodic users. Since it is built on existing 802.11 wireless networking, additional users could be added and removed at will, with only those currently accessing the system using the system's resources.

4. Experimental Results and Discussion

For our proof-of-concept device, we used a commercially available IP-based camera, the Airlink SkyIPCam500W. This camera uses a CMOS sensor that can output individual 640 by 480 pixel images captured at 30 frames per second in the near-infrared spectrum with built in 802.11a/b connectivity and near-infrared LED illumination. A LG IrisAccess system was used to perform the iris matching, and the remaining functionality was programmed in C# using in-house developed software.

We have tested our system using a small test database of 6 enrolled individuals, and our system was able to correctly identify all the individual users. Additionally, we submitted an individual who was not enrolled in the system for matching and the system properly identified that the 'imposter' user did not match any of the enrolled users.

Figure 2 shows some sample enrollment images, acquired with the LG IrisAccess system, and Figure 3 shows some test images acquired with the proposed system.

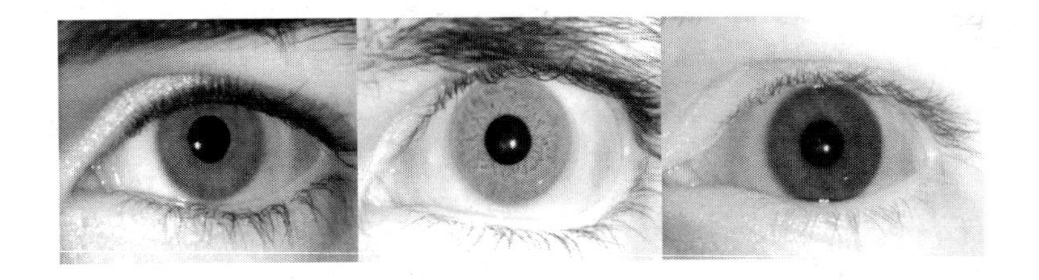

Figure 2. Example enrollment images.

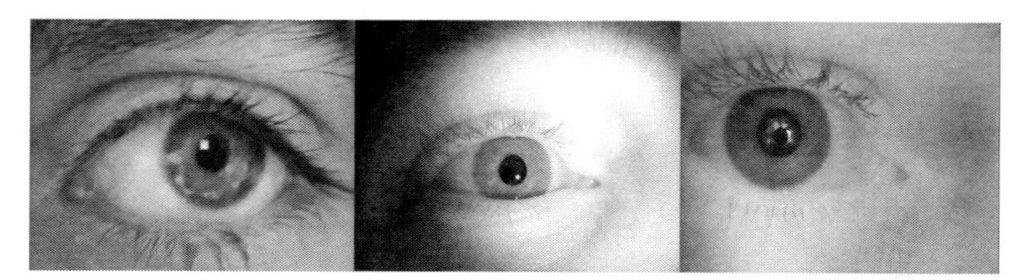

Figure 3. Example images acquired with proposed system.

The camera used for the project only outputs images in JPEG form, a lossy compression image format. JPEG (Joint Pictures Expert Group) format compresses images by compressing each 8 pixel by 8 pixel block of the target image using the discrete cosine transform and then independently quantizing the resultant coefficients. This compression technique is designed to retain the portions of the image that the human eye is most sensitive to while achieving a reasonable level of data compression. For iris recognition, the problem with this scheme is that the compression introduces artifacts, or features in the acquired image that were not present in the target image. Since iris recognition, at its most basic, measures and compares the features within an iris image, this compression scheme can introduce 'false' features into the image that can change its matching result. A simple example of this effect is when the boundaries between the JPEG encoded blocks are not as contiguous as the original image and create false edges in the image (fig 4).

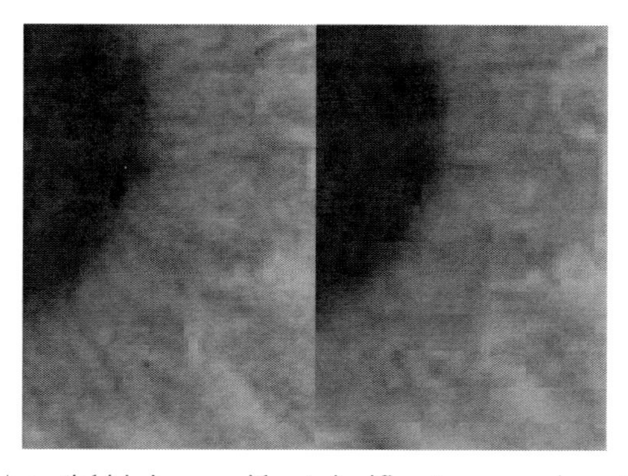

Figure 4. A partial iris image without significant compression artifacts (left) and with obvious compression artifacts (right).

In [18] Daugman discussed the effect of image compression on iris recognition, and showed that using the JPEG2000 compression technique with ROI could achieve a high compression rate without compromising much recognition accuracy. However, this requires the compression system to segment the iris image before compression, which is not the case for our application. In [19], Belcher *et al.*, showed regular JPEG2000 compression without ROI processing could reduce recognition accuracy. In this research, due to the limitation of our database size, we do not have conclusive results, but anecdotally we do see the reduction of recognition accuracy due to compression. For our system, we limited this effect by matching several images and producing a single matching score from all the matching results. While the compression increased the hamming distance as compared to images acquired and matched with the LG IrisAccess system, the system was able to accurately identify each user with the score fusion and multiple test images from a single video acquisition.

5. Conclusion and Future Work

Overall, the system is a proof-of-concept for a simple, low-cost, video-based, wireless iris recognition system. While certainly not implementing the concept at the same level that a production device would be implemented at, we believe that the final system demonstrates many of the advantages that this type of iris recognition system could have over existing systems—low cost, centralized processing and matching, decentralized data acquisition, simple expansion and maintenance, and a high matching accuracy.

This type of system could be used in medical settings—such as verification of patient identity prior to surgery, detention settings—such as in a prison or detention center, or access control—such as an office building. In each of these settings, cost is the major prohibition to broader biometric use, the security requirements are not extremely high, and the database size would be manageable for a system using compressed images.

For future work, the data security and encryption should be considered in the system for privacy and security. The iris recognition system can also be replaced with another recognition system for multiple functions, such as face recognition, ear recognition, etc.

Acknowledgement

This research is a result of an undergraduate senior design project. Part of this research was funded by the Electrical and Computer Engineering Department at Indiana University Purdue University Indianapolis, and ONR Young Investigator Award #: N00014-07-1-0788. The authors would like to acknowledge Mr. R.P. Kirchner from the Department of Defense for his help and support.

References

[1] A. K. Jain, S. Pankanti, S. Prabhakar, H. Lin, and A. Ross, "Biometrics: a grand challenge," in *Pattern Recognition, 2004. ICPR 2004. Proceedings of the 17th International Conference on*, 2004, pp. 935-942 Vol.2.

[2] J. Daugman, "How iris recognition works," *Circuits and Systems for Video Technology, IEEE Transactions on,* vol. 14, pp. 21-30, 2004.

[3] J. Daugman, "New Methods in Iris Recognition," *Systems, Man, and Cybernetics, Part B, IEEE Transactions on,* vol. 37, pp. 1167-1175, 2007.

[4] Y. Du, "Review of Iris Recognition: Cameras, Systems, and Their Applications," *Sensor Review,* vol. 26, pp. 66-69, 2006.

[5] J. Daugman, "Probing the Uniqueness and Randomness of IrisCodes: Results From 200 Billion Iris Pair Comparisons," *Proceedings of the IEEE,* vol. 94, pp. 1927-1935, 2006.

[6] M. Imaizumi and T. Kuwabara, "Development of the rat iris," *Invest Ophthalmol,* vol. 10, pp. 733-44, Oct 1971.

[7] R. M. Gladstone, "Development and significance of heterochromia of the iris," *Arch Neurol,* vol. 21, pp. 184-91, Aug 1969.

[8] I. C. Mann, "The Development of the Human Iris," *Br J Ophthalmol,* vol. 9, pp. 495-512, Oct 1925.

[9] R. W. Ives, Y. Du, D. M. Etter, and T. B. Welch, "A Multidisciplinary Approach to Biometrics," *Education, IEEE Transactions on,* vol. 48, pp. 462-471, 2005.

[10] R. P. Wildes, "Iris recognition: an emerging biometric technology," *Proceedings of the IEEE,* vol. 85, pp. 1348-1363, 1997.

[11] D. Hubler, "L-1 Supplies Army with portable recognition devices." vol. 2009, 2009.

[12] C. Dal-ho, P. Kang Ryoung, R. Dae Woong, K. Yanggon, and Y. Jonghoon, "Pupil and Iris Localization for Iris Recognition in Mobile Phones," in *Software Engineering, Artificial Intelligence, Networking, and Parallel/Distributed Computing, 2006. SNPD 2006. Seventh ACIS International Conference on*, 2006, pp. 197-201.

[13] S. J. Yang, "A New Look in Security," in *PcMag.com*, 2002.

[14] "http://www.lgiris.com/ps/products/irisaccess4000.htm."

[15] "http://www.panasonic.com/business/security/products/biometrics.asp."

[16] J. R. Matey, O. Naroditsky, K. Hanna, R. Kolczynski, D. J. LoIacono, S. Mangru, M. Tinker, T. M. Zappia, and W. Y. Zhao, "Iris on the Move: Acquisition of Images for Iris Recognition in Less Constrained Environments," *Proceedings of the IEEE,* vol. 94, pp. 1936-1947, 2006.

[17] C. Belcher and D. Yingzi, "Feature information based quality measure for iris recognition," in *Systems, Man and Cybernetics, 2007. ISIC. IEEE International Conference on*, 2007, pp. 3339-3345.

[18] J. Daugman and C. Downing, "Effect of Severe Image Compression on Iris Recognition Performance," *Information Forensics and Security, IEEE Transactions on,* vol. 3, pp. 52-61, 2008.

[19] C. Belcher, Y. Du, and R. Ives, "Feature Correlation Evaluation Approach for Iris Image Quality Measure," *submitted to Signal Processing,* 2009.

An Orthogonal Subspace Projection Approach for Face Recognition

Zhi Zhou[1] *Yingzi Du*[1]* *Chein-I Chang*[2]

[1]Department of Electrical and Computer Engineering
Purdue School of Engineering and Technology
Indiana University-Purdue University Indianapolis
Indianapolis, IN 46202
Email {zhizhou, yidu}@iupui.edu

[2] Department of Computer Science and Electrical Engineering
University of Maryland, Baltimore County
Baltimore, Maryland 21250
Email cchang@umbc.edu

ABSTRACT

Face recognition has been widely used to automatically identify and verify a person. In this paper, we proposed a new approach based on orthogonal subspace projection (OSP) to identification of human faces. In linear mixture model of the face images, the OSP faces of the training images are calculated by using orthogonal subspace projection approach and the signal-to noise ratio maximization. And the weight parameter of the input image is obtained to do face recognition.

Index Terms— face recognition; PCA; LDA; orthogonal subspace projection (OSP)

1. INTRODUCTION

Over the past decade, many face recognition algorithms were designed [1-17]. The principle components analysis (PCA), also called eigenface method, is considered as the classic approach [1-3, 11]. Its principle is to find an orthogonal coordinate system such that data is approximated best and the correlation between different axes is minimized. Different from PCA, the linear discriminate analysis (LDA) searches the directions for maximum discrimination of classes[4, 15]. And within-class and between-class matrices are defined in order to achieve this goal. The dimensionality of face image is reduced both in these two approaches. Moreover, the integral operator kernel function was used to perform a nonlinear form of principle components analysis in face recognition [7, 13, 14, 17].

In this paper, we propose a new approach for face recognition using orthogonal subspace projection (OSP). It is different from previous approaches that focus on how to reduce the dimensionality, but rather to project the face is the dimension that best separate them. Histogram equalization and histogram specification are applied in the preprocessing step to enhance illumination. Based on linear mixture model of the face images, the OSP faces are calculated by using orthogonal subspace projection approach and the Signal-to Noise Ratio maximization. The weight parameter of the input image is obtained to identification of human faces. In our experimental result, we compared our proposed method with two traditional methods (PCA and LDA). The result shows our OSP approach can achieve better accuracy than other two.

Mobile Multimedia/Image Processing, Security, and Applications 2009, edited by Sos S. Agaian, Sabah A. Jassim,
Proc. of SPIE Vol. 7351, 73510E · © 2009 SPIE · CCC code: 0277-786X/09/$18 · doi: 10.1117/12.818037

2. METHODOLOGY

2.1 Illumination Enhancement

As we mentioned before, face recognition accuracy will decrease dramatically in uncontrolled environmental illumination. In order to obtain better performance of the system, we do preprocessing of the face images before feature extraction. Fig. 1 (a) shows the original image from Yale database B [18].

2.1.1. Histogram Equalization

In histogram equalization, the image's histogram is used to do contrast adjustment[19]. The general histogram equalization formula is:

(1)

Where is the minimum value of the cumulative distribution function. is the size of the image (M is width and N is height) and L is the number of grey levels used(in here $L=256$). After this adjustment, the image's intensities can be better distributed on the histogram. Fig.1(b) shows the image after histogram equalization. We can see the intensity of the image is much better than the original image. However, the illumination of the left side in Fig.1 (b) is still very dark because of the shadow.

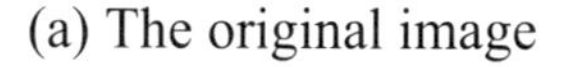

(a) The original image (b) The enhanced image

Figure 1. Histogram equalization

2.1.2. Histogram Specification

In Fig.1 (b), since the left side's illumination is still not good enough, we introduced the histogram specification method to do the second step adjustment. Histogram specification is a technique that transforms the histogram of one image into the histogram of another image[19]. This method is mapping pixels from one image to another according to the rules of histogram equalization. The primary rule is that the actual cumulative sum can be no less than the template cumulative sum.

In here, we used the enroll image's histogram as the template histogram (Fig.2 (a)), and then transform the histogram of the testing image into the template histogram. Observed the shadow in the testing image, we divided the template histogram into two sides (left face and right face). Each side of testing image is matching with the corresponding side in template histogram. Fig.2 shows the result of histogram specification. We can see the contrast of left side in Fig.2(c) is much better than the original image (Fig.1 (a)) and the first enhance image (Fig.1(b) or Fig.2(b)).

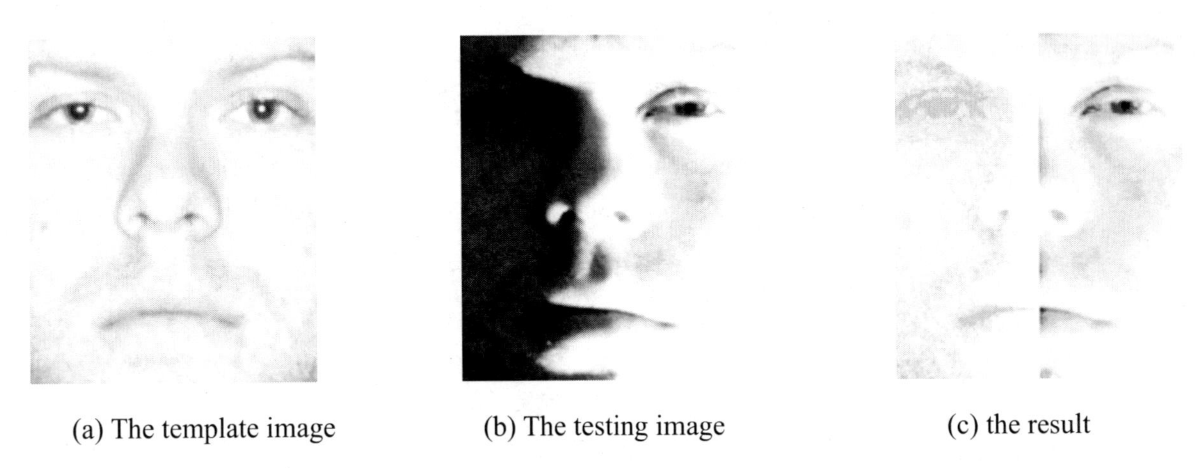

(a) The template image (b) The testing image (c) the result

Figure 2. Histogram specification

2.2. Orthogonal Subspace Projection Approach

2.2.1. Linear Mixture Model for Face Images

Let ***r*** to be a $l \times 1$ column vector and defined as the input image where l is the number of pixel [20-22]. Assume the M is a $l \times p$ matrix and made up of p training images denoted by (, , … ,) and is a $l \times 1$ column vector of each training image. Let be a $p \times 1$ column vector given by and denote the fraction of the jth training image in the input image ***r***. A liner mixture model for face images assume that input image ***r*** is a linear combination of p training images with the weight coefficients designated by vector and noise ***n***. Eq.(1) shows the detail.

$$\boldsymbol{r} = M \quad + \boldsymbol{n} \tag{1}$$

For training images M, it includes the desired signature and undesired signatures. In order to reduce image dimensionality and enhance the effect of desired signature while decrease undesired signatures, we expand in Eq.(2).

(2)

where ***d*** is a $l \times 1$ column vector of the desired signature with weight coefficient . ***U*** is a $l \times (p-1)$matrix of the undesired signatures with weight coefficients which is a $(l-1) \times 1$ column vector. Let and to be the estimate of and based on the observation input image .

(3)

2.2.2. Orthogonal Subspace Projection Approach.

In order to obtain accurate $\widehat{\alpha}_p$, we have to eliminate the undesired signatures U and noise $\boldsymbol{n}$. At first, we project $\boldsymbol{r}$ onto a subspace which is orthogonal to the matrix U using $l \times l$ projection matrix $\boldsymbol{P}$ given by the following:

$$\boldsymbol{P}=\boldsymbol{I}-\boldsymbol{U}(\boldsymbol{U}^{\boldsymbol{T}}\boldsymbol{U})^{-1}\boldsymbol{U}^{\boldsymbol{T}} \tag{4}$$

$$\boldsymbol{Pr} = \boldsymbol{Pd}\hat{\alpha}_p + \boldsymbol{PU}\hat{\boldsymbol{\gamma}} + \boldsymbol{Pn} \tag{5}$$

Since $\boldsymbol{P}$ is a projection matrix of the orthogonal subspace to the matrix U, $\boldsymbol{PU}\hat{\boldsymbol{\gamma}} = \boldsymbol{0}$ which means the contribution of $\boldsymbol{U}$ is reduced to zero. We have:

$$\boldsymbol{Pr} = \boldsymbol{Pd}\hat{\alpha}_p + \boldsymbol{Pn} \tag{6}$$

After undesired signatures eliminated, next step is to find $\boldsymbol{x}^{\boldsymbol{T}}$ (Eq.(7)) to maximize the *Signal-to Noise Ratio* (*SNR*).

$$\boldsymbol{x}^{\boldsymbol{T}}\boldsymbol{Pr} = \boldsymbol{x}^{\boldsymbol{T}}\boldsymbol{Pd}\hat{\alpha}_p + \boldsymbol{x}^{\boldsymbol{T}}\boldsymbol{Pn} \tag{7}$$

According to Ref.[20-22], the $1 \times l$ operator $\boldsymbol{x}^{\boldsymbol{T}} = k\boldsymbol{d}^{\boldsymbol{T}}$, k is an arbitrary scalar. Let's choose k=1 in general. The final equation is given by:

$$\boldsymbol{d}^{\boldsymbol{T}}\boldsymbol{Pr} = \boldsymbol{d}^{\boldsymbol{T}}\boldsymbol{Pd}\hat{\alpha}_p \tag{8}$$

According to Eq.(8), $\hat{\alpha}_p$, the estimate of α_p , can be obtained as follows:

$$\hat{\alpha}_p = \frac{\boldsymbol{d}^{\boldsymbol{T}}\boldsymbol{P}}{\boldsymbol{d}^{\boldsymbol{T}}\boldsymbol{Pd}}\boldsymbol{r} \tag{9}$$

where $\boldsymbol{d}$ is the desired signature of the training image, $\boldsymbol{P}$ is a projection matrix of the orthogonal subspace to the undesired signatures of the training images $\boldsymbol{U}$. $\boldsymbol{r}$ is the input image. $\hat{\alpha}_p$ is the similarity score of the input image $\boldsymbol{r}$ with the desired signature of training image $\boldsymbol{d}$. We saved $\frac{\boldsymbol{d}^{\boldsymbol{T}}\boldsymbol{P}}{\boldsymbol{d}^{\boldsymbol{T}}\boldsymbol{Pd}}$ as the OSP face of the training image.

The following steps summarize the face recognition process using OSP approach.

1. Initialization: Acquire the training set of face images and calculate the OSP faces for each training image.

2. When an input face image is encountered, calculate the input image into a $l \times 1$ column vector $\boldsymbol{r}$.

3. Obtaining the weight parameter $\hat{\alpha}_p$ by multiplying each training image's OSP face with the vector $\boldsymbol{r}$.

4. Classify the weight parameter $\hat{\alpha}_p$ to determine if the image is a known person or an unknown person. The higher value of $\hat{\alpha}_p$ implies the similarity.

In PCA approach, since we discard those eigenvectors with 0 or very small eigenvalues, the number of eigenface is smaller than the training image. Compared to the eigenface, we can obtain all of the OSP face according to the training image.

3. EXPERIMENTAL RESULTS

We evaluated the performances of the proposed orthogonal subspace projection (OSP) approach for face recognition based on Yale face database B [18]. The Yale face database B contains 2432 images of 38 human subjects under 9 poses and 64 illumination conditions. Fig.3(a) shows some face images in the database. Some of the images have shadow in one side, and some of them are totally dark in entire image.

In our experiment, we chose the first image in every human subject as the enrollment image, and the rest are the testing image. Since there are 64 illumination conditions in the database, we did illumination enhancement using histogram equalization and histogram specification to all of the images in the database. Fig.3 shows some results. The face images in Fig.3(b) are enhance images corresponding to the original image in Fig.3(a).

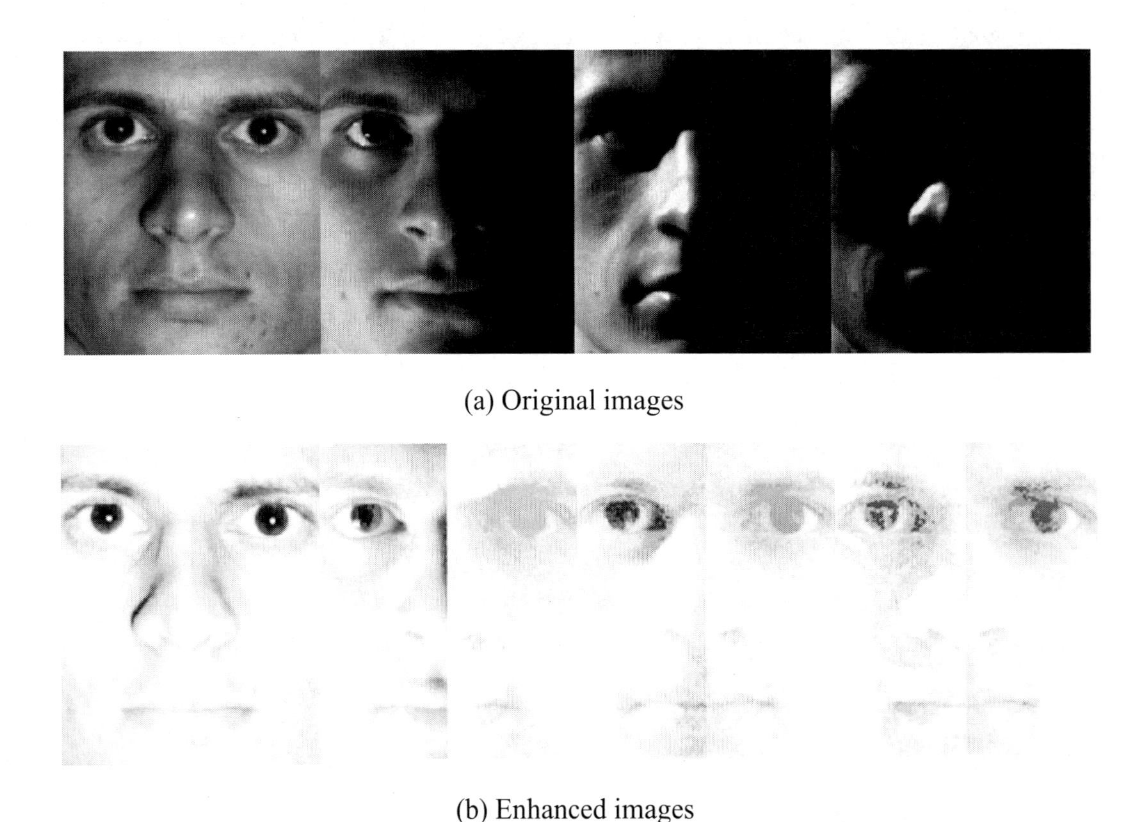

(a) Original images

(b) Enhanced images

Figure 3. Some results of illumination enhancement

After preprocessing, we compared our proposed OSP approach with two traditional approaches (PCA and LDA) to check if our OSP approach can achieve better accuracy or not.Fig.4 shows the comparison result.

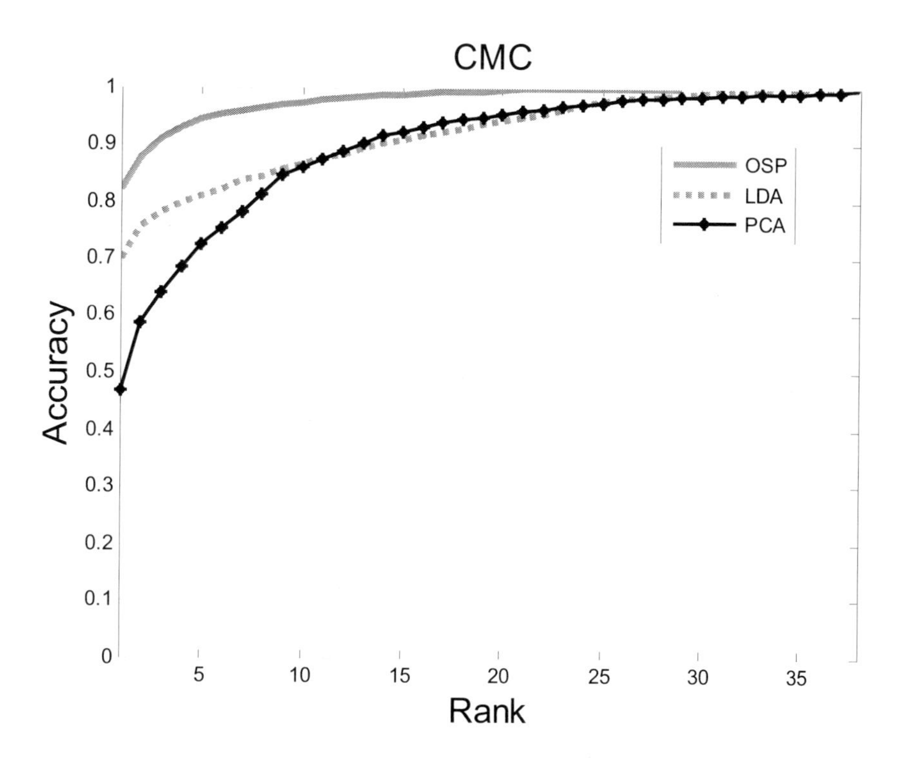

Figure 4. The comparison result

From Fig.4 we can obtain, the accuracy of our proposed OSP approach can achieve around 82% in rank 1, which is the highest one compared with other two approaches. PCA is just 50% in rank 1, and LDA is also lower than OSP (70% in rank 1). Fig.5 and Fig. 6 shows an OSP face based on the original image and the entire OSP faces of all training images respectively.

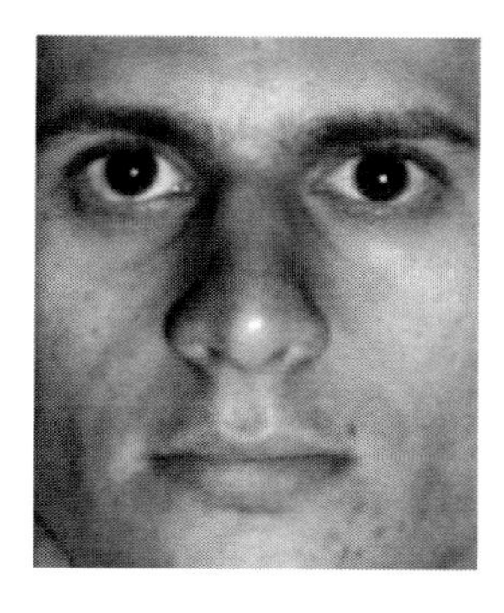

(a) Original image

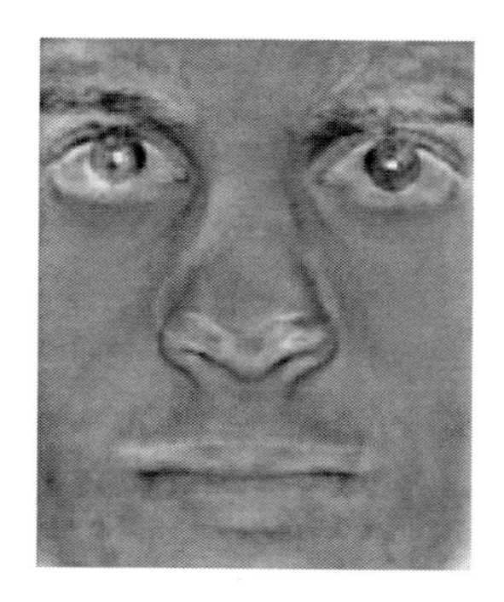

(b) OSP face

Figure 5. The OSP face

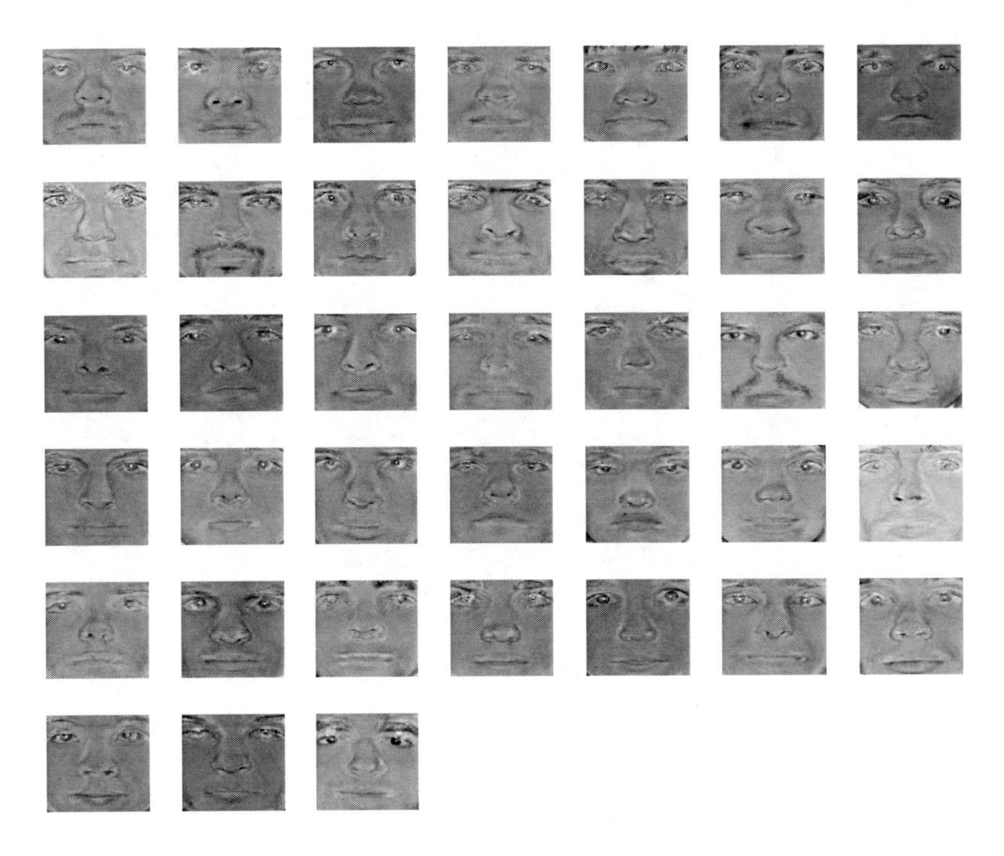

Figure 6. The OSP faces for all of the training images.

4. CONCLUSIONS

In this paper, we proposed a robust a new approach for face recognition using orthogonal subspace projection (OSP). In the preprocessing, we applied histogram equalization and histogram specification to do illumination enhancement. Furthermore, based on linear mixture model of the face images, the OSP faces are calculated by using orthogonal subspace projection approach and the Signal-to Noise Ratio maximization. The weight parameter of the input image is obtained to do face recognition. We compared our proposed method with two traditional methods (PCA and LDA) using Yale database B. The result shows our OSP approach can achieve better accuracy than other two.

5. REFERENCES

[1] M. Turk and A. Pentland, "Eigenfaces for Recognition," *Journal of Cognitive Neuroscience*, vol. 3, no. 1, pp. 71-86;1991

[2] M.A. Turk and A.P. Pentland, "Face recognition using eigenfaces," *IEEE Computer Society Conference on Computer Vision and Pattern Recognition*, pp. 586-591;1991

[3] A. Pentland, B. Moghaddam and T. Starner, "View-based and modular eigenspaces for face recognition," *Proc. IEEE Computer Society Conference on Computer Vision and Pattern Recognition*, pp. 84-91. 1994

[4] K. Etemad and R. Chellappa, "Discriminant analysis for recognition of human face images," Springer. 1997

[5] L. Wiskott, J. Fellous, N. Krer and C. von der Malsburg, "Face Recognition by Elastic Bunch Graph Matching," *IEEE Transactions on Pattern Analysis and Machine Intelligence*, pp. 775-779;1997

[6] C. Liu and H. Wechsler, "Face recognition using evolutionary pursuit," *Lecture Notes in Computer Science*, pp. 596-612;1998

[7] B. Scholkopf, A. Smola and K. Muller, "Nonlinear Component Analysis as a Kernel Eigenvalue Problem," *Neural Computation*, vol. 10, no. 5, pp. 1299-1319;1998
[8] C. Liu and H. Wechsler, "Comparative Assessment of Independent Component Analysis (ICA) for Face Recognition." 1999
[9] C. Liu and H. Wechsler, "Evolutionary Pursuit and Its Application to Face Recognition," *IEEE Transactions on Pattern Analysis and Machine Intelligence*, pp. 570-582;2000
[10] A. Kadyrov and M. Petrou, "The Trace Transform and Its Applications," *IEEE Transactions on Pattern Analysis and Machine Intelligence*, pp. 811-828;2001
[11] H. Moon and P. Phillips, "Computational and performance aspects of PCA-based face-recognition algorithms," *Perception-London-*, vol. 30, no. 3, pp. 303-322;2001
[12] M.S. Bartlett, J.R. Movellan and T.J. Sejnowski, "Face recognition by independent component analysis," *IEEE Transactions on Neural Networks*, vol. 13, no. 6, pp. 1450-1464;2002
[13] M. Yang, "Face Recognition using Kernel Methods," *Advances In Neural Information Processing Systems*, vol. 2, pp. 1457-1464;2002
[14] F. Bach and M. Jordan, "Kernel independent component analysis," *The Journal of Machine Learning Research*, vol. 3, pp. 1-48;2003
[15] J. Lu, K. Plataniotis and A. Venetsanopoulos, "Face recognition using LDA-based algorithms," *IEEE Transactions on Neural Networks*, vol. 14, no. 1, pp. 195-200;2003
[16] S. Srisuk, M. Petrou, W. Kurutach and A. Kadyrov, "Face authentication using the trace transform," *Proc. IEEE Computer Society Conference on Computer Vision and Pattern Recognition*, pp. I-305-I-312 vol.301. 2003
[17] S. Zhou and R. Chellappa, "Multiple-exemplar discriminant analysis for face recognition." 2004
[18] A.S. Georghiades, P.N. Belhumeur and D.J. Kriegman, "From few to many: illumination cone models for face recognition under variable lighting and pose," *IEEE Transactions on Pattern Analysis and Machine Intelligence*, vol. 23, no. 6, pp. 643-660;2001
[19] R. Gonzalez and R. Woods, *Digital Image Processing*, Prentice Hall, 2007.
[20] I.C. Cheng, Z. Xiao-Li, M.L.G. Althouse and P. Jeng Jong, "Least squares subspace projection approach to mixed pixel classification for hyperspectral images," *IEEE Transactions on Geoscience and Remote Sensing*, vol. 36, no. 3, pp. 898-912;1998
[21] J.C. Harsanyi and C.I. Chang, "Hyperspectral image classification and dimensionality reduction: an orthogonal subspace projection approach," *IEEE Transactions on Geoscience and Remote Sensing*, vol. 32, no. 4, pp. 779-785;1994
[22] T. Te-Ming, C. Chin-Hsing and I.C. Chein, "A posteriori least squares orthogonal subspace projection approach to desired signature extraction and detection," *IEEE Transactions on Geoscience and Remote Sensing*, vol. 35, no. 1, pp. 127-139;1997

Selective Object Encryption for Privacy Protection

Yicong Zhou*[a], Karen Panetta[a], Ravindranath Cherukuri[b], Sos Agaian[b]
[a]Department of Electrical and Computer Engineering, Tufts University, Medford, MA USA 02155;
[b]Department of Electrical and Computer Engineering, University of Texas at San Antonio, San Antonio, TX USA 78249

ABSTRACT

This paper introduces a new recursive sequence called the truncated P-Fibonacci sequence, its corresponding binary code called the truncated Fibonacci p-code and a new bit-plane decomposition method using the truncated Fibonacci p-code. In addition, a new lossless image encryption algorithm is presented that can encrypt a selected object using this new decomposition method for privacy protection. The user has the flexibility (1) to define the object to be protected as an object in an image or in a specific part of the image, a selected region of an image, or an entire image, (2) to utilize any new or existing method for edge detection or segmentation to extract the selected object from an image or a specific part/region of the image, (3) to select any new or existing method for the shuffling process. The algorithm can be used in many different areas such as wireless networking, mobile phone services and applications in homeland security and medical imaging. Simulation results and analysis verify that the algorithm shows good performance in object/image encryption and can withstand plaintext attacks.

Keywords: Image encryption, object encryption, truncated P-Fibonacci sequence, truncated Fibonacci p-code, truncated Fibonacci p-code decomposition

1. INTRODUCTION

Video surveillance systems for homeland security purposes are used to monitor many strategic places such as public transportation hubs, airport security, commercial and financial centers. Large amounts of videos and images with private information are generated. The potential exists for the loss of privacy and information abuse. Privacy protection becomes an important issue for these monitoring systems. One solution is to encrypt the selected object in a video/image to ensure privacy protection while allowing decryption for legitimate security needs at anytime. Privacy is preserved since the selective object encryption allows monitoring the activities without knowing the identities monitored in the videos. When a suspicious activity needs to be investigated, the identities can be uncovered with proper authorization [1].

Privacy protection is also important in many other areas. Biometrics authentication systems use personal biological or behavioral characteristics, such as face, fingerprint and signatures, to verify the user identity. The security of biometric data is particularly important because these biometric data not only serve as security keys for the authentication systems but also contain private information. Medical image transmission through wired and wireless networks allows different doctors the ability to digitally access the medical records of a patient. Hence, there is an urgent need to provide the confidentiality of medical image data related to the patient when medical image data is stored in databases and transmitted over networks. [2-4].

To protect privacy, several interesting approaches have been developed such as concealing regions of interest (ROIs) by scrambling the sign of selected transform coefficients in the transform-domain [5], Encrypting biometric images using Fractional Fourier transform [6], using a track-based system for human movement analysis and privacy protection adaptive to environmental contexts [7], de-identifying face images[8] and many others. These methods are good contributions to privacy protection for specific applications. However, a universal security method is required.

In this paper, we introduce a new lossless image encryption algorithm based on the truncated Fibonacci p-code decomposition. It is suitable for all types of images or objects in any kind of images for privacy protection in real-time applications. It protects privacy by encrypting a selected object in either a part of an image or the entire image, or encrypting the whole image, achieving different security needs in practical applications.

* Yicong.Zhou@tufts.edu; phone 1-617-627-5183; fax 1-617-627-3220

Mobile Multimedia/Image Processing, Security, and Applications 2009, edited by Sos S. Agaian, Sabah A. Jassim, Proc. of SPIE Vol. 7351, 73510F · © 2009 SPIE · CCC code: 0277-786X/09/$18 · doi: 10.1117/12.817699

The algorithm first generates a boundary mask of the selected object by using a segmentation algorithm or an edge detection method such as Canny, Sobel, or the shape-adaptive DCT [9, 10]. It then separates the image into the selected object and the image without the selected object based on this boundary mask. The selected object is encrypted using the truncated Fibonacci p-code decomposition method and bit plane shuffling. Finally, the encrypted object is combined with the image without the selected object to obtain the resulting encrypted image.

2. TRUNCATED FIBONACCI P-CODE AND ITS DECOMPOSITION

In this section, we introduce a new recursive sequence called truncated P-Fibonacci sequence (TPFS). We also introduce its corresponding binary code, called truncated Fibonacci p-code (TFPC), and bit-plane decomposition, namely truncated Fibonacci p-code decomposition. This decomposition method is well suitable for image encryption because the truncated Fibonacci p-code and its decomposition results are parameter dependent.

2.1 Truncated P-Fibonacci sequence

Definition 2.1: The P-Fibonacci sequence is a recursive sequence defined by [11],

$$F_p(n) = \begin{cases} 0 & n < 0 \\ 1 & n = 0 \\ F_p(n-1) + F_p(n-p-1) & n > 0 \end{cases} \tag{1}$$

where p is a non-negative integer.

Based on the definition above, P-Fibonacci sequence will differ based on the p value. Specially,

1) $p = 0$, the P-Fibonacci sequence is geometric progression increasing by two, 1, 2, 4, 8, 16…;
2) $p = 1$, the P-Fibonacci sequence is the classical Fibonacci sequence is 1, 1, 2, 3, 5, 8, 13, 21…;
3) For the large values of p, the P-Fibonacci sequence starts with successive 1's and immediately after that 1, 2, 3, 4 … p …

Some examples are given in Table 1.

Table 1. P-Fibonacci sequences with different p values.

P \ n	0	1	2	3	4	5	6	7	8	…
0	1	2	4	8	16	32	64	128	256	…
1	1	1	2	3	5	8	13	21	34	…
2	1	1	1	2	3	4	6	9	13	…
3	1	1	1	1	2	3	4	5	7	…
4	1	1	1	1	1	2	3	4	5	…
…	…									
∞	1	1	1	1	1	1	1	1	1	…

Table 2. Truncated P-Fibonacci sequences with different p values.

P \ n	0	1	2	3	4	5	6	7	8	…
0	1	2	4	8	16	32	64	128	256	…
1	1	2	3	5	8	13	21	34	55	…
2	1	2	3	4	6	9	13	19	28	…
3	1	2	3	4	5	7	10	14	19	…
4	1	2	3	4	5	6	8	11	15	…
…	…									

Definition 2.2: A recursive sequence called truncated P-Fibonacci sequence (TPFS) is defined as,

$$T_p(n) = \begin{cases} 0 & n < 0 \\ 1 & n = 0 \\ F_p(n+p) & n > 0 \end{cases} \quad (2)$$

where $F_p(n+p)$ is P-Fibonacci sequence defined in equation (1).

The truncated P-Fibonacci sequence also changes with different p values. For example,

1) $p = 0$, the truncated P-Fibonacci sequence is geometric progression increasing by two, 1, 2, 4, 8, 16…;

2) $p = 1$, the truncated P-Fibonacci sequence is the truncated classical Fibonacci sequence is 1, 2, 3, 5, 8, 13, 21…;

3) $p = \infty$, the truncated P-Fibonacci sequence is a positive integer sequence, 1, 2, 3, 4, 5, 6, 7…

Some TPFS examples are also given in Table 2.

2.2 Truncated Fibonacci P-code

Definition 2.3: A non-negative decimal number can be represented by the following format,

$$A = c_0 T_p(0) + c_1 T_p(1) + ... + c_{n-1} T_p(n-1) \quad (3)$$

where n and p are nonnegative integers, $i = 0,1,...,n-1$, $c_i \in (0,1)$, $T_p(i)$ is the TPFS defined in equation (2). The binary coefficient sequence $(c_{n-1},...,c_1,c_0)$ is called the truncated Fibonacci P-code (TFPC) of A, namely,

$$A = (c_{n-1},...,c_1,c_0)_p \quad (4)$$

For a certain p value, the TFPC of a specific decimal number is shorter than the Fibonacci p-code in [11]. This makes TFPC more efficient to be generated. Similar to the Fibonacci p-code, TFPC is also not unique. For example, if $A = 35$, $p = 4$, the TFPC of A will be,

$$35 = (1,0,0,0,0,0,0,0,0,0,0,1)_4 = (0,1,0,0,0,1,0,0,0,0,0,1)_4 = (0,1,0,0,0,0,1,0,0,1,0,0)_4 = ...$$

To obtain a unique TFPC of a non-negative decimal number for specific parameter p, the following condition should be satisfied,

$$A = T_p(n) + s \quad (5)$$

where $0 \le s < T_p(n-p)$.

The above condition is same as the constraint of the Fibonacci p-code in [11]. There are at least p 0's between two consecutive 1's in the unique TFPC of any non-negative decimal number. For the above example, its unique TFPC will be $35 = (1,0,0,0,0,0,0,0,0,0,0,1)_4$.

2.3 Truncated Fibonacci P-code Decomposition

For a certain p value, every non-negative decimal number has a unique TFPC representation if the condition in equation (5) is satisfied. Its TFPC will differ only based on different p values because the TFPS is specified by the parameter p values.

Based on the definition 2.3, a grayscale image can also be decomposed into the TFPC bit-planes called the Truncated Fibonacci P-code decomposition. The traditional image decomposition is a special case of the TFPC decomposition when $p = 0$.

A grayscale image with gray levels within 0-255 is decomposed into 12 TFPC bit planes when $p = 1$. A TFPC decomposition example is shown in Fig.1. For a specific grayscale image, the number of its TFPC bit-planes changes with parameter p values. For instance, the number of its TFPC bit-plane is 17 for $p = 3$, and 21 for $p = 5$ respectively.

Moreover, the contents of its TFPC decomposition results are also different based on different p values. This makes the TFPC decomposition a suitable method for image encryption. To show the difference between the TFPC decomposition and the Fibonacci p-code decomposition, Fig. 2 provides the decomposition result of the same grayscale image using the Fibonacci p-code with $p=1$.

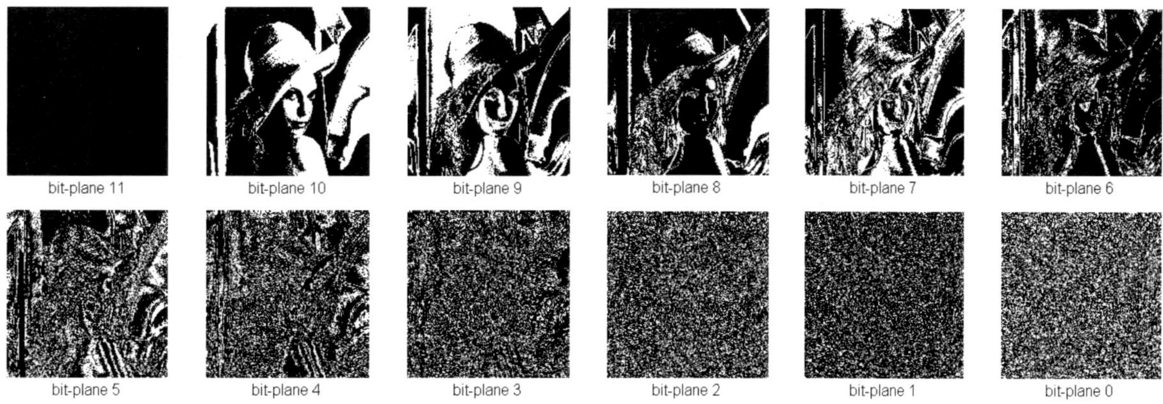

Fig. 1. Truncated Fibonacci P-code decomposition of the grayscale Lena image, p=1.

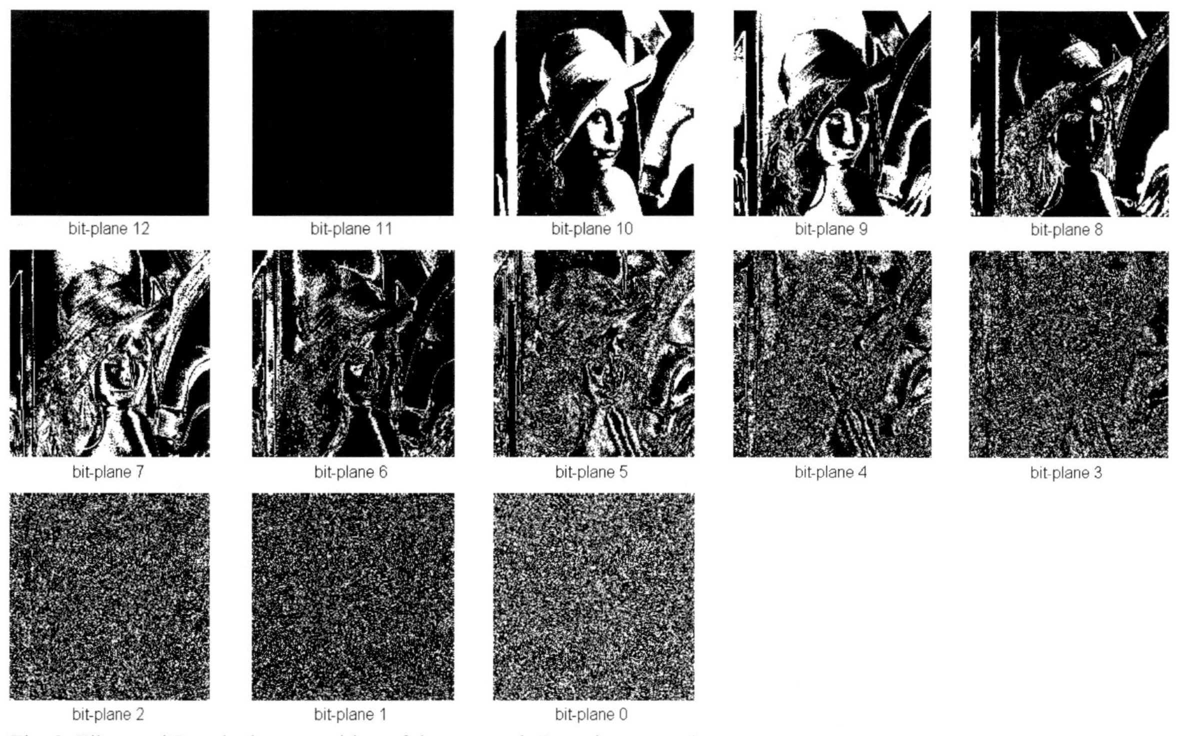

Fig. 2. Fibonacci P-code decomposition of the grayscale Lena image, p=1.

3. IMAGE ENCRYPTION ALGORITHM

The 2D image such as grayscale image, biometric image and medical image contain a 2D data matrix. To improve the speed of encryption process while protecting private information, one solution is to selectively encrypt the important part or region of an image. In this section, we introduce a new lossless encryption algorithm called "ObjectEcrypt" to encrypt a selected object. This selected object can be defined as either an object in an image or in a specific region of the image, a selected part/region of an image, or an entire image. The ObjectEcrypt algorithm is shown in Fig. 3. It can be used in real-time applications such as wireless network and mobile phone services.

The ObjectEcrypt algorithm first creates a boundary mask of the selected object using an edge detector or segmentation algorithm. In this paper, we use Canny edge detector to generate the object boundary mask. The algorithm then uses this mask to separate the original image into the selected object and image without the object. The selected object is decomposed into several TFPC bit planes based on the specified parameter p. The parameter p which has infinite number of possible choices can act as one of the security keys for the ObjectEcrypt algorithm. The order of TFPC bit planes is shuffled by an existing or new shuffling algorithm. The encrypted object can be obtained by combining all the shuffled bit planes and scaling down all pixel values back to the range of gray levels. Finally, the encrypted object is combined with the image without the object to acquire the resulting encrypted image.

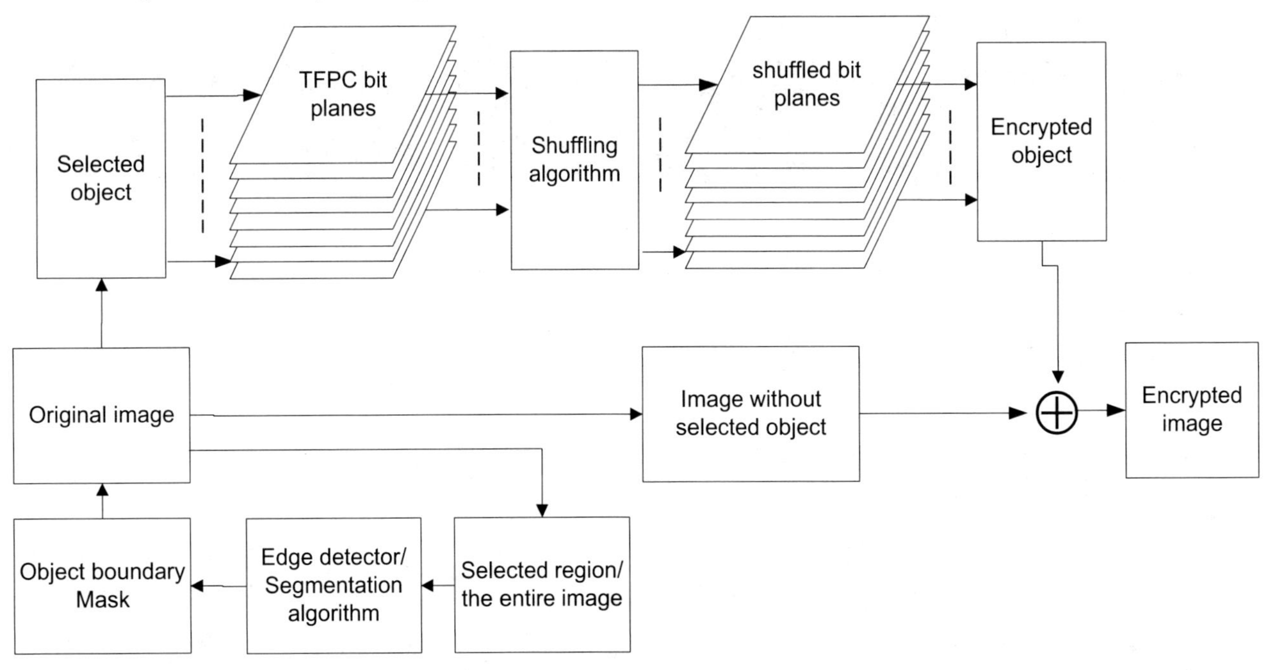

Fig. 3. Block diagram of the ObjectEcrypt algorithm.

The users also have flexibility to choose any new or existing method for the bit plane shuffling process such as inverting the order of the bit planes. In this paper, we choose right-round shift to shuffle the order of the TFPC bit planes. To make the shifted TFPC of each image pixel satisfy the constraint in equation (5), p zero bit-planes are added in front of the most significant bit plane before shifting all the TFPC bit planes. The shift algorithm is shown in Fig. 4. The shifting process will move all bit planes one bit position to their right side and the pth zero bit plane will be shifted to the position of the least significant bit plane in the TFPC bit planes. If the shifting times $r > 1$, the shifting process will more all bit plane r bit positions.

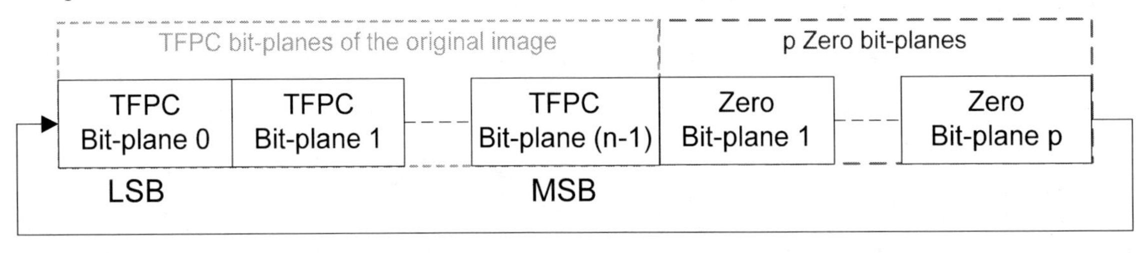

Fig. 4. Block diagram of the shifting algorithm.

The security keys of the ObjectEcrypt algorithm consist of the parameter p in the TFPC, the security keys in the shuffling algorithm, and the information of the encrypted object including the boundary mask and original data range before scaling down. In this paper, the number of times to shift, parameter r, is the security key of the shifting algorithm. The shifting times r should be less than the sum of the parameter p and the number of TFPC bit planes, i.e., $r < p+n$. These are required to be provided to the authorized users for their decryption process.

To reconstruct the original object/image in the decryption process, the encrypted object is extracted from the encrypted images using the object boundary mask. It is first scaled up to the original data range and decomposed to TFPC bit planes. The order of these TFPC bit planes is reverted back to its original order by using left-round shift. The reconstructed image can be obtained by converting the TFPC bit planes back to gray levels.

The 3D images such as color images and 3D medical images contain three 2D data matrices called 2D components. The object in 3D images can be encrypted by applying the ObjectEncrypt algorithm to its 2D components one by one.

4. SIMULATION RESULTS

The selected object can be defined as either an entire image or an object in a specific region of the image. In this section, we provide some experimental examples for these two cases to show the performance of the ObjectEcrypt algorithm for encrypting the selected objects in 2D and 3D images.

4.1 Object encryption in 2D images

Fig. 5 shows an example of grayscale image encryption using the ObjectEcrypt algorithm with security key, $p=2, r=5$. Fig. 6 provides a medical image encryption example, $p=2, r=4$. The objects in these two examples are defined as the entire images. In both examples, the original images are fully encrypted by the ObjectEcrypt algorithm. The encrypted images shown in Fig. 5(b) and Fig. 6(b) are unrecognizable. The original images are also completely reconstructed without any distortion. This can be verified by the reconstructed images in Fig. 5(c) and Fig. 6(c) which visually look the same as their original images. Their corresponding histograms in Fig. 5(f) and Fig. 6(f) also witness this perfect reconstruction.

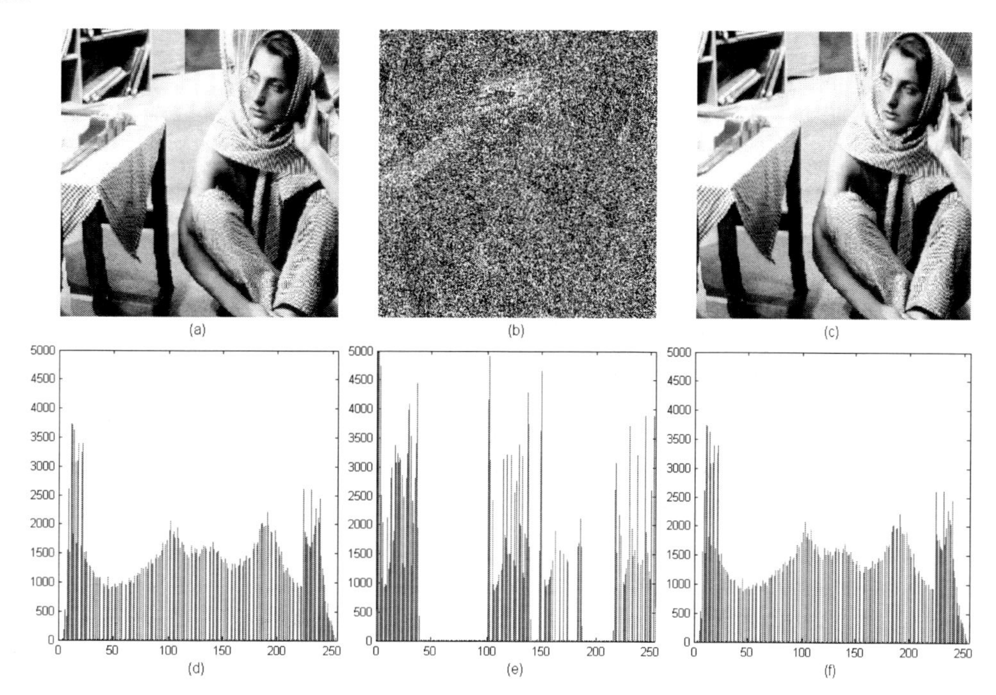

Fig. 5. Grayscale image encryption, $p=2, r=5$. (a) Original grayscale image; (b) Encrypted image; (c) Reconstructed image; (d) Histogram of the original image; (e) Histogram of the encrypted image; (f) Histogram of the encrypted image.

To achieve the goal of privacy protection in real-time application, it is not necessary to encrypt the entire image/video. Instead, selectively encrypting some important objects/regions in the image/video is an effective scheme. These important objects/regions usually contain private information such as human faces, fingerprints or patient's medical records along with text based personal information. Fig. 7 gives an example of the ObjectEcrypt algorithm for selected object encryption. Only the selected object has been fully protected. As a result, the computation cost of the encryption process will be significant decreased. This makes that the ObjectEcrypt algorithm is well suitable for privacy protection in real-time applications. The example in Fig. 7 shows that the selected object and the original image are also fully recovered. All these above examples demonstrate that the ObjectEcrypt algorithm is a lossless encryption method.

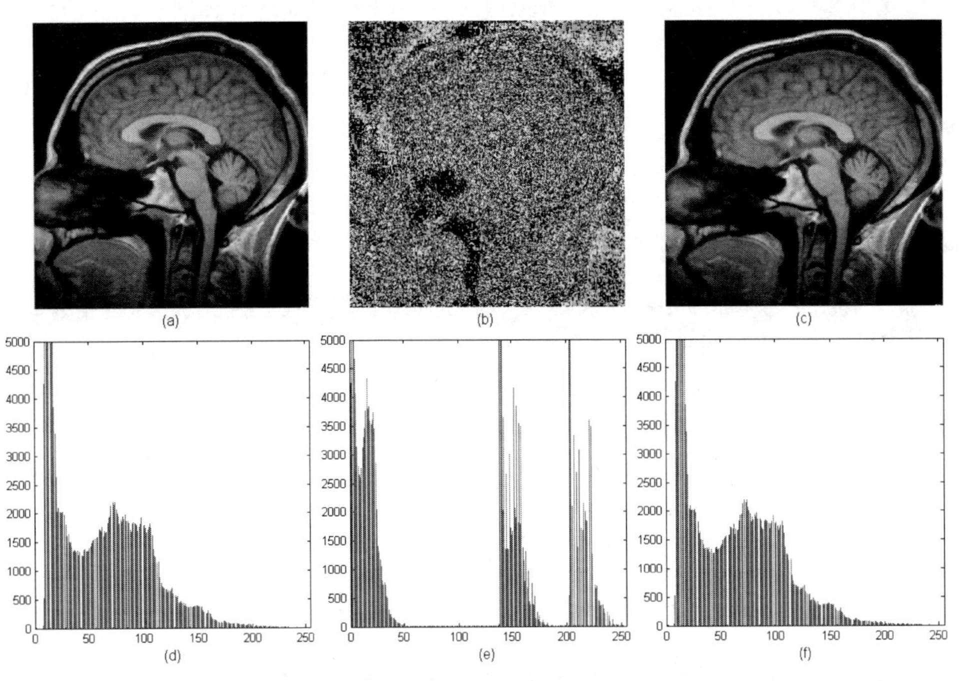

Fig. 6. Medical image encryption, $p = 2, r = 4$. (a) Original medical image; (b) Encrypted image; (c) Reconstructed image; (d) Histogram of the original image; (e) Histogram of the encrypted image; (f) Histogram of the encrypted image.

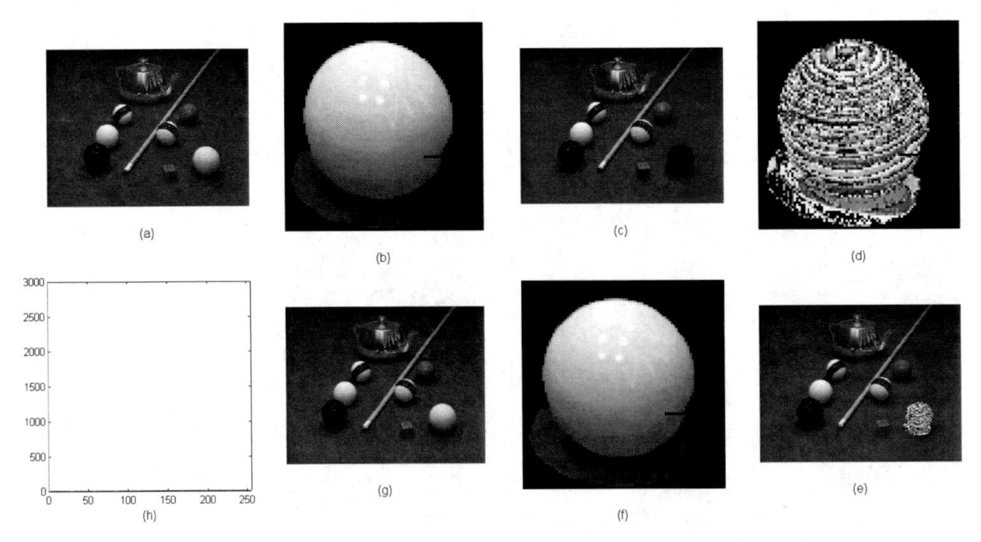

Fig. 7. Selected object encryption in a grayscale image, $p = 1, r = 5$. (a) Original grayscale image; (b) Selected object; (c) Image without the selected object; (d) Encrypted image of the selected object; (e) Encrypted image; (f) Reconstructed object; (g) Reconstructed image; (h) Histogram of the difference between (a) and (g).

4.2 Object encryption in 3D images

The selected 3D objects can be encrypted by applying the ObjectEcrypt algorithm to their corresponding 2D components one by one. Fig. 8 provides an example to encrypt a color image defined as an object. Fig. 9 shows the performance of the ObjectEcrypt algorithm for the selected 3D object encryption. The results show that the ObjectEcrypt algorithm can fully protect the private information and the original image/object can be also completely recovered.

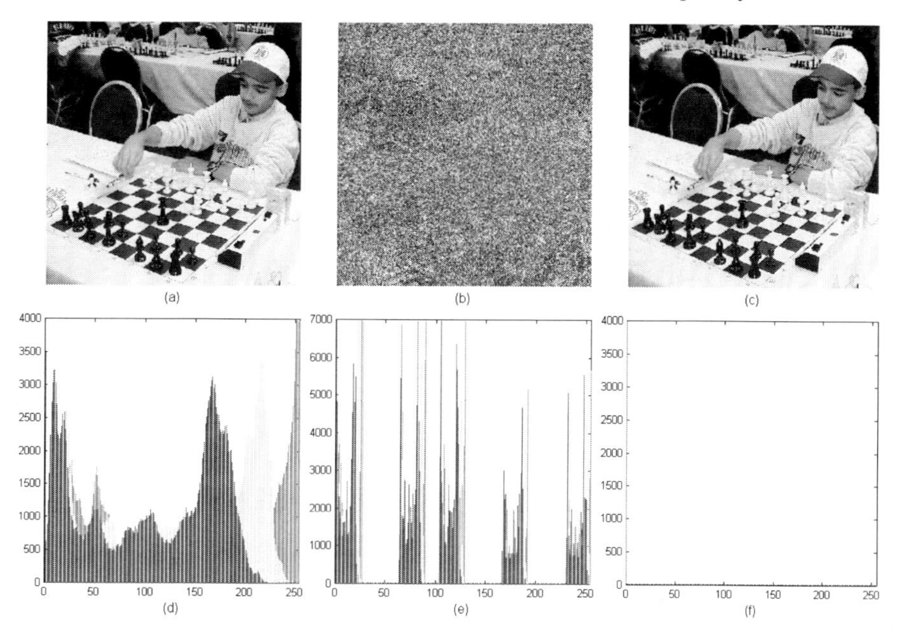

Fig. 8. Color image encryption, $p=1,r=4$. (a) Original color image; (b) Encrypted image; (c) Reconstructed image; (d) Histogram of the original image; (e) Histogram of the encrypted image; (f) Histogram of the difference between (a) and (c).

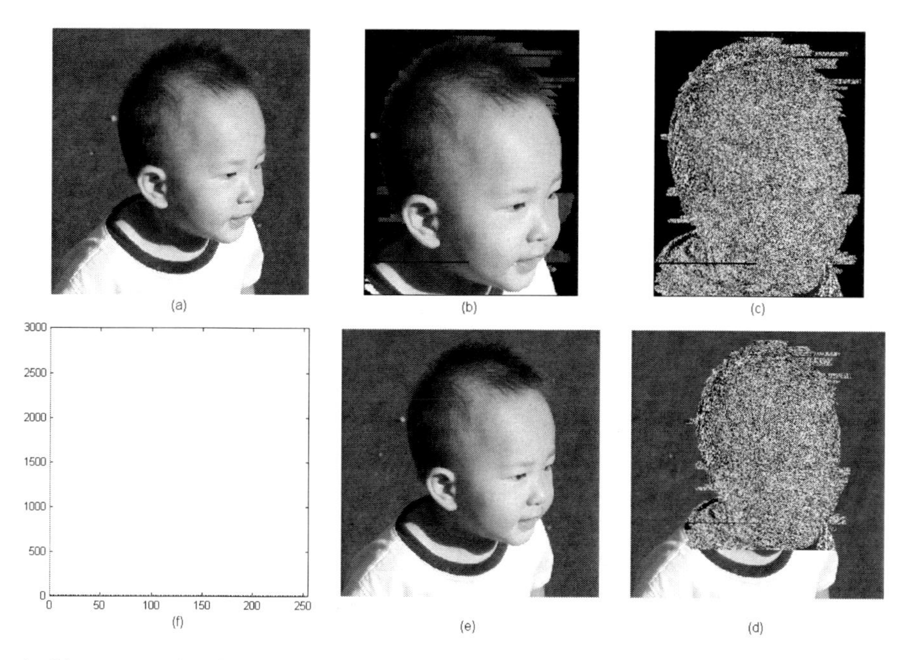

Fig. 9. Selected object encryption in a color image, $p=3,r=5$. (a) Original grayscale image; (b) Selected object; (c) Encrypted image of the selected object; (d) Encrypted image; (e) Reconstructed image; (f) Histogram of the difference between (a) and (f).

5. SECURITY ANALYSIS

In this section, we discuss the security issue of the ObjectEcrypt algorithm such as security key space and plaintext attacks.

5.1 Security key space

The security key of the ObjectEcrypt algorithm is the combination of the parameter p of TFPC, the shifting times r of the shifting algorithm in this paper, and information of the encrypted object. The parameter p of TFPC has infinite number of possible choices. The number of the TFPC bit planes changes with different p values. Based on discussion in section 3, the shifting times r is less than the sum of the parameter p and the number of TFPC bit planes ($r < p+n$). Thus, the parameter r also has limitless number of possible choice. Furthermore, the selected object changes with its regions and different methods to generate its boundary mask. All these ensure the possible number of combination of the security keys for the ObjectEcrypt algorithm to be infinite. As a result, the algorithm has unlimited security key space. It can resist the brute force attack.

5.2 Plaintext attacks

The users have flexibility to define the object to be encrypted as an entire image, any specific part/region of an image, or an object in an image or in a selected region in an image. They are also allowed to select any desired edge detection method or segmentation algorithm to obtain the object boundary mask. Therefore, the selected object is unpredictable and user-dependent. To encrypt the selected object, the ObjectEcrypt algorithm changes all pixel data in the select object by shuffling the order of its TFPC bit planes. The data of the encrypted object is not useful for the purpose of plaintext attacks. As a result, the selected object is protected with a high level of security. The ObjectEcrypt algorithm can withstand the plaintext attacks.

6. CONCLUSION

We have introduced a new recursive sequence called the truncated P-Fibonacci sequence. We also presented its corresponding binary code, the truncated Fibonacci p-code (TFPC), and the TFPC based decomposition method. They are parameter-dependent.

Based on these, we have introduced a new lossless encryption algorithm to encrypt the selected object. The object to be encrypted is either a full image, a part of an image, or a selected object in an image or in a specific region in an image. Any new or existing method of edge detection and image segmentation can be used to generate the object boundary mask in the ObjectEncrypt algorithm. The users have flexibility to apply any new or existing method for shuffling process.

Experimental results and security analysis have demonstrated that the ObjectEcrypt algorithm can protect the selected object with a high security level and withstand the brute force attack and the plaintext attacks. It can be used for privacy protection in the real-time applications such as homeland security, medical imaging, wireless network, and mobile phone services

REFERENCES

[1] P. Carrillo, H. Kalva, and S. Magliveras, "Compression independent object encryption for ensuring privacy in video surveillance," in *Multimedia and Expo, 2008 IEEE International Conference on*, 2008, pp. 273-276.

[2] K. Usman, H. Juzoji, I. Nakajima, S. Soegidjoko, M. Ramdhani, T. Hori, and S. Igi, "Medical Image Encryption Based on Pixel Arrangement and Random Permutation for Transmission Security," in *e-Health Networking, Application and Services, 2007 9th International Conference on*, 2007, pp. 244-247.

[3] R. Norcen, M. Podesser, A. Pommer, H. P. Schmidt, and A. Uhl, "Confidential storage and transmission of medical image data," *Computers in Biology and Medicine,* vol. 33, no. 3, pp. 277-292, 2003.

[4] Yang Ou, Chul Sur, and Kyung Rhee, "Region-Based Selective Encryption for Medical Imaging," in *Frontiers in Algorithmics*, 2007, pp. 62-73.
[5] F. Dufaux and T. Ebrahimi, "Scrambling for Privacy Protection in Video Surveillance Systems," *Circuits and Systems for Video Technology, IEEE Transactions on,* vol. 18, no. 8, pp. 1168-1174, 2008.
[6] Muhammad Khurram Khan and Jiashu Zhang, "An Intelligent Fingerprint-Biometric Image Scrambling Scheme," in *Advanced Intelligent Computing Theories and Applications. With Aspects of Artificial Intelligence.* vol. 4682/2007: Springer Berlin / Heidelberg, 2007, pp. 1141-1151.
[7] Park Sangho and M. M. Trivedi, "A track-based human movement analysis and privacy protection system adaptive to environmental contexts," in *Advanced Video and Signal Based Surveillance, 2005. AVSS 2005. IEEE Conference on*, 2005, pp. 171-176.
[8] Elaine M. Newton, Latanya Sweeney, and Bradley Malin, "Preserving privacy by de-identifying face images," *Knowledge and Data Engineering, IEEE Transactions on,* vol. 17, no. 2, pp. 232-243, 2005.
[9] A. Foi, V. Katkovnik, and K. Egiazarian, "Pointwise Shape-Adaptive DCT for High-Quality Denoising and Deblocking of Grayscale and Color Images," *Image Processing, IEEE Transactions on,* vol. 16, no. 5, pp. 1395-1411, 2007.
[10] Hui-Cheng Hsu, Kun-Bin Lee, N. Y. C. Chang, and Tian-Sheuan Chang, "Architecture Design of Shape-Adaptive Discrete Cosine Transform and Its Inverse for MPEG-4 Video Coding," *Circuits and Systems for Video Technology, IEEE Transactions on,* vol. 18, no. 3, pp. 375-386, 2008.
[11] David Z. Gevorkian, Karen O. Egiazarian, Sos S. Agaian, Jaakko T. Astola, and Olli Vainio, "Parallel algorithms and VLSI architectures for stack filtering using Fibonacci p-codes," *Signal Processing, IEEE Transactions on,* vol. 43, no. 1, pp. 286-295, 1995.

Characterizing cryptographic primitives for lightweight digital image encryption

Farid Ahmed and Cheryl L Resch
Johns Hopkins University Applied Physics Laboratory
11100 Johns Hopkins Road, Laurel, MD 20723

ABSTRACT

We present a statistical footprint-based method to characterize several symmetric cryptographic primitives as they are used in lightweight digital image encryption. In particular, using spatial-domain histogram and frequency-domain image analysis techniques, we identify a number of metrics from the encrypted images and use them to contrast the security performance of different cryptographic primitives. For each of the metrics, the best performing cryptographic primitive is identified. Complementary primitives are then combined to result in a product cipher with better cryptographic performance.

Keywords: Cryptographic primitives, image encryption, security performance, lightweight encryption

1. INTRODUCTION

The number of multimedia applications involving the exchange of digital image, audio, and video has recently grown with the advent of faster processing capabilities and increased connectivity due to the use of the internet. Secure transmission of digital images in some of these applications has become increasingly challenging, due to a number of constraints such as limited bandwidth, limited storage memory, unreliable network connectivity, and stringent timing requirements. In addition, multimedia data usually need to be compressed for efficient storage and transmission, resulting in a possible interoperability problem between encryption and compression. Due to these constraints, application of standard cryptographic solutions to digital multimedia leads to some provisions, including: i) the computational complexity of encryption and decryption should be as low as possible, ii) the encryption procedure should not significantly decrease the compression efficiency if encryption is done first, and iii) the encryption should be relatively robust against other common image processing techniques.

The stringent quality of service requirement (as typically measured by delay, jitter, and data loss) sometimes leads to a relaxation of security requirements in multimedia transmission. Security requirements may be relaxed because multimedia data can tolerate a degree of 'flexible protection' as opposed to all-or-none protection, based on relative value, quality, and timeliness. As an example, a highly significant high-resolution video of a live event will require more security protection compared to a low-resolution, offline, on-demand version. In [1], the authors propose the term 'entertainment security' to refer to the level of security that protects only the significant portion of the digital image. This relaxation of security is one of the important motivations of characterizing cryptographic primitives for image encryption.

Another motivation for characterizing and understanding signal processing in encrypted images is that non-traditional attacks may be launched in the encrypted media using signal processing algorithms in a variety of attack modes including cipher-text-only attack [2], chosen cipher-text attack, and known plaintext attack. Signal processing, data communication, and coding are very well-researched areas in analyzing digital image data. In this paper we present a characterization of cryptographic performances of some classical and advanced symmetric-key cryptographic primitives using some signal processing operations and metrics. Specifically, the objective of our analysis is to characterize the *confusion* and *diffusion* function [3] of different cryptographic operations and techniques that are used as building blocks of multimedia encryption. In the paper, we shall use the term 'original' or 'plaintext' for the unencrypted version of the image, and 'encrypted', 'cipher', or 'cipher-text' for the encrypted version.

Mobile Multimedia/Image Processing, Security, and Applications 2009, edited by Sos S. Agaian, Sabah A. Jassim,
Proc. of SPIE Vol. 7351, 73510G · © 2009 SPIE · CCC code: 0277-786X/09/$18 · doi: 10.1117/12.817907

2. BACKGROUND AND SCOPE

Research of cryptographic techniques as applied to multimedia security has been pursued in two different tracks. The first track aims at developing a lightweight cryptographic algorithm for real-time rendering of multimedia data. Lightweight cryptographic techniques generally result in less computationally demanding encryption/decryption at the cost of 'acceptable loss' of security margin. Partial encryption techniques [4-9] and chaotic and cellular automata-based encryption [10-13] fall into this category. Of particular relevance is the work by Jipping [9], who demonstrated AES [14] S-Box based image encryption technique. The second track addresses the integration of encryption with compression, coding, and information embedding of data [15-17]. Andreas *et. al.* [5] addresses the relation of compression and coding with the encryption of image and video. Dang *et. al.* [16] proposes a new joint encryption and compression algorithm suitable for internet multimedia.

One important aspect missing from these and other related research works is the 'signal processing characterization' of different cryptographic primitives and the robustness or fragility of those characterizations against typical information loss incurred in multimedia communication. A number of metrics have been reported in literature to benchmark the performance of cryptographic techniques. But most of them deal with the standardized cryptographic technique, and the performance metrics are usually computed relative to the original un-encrypted image [10]. Correlation, mostly captured as a normalized correlation coefficient (NCC) is a widely used metric [8]. Metrics based on histogram differences of the original and encrypted images have also been used. In contrast, our approach compares the performance of different classical cryptographic techniques as well as a few advanced primitives used in the current Advanced Encryption Standard (AES) [14].

One unique aspect of our approach is blind characterization, which measures the *confusion* and *diffusion* [2, 3] functions of different classical and symmetric cryptographic techniques in terms of only the cipher-text image. To the best of our knowledge there has not been any reported work undertaking this approach in digital imaging area. This work will help understand the strengths and weaknesses of specific cryptographic operations necessary for the design and analysis of a group of lightweight multimedia encryption techniques.

3. CHARACTERIZING DIGITAL IMAGE ENCRYPTION

3.1 Characterizing Confusion and Diffusion

Symmetric key cryptographic algorithms are based on the Feistel Cipher structure [2] that encrypts data by processing a multiplication of complementary cryptographic primitives through a number of scrambling cycles such that the desired *confusion* and *diffusion* of data is achieved. *Confusion* is the outcome of cryptographic processing that makes the statistical relation between the cipher-text and the key difficult. In other words, given the cipher-text, it is difficult to know the key. This is partly responsible for the avalanche effect [19], which says every bit of the key affects every bit of the cipher-text. *Confusion* is usually achieved by substitution/permutation that makes the relationship between the key and cipher-text uncertain. From an information-theoretic point of view, *confusion* indicates how much information the cipher-text reveals about the key. More *confusion* means more uncertainty, or entropy, in the cipher value [2]. The term entropy as it is applied in the coding of digital images implies how many bits are required to encode each pixel value of the image.

Diffusion dissipates the statistical relation between the cipher-text and the plaintext and results in a statistically similar plaintext image having a completely different cipher-text image, even when encrypted with the same key. The use of *diffusion* makes any cryptanalysis based on statistical relations very difficult. The avalanche effect is also applicable here in the sense that any element of the input block changes every element of the output block. The use of diffusion tends to reduce the redundancy of the image representation and thus increases the entropy.

3.2 Cryptographic primitives under study

We start with a number of selected cryptographic primitives, which are the building blocks of classical and modern cryptography and have been used in the pursuit of lightweight cryptographic solutions for secure multimedia communication. We first consider the classical Affine cipher. Let us consider an 8-bit grayscale image x, where the pixel values range from 0 to 255. An Affine cipher given by

$$y = (ax + b) \bmod 256 \quad (1)$$

transforms the 256 gray-level input image x into the encrypted image y. Here the coefficient pair (a,b) is the key. *'b'* can have any integer value in the range {0,255}. The only restriction on *'a'* is that it has to be relatively prime [2] to 256. All the odd values in the range {1,255} meet that restriction. A special case of $a=1$, results in a more rudimentary Shift cipher. This is a *confusion* process, as described above.

Next we consider the Hill cipher, which is classified as one of a 'Poly-alphabetic cipher' in classical cryptography [3]. It is essentially given by a matrix multiplication,

$$C = K * P \bmod 256 \quad (2)$$

where we consider a MxM Key matrix K, and $Mx1$ plaintext vector P, resulting in a $Mx1$ cipher-text vector C. For a 4x4 key matrix, every element of the encrypted version, is a linear combination of 4 neighboring pixels, which in turn diffuses the statistics of the original image. Hence this is essentially a *diffusion* process.

We then consider the four core operations in the AES [14]: ByteSub, ShiftRow, MixColumns, and AddRoundKey. All these operations operate on a 'state' of plaintext, which is 16 pixels (equivalent to 128 bits) long. The 16 pixels are arranged in a 4X4 matrix named as the 'state'. The ByteSub operation is essentially a substitution cipher, which uses an invertible substitution box, followed by an XOR operation. In essence, it is the realization of *confusion* by substituting the value of one pixel with another. The ShiftRow step performs an asymmetric shift on each row of pixel values in the state. Therefore, it does not really change the pixel values, only permutes or transposes them. It is thus a transposition cipher, resulting in confusion. The MixColumns operation performs a linear combination of columns of the state, and thus realizes a diffusion operation. Finally, we consider the AddRoundkey operation, which performs an XOR operation of the state with the round key. The round key is obtained from the master key using a key generation algorithm, which essentially adds *diffusion* to the process.

3.3 Metrics used for characterization

As mentioned in section 2, our blind characterization is based on the statistics obtained from the spatial pixel domain, in addition to a feature from the Fourier frequency for the characterization. Specifically, we employ image histogram statistics and energy content at different frequencies of the encrypted image, yielding the following metrics.

a. ***SAD (Sum of Absolute Difference) --*** The *SAD* of a histogram is a measure of how different the distribution is from a uniform distribution. SAD is defined as follows:

$$SAD = \frac{\sum_{i=1}^{255} |h(i) - h(i-1)|}{\sum_{i=1}^{255} h(i)} \quad (3)$$

Let $h(i)$ denote the histogram of an image, where i represents the gray value of a pixel. With normalization, $h(i)$ represents the fraction of total pixels with the gray value i. The histogram of typical natural images takes the shape of normal, bi-modal, or tri-modal distribution [14]. Moreover, the histogram of natural image is usually continuous with little change between neighboring pixels except near the modes.

It can be shown that for a uniformly distributed histogram (which is desirable in high-*diffusion* encryption algorithms), SAD is 0. On the other hand, for a saw-tooth distribution, where the alternate pixel levels will have a value of 1/128 and 0, the SAD can be shown to be 255/128, which is slightly less than 2. Substitution-based cryptographic processing usually results in a saw-toothed histogram. For all other forms of histogram of encrypted image, SAD should assume some value in between.

b. ***Range*** -- The *Range* refers to the range of the histogram density. It is defined as the difference between the maximum and minimum value of the histogram density, as follows:

$$r = \max(h(i)) - \min(h(i)) \tag{4}$$

Where the index i varies from 0 to 255. The maximum value is 1.0. For typical natural images it is around 0.1. Cryptographic primitives generally keep the range value constant or reduce it. For example substitution and permutation does not change the range value because the maximum and minimum frequency count remains the same, only the actual gray value changes. On the other hand, *diffusion* processes will likely decrease the range value.

c. ***Entropy*** -- The *Entropy* of the histogram is defined as follows:

$$entropy = -\sum_{i=0}^{255} h(i)\log_2(h(i)) \tag{5}$$

It can be shown easily that for a uniformly distributed histogram (with $h(i) = 1/256$, *for all values of* i), the entropy is 8, which conforms to the information-theoretic definition of entropy. The objective of *diffusion*-type cryptographic primitives is to achieve this value of 8. For natural images and cryptographic operations with less *diffusion*, entropy is less than 8. It can also be shown that for a saw-toothed histogram as mentioned above the value is 7.

d. ***Contrast*** -- *Contrast* is defined as the standard deviation of the histogram values. Since, the goal of *diffusion*-type processing is to diffuse the histogram differences, it turns out that *diffusion* or mixing processes reduce this value. Therefore the Hill cipher, AES Mixcolumns, and Addroundkey are expected to decrease this value. Substitution-type operations generally should not increase this value.

e. ***Retained Energy*** -- For frequency domain characterization, a metric based on the Fourier domain energy distribution is defined. We represent the Fourier transform of an image in polar form, compute the radial frequency, normalize it and compute the energy histogram at different radial frequency. *Retained Energy* is defined as the ratio of energy in the low radial frequency to the total energy. In the simulation results furnished in section 4, we choose a frequency value of 0.125 for the normalized radial frequency. A value of 75 for the *Retained Energy* metric then means that 75% of the total energy is retained in the lowest 12.5% of all the frequency components.

4. SIMULATION/EXPERIMENTAL RESULTS

We consider the encryption of 8-bit gray image represented by 256 shades of gray values. The images are taken from the University of Southern California Signal and Image Processing Institute (USC-SIPI) image database [21]. Three classical cryptosystems (Shift cipher, general Affine cipher, and Hill cipher) in addition to the four core operations (ByteSub, ShiftRow, MixColumns, and AddRoundKey) of the current symmetric key-cryptography standard, AES, are considered in the simulation.

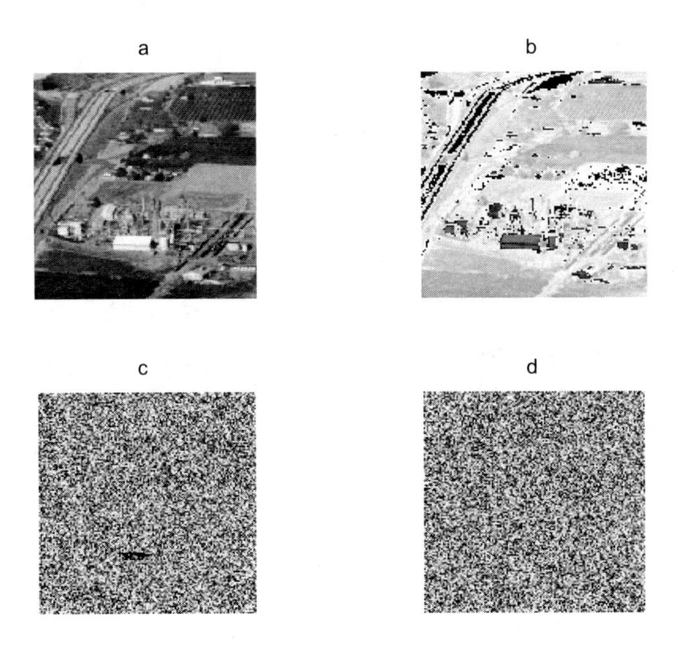

Fig. 1: a) Original image, b) Shift ciphered, c) Affine ciphered, d) Hill ciphered

Figure 1 shows the original of a 'Chemical Plant' image along with the three cipher versions encrypted by classical ciphers of Shift, Affine, and Hill. Note the impact of *confusion* and *diffusion* added in the Affine and Hill ciphered versions. Figure 2 shows the versions encrypted by the four core operations of AES.

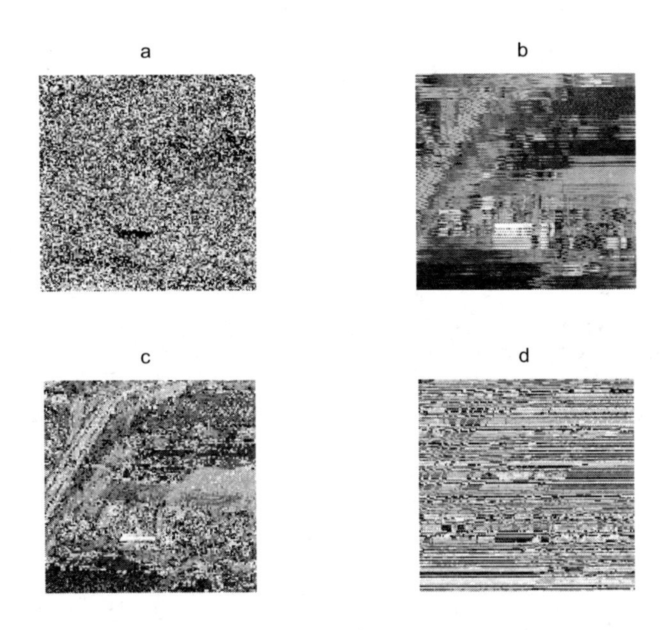

Fig. 2: Encrypted image after a) AES Sub, b) AES ShiftRow, c) MixColumns, d) AddRoundKey

Figure 3 shows the histogram of the images shown in Fig. 1. The x- axis represents gray level and y-axis represents histogram density. As expected, histogram of the Shift cipher is simply the shift of the original histogram. The histogram corresponding to the Affine cipher is seen to change very rapidly while keeping the range bounded by the same value as the original histogram. The *diffusion* effect in Hill cipher is seen to distribute the pixel values all around, such that all the pixel values have non-zero counts. The range of histogram is seen to decrease correspondingly.

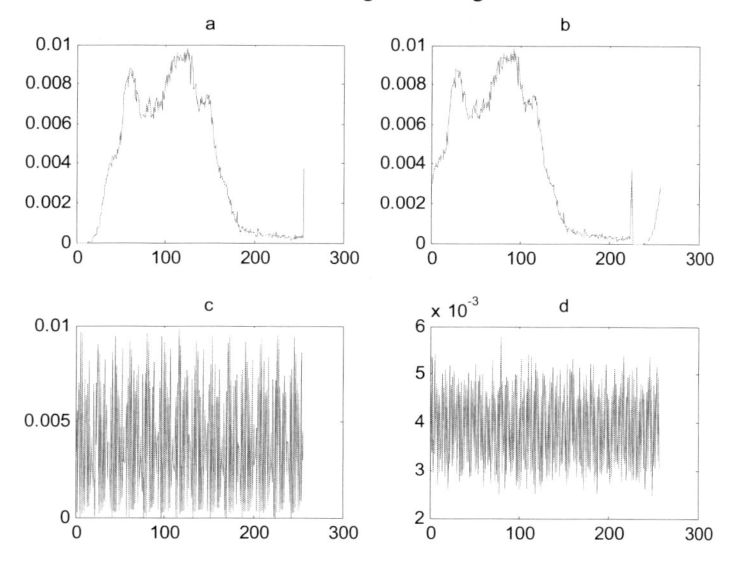

Fig. 3: Histogram Density a) Original Image, b) Shift Ciphered, c) Affine Ciphered, d) Hill Ciphered.

Figure 4 shows the histogram of the four AES operations. As before, the x- axis represents gray level, y-axis represents histogram density. AESSub is seen to behave similar to the Affine cipher, as in Fig. 4(a). The MixColumns and ADDRoundKey yield very interesting multi-mode histograms that show their contributions to both *confusion* and *diffusion* outcomes.

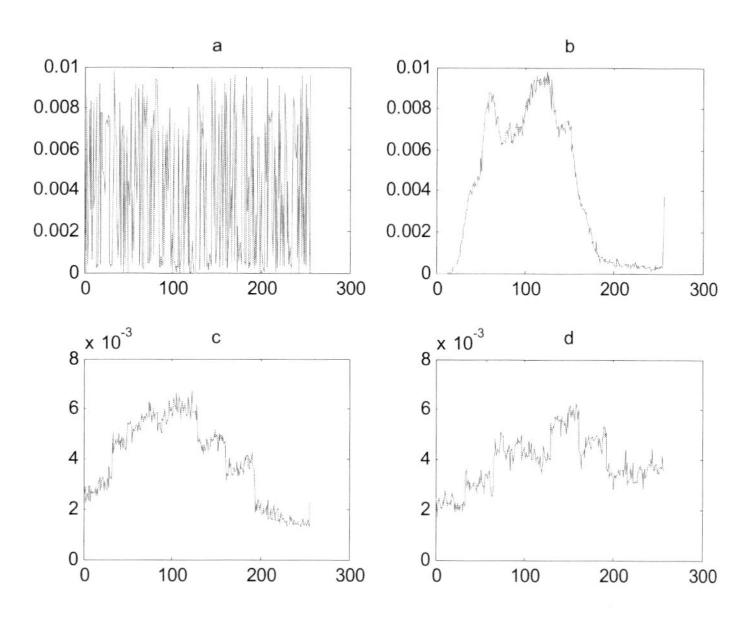

Fig. 4: Histogram Density for AES operations-based encrypted images a) Substitution, b) ShiftRow, c) Mix Columns, d) AddRoundKey.

Figure 5 shows the energy histograms in the Fourier domain. The x-axis represents the normalized radial frequency and y-axis represents normalized energy. For all the ciphers, the low-frequency bin does not change much. But the high-frequency distribution of the encrypted image changes significantly. For the Shift cipher, the pattern is a sort of monotonic decrease in the energy content, with increasing frequency, as is expected from the original unencrypted natural image. The Affine cipher which contributes to *confusion* appears to have a reversal of the distribution of energy, as shown in Fig 5c. There is a similar trend for the Hill cipher.

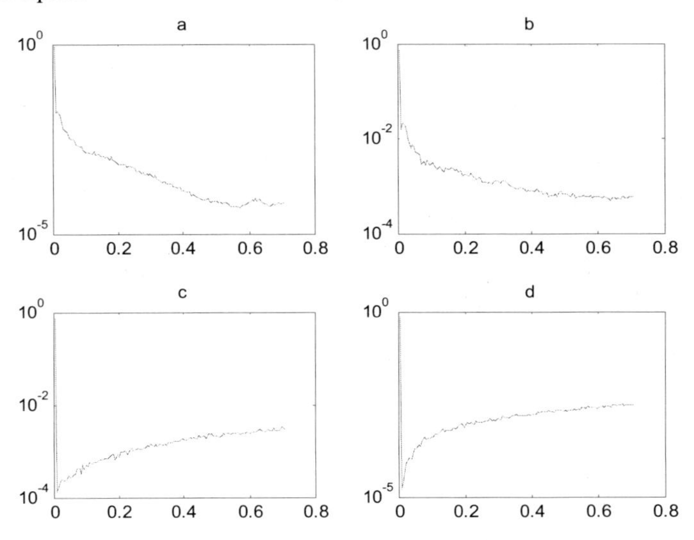

Fig. 5: Energy Histogram in the Fourier domain. a) Original Image, b) Shift Ciphered, c) Affine Ciphered, d) Hill Ciphered.

Figures 5 and 6 shows why compression done in encrypted domain will not result in a very good quality decrypted image, since the compression has the potential of taking away a significant amount of mid-to-low frequency components, which are important for rendering the image. This result has a significant implication in applications like digital watermarking, which tends to hide energy in mid-frequency region. The situation is particularly worse for AES MixColumns which diffuses the energy distribution the most, as shown in Fig 6c.

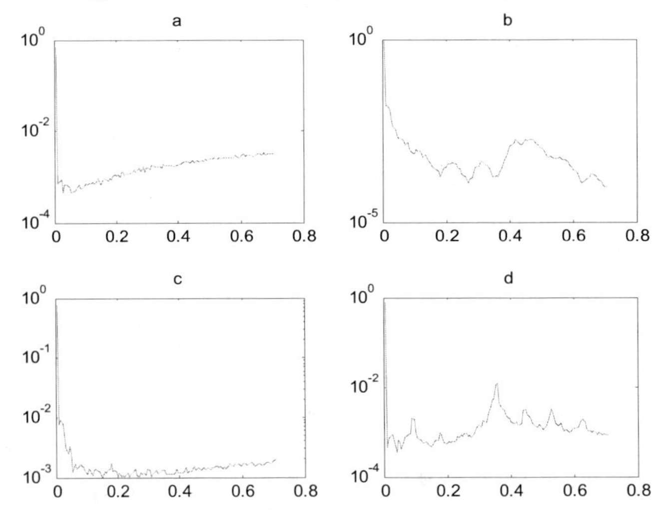

Fig. 6: Energy Histogram in the Fourier domain for AES operations-based encrypted images a) Substitution, b) ShiftRow, c) Mix Columns, d) AddRoundKey. x-axis represents the normalized radial frequency and y-axis represents normalized energy

Next, in order to grasp a better understanding of the underlying patterns, we performed each encryption primitive on 10 different 256X256 images with 10 different keys, thus gathering 100 different scores for each metric, corresponding to each primitive operation. In the subsequent discussion, we report the average of these 100 scores. Table 1 summarizes the characterization results for the selected encryption primitives.

Table 1: Characterization of the Encryption Primitives

Encryption Primitive	Operation Type	Characterization Metric					
		NCC	***SAD***	***Range***	***Contrast***	***Entropy***	***Retained Energy (%)***
None	None	1	0.14	0.042	0.0063	6.84	97.63
Shift	Substitution	0.13231	0.14	0.042	0.0063	6.84	91.61
Affine	Substitution	0.009633	1.18	0.042	0.0063	6.84	77.48
Hill	Mix	0.001136	0.16	0.004	0.0006	7.97	75.23
AESSub	Substitution	0.02244	1.20	0.042	0.0063	6.84	76.97
AESShift	Permutation	0.7399	0.14	0.042	0.0063	6.84	96.22
AESMix	Mix	0.6039	0.13	0.022	0.0039	7.41	90.04
ARK	Mix	-0.169	0.17	0.008	0.0015	7.86	77.63

Coulmn 2 of the Table 1 lists the type of primitive cryptographic operation used in the tests. Generally, encrypted images have very small correlation with the original image as shown in column 3. Among them, Shift operations usually have higher correlation value as evident in the cases of Shift and AES Shift. In addition, AESMix, which performs a linear combination of the 4 columns of AES state matrix, also generally maintains a higher correlation with the original image. Hill cipher is always seen to have a correlation close to zero.

The most interesting metric is found to be the *SAD*. *SAD* is a measure of the first derivative of the histogram. A higher value of this signifies neighborhood *confusion*, drifting away from the uniform distribution. Substitution operations result in the most *confusion*. As a result the Affine cipher and the AES SubBytes has resulted in higher values of SAD. This is followed by the Hill cipher, which is in between. But other ciphers result in a histogram that has some continuity and thus smaller value of SAD.

As for the range of histogram, since the Shift cipher, Affine, AES Sub, and AES Shift all are substitution or permutation cipher, the maximum and minimum values of histogram remain the same. Hence the range is same for these ciphers. The Hill cipher, AES MixColumns, and the AES AddRoundKey, on the other hand, are more of a *diffusion* processing. Hence they tend to flatten the histogram to reduce the gap between maximum and minimum value, and thus reducing the range. The smallest of these are found to be with the Hill cipher, which apparently does a very good job in *diffusion*.

Histogram *contrast* and *entropy* have similar trends. Note that the entropy here is defined from the information theoretic point of view, which is the average number of bits required to represent the images. The last column of Table 1 shows the energy retained in the encrypted image, when 87.5% of the high-frequency coefficients were zeroed out. The *diffusion* process again seems to have reduced the energy retained.

The top three ranked encryption primitives for each of the metrics are shown in Table 2. The next objective is to combine the primitives which are complementary, to result in a product encryption which has better cryptographic performance, as demonstrated in general Feistel cipher structures [2, 3].

Table 2: Rank order of the Cryptographic primitives for different metrics

Metric	Rank Order		
	1	2	3
SAD	AES Sub	Affine	Hill
NCC	Hill	Affine	AES Sub
Histogram Range	Hill	AES ARK	AES Mix
Entropy	Hill	AES ARK	AES Mix
Energy	Hill	AES Sub	Affine

The characterization results for the product crypto are shown in Table 3. Affine2 is the realization of the Affine encrypted image with another Affine cipher. Since the two operations are same, together they are equivalent to one single Affine transform, thus validating the concept of an idempotent [2] system, which says the product of two similar primitive operations does not enhance the security of the cryptographic system. All of our proposed characterization metrics, remain almost the same with Affine2, as they were with Affine transform as shown in Table 1 row 3 and Table 3 row 1. This clearly gives a gross validation of the metrics.

Table 3: Characterization of a few Product Crypto operations

Product Crypto	NCC	SAD	Range	Contrast	Entropy	Energy
Affine2	0.010144	1.17	0.042	0.0063	6.84	77.67
HillAffine	-0.00113	0.16	0.004	0.0006	7.97	75.13
MixAffine	0.006473	0.41	0.015	0.0021	7.76	75.93
AESSHill	-0.00096	0.15	0.004	0.0006	7.97	75.23
ARKAffine	-0.00219	0.45	0.009	0.0017	7.84	75.21
AES	0.00062	0.07	0.001	0.0003	8.00	75.21

The next four (HillAfine, MixAffine, AESSHill, and ARKAffine) are products of a substitution cipher followed by a mix primitive or vice versa, resulting in a non-idempotent system, and thus increasing the security margin as expected. Specifically, the Hill cipher, operating on an Affine encrypted version, does seem to improve the correlation (NCC) metric. It does not however, enhance the performance compared to the Hill-only performance. MixAffine, which uses an AES MixColumns on the Affine encrypted version, is seen to improve on every metric we used, compared to its constituent cipher Affine and MixColumns. AESSHill, which uses AES Sub operation on the Hill ciphered image, does not significantly improve on the Hill-only cipher. The reason could be that although the Hill cipher linearly combines the plaintext, it is treated as a poly-alphabetic substitution cipher, which is still a substitution cipher like Affine.

The strength of Hill Cipher followed by AESSub, is indicated by a smaller correlation coefficient (negative in this case). Since AESSub, being a substitution cipher, does not change the range, standard deviation, or the entropy of the histogram, hence these three values are same as only Hill. AES Sub increases the SAD value when it operates on the original image. This is because the range of the original image was higher than that of the Hill ciphered image by a factor of more than 10.

Next we explore the sensitivity of the products of cryptographic primitives (as shown in Table 3) and the standard AES against a typical distortion frequently encountered in lossy communication. We assume the images encrypted by different cryptographic products are transmitted through a lossy channel, where they are under attack of 'salt and pepper noise' [20]. Figure 7 shows the quality of the decrypted image for 6 different product encryptions under a noise density of 0.02. X-axis shows the index of 10 different images and y-axis shows the quality of the decrypted image in terms of PSNR [20].

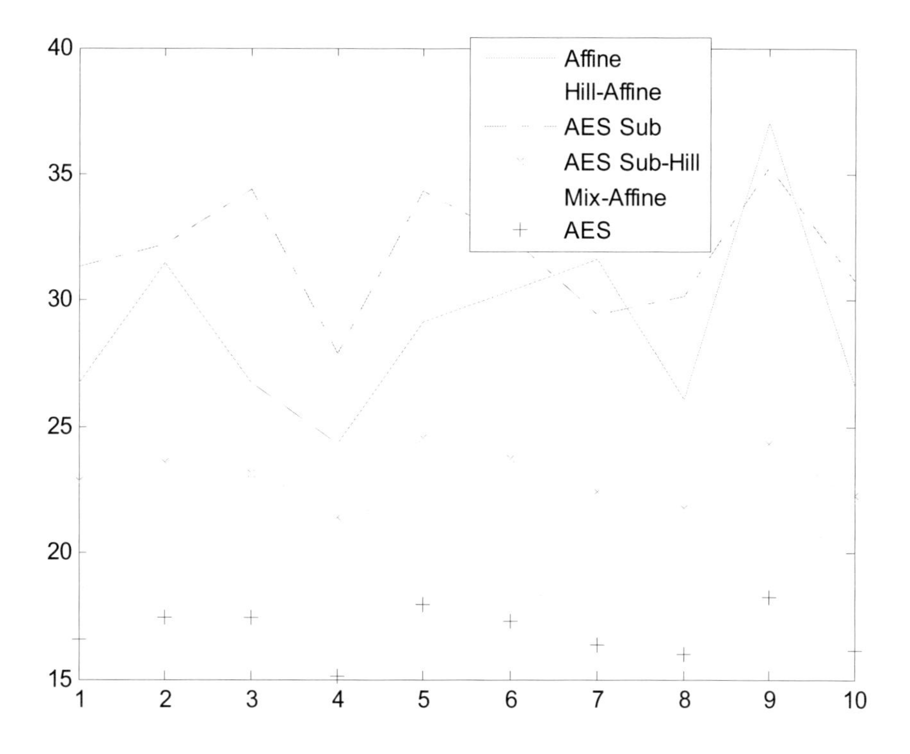

Fig. 7: Quality of the decrypted image for different product crypto with salt and pepper noise of density 0.02.

It is to be noted that the AES resulted in lowest quality. The two substitution type operations, Affine and AESSub, produced the highest quality. Interestingly, the quality produced by of the three product encryption (Mix-Affine, AES Sub-Hill, and Hill-Affine) are almost equivalent. Contrasting these results with those reported in Table 3 shows that although the standard AES outperforms all other product encryption in *confusion* and *diffusion* metrics, the *MixAffine* is a viable alternate, especially because it is manifold faster than the standard AES, and because it is more robust to the typical 'salt and pepper noise' occurring in lossy transmission as shown in Fig. 7.

5. CONCLUSION

This paper has presented the characterization of a number of cryptographic operations that were performed in the encrypted domain of an image using some signal processing techniques. The work is expected to be very useful in making an informed decision about the choice of a specific cryptographic primitive better suited for the networking and computational requirements at hand. Since our method ranks a number of cryptographic primitives and their combinations in terms of the proposed metrics, a potential application of this could be a multimedia system, where variable levels of security might be needed to adjust to the experienced quality of service. With the limited number of cryptographic primitives that we experimented with, the AESSub and Hill cipher are found to yield better cryptographic security in terms of the metrics we dealt with. This work will help understand the strengths and weaknesses of specific cryptographic operations necessary for the design and analysis of a group of lightweight multimedia encryption techniques. Future work will address the interoperability with compression and coding of the different cryptographic operations.

REFERENCES

[1] Zeng, W., "On Security Architecture and Functionality of Distributed Multimedia," 40th Annual Conf. on Information Sciences and Systems, 1164 – 1169 (2006).

[2] Stinson, D. R., Cryptopgraphy- Theory and Practice, Second ed., ISBN 1-58488-206-9, Chapman and Hall/CRC.

[3] Trappe, W. and Washington, L., C., Introduction to Cryptography with Coding Theory, Prentice Hall, 2006, ISBN 0-13-186239-1.

[4] Zheng Liu, Xue Li and Zhaoyang Dong' "A Lightweight Encryption Algorithm for Mobile Online Multimedia Devices," LNCS Springer Berlin, Web Information Systems, pp. 653-658.

[5] Uhl, A. and Pommer, A., "Image and Video Encryption- From Digital Rights Management to Secured Personal Communication', Springer, 2005, ISBN 0-387-23402-0.

[6] Podesser, M., Hans-Peter Schmidt, Uhl, A. "Selective Bitplane Encryption for Secure Transmission of Image Data in Mobile Environments," in CD-ROM Proceedings of the 5th IEEE Nordic Signal Processing Symposium (NORSIG) 2002.

[7] Cheng, H. and Xiaobo Li, "Partial encryption of compressed images and videos," IEEE Trans. Signal Processing, 48(8), 2439-2451 (2000).

[8] Ali, M., Younes, B., and Jantan, A., "Image Encryption Using Block-Based Transformation Algorithm," Intl J of Comp Sc, 35(1), IJCS_35_1_03

[9] Niu, Jiping;, "Image Encryption Algorithm Based on Rijndael S-boxes," Intl Conf on Computational Intelligence and Security, vol 1, 277 – 280 (2008).

[10] Peng, Jun; Shangzhu Jin,; Yongguo Liu,; Zhiming Yang,; Mingying You,; Yangjun Pei,;
, "A novel scheme for image encryption based on piecewise linear chaotic map," IEEE Conference on Cybernetics and Intelligent Systems, 1012 – 1016 (2008).

[11] Yu, X.Y., J Zhang, Ren, H., E., G S Xu, and Luo, X., Y., "Chaotic Image Scrambling Algorithm Based on S-DES," Journal of Physics: Conference Series, 48, 349-353 (2006).

[12] Yong-Hong Zhang; Bao-Sheng Kang; Xue-Feng Zhang;, "Image Encryption Algorithm Based on Chaotic Sequence," Artificial Reality and Telexistence—Workshops, ICAT '06. 221 – 223 (2006).

[13] Rong-Jian Chen; Yuan-Hsin Chen; Chao-Shen Chen; Jui-Lin Lai, "Image Encryption/Decryption System using 2-D Cellular Automata," IEEE Tenth International Symposium on Consumer Electronics, 28-01, 1 – 6 (2006).

[14] Advanced Encryption Standard, http://csrc.nist.gov/CryptoToolkit/aes/

[15] Dang, P., P. and Chau, P. M., "Image Encryption for Secure Internet Multimedia Applications," IEEE Trans. Consumer Electronics, 46(3), 395 – 403 (2000).

[16] Mao, Y. and Wu, M., "A Joint Signal Processing and Cryptographic Approach to Multimedia Encryption," IEEE Trans. On Image Processing, 15(7), 2061-2075 (2006).

[17] Wu, C.P. and Kuo, C.-C. J., "Design of integrated multimedia compression and encryption systems," IEEE Trans. Multimedia, 7(5), 828-839 (2005).

[18] Ismail I. A., Mohammed A., and Hossam, D., "How to repair the Hill Cipher," Journal of Zhejiang University Science A, 7(12), 2022-2030(2006).

[19] Stallings, W., Cryptography and Network Security-Principles and Practices, 3rd ed. Prentice Hall, 2003.

[20] Arthur R. Weeks, Jr., "Fundamentals of Electronic Image Processing," IEEE Press, ISBN 0-7803-3410-8, 1996.

[21] USC SIPI Image Database, http://sipi.usc.edu/database/

Fast Unitary Heap Transforms: Theory and Application in Cryptography

Artyom M. Grigoryan and Khalil Naghdali

Department of Electrical and Computer Engineering
The University of Texas at San Antonio
One UTSA Circle, San Antonio, TX 78249, USA

ABSTRACT

This paper presents a novel approach to compose discrete unitary transforms that are induced by input signals which are considered to be generators of the transforms. Properties and examples of such transforms, which we call *the discrete heap transforms* are given. The transforms are fast, because of a simple form of decomposition of their matrices, and they can be applied for signals of any length. Fast algorithms of calculation of the direct and inverse heap transforms do not depend on the length of the processed signals. In this paper, we demonstrate the applications of the heap transforms for transformation and reconstruction of one-dimensional signals and two-dimensional images. The heap transforms can be used in cryptography, since the generators can be selected in different ways to make the information invisible; these generators are keys for recovering information. Different examples of generating and applying heap transformations over signals and images are considered.

Keywords: The Fourier transform, heap transforms, Givens transforms, cryptography.

1. INTRODUCTION

The representations of the one- and two-dimensional signals and images in the frequency domain are well-known methods that are used for effectively processing signals and images in different applications, such as biomedical image processing, communication, and cryptography[1]-.[4,9] The representation is performed by orthogonal basis functions, which have different forms for existent linear unitary transforms, such as the Fourier, Hadamard, and cosine transforms[5]-.[7] However, all these complete systems are systems with non-variable functions (except the case of the Karhunen-Loeve transformation). In other words, these transforms are described by matrices with constant coefficients, which do not depend on the signals to be processed.

The discrete Fourier transform and other unitary transforms are used in image cryptography and stenography. We mention the importance of the phase of the spectrum,[8] which defines the information of a processed signal or image. A small change in phase leads to significant changes. This property can be used for image encryption and to make the image less visible. As an example, the random phase mask encoding method[10] uses the phase modification of the Fourier transform. There are many possible ways to modify the phase of the spectrum, and other effective transforms can be developed for these purposes.

In this paper, we discuss a new class of discrete unitary transforms, which we call the heap transforms, or transforms which are generated by input signals.[12] The complete systems of basis functions of heap transforms are referred to as waves generated by input signals, the waves with their specific motion in the space of functions. To demonstrate the main properties of the heap transforms, we here stand on the linear heap transforms defined by simple rotations, or Givens rotations. The heap transforms are defined by the signal-generators which allow us to tune the transforms when applying them for specific classes of applied signals and images. These generators play the role of keys without which it is not possible to reconstruct the initial signals or images.

Address all correspondence to Artyom M. Grigoryan. E-mail: amgrigoryan@utsa.edu, Tel/Fax: (210) 458-7518/5947.

Mobile Multimedia/Image Processing, Security, and Applications 2009, edited by Sos S. Agaian, Sabah A. Jassim, Proc. of SPIE Vol. 7351, 73510H · © 2009 SPIE · CCC code: 0277-786X/09/$18 · doi: 10.1117/12.818946

1.1. Heap Transforms

The discrete signal induced heap transforms, DsiHT, are defined by a given vector-generator, or signal and system of specific decision equations.[11] We here stand on the simple case, when the basic two-dimensional transformations of the heap transforms are defined by the Givens transformations, or elementary rotations which satisfy two decision equations.

The composition of the N-dimensional DsiHT by a given vector-generator $\mathbf{x} = (x_0, x_1, x_2, ..., x_{N-1})'$ is performed by the sequential calculation of two-dimensional transforms. This composition can be written in matrix form as

$$\mathbf{H} = \mathbf{T}_{N-1}\mathbf{T}_{N-2}\mathbf{T}_{N-3}\cdots\mathbf{T}_2\mathbf{T}_1 \tag{1}$$

where matrices $\mathbf{T}_k$, $k = 1 : (N-1)$, correspond to the transformations T_k that change only two components of the input. The components of an input $\mathbf{z} = (z_0, z_1, z_2, ..., z_{N-1})'$ are processed in order z_0, z_1, ..., z_{N-2}, and then z_{N-1}. This is a natural path P, and in general, such a path can be taken in many different ways. Transformations T_k are parameterized, and values of their parameters are defined by the given vector-generators. This is the main point, the generator $\mathbf{x}$ determines all transformations T_k which then are applying on the input $\mathbf{z}$. The vector-generator, when composing the heap transform, and the input signal are processed in the same order, or path P. Thus, the process of the $\mathbf{x}$-signal induced transformation of the signal $\mathbf{z}$ can be described as

$$\mathbf{x} \rightarrow \{T_k; k = 1 : (N-1)\} \rightarrow H = H_{\mathbf{x}}, \quad \text{and then} \quad \mathbf{z} \rightarrow H[\mathbf{z}]. \tag{2}$$

We consider the case when all basic transformations T_k are parameterized, and the parameter is referred to as an angle φ_k. The special selection of a set of these parameters is initiated by the vector-generator through the so-called *decision equations* and a given set of constants $A = \{a_1, a_2, \ldots, a_{N-1}\}$ in the following way. Let $f(x, y, \varphi)$ and $g(x, y, \varphi)$ be functions of three variables; x and y are referred to as the coordinates of a point (x, y) on the plane. It is assumed that, for each $a \in A$ and point (x, y) on the plane or a chosen subset of the plane, the equation $g(x, y, \varphi) = a$ has a unique solution with respect to φ. We denote such a solution by $\varphi = r(x, y, a)$. The system of equations

$$\begin{cases} f(x, y, \varphi) = y_0 \\ g(x, y, \varphi) = a \end{cases} \tag{3}$$

is called *the system of decision equations.* The value of φ is calculated from the second equation which is called *the angular equation.* The value of y_0 is calculated from the given input (x, y) and obtained φ.

The given generator is processed first and during this process all required angles, φ_k, are calculated as shown in the following diagram:

$$\left\{ \begin{array}{llll} \begin{bmatrix} x_0 \\ x_1 \end{bmatrix} \xrightarrow{T_{\hookleftarrow\varphi_1}} \begin{bmatrix} y_0^{(1)} \\ a_1 \end{bmatrix} & & & \\ & \begin{bmatrix} y_0^{(1)} \\ x_2 \end{bmatrix} \xrightarrow{T_{\hookleftarrow\varphi_2}} \begin{bmatrix} y_0^{(2)} \\ a_2 \end{bmatrix} & & \\ & & \ddots & \\ & & & \begin{bmatrix} y_0^{(N-2)} \\ x_{N-1} \end{bmatrix} \xrightarrow{T_{\hookleftarrow\varphi_{N-1}}} \begin{bmatrix} y_0^{(N-1)} \\ a_{N-1} \end{bmatrix} \end{array} \right\} \tag{4}$$

The rotation angles φ_k are calculated sequentially by

$$\varphi_k = r(y^{(k-1)}, x_k, a_k), \quad k = 1 : (N-1), \quad (y^{(0)} = x_0), \tag{5}$$

where the "heap", $y_0^{(\cdot)}$, is accumulated at the original point $t = 0$ as follows:

$$\begin{aligned} y_0^{(1)} &= f(x_0, x_1, \varphi_1), \\ y_0^{(2)} &= f(y_0^{(1)}, x_2, \varphi_2), \\ y_0^{(3)} &= f(y_0^{(2)}, x_3, \varphi_3), \quad \ldots, \\ y_0^{(N-1)} &= f(y_0^{(N-2)}, x_{N-1}, \varphi_{N-1}). \end{aligned} \tag{6}$$

The transform of the vector $\mathbf{x}$ results in a vector with constant components of the set A plus $y_0^{(N-1)}$ as the first component,

$$\mathcal{H}: \mathbf{x} \rightarrow H(\mathbf{x}) = (y_0^{(N-1)}, a_1, a_2, ..., a_{N-1})'. \tag{7}$$

The transformation $\mathcal{H}$ is called the *N-point discrete* $\mathbf{x}$*-signal-induced heap transformation* (DsiHT), and the vector $\mathbf{x}$ is *the generator* of this transformation. This transformation is linear and unitary in the space of N-dimensional vectors. The following should be mentioned about the heap transform. The transform is performed in a space of N-dimensional vectors $\mathbf{z}$, but all required angles φ_k are found sequentially and the transformation $\mathcal{H}$ is composed after solving the decision equations relative to a given vector-generator $\mathbf{x}$.

The process of transformation of the input vector $\mathbf{z}$ by the DsiHT can be described by the following diagram:

$$\left\{ \begin{array}{llll} \begin{bmatrix} z_0 \\ z_1 \end{bmatrix} \xrightarrow{T_{\varphi_1}} \begin{bmatrix} z_0^{(1)} \\ z_1^{(1)} \end{bmatrix} & & & \\ & \begin{bmatrix} z_0^{(1)} \\ z_2 \end{bmatrix} \xrightarrow{T_{\varphi_2}} \begin{bmatrix} z_0^{(2)} \\ z_2^{(1)} \end{bmatrix} & & \\ & & \ddots & \\ & & & \begin{bmatrix} z_0^{(N-2)} \\ z_{N-1} \end{bmatrix} \xrightarrow{T_{\varphi_{N-1}}} \begin{bmatrix} z_0^{(N-1)} \\ z_{N-1}^{(1)} \end{bmatrix} \end{array} \right\} \tag{8}$$

The components of $\mathbf{z}$ can also be processed sequentially together with components of the vector-generator, namely at once the angles φ_k are calculated.

The N-point heap transformation $\mathcal{H}$ in the space of N-dimensional vectors is thus defined as

$$\mathbf{z} = (z_0, z_1, z_2, ..., z_{N-1})' \rightarrow H(\mathbf{z}) = (z_0^{(N-1)}, z_1^{(1)}, z_2^{(1)}, ..., z_{N-1}^{(1)})' \tag{9}$$

where the components of the transform are calculated as

$$z_k^{(1)} = g(z_0^{(k-1)}, z_k, \varphi_k), \quad k = 1 : (N-1), \quad (z^{(0)} = z_0). \tag{10}$$

The first component, where the heap is accumulated, is transformed sequentially $(N-1)$ times,

$$z_0 \rightarrow z_0^{(1)} \rightarrow z_0^{(2)} \rightarrow \cdots \rightarrow z_0^{(N-1)}. \tag{11}$$

Values of components $z_0^{(k-1)}$ are defined as follows:

$$\begin{aligned} z_0^{(1)} &= f(z_0, z_1, \varphi_1), \\ z_0^{(2)} &= f(z_0^{(1)}, z_2, \varphi_2), \quad \ldots, \\ z_0^{(N-1)} &= f(z_0^{(N-2)}, z_{N-1}, \varphi_{N-1}). \end{aligned} \tag{12}$$

As an example, Figure 1 shows the signal-flow graph, or network of calculation of the five-point transform of a vector $\mathbf{z} = (z_0, z_1, z_2, z_3, z_4)'$. It is assumed that all parameters φ_k of the transformation have been calculated by the given vector $\mathbf{x} = (x_0, x_1, x_2, x_3, x_4)'$ and the set of constants $A = \{a_1, a_2, a_3, a_4\}$.

Example 1 (elementary rotation). For a given real number a, we consider the following functions defined on the unbounded set of points $\{(x, y);\ x^2 + y^2 \geq a^2\}$

$$\left\{ \begin{array}{l} f(x, y, \varphi) = x \cos\varphi - y \sin\varphi, \\ g(x, y, \varphi) = x \sin\varphi + y \cos\varphi. \end{array} \right. \tag{13}$$

The basic transformation is defined as a rotation of the point (x, y) to the horizontal $Y = a$,

$$T_\varphi : (x, y) \rightarrow (y_0, a) = (x \cos\varphi - y \sin\varphi, a) \tag{14}$$

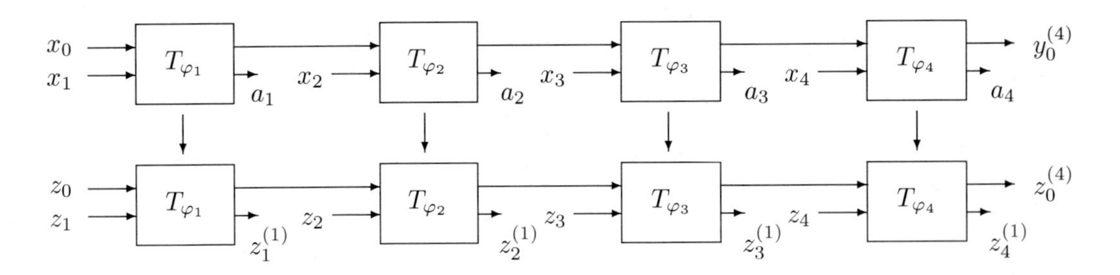

Figure 1. Signal-flow network of calculation of the five-point DsiHT of a vector $\mathbf{z}$.

where the angle φ is calculated by

$$\varphi = \arccos\left(\frac{a}{\sqrt{x^2+y^2}}\right) - \arctan\left(\frac{x}{y}\right), \qquad \left(\varphi = \arcsin\left(\frac{a}{x}\right), \text{ if } y = 0\right). \tag{15}$$

The transformation T_φ is defined in matrix form as

$$T_\varphi : \begin{bmatrix} x \\ y \end{bmatrix} \to \begin{bmatrix} \cos\varphi & -\sin\varphi \\ \sin\varphi & \cos\varphi \end{bmatrix} \begin{bmatrix} x \\ y \end{bmatrix}.$$

The graphs of the transformation T_φ are shown in Fig. 2, when inputs are the generator $(x_0, x_1)'$ and vector $(z_0, z_1)'$.

x_0, x_1 → T_φ → y_0, a

(a)

z_0, z_1 → T_φ → $z_0^{(1)}$, $z_1^{(1)}$

(b)

Figure 2. Graphs of the two-dimensional rotation T_φ : (a) composition and (b) application.

We consider only the case when $a = 0$ in (14), i.e. when the N-point DsiHT with matrix $\mathbf{T} = \mathbf{T}_{N-1} \cdots \mathbf{T}_2 \cdot \mathbf{T}_1$ is defined by the set $A = \{0, 0, ..., 0\}$. The angles of basic transformations $T_k = T_{\varphi_k}$ can be defined as

$$\begin{aligned} \tan(\varphi_k) &= -\frac{x_k}{y_0^{(k-1)}}, \qquad k = 1 : (N-1), \\ y_0^{(k)} &= (\mathbf{T}_k \cdots \mathbf{T}_1 \mathbf{x})_0 = y_0^{(k-1)} \cos(\varphi_k) + x_k \sin(\varphi_k) = \pm\sqrt{x_0^2 + x_1^2 + ... + x_k^2}, \end{aligned} \tag{16}$$

where $y_0^{(0)} = x_0$. As a result, the whole energy of the input signal is collected consequently and, then, transferred to the first component (heap)

$$y_0 = y_0^{(N-1)} = y_0^{(N-2)} \cos(\varphi_{N-1}) + x_{N-1} \sin(\varphi_{N-1}) = \pm\sqrt{x_0^2 + x_1^2 + ... + x_{N-1}^2} \tag{17}$$

$(y_0^{(N-1)})^2 = ||\mathbf{x}||^2$. Thus, $\mathcal{H} : \mathbf{x} \to H(\mathbf{x}) = (||\mathbf{x}||, 0, 0, ..., 0)'$. (It follows that $y_0^{(N-1)} = \pm||\mathbf{x}||$, but our code is designed in a way that $y_0^{(N-1)}$ is always positive. We can also propose $y_0^{(N-1)} = \text{sign}(x_0) \cdot ||\mathbf{x}||$.)

EXAMPLE 2. Let $\mathcal{H}$ be the heap transformation generated by the five-dimensional vector $\mathbf{x} = (1, 2, 3, 2, 1)'$. The generator itself is transformed into the scaled unit vector

$$H(\mathbf{x}) = ||\mathbf{x}||e_1 = (||\mathbf{x}||, 0, 0, ..., 0)' = (\sqrt{19}, 0, 0, ..., 0)' = \sqrt{19}(1, 0, 0, ..., 0)'. \tag{18}$$

The matrix of the heap transform generated by this vector equals

$$\mathbf{H} = \begin{bmatrix} 0.2294 & 0.4588 & 0.6882 & 0.4588 & 0.2294 \\ -0.8944 & 0.4472 & 0 & 0 & 0 \\ -0.3586 & -0.7171 & 0.5976 & 0 & 0 \\ -0.1260 & -0.2520 & -0.3780 & 0.8819 & 0 \\ -0.0541 & -0.1081 & -0.1622 & -0.1081 & 0.9733 \end{bmatrix}, \tag{19}$$

and the angels of rotations are $\varphi_1 = -63.4349°$, $\varphi_2 = -53.3008°$, $\varphi_3 = -28.1255°$, and $\varphi_4 = -13.2627°$. Figure 3 shows the mesh of this matrix in part a, along with the angles φ_k, $k = 1 : 4$, of rotations in b.

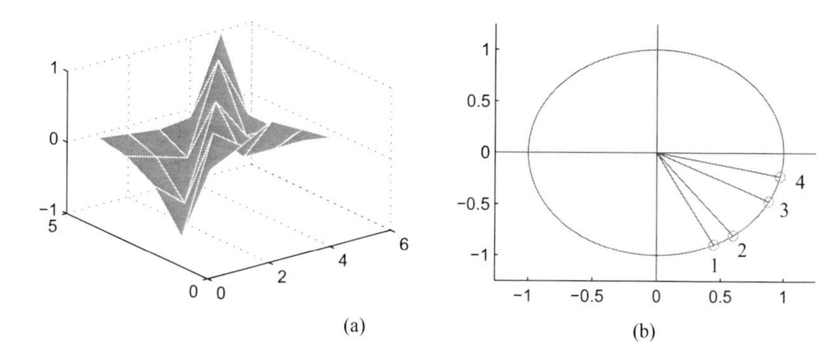

Figure 3. (a) Mesh of the matrix **H** of the DsiHT generated by the signal $\{1, 2, 3, 2, 1\}'$, and (b) angles of rotations.

The matrix of this five-point DsiHT can be represented as $\mathbf{H} = \mathbf{DM}$, where the matrix **M** and the diagonal matrix **D** are defined as

$$\mathbf{H} = \begin{bmatrix} 0.2294 & 0 & 0 & 0 & 0 \\ 0 & 0.4472 & 0 & 0 & 0 \\ 0 & 0 & 0.3586 & 0 & 0 \\ 0 & 0 & 0 & 0.1260 & 0 \\ 0 & 0 & 0 & 0 & 0.0541 \end{bmatrix} \begin{bmatrix} 1 & 2 & 3 & 2 & 1 \\ -2 & 1 & 0 & 0 & 0 \\ -1 & -2 & \frac{5}{3} & 0 & 0 \\ -1 & -2 & -3 & 7 & 0 \\ -1 & -2 & -3 & -2 & 18 \end{bmatrix}. \tag{20}$$

The five basis functions of this heap transform are shown in Fig. 4 in part a, along with the basis functions of another heap transform generated by the vector $(1, 2, 3, 4, 5)'$ in b. The first three basis functions are the same since the vector-generators coincide at three first points. We can see that the basis functions represents themselves the moving waves that change in their movement from the left to right. The first basis function $h_1(n)$ of the DsiHT coincides with the generator, up to the normalized coefficient $||\mathbf{x}||$. This property holds for the considered above examples, as well as in the general case of $\mathbf{x}$, i.e.

$$h_1(n) = \frac{x_n}{||\mathbf{x}||}, \quad n = 0 : (N-1). \tag{21}$$

Figure 5 shows the original discrete-time signal $\mathbf{z}$ in part a, which has been obtained by sampling the function

$$z(t) = 2\cos(8t) - 3\sin(t), \quad t \in [0, 2\pi],$$

at 512 equidistant time-points of the interval $[0, 2\pi]$. We consider the heap transforms generated by two waves of frequencies $\omega_1 = 8$ and $\omega_2 = 1$ (in rad/s) which are carrier frequencies of the original signal. Two generators of length 512 each, that correspond to the waves $\cos(8t)$ and $\sin(t)$ are shown in b. The 512-point discrete heap transform of the signal $\mathbf{z}$, which was generated by the discrete-time signal $\mathbf{x}_1$ sampled from the wave $\cos(8t)$ is shown in c, and the 512-point heap transform of the signal $\mathbf{z}$, which was generated by $\mathbf{x}_2$ sampled from the wave $\sin(t)$ in d. One can notice that the transform of the signal by generator $\mathbf{x}_1$ deletes the contribution of the cosine wave of frequency $\omega_1 = 8$. A wave of the high frequency is present in the beginning of the transform, but the remaining part is similar to the wave $3\sin(t)$ up to the sign. The application of the transform generated by $\mathbf{x}_2$ illuminates the frequency $\omega_2 = 1$ and results in a wave which is similar to the wave $-2\cos(8t)$.

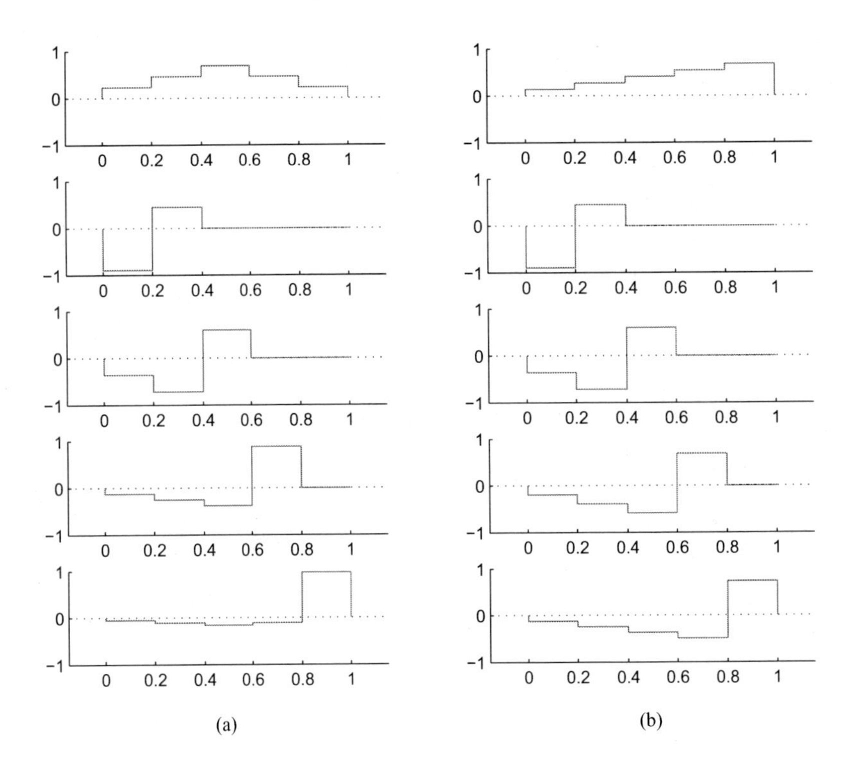

Figure 4. Basis functions of the five-point DsiHTs generated by the vectors (a) $(1, 2, 3, 2, 1)'$ and (b) $(1, 2, 3, 4, 5)'$.

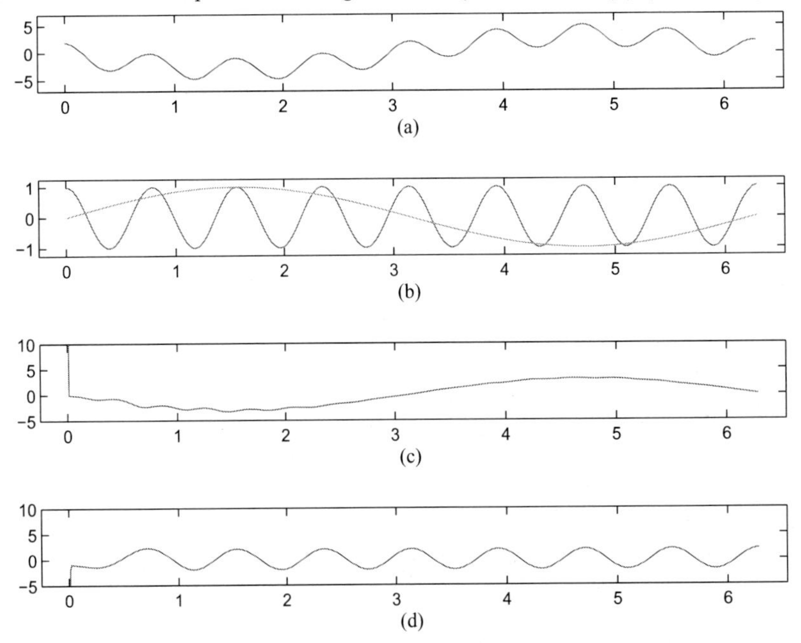

Figure 5. (a) Original 512-point discrete signal, (b) two generators $x_1(t) = \cos(8t)$ and $x_2(t) = \sin(t)$, (c) the 512-point $\mathbf{x}_1$-induced heap transform of the signal, and (d) the 512-point $\mathbf{x}_2$-induced heap transform of the signal.

2. ANGULAR REPRESENTATION

The discrete heap transformation $\mathcal{H}$ can be described uniquely by its *angular representation* which includes the square root energy of the generator and the set of rotation angles,

$$\mathbf{x} \to \mathcal{A}_{\mathbf{x}} = \{||\mathbf{x}||, \mathcal{A}_{\varphi}\} = \{||\mathbf{x}||, \varphi_1, \varphi_2, ..., \varphi_{N-1}\}. \tag{22}$$

The sequence $\mathcal{A}_\varphi$ is referred to as an angular signal $\mathbf{a} = \mathbf{a}(\mathbf{x})$. The signal generator is reconstructed by its angular representation as

$$\mathbf{x} = \mathcal{H}^{-1}(||\mathbf{x}||, 0, 0, ..., 0)' = ||\mathbf{x}|| \left(\mathcal{H}^{-1}(1, 0, 0, ..., 0)'\right). \tag{23}$$

Figure 6 shows the angular representation of the 512-point discrete heap transform generated by the signal-generator $\mathbf{x}_1$ in part a, along with the angular representation of the 512-point discrete heap transform generated by $\mathbf{x}_2$ in b. The signals $\mathbf{x}_1$ and $\mathbf{x}_2$ are signals shown in Fig. 5. It is interesting to note, that the angular representation is similar to the generator but in a very small range $[-\pi, \pi]$, or $[0, 2\pi]$. The major difference of signals occurs in the very beginning. This angular representation as a compressed form of the input data can be

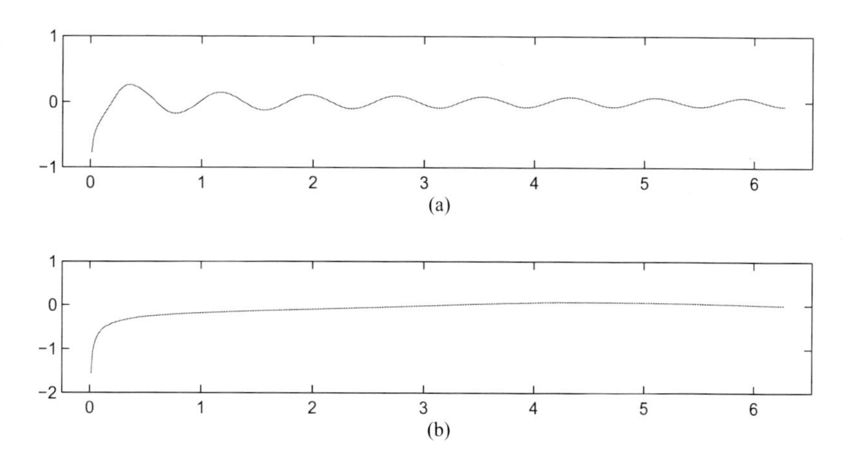

Figure 6. (a) The angular representation of the 512-point DsiHT generated by $x_1(t)$, and (b) angular representation of the 512-point DsiHT generated by $x_2(t)$.

used in signal and image processing. At the same time, the signal $\mathbf{x}$ can be considered as an angular signal of other signal. Indeed, the signal $\mathbf{x}$ in its normalized form $\bar{\mathbf{x}}$ defines the $(N+1)$-dimensional vector $\mathbf{u}$ by:

$$\mathbf{x} \to \bar{\mathbf{x}} = \left(\frac{x_0}{||\mathbf{x}||}, \frac{x_1}{||\mathbf{x}||}, ..., \frac{x_{N-1}}{||\mathbf{x}||}\right)' \to \mathbf{u} = (1, \bar{\mathbf{x}})' = \left(1, \frac{x_0}{||\mathbf{x}||}, \frac{x_1}{||\mathbf{x}||}, ..., \frac{x_{N-1}}{||\mathbf{x}||}\right)'. \tag{24}$$

The obtained signal $\mathbf{u}$ is the angular representation of a signal, which we denote by $\mathbf{y}$. In other words, $\mathcal{A}_\mathbf{y} = \mathbf{u}$. Thus, the following representation holds

$$\mathbf{x} \to \mathbf{u} \to \mathbf{y}. \tag{25}$$

We call the signal $\mathbf{y}$ to be *the dual signal* to $\mathbf{x}$. The duality is with respect to the angular representation.

We now analyze the heap transforms over a real speech signal. As a generator for such a signal, we consider a short part of the speech signal, and then, we will apply the transform generated by this part to other parts. As an example, Figure 7 shows a speech signal of duration of 38.7360 seconds, which was sampled with the rate $f_s = 22050$Hz. The generator for processing this signal is considered to be the signal composed of the first part of the signal, which has the duration of 2.9722 seconds (or 2^{16} points). Figure 8 shows this signal-generator in part a, along with the angular representation of this part in b. Thus, this is a set of rotation angles by means of which the part of the signal is transformed to the magnified unit vector. It can be seen, that the angular representation differs much from the generator $\mathbf{x}_1$ at the beginning, but in a short time, the form of the angular representation becomes similar to the generator.

3. EXPERIMENTAL RESULTS

In this section, we briefly describe a few applications of the heap transforms. It is very important to note, that we can hear the angular signal $\mathcal{A}_\varphi$; it reproduces all sounds and words as the original signal. The human ear is able thus to reproduce the speech signal in its angular representation; it needs only to be magnified. In the

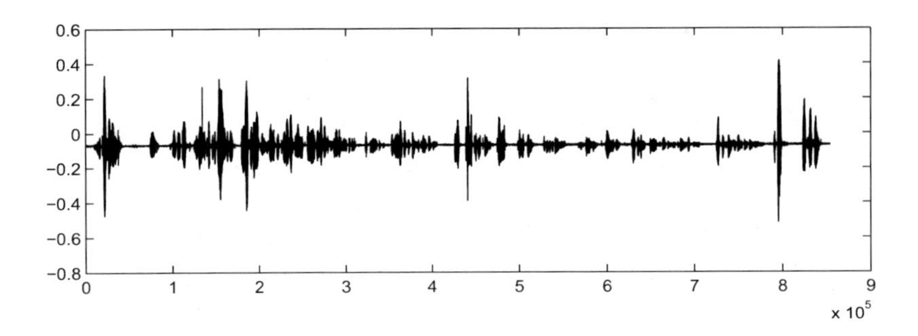

Figure 7. Discrete-time speech signal of duration of 38.786sec.

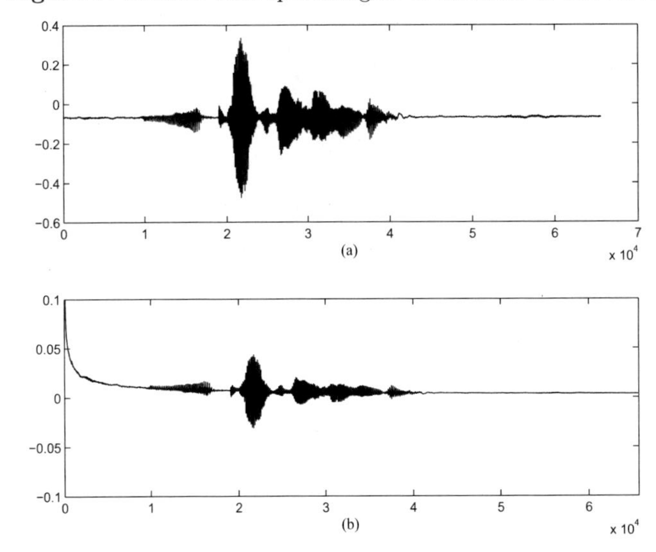

Figure 8. (a) The first part of the speech signal of duration of 2.9722sec and (b) the angular representation of this part.

above example, the angular signal $\mathcal{A}_\varphi$ has been magnified by factor of 32. Thus, together with the time and frequency representation of the signal, we obtain a new, angular representation.

Our experimental results show that the angular transformation has a root property. In other words, the next angular representations over the previous angular-signals do not change almost. As an example, Figure 9 shows the second angular representation of the angular signal, i.e. transformation which is defined as

$$\mathbf{x} \to \mathbf{u} = \mathcal{A}_{\mathbf{x}} = \{||\mathbf{x}||, \mathcal{A}_\varphi\} \to \mathcal{A}_{\mathbf{u}} = \{||\mathbf{u}||, \mathcal{A}_\phi\} \tag{26}$$

or $\mathbf{x} \to \mathcal{A}_\varphi \to \mathcal{A}_\phi$. If we perform this transformation one time more, the angular representation will change the magnitude, but not the form.

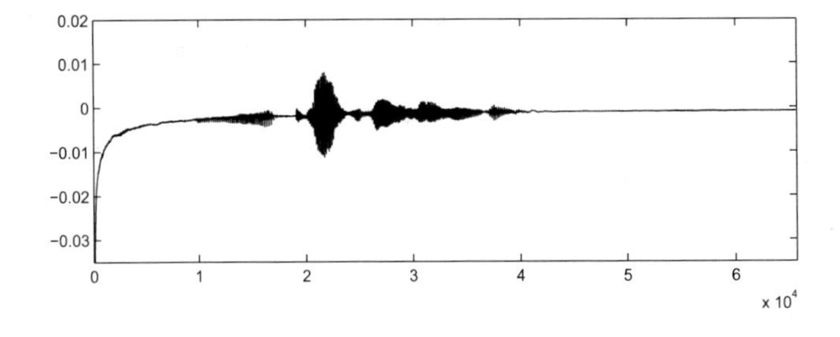

Figure 9. The second angular representation $\mathcal{A}_\phi$ of the speech signal.

When applying the heap transform generated by the selected part of the speech signal to the whole signal, we observe the heaps of the transforms at the beginning of each part of the signal. Figure 10 shows the application of the heap transform generated by the signal of Figure 8(a) to the whole speech signal. There are 13 parts of the signals that are processed separately. One can see that the first part contains only the heap (the generator coincides with this part), and all others parts of the transform contain waves similar to the generator. This interesting property can be hearted very clearly; within each part the transform reproduce the generator. It means that the heap transform, when processing the input signal, includes in the output much information about the signal-generator. In part b of this figure, the same transform of the speech signal is shown, but when

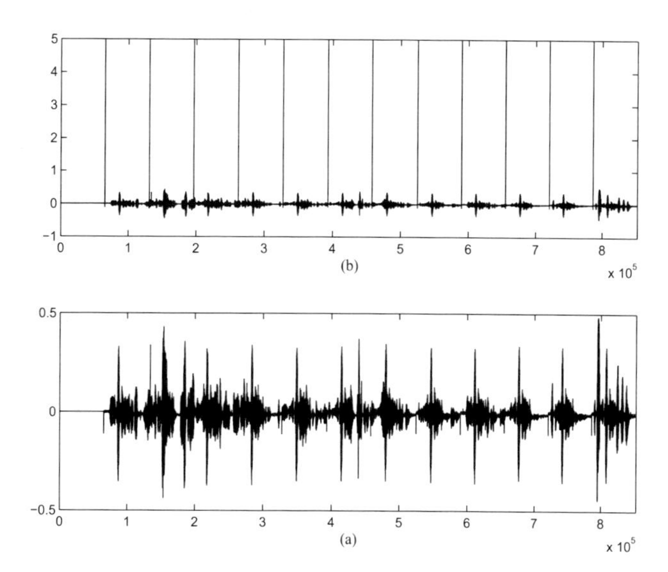

Figure 10. (a) Application by of the heap transform to all parts of the speech signal (2^{16} points each), and (b) the same transform after suppressing all heaps.

all heaps of parts are suppressed to zero. The amplitudes of all heaps are shown in Fig. 11. The first heap corresponds to the norm, or square root energy of the generator, and the other heaps have almost the same amplitudes.

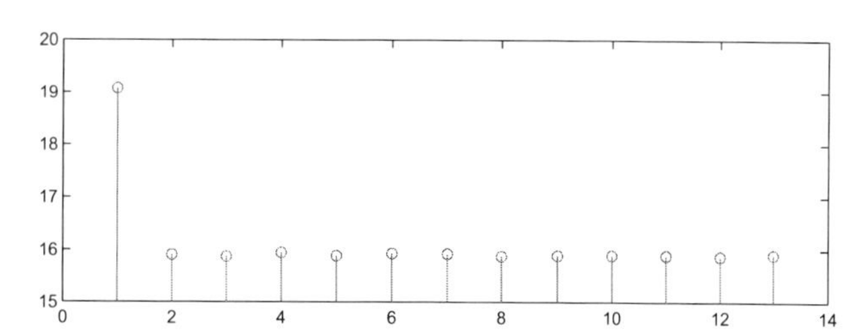

Figure 11. Amplitudes of all 13 heaps, when applying the DsiHT over the speech signal.

The above mentioned property of the DsiHT allows for changing the information of the signal into a very compact form and even to hide the signal within other one. Thus the signal can be enhancement or becomes imperceptible. To demonstrate this fact, we consider the simple example with the speech signal of Fig. 7. It was said, that when processing one signal by the DsiHT generated by a given signal, the information of the generator is imposed in the processed signal. As an example, Figure 12 shows the first part of the speech signal in part a, which is consider to be the signal-generator for the DsiHT. As the input signal, we take the third parts of the speech signal, which is shown in b. The heap transform of this signal is given in c.

The transform contains the information about both signals, and during the reproduction of this transform as a speech signal, in many cases, both signals can be recognized. This effect can be used in different ways. As an

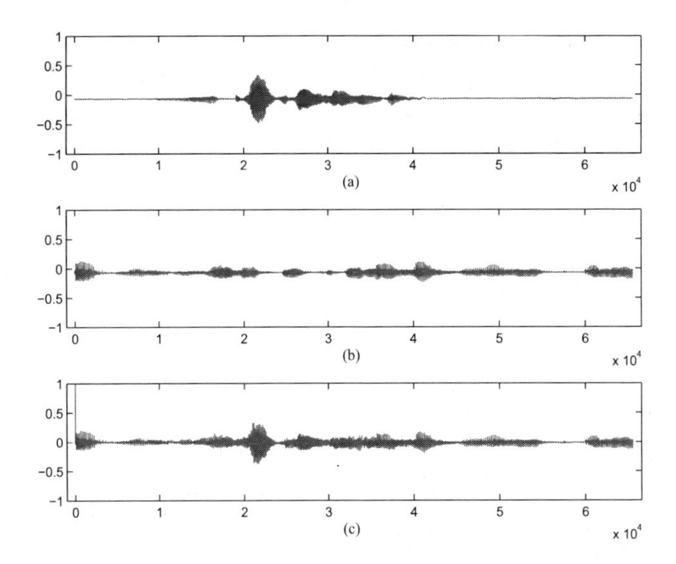

Figure 12. (a) The first part of the speech signal, (b) the third part of the speech signal, and (c) the DsiHT of the third part, when the generator is the first part of the signal.

example, Figure 13 shows the first fragment of the speech signal in part a, along with the heap transform of this signal, when the generator of the transform is the constant signal $(1, 1, 1, ..., 1)'$ in b. One can see that the signal has been enhanced. The part of the heap transform of the same signal, when the generator of the transform is the binary signal $(1, -1, 1, -1, ..., 1, -1)'$ is shown in c. It is possible also to superpose similarly the speech signal

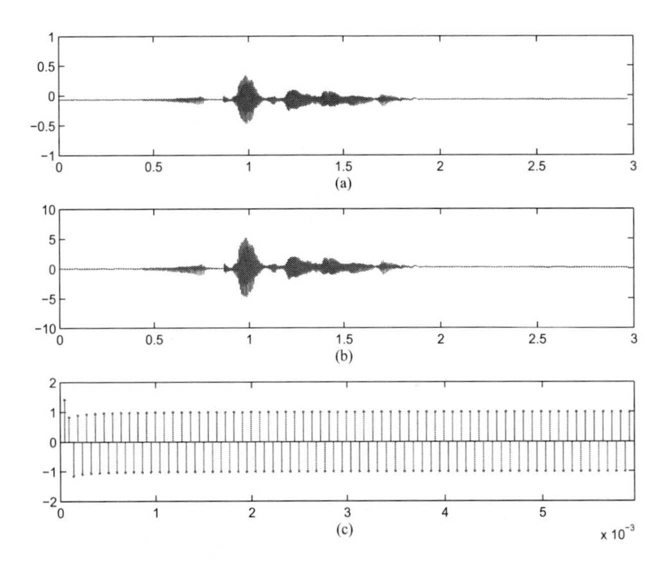

Figure 13. (a) The first part of the speech signal, (b) the DsiHT of this signal, when the generator is the vector $(1, 1, 1, ..., 1)'$, and (c) the DsiHT of this signal, when the generator is the vector $(1, -1, 1, -1, ..., 1, -1)'$.

in a cosine wave and make the speech signal imperceptible.

3.1. Heap transform of images

The selection of generators plays important role when applying the heap transform in image processing. We here consider the simple row-wise method of image processing, when each row of the image is processed by the 1-D heap transform generated by the same row or a different generator.

We first consider two different images processed by DsiHTs generated by rows of these images. As an example, Figure 14 shows the tree image (256×256) processed row-wise by the 1-D heap transforms which are generated by rows of the moon image in part a, along with the image processed by the 1-D heap transforms generated by rows of the girl image in b. One can note how tree image changes; the moon image cannot be seen in the image,

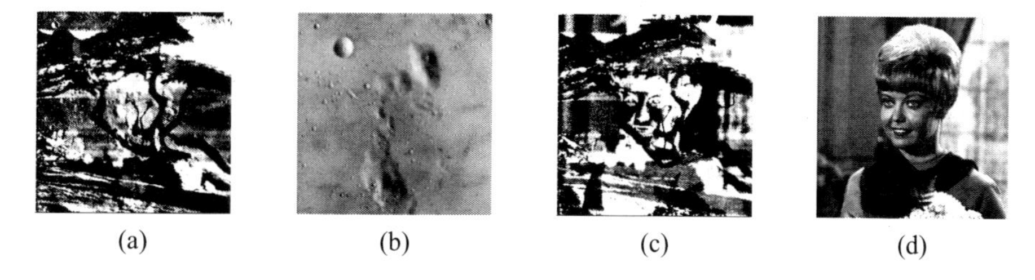

Figure 14. (a) The tree image processed by the DsiHTs generated by rows of (b) the moon image, and (c) the tree image processed by DsiHTs generated by rows of (d) the girl image.

but some details of the girl image can be observed. This figure illustrates the importance of the generators and show that the key-generator presents in the transform together with the image.

It should be noted, that the generator can be extracted from the transform and processed image. In other words, in general, if a signal $\mathbf{z}$ has been processed by the DsiHT generated by a signal $\mathbf{x}$, then there is the unique transformation restoring $\mathbf{x}$,

$$\mathcal{R} : \{H_{\mathbf{x}}[\mathbf{z}], \mathbf{z}\} \to \mathbf{x}.$$

Thus $\mathbf{z}$ is the key to obtain $\mathbf{x}$. As an application of this property, we consider the generator to be the information we want to make invisible. For instance, we consider the tree image be such information or a 2-D generator, and the signal which makes this image invisible is a cosine wave, for instance, $\cos(32t)$ sampled with 256 equidistant points in the interval $[-\pi, +\pi)$. Figure 15 show the result of such processing of the tree image.

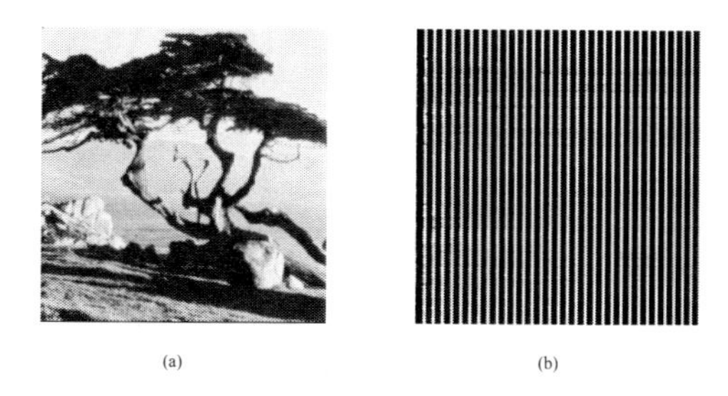

Figure 15. (a) The original tree image and (b) the image processed by rows by the DsiHT of the cosine wave.

Figure 16 shows the first fragment (of length $2^{16} = 256 \times 256$) of the speech signal in part a, along with the DsiHTs of this signal, when 256 generators represent the rows of the tree image in b. Each DsiHT was applied to one part of the speech signal. Figure 17 shows the original tree image in part a, processed row-wise and column-wise by the 256 DsiHTs generated sequentially by parts of the speech signal in b and c, respectively.

4. CONCLUSIONS

The described heap transformations represent a subclass of discrete unitary signal-induced transformations which are generated by input signals and defined by complete systems of moving functions. Such transformations are linear, but the method of their composition is not linear. The basis functions of these transformations are variable waves, but not simple sliding windows as in the wavelet theory. We have considered the class of transformations that are defined by one generator and two decision equations, however more general cases with two or more generators with a few decision equations can be also considered. The matrices of the heap transformations are triangle from the second row, and the first rows represent the generators themselves. The transforms are fast

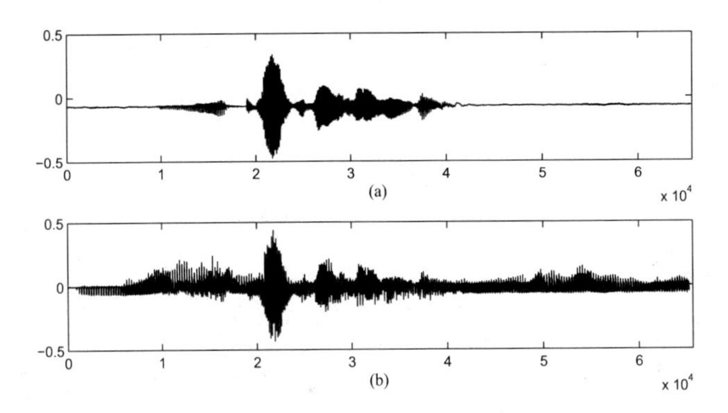

Figure 16. (a) The speech signal, (b) the signal after processing by the 256 DsiHTs generated by rows of the tree image.

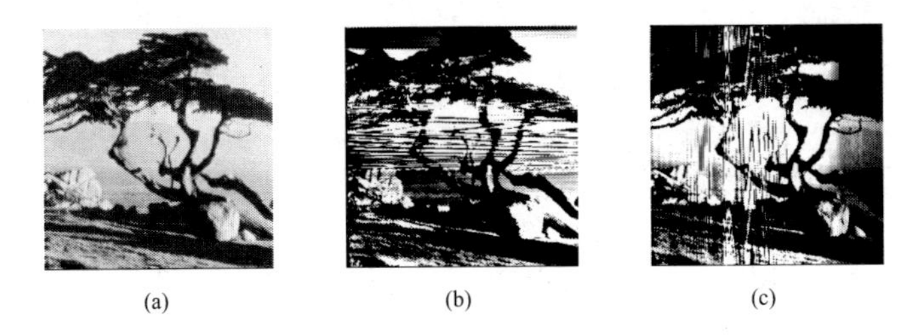

Figure 17. (a) The original tree image and the images processed row-wise (b) and column-wise (c) by the DsiHTs generated by parts of the speech signal. (Each part is of length 256.)

and have place for any length of signals, and we believe that these transforms can be used in signal and image processing, communication, cryptography, and other areas as well.

REFERENCES

1. N. Ahmed and K.R. Rao, *Orthogonal transforms for digital signal processing,* Springer-Verlag, Berlin, 1975.
2. D. Gabor, "Theory of communication," *J. Inst. Elec. Eng.,* vol. 93, pp. 429-457, 1946.
3. R.C. Gonzalez and R.E. Woods, *Digital Image Processing,* Addison-Wesley, Reading 1992.
4. M. Akay, "Wavelet applications in medicine," *IEEE Spectrum,* vol. 34, no. 5, pp. 50-56, 1997.
5. R.E. Blahut, *Fast Algorithms for Digital Signal Processing*. Reading, MA: Addison-Wesley, 1984.
6. K.R. Rao and P. Yip, *Discrete cosine transform–algorithms, advantages, applications,* Academic Press, Boston, 1990.
7. A.M. Grigoryan and S.S. Agaian, *Multidimensional Discrete Unitary Transforms: Representation, Partitioning and Algorithms,"* Marcel Dekker Inc., New York, 2003.
8. A.V. Oppenheim and J.S. Lim, "The importance of phase in signals," *Proc. of the IEEE,* vol. 69, pp. 529-541, 1981.
9. C.J. Kuo and C.S. Huang, "Robust coding technique-transform encryption coding for noisy communication," *Optical Engineering,* vol. 32, pp. 150-156, Jan. 1993.
10. Z.Hrytskiv, S. Voloshynovskiy, and Y. Rytsar, "Cryptography and steganography of video information in modern communications," *Facta Universitatis (NIS), Series: Electronics and Energetics,* vol. 11, no. 1, pp. 115-125, 1998.
11. A.M. Grigoryan and M.M. Grigoryan, "Nonlinear approach of construction of fast unitary transforms," *Proc. of the 40th Annual Conference on Information Sciences and Systems (CISS 2006),* Princeton University, pp. 1073-1078, March 22-24, 2006, Princeton.
12. A.M. Grigoryan and M.M. Grigoryan, "Discrete unitary transforms generated by moving waves," *Proc. of the International Conference: Wavelets XII, SPIE: Optics+Photonics 2007,* vol. 6701, 670125, 27-29 August, 2007, San Diego, CA.

Improved accuracy with higher protection of a biometric system using image and decision fusion techniques

Salim Alsharif, Aed El-Saba, and Syed Bokhari
Department of Electrical and Computer Engineering
University of South Alabama, Mobile, AL 36688
salsharif@usouthal.edu

ABSTRACT

Fingerprint recognition applications as means of identity authentication that deals with accuracy and security are becoming more acceptable in areas such as financial transactions, access to secured buildings, commercial driver license and identity check at entry borders, to mention a few. This paper presented a new approach of using two patterns of the same person, intelligently fused, to form a new unique pattern of the same person. The Laplacian pyramid (LP) level 7 image fusion approach and the logical "OR" and logical "AND" operators for the decision fusion approach were tested with respect to their performance in accuracy, security and processing speed of the recognition system. The concept of receiver operator characteristic (ROC) curve to indicate any improvement in accuracy and security of the process was used.

Finally, an overall comparison and analysis of performance between traditional systems that used a single pattern and our proposed system that used two fused fingerprint patterns in the biometric system was presented.

Keywords: Biometric system, image fusion, decision fusion

1. INTRODUCTION

Our aim in this paper is to establish a more secure and more accurate system, which is done by fusing two impressions of a particular person to form a new unique pattern. Image fusion is done using LP at level 7 fusion [1, 2], and decision fusion is achieved using logical AND and logical OR operators. Logical operators were used in detecting landmines [3, 4] and will be used here on the biometric system. Images are fused and saved as a new pattern, then image enhancement is performed using Filterbank-based algorithm [5], and tested to determine its security and accuracy for fused image before and after enhancement, from which ROC curve is established. The biometric fusion takes an impression from two fingerprints and fuses them together to form a single pattern using decision or image fusion techniques [1, 2, 6, 7] as shown in Fig. 1. The presence of dirt or any other form of error could be replaced after the patterns have been fused together, hence, greatly reducing processing time [8].

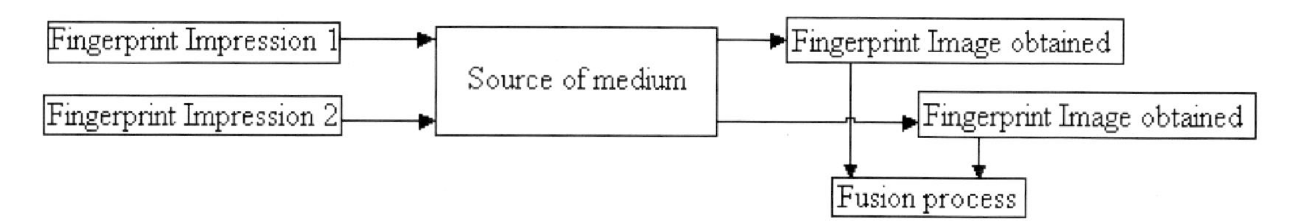

Fig. 1. Two fingerprint patterns fusion process [7].

This approach has been applied to several impressions. Several trials have been applied and in this paper we present an example showing impression of these images as in Fig. 1. These images are fused using image and decision fusion techniques.

Mobile Multimedia/Image Processing, Security, and Applications 2009, edited by Sos S. Agaian, Sabah A. Jassim,
Proc. of SPIE Vol. 7351, 73510I · © 2009 SPIE · CCC code: 0277-786X/09/$18 · doi: 10.1117/12.829896

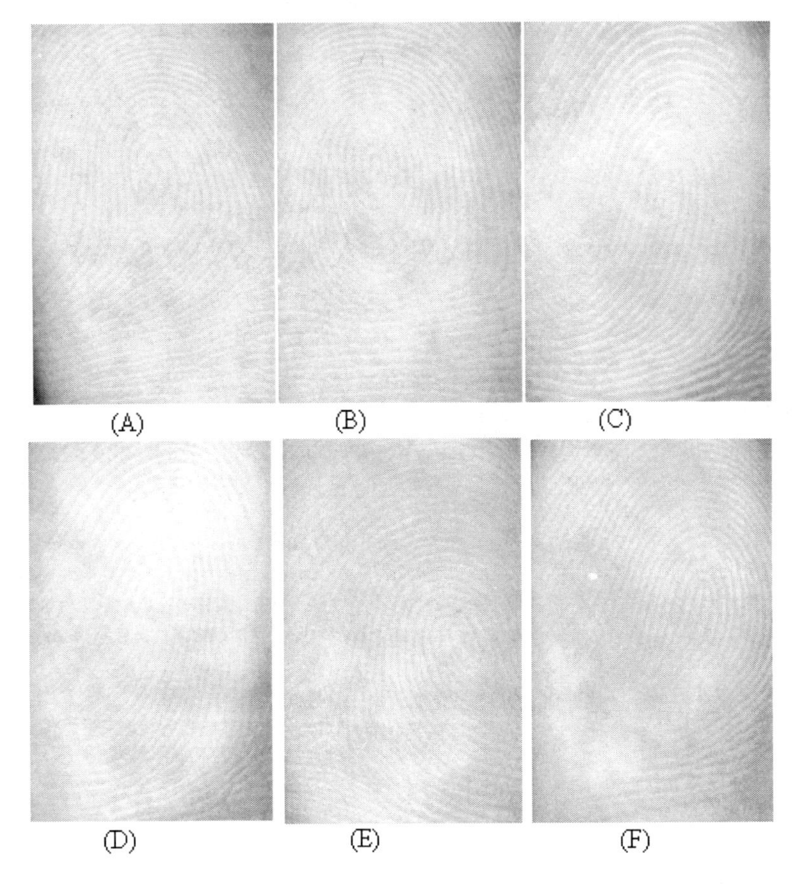

Fig. 2. Input images: (A, B) First person impressions 1 and 2; (C, D) Second person impressions 1 and 2 (E, F) Third person impressions 1 and 2, respectively.

2. IMAGE FUSION AND ENHANCEMENT APPROACH

The combination of the two patterns would make a unique set that is different from the other individual two. The fused pattern is expected to yield a more accurate and secure biometric system. Fig. 3 shows an example of single pattern image fusion with and without enhancement. The image enhancement takes two patterns of a single person where a clearer pattern is expected but it has a complex structure that is hard to manipulate and copy, making the process more secure and accurate as shown in Fig. 3.

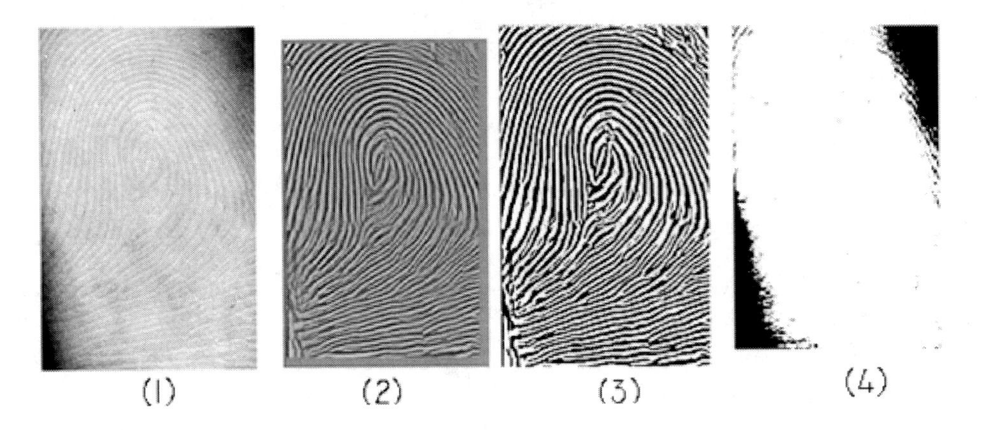

Fig. 3. (1) Unenhanced fused pattern (2) Enhanced fused pattern (3) Extracted minutiae from 2 (4) Extracted minutiae from 1 directly.

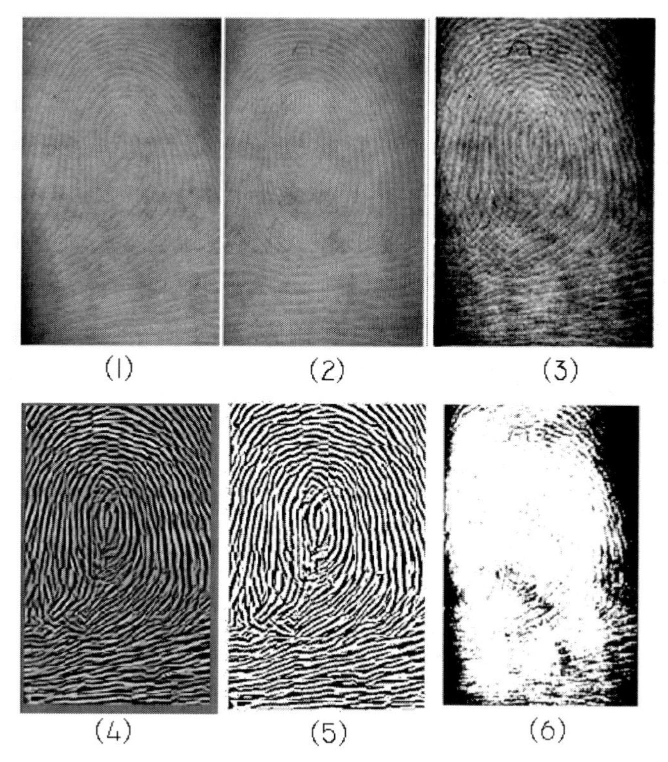

Fig. 4. (1) Pattern 1, (2) Pattern 2, (3) Fused pattern, (4) Enhanced fused pattern (5) Extracted minutiae from (4), (6) Extracted minutiae from (3)

Figs. 3 and 4 show that the minutiae threshold for patterns before image enhancement has low threshold making them easily accessible, but the minutiae threshold for a fused patterns is higher and thus less accessible for imposters. Image enhancement Filterbank-based fingerprint matching is a combination of wavelet transform and filter-based [5]. The noise is removed using Gabor filters and then the image is enhanced using wavelet transform. Wavelet transform is explained in five steps: (1) Normalization where the input images are adjusted to have a specific mean and variance of the input image. (2) Wavelet decomposition is where the image is decomposed into sub-images in the spatial domain using Gabor filters of its sub-images, and then the decomposed image is established by approximating the region. (3) Global texture filtering where the sub-images that are obtained in step (2) are converted into texture spectrum domain, which contains all the texture unit of all the pixels. (4) Local directional compensation is where the local orientation of the pixels is estimated by using predefined directional masks. (5) Wavelet decomposition is where the sub-image is reconstructed to form a new improved image.

LP is described as a multi-resolution image transform [2], where a collection of low or band pass copies of an original image containing both the band limit and sample density are reduced in regular steps. A multi-resolution transform decomposes an image into different images of different resolutions. This transform forms a sequence of images in which each level is filtered and a sub-sampled copy of its pervious image. Thus the lower level of the pyramid has the lowest resolution of the original pattern, whereas the highest level has the opposite characteristics. Fusion of two patterns in LP requires four steps, (1) Obtain two patterns and check for dimensions (2) Construct the pyramid of n-level dimensions (3) Fuse the patterns in n-level dimension separately, known as pyramid level fusion [2] (4) Reconstruct the fused level image into a single image. Fig. 5 shows an example of one pattern being converted into level 5. Fig. 6 shows the image fusion for level 5 and level 7 respectively using LP.

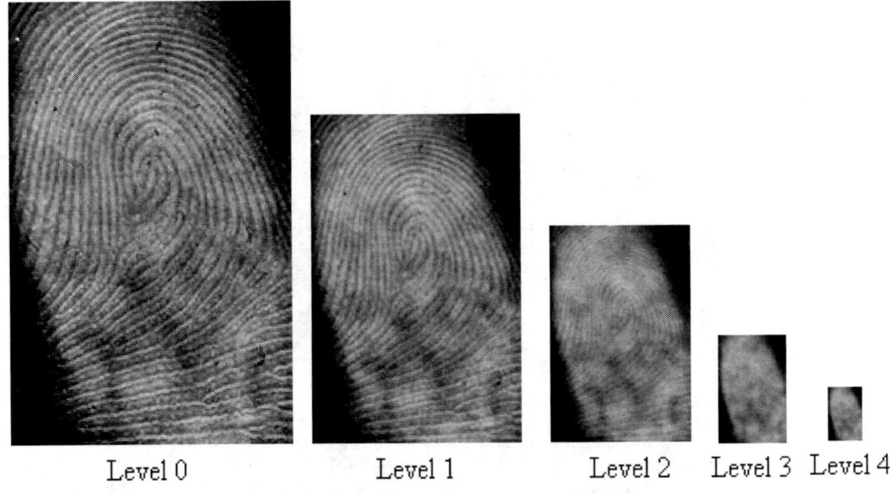

Fig. 5. Level 5 LP layers

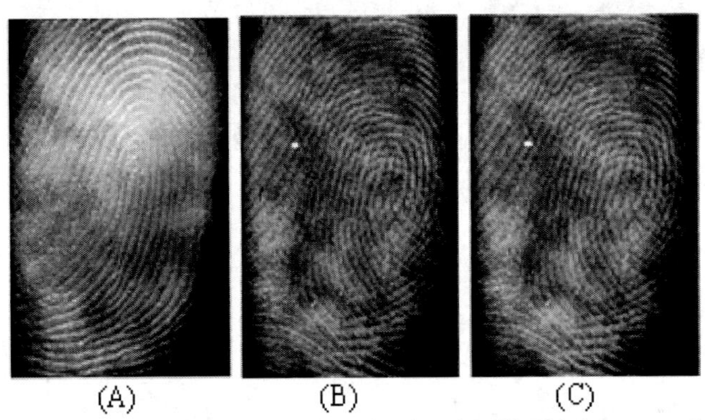

Fig. 6. LP fusion. (A) First person using level 7 (B) Second person using level 7 (C) Third person using level 7.

3. DECISION FUSION

Logical operators "AND" or "OR" are used to fuse two images together [3]. Referring to Fig. 7 below, the two logical states (TRUE or FALSE) combine together to form the decision on the fusion of the image. The AND operator produces TRUE outcome if both of the inputs are true, otherwise the value is FALSE. Each image has a set of image pixel values that are converted into grayscale for better decision fusion process. For example, if the inputs for two images at the same position have pixel decimal values of 157 and 234 (binary 10011101 and 11101010) respectively. Their binary output when ANDed will be 10001000 or 136 decimal. The process is repeated until all the 148635 pixels have been processed.

For the OR operation, fusion occurs when one state is true and the second state is false, or when both are true, the resulting decision is TRUE decision. Even when the images are false, the logical OR fusion occurs the same way as logical AND operation (between two true states). However, the resulting fused pattern will be different in intensity than the logical AND fused pattern, i.e. taking the same value of pixels as 157 and 234; the output would be decimal 255 or binary 11111111.

For the AND operation we will take the true states of all persons and compare it with the combination of all the other false states. Since AND decision is yes for only two true states, the other conditions give us a false or no decision to fuse, so the process of AND is only performed if two true states of a person are present. Figure 8 represents the AND fused and enhanced patterns of two true states for different patterns of the same person.

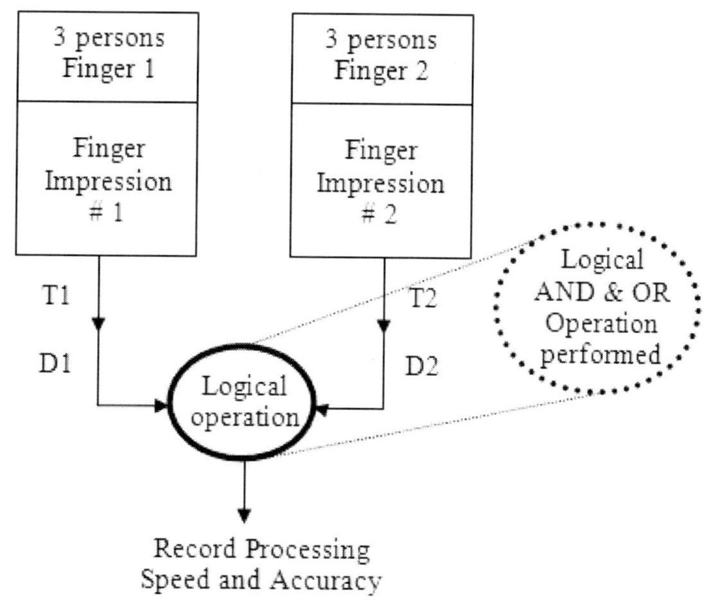

Fig. 7. Decision fusion of logical operations using two fingerprint patterns

For the logical OR operation, the fusion on the decision of false patterns is acceptable, since in logical OR the decision between a true and a false state is true. The logical OR operation has the same outcome as the logical AND but since we are looking at different conditions, the outcomes are more for a person than three persons.

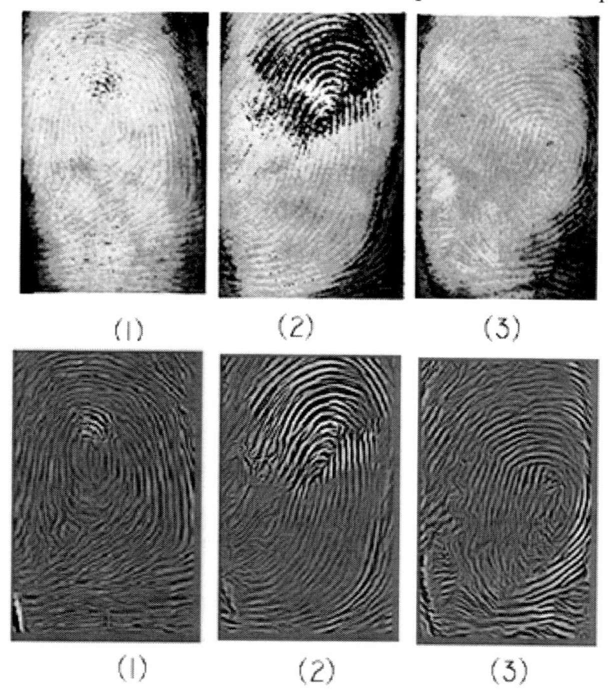

Fig. 8. Logical AND fused (top three) enhanced fused patterns (lower three) for (1) First person, (2) Second person and (3) Third person

A total of 12 new set of patterns are produced with the combination of three true sets of three fingerprint fusion. Fig. 9 below shows the logical OR fused patterns and enhanced patterns.

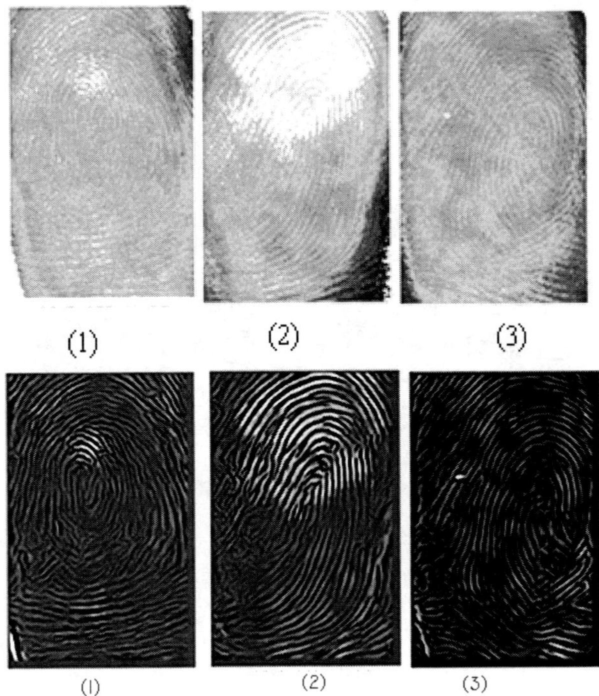

Fig. 9. Logical AND fused (top three) enhanced fused patterns (lower three) for (1) First person, (2) Second person and (3) Third person

4. ROC CURVES

A ROC curve determines the responsivness of the system and measures the percentage error between a real image and set of experimental data [9, 10], and estimates the values of TPR (True positive rate) and FPR (False positive rate). Other parameters such as FNR (False negative rate) and TNR (True negative rate) are also calculated. TPR occurs when a real image A is tested across a set of stored images and gives a correct answer, whereas FNR occurs when a real image A gives an inaccurate result. When a another tested real image B (different from real image A) gives a false answer, it is recorded as FNR, and when it identifies real image A, then that parameter is known as FPR. The outcomes are examined to see if the system recognizes the patterns stored correctly. The values are then fed into the ROC parameter analyzer to find out the accuracy of logical operators fusion. Decision fusion is expected to show more accurate and secure ROC curves than image fusion. ROC curves that are generated in the verification and identification stage after image enhancement are shown below. The verification step requires first a decision to be made, so our analysis in this step is to accurately make the decision if it is true and then draw the ROC curve. The second step is to make sure that the fused patterns have been stored correctly by repeating the process and ensure that the overall system is accurate.

Similarly, the identification operation is performed in which a new set of patterns of three persons are entered to the decision fusion process to evaluate how much identification has been achieved. In case of OR decision, even when the decision is false the fusion can proceed, so we expect to clearly see less security.

Figure 10 shows that logical operator AND is way better in recognizing a true pattern and is more accurate. But the logical operator OR is better in understanding a false state, and that is why it has low classification. The area under the curve represents how accurate the system actually is, and the corresponding accuracy and security values which are recorded in Table 1.

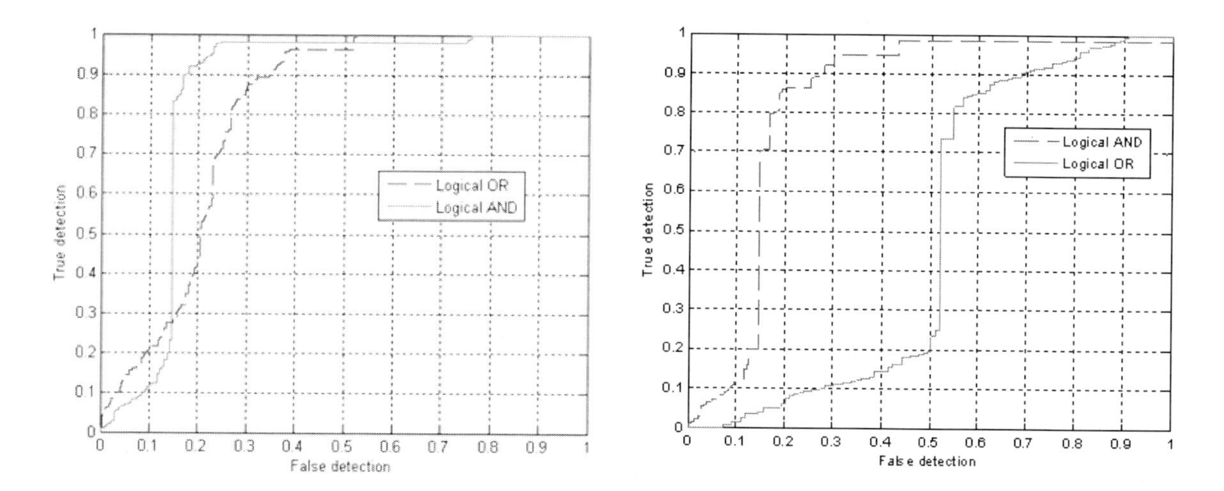

Fig. 10. ROC curves for verification and identification, respectively, using logical AND and OR operations.

Table 1. Accuracy and security values for logical operations

	Accuracy	Security
Logical AND	**81.65%**	**79.31%**
Logical OR	77.56%	53.12%

5. IMAGE AND DECISION FUSION COMPARISON

This section compares the LP image fusion method and logical operators AND and OR decision fusion methods in terms of accuracy and security. For accuracy, we looked at the verification process in which a system correctly verifies that the fused pattern is correctly stored in the database with the same fingerprint impressions of the same person using the ROC curve analysis.

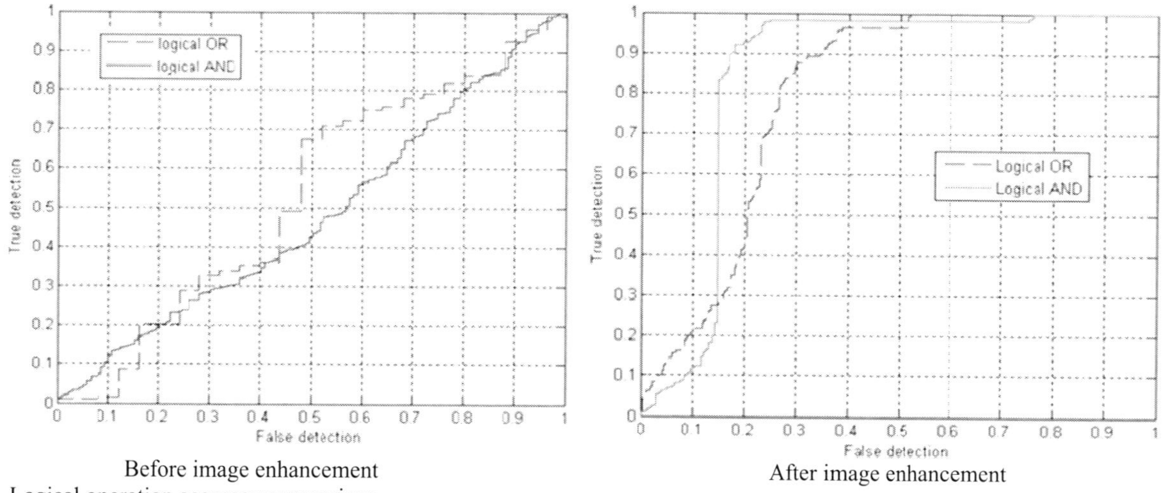

Before image enhancement

After image enhancement

Fig. 11. Logical operation accuracy comparison

It is clear from the figure above that image enhancement improved the accuracy considerably. Logical AND showed almost perfect identification and gave better result than logical OR as expected.

Security is the introduction of new samples into the system for the same person. The results are similar to the analysis for accuracy. The identification process tends to have a less expected curve classification than the verification process. Fig. 12 shows the comparison between each method.

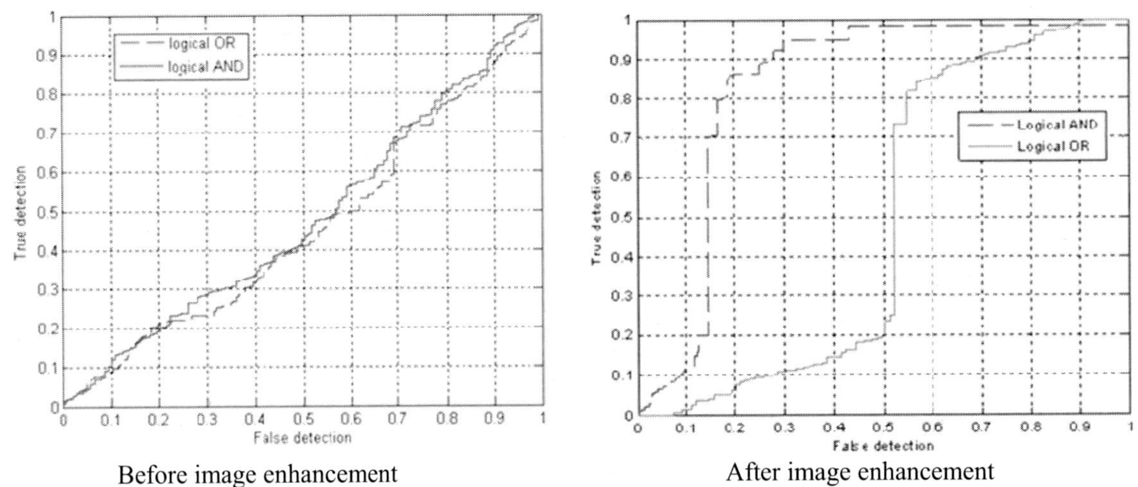

Fig. 12. Logical operation security comparison

Table 2 below shows the corresponding parameter values for accuracy and security before and after image enhancement. It clear from the table that the accuracy and security is much improved after image enhancement as expected.

Table 2. Accuracy and security values for logical operations

	Before image enhancement		After image enhancement	
	Accuracy	Security	Accuracy	Security
Logical AND	54.07%	52.1%	81.65%	79.31%
Logical OR	52.89%	52.2%	77.56%	53.12%

From Fig. 13, it can be seen from the ROC curves that decision fusion after enhancement performs better then image fusion in identification, as it correctly identifies much more users. The ROC curve for logical AND shows an almost perfect system, whereas the ROC curve for LP is lower, making it obvious that logical AND has better performance when compared to LP. ROC curve for logical AND shows more accuracy in verification as well as in security, The more perfect the ROC curve becomes the more secure the system becomes.

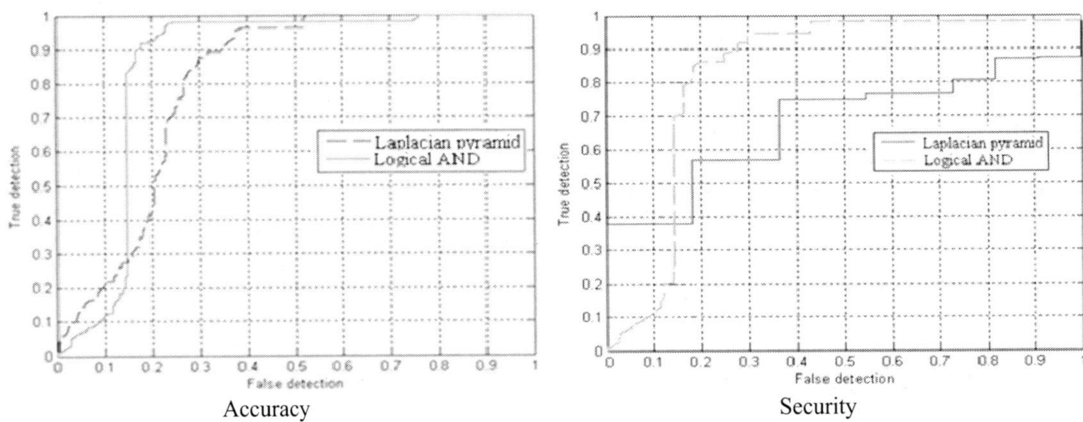

Fig. 13. Image and decision fusion comparison for LP and logical AND.

Table 3 clearly shows that the fusion of two patterns increases the accuracy and security than ordinary patterns by a significant percentage value. Decision fusion logical operator AND performs better as compared to all other methods of fusion, thus suggesting that decision fusion of two ordinary patterns are better than single pattern.

Table 3. Accuracy and security comparison between single and double pattern after image enhancement

	Single Pattern 1		Single Pattern 2	
	Accuracy	**Security**	**Accuracy**	**Security**
	61.12%	50.29%	61.27%	55.65%
Fusion Method	**Accuracy**		**Security**	
Logical AND	**81.65%**		**79.31%**	
Logical OR	77.56%		53.12%	
LP	73.00%		70.00%	

6. CONCLUSION

In conclusion, we calculated the three parameters for fusing different patterns for different persons, using image and decision fusion. Accuracy and security where calculated for both fusion techniques. Image fusion used Laplacian pyramid to fuse the two patterns whereas decision fusion used simple logical operators, AND and OR. Image enhancement proved to have a better response for all the fusion techniques. Performance of image and decision fusion where compared after image enhancement. Overall logical AND had the best accuracy of 81.65% when compared with 73.00% of Laplacian pyramid and 61.12% of a single pattern. Security performance for logical AND was 79.31% when compared to 70.00% of Laplacian pyramid. Decision fusion was much better than image fusion from the security and accuracy point of view of the system. A better choice is to use two patterns for a particular person to increase the security and accuracy.

Future work can be implemented by using three or more patterns and then fuse them, from which we can expect to see an improvement in the performance for security and accuracy, but at a little cost of processing time.

REFERENCES

[1] Hill, P., Canagarajah, N. and Bul, D., "Image fusion using complex wavelets," Dept. of Electrical and Electronic Engineering, University of Bristol, UK, 2002.

[2] Jalili-Moghaddam, M., "Real-time Multi-focus Image fusion using Discrete Wavelet Transform and Laplacian pyramid transform." *The image analysis group Department of signals and systems,* Chalmers University of technology, Sweden, 2005.

[3] EL-Saba, A. M. and Bezuayehu, T., "Fusion of stokes vector imagery using simple local operators: application to the problem of surface landmine detection," *Proc. of SPIE defense and security Conf.*, Vol. 6972, 697228, Orlando, FL., Mar. 2008.

[4] EL-Saba, A. M. and Bezuayehu, T., "Higher probability of detection of substance landmines with a single sensor using multiple polarized and un-polarized image fusion," *Proc. of SPIE defense and security Conf.*, Vol. 6972, 697232, Orlando, FL., Mar. 2008.

[5] Hsieh, C.-T., Lai, E., and Eang, Y.-C., "An effective algorithm for fingerprint image enhancement based on wavelet transform," *Patter Recognition, Vol. 36,* pp. 303-312, 2003.

[6] Mueller, D., Maeder, A. and O'Shea, P., "The Generalized Image Fusion Toolkit (GIFT) Release 4.0", Queensland University of Technology, Brisbane, Australia, e-Health Research Centre, CSIRO ICT Centre, Brisbane, Australia, Sept. 13, 2006.

[7] Canga, E. F., "Image fusion", project report for the degree of M. Eng. in electrical and electronics, University of Bath, signal and image processing group, Jun. 2002.

[8] Duraisamy, P., "Visual contrast enhancement of optical coherence tomography images using combined digital and image fusion methods", MS thesis, University of South Alabama, Jul. 2008.

[9] Fawcett, T., "An introduction to ROC analysis," *The Journal of the Pattern Recognition Society, Pattern Recognition Lett.*, Vol. 27, pp. 861-874, 2006.

[10] Marzban, C., "A Comment on the ROC Curve and the Area Under it as Performance Measures," The Applied Physics Laboratory and the Department of Statistics, University of Washington, and Center for Analysis and Prediction of Storms University of Oklahoma, Norman, Jun. 28, 2004.

A New Reference-Based Measure For Objective Edge Map Evaluation

Shahan C. Nercessian[a], Sos S. Agaian[b], and Karen A. Panetta[a*]

[a] Tufts University, Department of Electrical and Computer Engineering, 161 College Ave., Medford, MA, USA

[b] University of Texas at San Antonio, College of Engineering, 6900 North Loop 1604 West, San Antonio, TX, USA

ABSTRACT

Edge detection is an important preprocessing task which has been used extensively in image processing. As many applications heavily rely on edge detection, effective and objective edge detection evaluation is crucial. Objective edge map evaluation measures are an important means of assessing the performance of edge detectors under various circumstances and in determining the most suitable edge detector or edge detector parameters. Quantifiable criteria for objective edge map evaluation are established relative to a ground truth, and the weaknesses and limitations in the Pratt's Figure of Merit (FOM), the objective reference-based edge map evaluation standard, are discussed. Based on the established criteria, a new reference-based measure for objective edge map evaluation is presented. Experimental results using synthetic images and their ground truths show that the new measure for objective edge map evaluation outperforms Pratt's FOM visually as it takes into account more features in its evaluation.

Keywords: Edge detection, edge map evaluation, performance measures

1. INTRODUCTION

Edge detection plays a pivotal role in determining structure in images. It aids image processing tasks by substantially reducing the amount of information needing to be processed [1, 2]. For this reason, it has served as a basis for many feature extraction, object detection, object recognition, image enhancement, and image segmentation algorithms [3, 4, 5]. It is felt that many applications share a common set of requirements for edge detection. For most image analysis processes that use edge information, it is required that

(1) The edges present should be correct. That is, edge points in the image must represent actual (true) edges. If a pixel does not correspond to a true edge point then it should not be included in the edge point set. If the pixel corresponds to a true edge point then it should be detected as an edge point with a certain degree of strength.

(2) Edges should be properly represented. A boundary model of objects requires that edges be extracted as continual lines that satisfy the following requirements:

(i) Edges should be represented by a certain width, neither too thin nor too thick. It is particularly undesirable to have edges of varying widths at different locations in the image.
(ii) Edges should have a certain degree of continuity. The set of edge points belonging to one edge segment should form a continuous line. The edge points should also be easily formulated to a functional description of the line.
(iii) Edges should have certain degree of connectivity. An edge segment should have connections with some other edge segments. These connections form closed boundaries of the objects in the image.
(iv) Edge segments should be separable from each other. This is to say parallel edges should not have large amounts of pixels touching each other. In other words, the intersection of the point sets for two edge segments should be limited [6].

[*] Further author information: (Send correspondence to S.C.N.)
S.C.N.: E-mail: shahan.nercessian@tufts.edu
S.S.A.: E-mail: sagaian@utsa.edu
K.A.P.: Email: karen@eecs.tufts.edu

Mobile Multimedia/Image Processing, Security, and Applications 2009, edited by Sos S. Agaian, Sabah A. Jassim, Proc. of SPIE Vol. 7351, 73510J · © 2009 SPIE · CCC code: 0277-786X/09/$18 · doi: 10.1117/12.816519

As a result, effective and objective edge map evaluation measures must be developed to assess edge detector performance. Objective edge map evaluation measures have many important uses. Obviously, after the development of so many edge detection algorithms over the years, there should be an objective way of determining which edge detection algorithm generally works best or which edge detection algorithm should be used for a certain application. Secondly, an objective edge map evaluation measure can select the tunable parameters that exist in many edge detection algorithms. Assessing the performance of edge detection algorithms is difficult because the performance depends on several factors. These factors include, but are not limited to,

1) the algorithm itself
2) the type of images used to measure the performance of the algorithm
3) the edge detector parameters used in the evaluation
4) the method for evaluating the edge detectors [7].

Three types of edge map evaluation approaches exists: human evaluation, reference-based measure, non-reference-based measure. Methods based on human evaluation allow us to make quantitative comparisons using subjective ratings made by people. This approach avoids the issue of pixel-level ground truth. As a result, however, it does not allow us to make statements about the frequency of false positive and false negative errors at the pixel level and cannot be automated for human vision systems [8]. Two types of edge map evaluation methods exists: Reference-based and non-reference-based measures. A non-reference-based edge map evaluation measure should only use information from the resultant edge map and the original image itself to make its evaluation. Examples of non-reference-based edge map evaluation measures include probabilistic measures, and edge connectivity and width uniformity measures [9, 10]. In their current stage, such non-reference-based edge map evaluation measures suffer from many biases and are therefore highly unreliable. They also tend to make little to no use of the original image data. For example, probabilistic measures using receiver operating characteristics attempt to estimate an ideal edge map, or ground truth, from multiple edge detector outputs, while edge width and connectivity based measures characterize the quality of edge fragments paying no respect to the original image data. As a result, the most reliable types of non-reference-based edge map evaluation still appear to be visual comparison, which is highly subjective, or task-based evaluation [11], which does not fit under a universal model and cannot always be automated [9]. Alternatively, reference-based edge map evaluation measures require a ground truth in addition to the resultant edge map. They can be used to determine edge detector performance within a controlled environment. This is because for real images, determining the ground truth is non-trivial. Therefore, reference-based edge map evaluation measures are used predominantly for evaluating edge maps of synthetic images. Pratt's Figure of Merit (FOM) [12] is considered to be the reference-based objective edge map evaluation standard. Its formulation is very easy to understand and in some instances, it has shown to be an adequate means for reference-based edge map evaluation. However, Pratt's FOM fails to account for some very important features that are seen visually in assessing the quality of an edge map, or only accounts for them indirectly. Therefore, a more extensive list of criteria must first be established in developing a new measure.

In this paper, a new reference-based objective edge map evaluation measure is presented. Pratt's FOM is briefly reviewed, and weaknesses and limitations in Pratt's FOM are discussed. An extensive list of criteria for edge map evaluation is then established, followed by the development of the new measure which intends on addressing the established criteria directly. The new measure consists of an edge pixel presence/localization term, an edge corner presence/localization term, and a double edge occurrence term. The edge pixel presence/localization term contains a false positive edge pixel term taking into account the number of false positive edge pixels and their distance from the closest ideal edge pixel, as well as a false negative edge pixel term taking into account the number of false negative edge pixels and their distance from the closest correctly detected edge pixel. Similarly, the edge corner presence/localization term contains false positive and false negative edge corner terms. Lastly, a double edge occurrence term penalizes the presence of double edges. These terms are fused based on their importance for a given application. The use of the new edge map evaluation method both as a means of comparing the performance of many edge detectors as well as a means of selecting parameters within a single edge detector is then examined via computer simulations.

The remainder of this paper is organized as follows: Section II provides background information about Pratt's FOM and analyzes the limitations of Pratt's FOM in terms of a set of established criteria. Section III introduces the new measure, Section IV illustrates experimental results, and Section V draws conclusions regarding the presented measure.

2. BACKGROUND INFORMATION

Pratt's FOM is given by

$$FOM_{Pratt} = \frac{1}{\max\{N_O, N_T\}} \sum_{i=1}^{N_T} \frac{1}{1+\alpha d_i^{\,2}} \tag{1}$$

where N_O is the number of edge pixels in the ground truth, N_T is the number of pixels in the edge map being evaluated, d_i is the distance between an ideal edge pixel and its nearest correctly detected edge pixel, and α is a penalty parameter. Pratt's FOM ranges from 0 to 1, where a 1 corresponds to a perfect match between the edge map being tested and the ground truth. For consistency's sake, the α value used in all tests with both Pratt's FOM and the presented measure is set to 1/9, as suggested in the literature [10, 11].

In developing a new edge map evaluation measure, we introduce a list of criteria for assessing edge map quality. Given a ground truth, edge maps detected by an edge detector are evaluated given the following criteria:

Edge pixel presence/localization: Edge pixel presence regards the number of false positive and false negative edge pixels in a detected edge map relative to a ground truth. Edge pixel localization regards the distance between a false negative edge pixel and its nearest correctly detected edge pixel as well as the distance between a false positive edge pixel and its nearest ideal edge pixel.

Edge corner presence/localization: A similar criterion is the presence and localization of edge corners. Often times, the edge corners in detected edge maps are displaced relative to ground truth or are non-existent. This is particularly prevalent in edge maps detected using the Canny edge detector due to its Gaussian smoothing pre-processing step. This said, the Canny edge detector was designed to detect "real-world" edges rather than step edges. Nonetheless, an objective edge evaluation measure should take such flaws regarding edge corner presence and localization into account.

Double edge occurrences: Many edge detectors, including the Canny, Sobel, and Laplacian of Gaussian (LoG) edge detectors, suffer from double edge occurrences and other artifacts due to their smoothing effects. It is understood that double edge occurrences are a special case of false positives, but they are regarded as relevant and undesirable enough to be considered on their own.

Some of the established criteria are directly accounted for in Pratt's FOM. For example, the distance associated with each false negative edge pixel and its nearest correctly detected edge pixel is accounted for in Pratt's FOM. Many of the established criteria are only indirectly accounted for in Pratt's FOM. For example, the presence of edge corners may correspond to fewer false negative edge pixels. However, this is in no way an accurate means of determining the quality of edge maps in terms of edge corner presence/localization. In Figure 1, it is obvious that the edge corners produced by the Roberts edge detector are better than those produced by the Canny edge detector. In general, the Roberts edge map visually outperforms the Canny edge map relative to the ground truth. However, Pratt's FOM for the Canny edge map is .9967, while Pratt's FOM for the Roberts edge map is .9501.

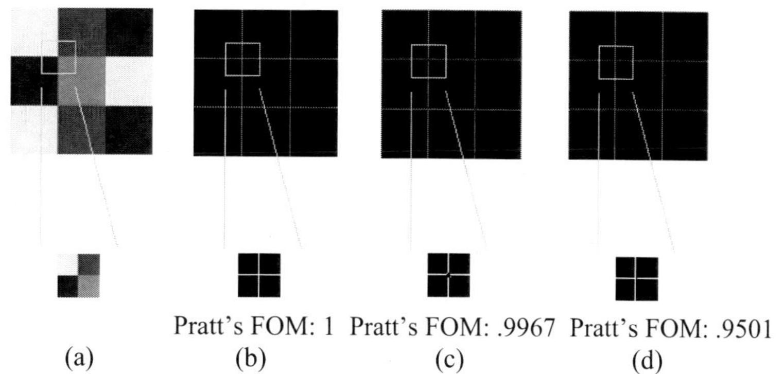

Figure 1 – (a) Original, (b) the ground truth, edge maps obtained using (c) Canny and (d) Roberts

Additionally, double edge occurrences should technically correspond to more false positive edge pixels. However, this is by no means an accurate way of determining the quality of edge maps in terms of double edge occurrences. In Figure 2, double edges in the edge map produced by the Canny edge detector are prevalent. In general, the Roberts edge map visually outperforms the Canny edge map relative to the ground truth. However, Pratt's FOM for the Canny edge map is .8886, while Pratt's FOM for the Roberts edge map is .8794.

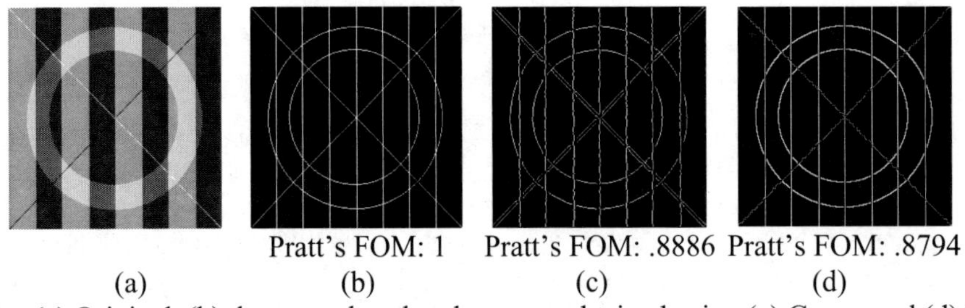

Pratt's FOM: 1 Pratt's FOM: .8886 Pratt's FOM: .8794

(a) (b) (c) (d)

Figure 2 – (a) Original, (b) the ground truth, edge maps obtained using (c) Canny and (d) Roberts

Other established criteria are completely neglected by Pratt's FOM. In general, the total number of edge pixels is considered rather than the number of false positive and false negative edge pixels. Also, the distance associated with each false positive and its nearest ideal edge pixel is not accounted for in Pratt's FOM. This point is illustrated with a simple example. Consider the ground truth and the two hypothetical edge maps shown in Figure 3. Both edge maps contain no false negatives edge pixels and 1 false positive edge pixel. That said, the edge map in Figure 3b is more favorable than the one in Figure 3c. In the extreme sense, this is to say that having a false positive edge near an ideal edge pixel is favored over having a false positive edge nowhere near an ideal edge pixel. However, Pratt's FOMs for both Figure 3b and Figure 3c are equal to .9697. The new measure overcomes this problem by including a term which accounts for the distance associated with each false positive and its nearest ideal edge pixel.

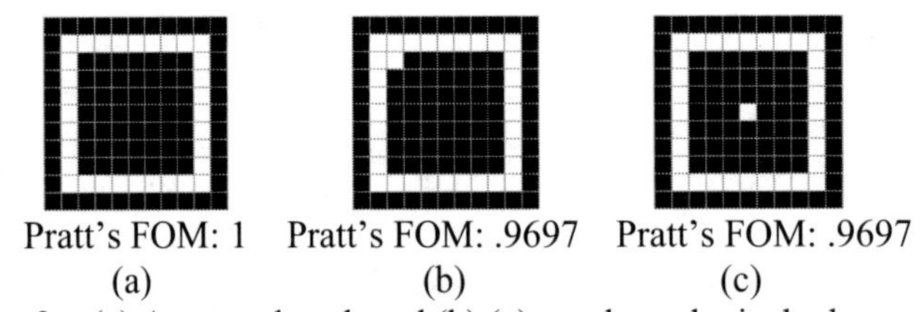

Pratt's FOM: 1 Pratt's FOM: .9697 Pratt's FOM: .9697

(a) (b) (c)

Figure 3 – (a) A ground truth and (b),(c) two hypothetical edge maps

Displaced or shifted edges are generally accounted by Pratt's FOM as there is some sort of distance term present. However, by using another hypothetical example, a weakness regarding displaced edges in Pratt's FOM is discussed. Consider the ground truth and the two hypothetical edge maps shown in Figure 4. Both edge maps contain 2 false negatives edge pixels and 2 false positive edge pixels. That said, the edge map in Figure b is more favorable than the one in Figure 4c. However, Pratt's FOMs for both Figure 4b and Figure 4c are equal to .9600. The new measure alleviates this problem as it considers both the distance between a false negative edge pixel and its nearest correctly detected edge pixel as well as the distance between a false positive edge pixel and its nearest ideal edge pixel.

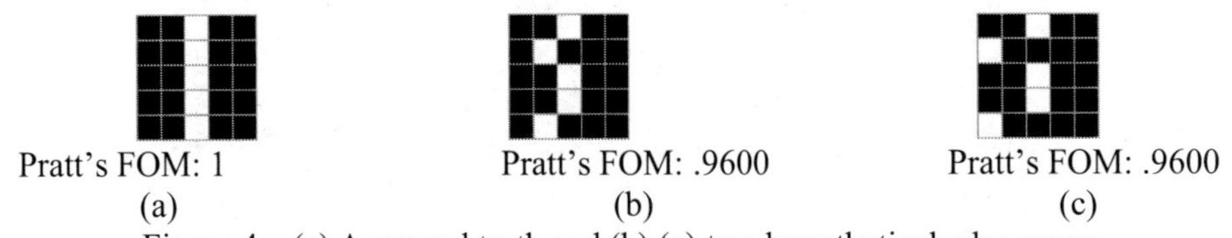

Pratt's FOM: 1 Pratt's FOM: .9600 Pratt's FOM: .9600

(a) (b) (c)

Figure 4 – (a) A ground truth and (b),(c) two hypothetical edge maps

3. THE NEW REFERENCE-BASED EDGE MAP EVALUATION MEASURE

Figure 5 shows a block diagram of the new reference-based edge map evaluation system. Measures regarding edge pixel presence/localization, edge corner presence/localization, and double edge occurrence are calculated. These values are then fused via weighted averaging based on their importance for a given application to yield the final evaluation. If desired, one may choose to completely neglect one of the terms if it is not a needed criteria for the given application.

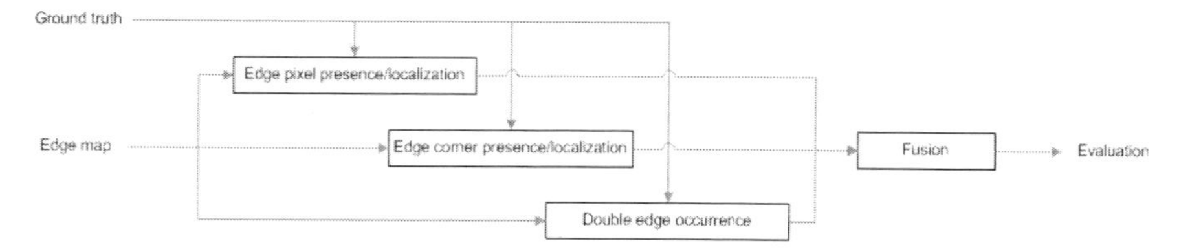

Figure 5 – Block diagram of the new reference-based edge map evaluation system

Edge pixel presence/localization: The edge pixel presence/localization term contains a false positive edge pixel term taking into account the number of false positive edge pixels and their distance from the closest ideal edge pixel. For an mxn edge map, it is given by

$$D_{PFP} = \frac{1}{mn - P_O} \sum_{i=1}^{P_{FP}} 1 - \frac{1}{1 + \alpha d_{PFP_i}^{\;2}} \tag{2}$$

where P_O is the number of edge pixels in the ground truth, P_{FP} is the number of false positive edge pixels in the edge map being evaluated, and d_{PFPi} is the distance between the ith false positive edge pixel from the closest ideal edge pixel. The edge pixel presence/localization term also contains a false negative edge pixel term taking into account the number of false negative edge pixels and their distance from the closest correctly detected edge pixel. It is given by

$$D_{PFN} = \frac{1}{P_O} \sum_{i=1}^{P_{FN}} 1 - \frac{1}{1 + \alpha d_{PFN_i}^{\;2}} \tag{3}$$

where P_{FN} is the number of false negative edge pixels in the edge map being evaluated, and d_{PFNi} is the distance between the ith false negative edge pixel and the closest correctly detected edge pixel. The two terms can be fused by weighted averaging based on a user specified tolerance for false positive or false negative edge pixels. That said, D_{PFP} will usually be substantially less than D_{PFN} due to the fact that the normalizing term in D_{PFP} is smaller than the normalizing term in D_{PFN}. For simplicity's sake, the terms are fused by the uniform average

$$D_P = \frac{D_{PFP} + D_{PFN}}{2} \tag{4}$$

The effectiveness of the D_P term is shown in Figure 6. The Roberts edge map generates more localized edge pixels than the Canny edge map. The new edge pixel measure correctly labels the Roberts edge output as the better of the two edge maps.

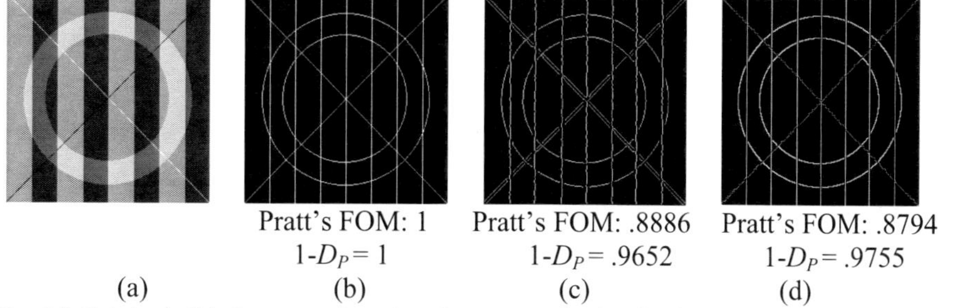

(a)	(b)	(c)	(d)
	Pratt's FOM: 1	Pratt's FOM: .8886	Pratt's FOM: .8794
	$1\text{-}D_P = 1$	$1\text{-}D_P = .9652$	$1\text{-}D_P = .9755$

Figure 6 – (a) Original, (b) the ground truth, edge maps obtained using (c) Canny and (d) Roberts

Edge corner presence/localization: Edge corners are detected using a simple template matching technique. If a given input window matches one of the templates, it is considered to be an instance of an edge corner. The four templates used to detect edge corners are shown in Figure 7. Note that these four templates detect not only all possible 90 degree edge corners, but also all possible 45 degree edge corners.

Figure 7 - Templates used for detecting edge corners

The edge corner presence/localization term contains a false positive edge corner term taking into account the number of false positive edge corners and their distance from the closest ideal edge pixel. For an mxn edge map, it is given by

$$D_{CFP} = \frac{1}{mn - C_O} \sum_{i=1}^{C_{FP}} 1 - \frac{1}{1 + \alpha d_{CFP\,i}^{\;2}} \tag{5}$$

where C_O is the number of edge pixels in the ground truth, C_{FP} is the number of false positive edge corners in the edge map being evaluated, and d_{CFPi} is the distance between the ith false positive edge corner from the closest ideal edge corner. The edge corner presence/localization term also contains a false negative edge corner term taking into account the number of false negative edge corners and their distance from the closest correctly detected edge corner. It is given by

$$D_{CFN} = \frac{1}{C_O} \sum_{i=1}^{C_{FN}} 1 - \frac{1}{1 + \alpha d_{CFN\,i}^{\;2}} \tag{6}$$

where C_{FN} is the number of false negative edge corners in the edge map being evaluated, and d_{CFNi} is the distance between the ith false negative edge corner and the closest correctly detected edge corner. The two terms can be fused by weighted averaging based on a user specified tolerance for false positive or false negative edge corners. That said, D_{CFP} will usually be substantially less than D_{CFN} due to the fact that the normalizing term in D_{CFP} is smaller than the normalizing term in D_{CFN}. For simplicity's sake, the terms are fused by the uniform average

$$D_C = \frac{D_{CFP} + D_{CFN}}{2} \tag{7}$$

The effectiveness of the D_C term is shown in Figure 8. The Roberts edge map generates more precise edge corners than the Canny edge map. The new edge corner measure correctly labels the Roberts edge output as the better of the two edge maps.

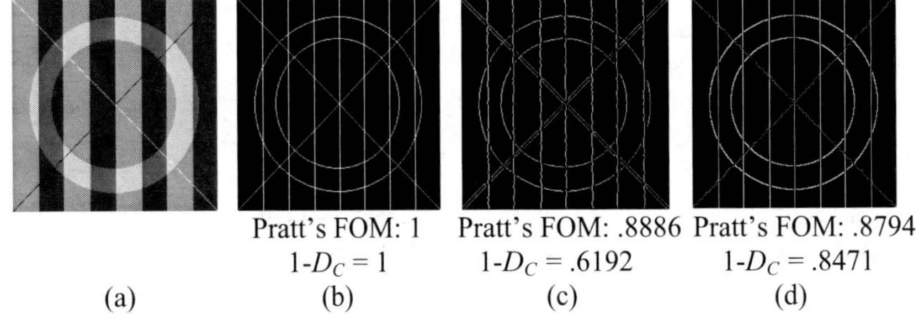

Pratt's FOM: 1 | Pratt's FOM: .8886 | Pratt's FOM: .8794

$1\text{-}D_C = 1$ | $1\text{-}D_C = .6192$ | $1\text{-}D_C = .8471$

(a) (b) (c) (d)

Figure 8 – (a) Original, (b) the ground truth, edge maps obtained using (c) Canny and (d) Roberts

Double edge occurrences: Downsampling is used to expose double edge occurrences in edge maps. The four types of horizontal/vertical downsampling methods that are performed are given by the functions y_z as

$$y_1[m,n] = x[2m,2n] \tag{8}$$

$$y_2[m,n] = x[2m,2n+1] \tag{9}$$

$$y_3[m,n] = x[2m+1,2n] \tag{10}$$

$$y_4[m,n] = x[2m+1,2n+1] \tag{11}$$

Edge maps are padded to ensure even row and column dimensions. The two types of diagonal downsampling methods that are then performed are given by the functions y_z as

$$y_5[m,n] = x[m,2n + n\%2] \tag{12}$$

$$y_6[m,n] = x[m,2n + (n+1)\%2] \tag{13}$$

For each downsampled image, the number of edge pixels in non-overlapping 2x2 windows is determined. By only considering the 2x2 windows in which the number of edge pixels is equal to 1, double edge locations are better exposed. This is illustrated in Figure 9 with the ground truth and edge maps obtained using Roberts and LoG. The edge map obtained using Roberts does not contain any distinct double edges. Conversely, the edge map obtained using the LoG contains very distinct double edges. The two edge maps are downsampled by y_1, and 2x2 windows in which the number of edge pixels is equal to 1 are set to 1. Note the similarity in this result between the Roberts edge map and the ground truth, and the difference between these two and LoG edge map.

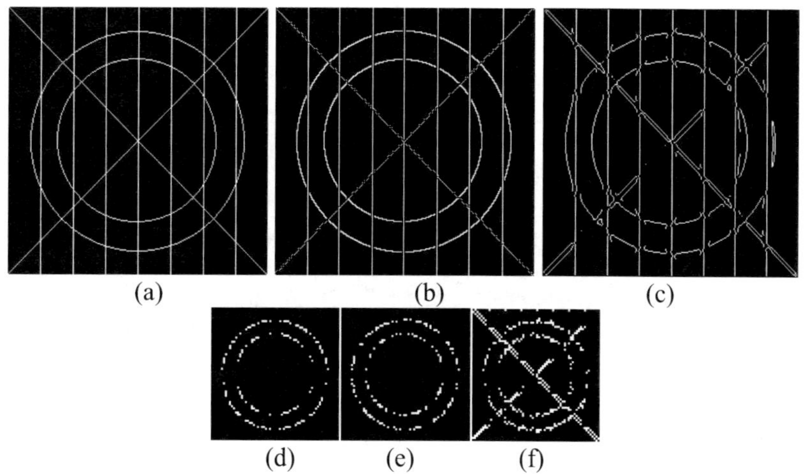

Figure 9 – (a) Ground truth, edge map obtained using (b) Roberts, (c) LoG, 2x2 windows in which the number of edge pixels is equal to 1 after downsampling with y_1 on (d) the ground truth, edge map obtained using (e) Roberts, and (f) LoG

Referring to this procedure for a given y_z as $box(y_z)$, distances d_{DE1} and d_{DE2} are defined as

$$d_{DE1} = \left(box(y_1)_O + box(y_2)_O + box(y_3)_O + box(y_4)_O\right) - \left(box(y_1)_T + box(y_2)_T + box(y_3)_T + box(y_4)_T\right) \tag{13}$$

$$d_{DE2} = \left(box(y_5)_O + box(y_6)_O\right) - \left(box(y_5)_T + box(y_6)_T\right) \tag{14}$$

where O denotes the ground truth and T denotes the edge map to be tested. The double edge term is then given by

$$D_{DE} = \frac{\frac{|d_{DE1}|}{4m_1n_1} + \frac{|d_{DE2}|}{2m_2n_2}}{2} \tag{15}$$

The effectiveness of the D_C term is shown in Figure 10. The Roberts edge map generates more precise edge corners than the Canny edge map. The new edge corner measure correctly labels the Roberts edge output as the better of the two edge maps.

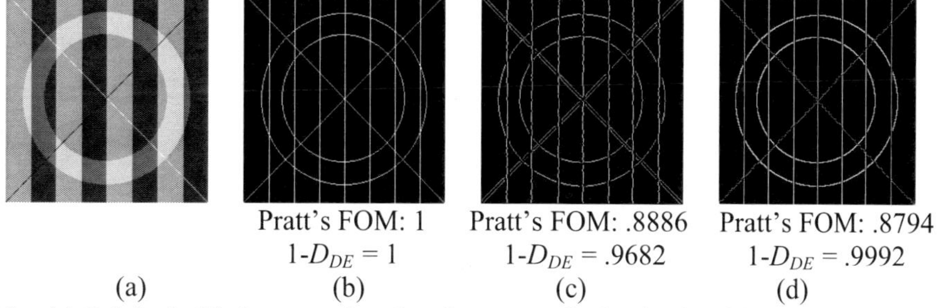

Figure 10 – (a) Original, (b) the ground truth, edge maps obtained using (c) Canny and (d) Roberts

where m_1 and n_1 are the dimensions of an edge map after horizontal/vertical downsampling and m_2 and n_2 are the dimensions of an edge map after diagonal downsampling. For even dimensions m_1 and n_1, $m_2n_2=2m_1n_1$, and in general, $m_2n_2 \approx 2m_1n_1$. Finally, the new measure is given by the weighted average of the individual terms as

$$measure = 1 - \frac{\beta_P D_P + \beta_C D_C + \beta_{DE} D_{DE}}{\beta_P + \beta_C + \beta_{DE}} \tag{16}$$

5. EXPERIMENTAL RESULTS

The effectiveness of the new edge map evaluation method both as a means of comparing the performance of many edge detectors as well as a means of selecting parameters within a single edge detector is examined. To assess the performance of the new edge map evaluation method, a number of synthetic images and their ground truths were generated. Some of these results are presented here. The synthetic images were designed to include many possible line and shape types. This includes straight lines oriented at various angles, curved lines such as those in ovals and circles, and various edge corner types. To test the use of the new edge map evaluation methods as a means of comparing edge detector performance, edges are detected using the Canny, Sobel, and LoG edge detectors. The criterion for selecting edge detectors was to evaluate only algorithms for which code was readily available. Note the Canny edge detection algorithm is considered to be the edge detection standard by many researchers. Evaluation measures using both the new measure and Pratt's FOM are then calculated for each edge map and compared to subjective evaluation. The same test is performed on each of the synthetic images. For simplicity and consistency's sake, all tests were performed with α=1/9, β_P=15, β_C=1, β_{DE}=10. Figure 11 summarize the results for some of the synthetic images that were tested, with the best performers according to the given measure highlighted in bold.

To test the use of the new edge map evaluation method as a means of selecting parameters within a single edge detector, two different experiments are performed. Both experiments are performed with the same parameters used previously, namely with α=1/9, β_P=15, β_C=1, β_{DE}=10. In the first experiment, Gaussian noise is added to an image, and edge detection is performed on the image using the Canny edge detector using various values for the Gaussian standard deviation σ. The Gaussian smoothing in the Canny edge detector is known to suppress Gaussian noise, but also eliminates and displaces edge pixels and causes double edge occurrences. The best value of σ successfully finds a balance between these two properties, and should correspond to a maximum in the presented measure. Note that in this experiment, the ground truth used is the ground truth edge map of the image before noise has been added. The results of this first experiment are shown in Figures 12 and 13, with the best performer highlighted in bold.

The results show that the presented measure outperforms Pratt's FOM as a means of comparing the performance of different edge detectors. This is to say that the relative values of the presented measure generally correspond better to subjective opinions. This is due to the fact that the new measure takes into account many more important features which are seen visually in making its evaluation. As a result, the best edge detector for each image determined by the presented measure and Pratt's FOM were dramatically different. One may note that Pratt's FOM often times falsely determined that edge maps containing double edges and smoothed corners were the best edge map for the given original image. The results also show that the presented measure can effectively serve as a means of selecting parameters within a single edge detector. In the instances where the Canny edge detector was used with various values of σ, the maximum in the measure corresponded to the best value of σ determined visually. In the instances where the Sobel edge detector was used with various values of T, the maximum in the measure corresponded to the best value of T determined visually.

Both subjective evaluation and the presented measure hint that the Roberts edge detector is the best edge detector for simple synthetic images when no noise is present. This makes sense as there is no embedded smoothing in the kernels used by the Roberts edge detector, and consequently fewer displaced edges, and smoothed corners, and double edge occurrences. Note that, according to the results, Pratt's FOM generally does not agree with this observation. However, it goes without saying that embedded smoothing is necessary for detecting edges in natural images.

6. CONCLUSIONS

A new reference-based edge map evaluation measure has been presented based on a set of established criteria. The presented measure incorporates many important features in its evaluation. The effectiveness of the new measure has been shown both as a means of comparing the performance of many edge detectors as well as a means of selecting parameters within a single edge detector. On experiments performed with synthetic images and their ground truths, the new measure outperformed Pratt's FOM as a means of comparing the performance of many edge detectors, as its results corresponded better to subjective opinions of many experts in the field. It also successfully selected parameters within a single edge detector, namely the standard deviation σ parameter in the Canny edge detector and the threshold T parameter in the Sobel edge detector.

REFERENCES

1. J. Parker, *Algorithms for Image Processing and Computer Vision*, John Wiley & Sons, New York, 1996.
2. A. Koschan and M. Abidi, "Detection and Classification of Edges in Color Images," *IEEE Signal Processing Magazine, Special Issue on Color Image Processing*, Vol. 22, No. 1, p. 64-73, 2005.
3. E. Danahy, S. Agaian, and K. Panetta, "Directional edge detection using the logical transform for binary and grayscale images," *Mobile Multimedia/Image Processing for Military and Security Applications, SPIE Defense and Security Symposium* Vol. 6250, April 2006.
4. M. Bennamoun and B. Boashash, "A Structural-Description-Based Vision System for Automatic Object Recognition," *IEEE Transactions on Systems, Man, and Cybernetics – Part B: Cybernetics*, Vol. 27, No. 6, December 1997.
5. P.L. Rosin, "Robust Pose Estimation," *IEEE Transactions on Systems, Man, and Cybernetics – Part B: Cybernetics*, Vol. 29, Issue 2, April 1999.
6. Q. Zhu, "Improving Edge Detection by an Objective Edge Evaluation," *Proceedings of the ACM/SIGAPP Symposium on Applied Computing: Technological Challenges of the 1990's*, pp. 459-468, 1992.
7. M.Heath, S.Sarkar, T. Sanocki, and K. Bowyer, "Comparison of Edge Detectors A Methodology and Initial Study" Computer Vision and Image Understanding, Vol. 69, No. 1, January, pp. 38–54, 1998
8. T. Kanungo, M. Jaisimha, J. Palmer, and R. Haralick, "A Methodology for Quantitative Performance Evaluation of Detection Algorithms," *IEEE Transactions on Image Processing*, Vol. 4, No. 12, December 1995.
9. Y. Yitzhaky and E. Peli, "A Method for Objective Edge Detection Evaluation and Detector Parameter Selection," *IEEE Transactions on Pattern Analysis and Machine Intelligence*, Vol. 25, No. 10, October 2003.
10. Q. Zhu, "Efficient evaluations of edge connectivity and width uniformity," *Image Vision Computing*, Vol. 14, p. 21-34, February 1996.
11. M. Shin, D. Goldgof, K Bowyer, and S. Nikiforou, "Comparison of Edge Detection Algorithms Using a Structure from Motion Task," *IEEE Transactions on Systems, Man, and Cybernetics – Part B: Cybernetics*, Vol. 31, No. 4, August 2001.
12. W. Pratt, *Digital Image Processing*, John Wiley & Sons, New York, 2nd edition, 1991.
13. P. Bao, L. Zhang, and X. Wu, "Canny Edge Detection Enhancement by Scale Multiplication," *IEEE Transactions on Pattern Analysis and Machine Intelligence*, Vol. 27, No. 9, p. 1485-1490, September 2005.
14. P. Trahanias and A. Venetsanopoulos, "Vector Order Statistics Operators as Color Edge Detectors," *IEEE Transactions on Systems, Man, and Cybernetics – Part B: Cybernetics*, Vol. 26, No. 1, February 1996.

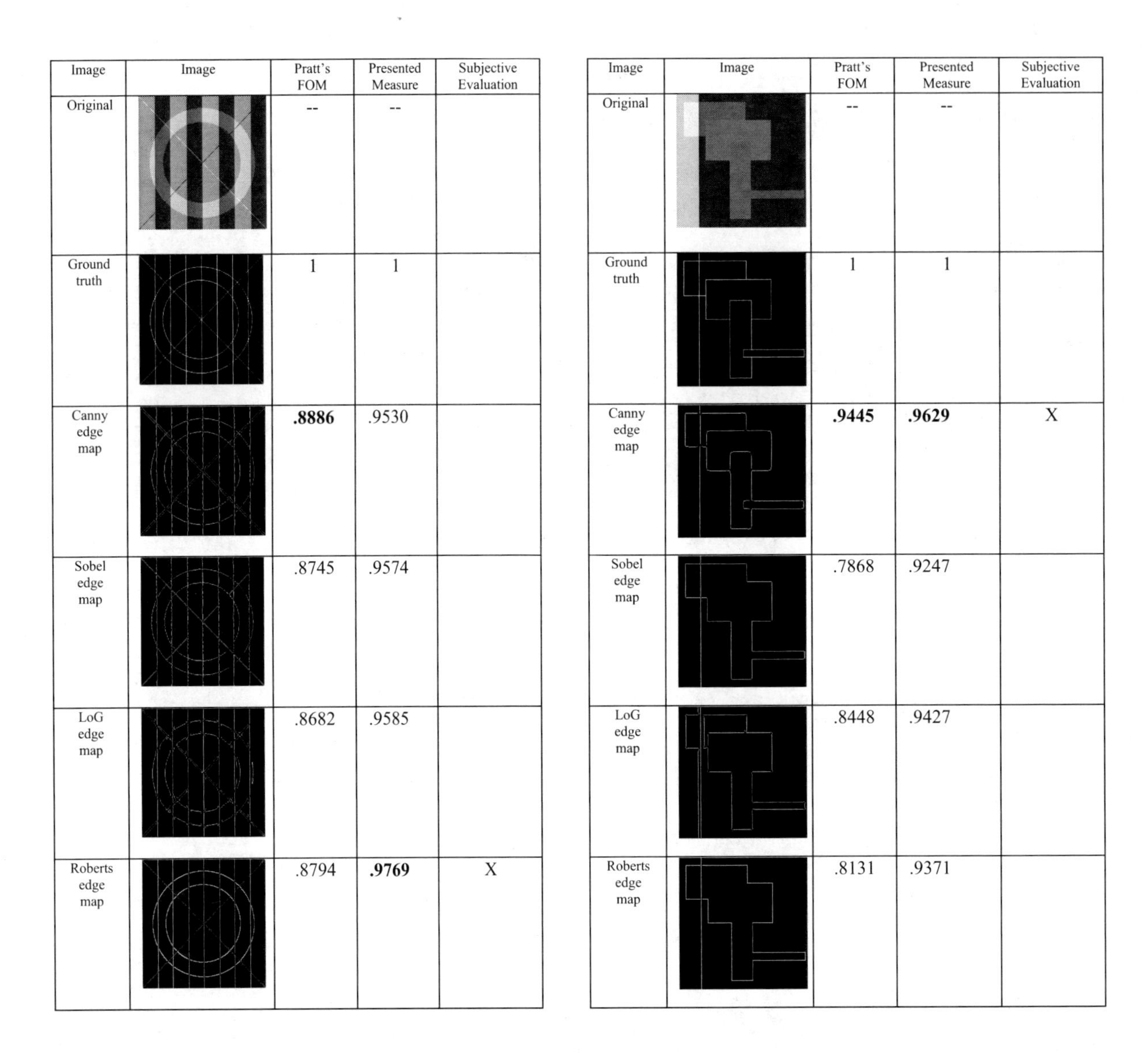

Image	Image	Pratt's FOM	Presented Measure	Subjective Evaluation
Original		--	--	
Ground truth		1	1	
Canny edge map		**.8886**	.9530	
Sobel edge map		.8745	.9574	
LoG edge map		.8682	.9585	
Roberts edge map		.8794	**.9769**	X

Image	Image	Pratt's FOM	Presented Measure	Subjective Evaluation
Original		--	--	
Ground truth		1	1	
Canny edge map		**.9445**	**.9629**	X
Sobel edge map		.7868	.9247	
LoG edge map		.8448	.9427	
Roberts edge map		.8131	.9371	

Figure 11 – Comparison of Pratt's FOM and the presented measure to subjective evaluation to determine the best edge detector output for a given synthetic image with ground truth

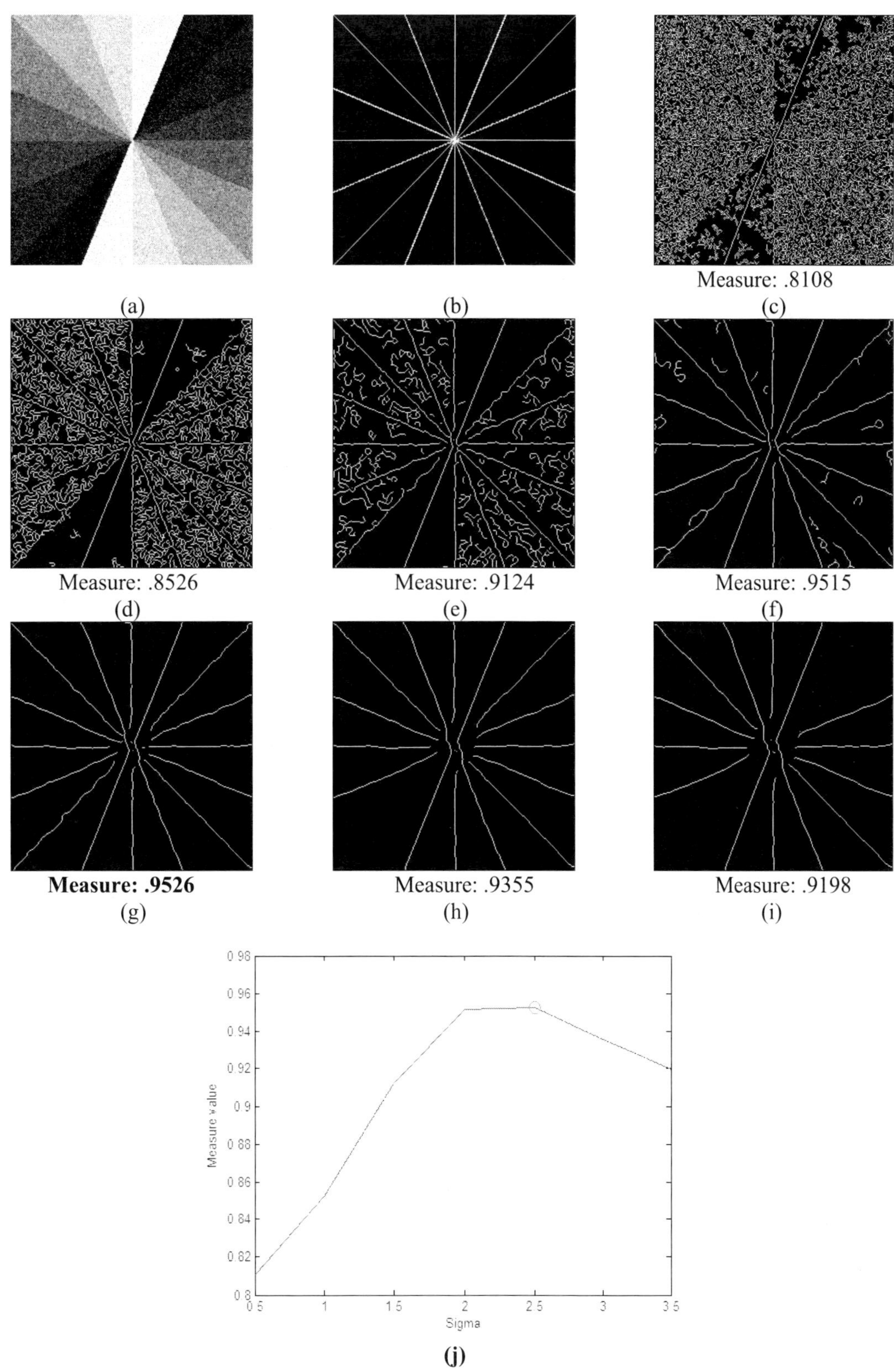

Figure 12 – (a) Image with Gaussian noise added, (b) ground truth, edge maps obtained using Canny with (c) $\sigma = .5$, (d) $\sigma = 1$, (e) $\sigma = 1.5$, (f) $\sigma = 2$, (g) $\sigma = 2.5$, (h) $\sigma = 3$, (i) $\sigma = 3.5$, (j) measure plot indicating a maximum at $\sigma = 2.5$

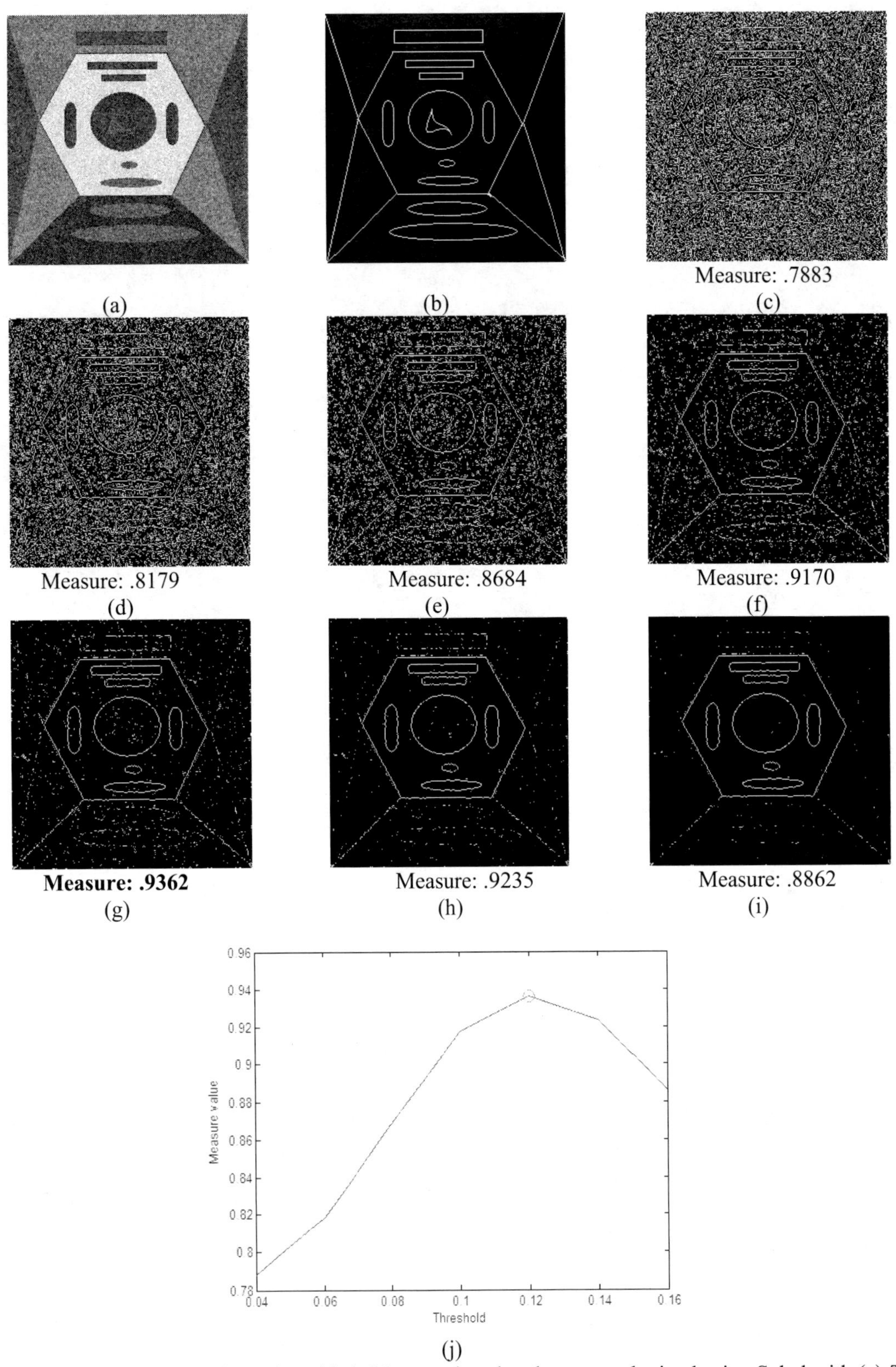

Figure 13 – (a) Image with Gaussian noise added, (b) ground truth, edge maps obtained using Sobel with (c) $T = .04$, (d) $T = .06$, (e) $T = .08$, (f) $T = .10$, (g) $T = .12$, (h) $T = .14$, (i) $T = .16$, (j) measure plot indicating a maximum at $T = .12$

3D fuzzy-directional processing to impulse video color denoising in real time environment

Alberto J. Rosales-Silva[*a], Volodymyr Ponomaryov[*a], Francisco Gallegos-Funes[*b]
[a]National Polytechnic Institute of Mexico, ESIME Culhuacan.
[b]National Polytechnic Institute of Mexico, ESIME Zacatenco.

ABSTRACT

It is presented a robust three dimensional scheme using fuzzy and directional techniques in denoising video color images contaminated by impulsive random noise. This scheme estimates a noise and movement level in local area, detecting edges and fine details in an image video sequence. The proposed approach cares the chromaticity properties in multidimensional and multichannel images. The algorithm was specially designed to reduce computational charge, and its performance is quantified using objective criteria, such as *Pick Signal Noise Relation*, *Mean Absolute Error* and *Normalized Color Difference*, as well visual subjective views. Novel filter shows superiority rendering against other well known algorithms found in the literature. Real-time analysis is realized on Digital Signal Processor to outperform processing capability. The DSP was designed by Texas Instruments for multichannel processing in the multitask process, and permits to improve the performance of several tasks, and at the same time enhancing processing time and reducing computational charge in such a dedicated hardware.

Keywords: Fuzzy, Directional Processing, Video Color, Impulsive Noise, Real-Time.

1. INTRODUCTION

There are several algorithms used for impulsive noise suppression in two dimensions in literature [1-6]. Here, a robust scheme is presented that process in two and three dimensions (2D, 3D), providing good numerical results. 2D developed algorithm outperforms other 2D filters used as comparatives ones that confirms in Numerical Results Section. With good capabilities results obtained under proposed 2D filter, it is possible to use it as a first step (*t-1*) in a temporal algorithm to provide good reference values to be processed joint with a subsequent frame (*t*). Complete 3D algorithm is described in Section 2.3, where it is used the 2D filter as a final step to suppress the non-stationary impulsive noise left by the temporal algorithm. The designed techniques use fuzzy theory and directional processing to suppress random impulsive noise. They demonstrate good preservation capabilities, such as edges, fine details and chromaticity properties.

One of the 2D filters used as comparative was the developed by Schulte et. al (INR filter) [1]. This technique uses fuzzy logic theory to calculate the color information detecting and removing impulsive noise, preserving inherent capabilities of the color images. Proposal scheme FTSCF-2D (Fuzzy Two Step Color Filter) works using not only fuzzy logic techniques but also directional processing too, in this way it is outperformed the rendering of Schulte's algorithm as can be seen in Numerical Results Section. Other 2D implemented algorithms used as comparative ones are: AMNF, AMNF2 (Adaptive Non-Parametric Filters) [2]. They employ nonparametric methodologies to adapt to local data in the color image; AMN-VRMKNNF(AMNF3) (Adaptive Multichannel Non-Parametric Vector Rank M-type K-nearest Neighbor Filter) [3]; GVDF (Generalized Vector Directional Filter) [4]. The presented above filters use directional techniques efficiently in 2D images. Another technique applied in CWVDF (Centred Weighted Vector Directional Filters) [5] minimizes the sum of weighted angular distances to other input samples from the filtering window. This filter takes advantage of adaptive stack filters design and weighted median filtering framework, finally VMF_FAS (Vector Median Filter Fast Adaptive Similarity) [6], this method is based in order statistics techniques.

Respects to 3D, some 2D algorithms were adapted to process in 3D, such as: MF-3F (Median Filter 3-Frames), VGVDF [7], VVMF (Video Vector Median Filter) [7], that uses directional techniques as a ordering criteria, and

* arosaless@ipn.mx or vponomar@ipn.mx: phobe/fax (525)6562058; IPN. Escuela Superior de Ingeniería Mecánica y Eléctrica, U.P. Culhuacan; Av. Santa Ana 1000, Col. San Francisco Culhuacan, 04430, México D.F.

Mobile Multimedia/Image Processing, Security, and Applications 2009, edited by Sos S. Agaian, Sabah A. Jassim,
Proc. of SPIE Vol. 7351, 73510M · © 2009 SPIE · CCC code: 0277-786X/09/$18 · doi: 10.1117/12.814982

VVDKNNVMF (Video Vector Directional K-nearest Neighbor Vector Median Filter) [8], where there are exploited the ideas of order statistics and directional processing. Algorithms designed directly in 3D are: KNNF (K-nearest neighbor filter) [9], where the nearest neighbor pixels provide an efficient solution in 3D; VATM (Video Alpha Trimmed Mean) [9], that presents a modification in 3D of the algorithm in 2D, and VAVDATM (Video Adaptive Vector Directional Alpha Trimmed Mean) [10], that connects the directional techniques ideas with adaptive procedures and order statistics techniques. All these filters use well known techniques: directional and order statistics techniques.

Images of video sequences used are in RGB color band space. In simulation, each frame is contaminated by impulsive random noise applied in each channel in independent way. After, it is processed in each R, G, and B channel using fuzzy and directional robust techniques to obtain parameters employed in spatial filtering. Values from spatial filtered image are used in temporal denoising algorithm to compute noise and movement levels under *fuzzy-directional* techniques using filtered spatial image (*t-1*) and corrupted image (*t*). Finally, the present frame is processed using the *fuzzy-directional* parameters computed to denoising the pixels in present frame.

Extensive simulations demonstrate that the proposed filter consistently outperforms well filtering video color sequences in different noise levels. We tested color images, and video sequences corrupted by different levels of impulsive random noise. Color Images “Lena”, “Peppers” and “Baboon”, and video color sequences “Flowers” and “Miss America” are well known data bases presenting different nature in pixel distributions and different color distribution. Color images are characterized by 320x320 pixels in RGB color space in true color (24 bits), and for video sequences we use video conference format QCIF with 176x144 pixels for each a frame in true color. Real-Time analysis was realized on the DSP (TMS320DM642) manufactured by Texas Instruments (TI). It was designed to process big data quantities in a real time environment. We based our approach using a Reference Framework defined as RF5 [11-12] designed by TI too. Here we use RGB color space in *false color* (5-bits in Red, 6-bits in Green, and 5-bits in Blue channels) representation for real-time video sequences in QCIF format.

2. SPATIAL METHOD DESCRIPTION

2.1 Fuzzy Two Step Color Filter Two Dimension (FTSCF-2D)

Because of the absolute differences (*gradients*), and the angle deviations (*directions*) characterize similarity among pixels, it is designed a procedure to exploit these values to estimate if the central pixel is a noisy or a free noise one. Procedure to achieve this ideas is to calculate for each a cardinal direction (*N, E, S, W, NW, NE, SE, SW*), respect to the central pixel. The parameter values denoted as “gradient values” defined as $\left|x_c^{\beta}(i,j)-x^{\beta}(i+k,j+l)\right| = \nabla_{(k,l)}^{\beta}x(i,j)$ with $(i,j)=(0,0)$, for each sliding window, where β represents the *R*, *G*, and *B* channel image components, (k,l) pair firstly corresponds for each one of the eight cardinal directions respect to the central pixel value, and their values are $\{-1,0,1\}$ as it was illustrated in Fig. 1. The eight gradient values associated with different directions are denominated as “basic gradient values”. As it is established in [13-14] (for only “two related gradient” values), to avoid smoothing, it is proposed to use not only one “basic gradient value” for each direction, but also four “related gradient values” too, where (k,l) pair takes values of $\{-2,-1,0,1,2\}$. Parameter ∇_{γ}^{β} is used to refer all the gradient values obtained for each a cardinal direction, where γ represents each one of the computed cardinal directions. The detailed procedure only for the “SE” direction is illustrated in Fig. 1. In this way, the gradients obtained for “SE” are computed as follows: $\nabla_{(1,1)}^{\beta}x(0,0)=\nabla_{SE(basic)}^{\beta}$, $\nabla_{(0,2)}^{\beta}x(i-1,j+1)=\nabla_{SE(rel1)}^{\beta}$, $\nabla_{(2,0)}^{\beta}x(i+1,j-1)=\nabla_{SE(rel2)}^{\beta}$, $\nabla_{(0,0)}^{\beta}x(i-1,j+1)=\nabla_{SE(rel3)}^{\beta}$, and $\nabla_{(0,0)}^{\beta}x(i+1,j-1)=\nabla_{SE(rel4)}^{\beta}$. The same procedure is used to obtain the rest of the cardinal direction gradient values, where ∇ is defined as:

$$\nabla_{\gamma=SE(basic)}^{\beta}=\left|x_{(0,0)}^{\beta}-x_{(1,1)}^{\beta}\right|. \tag{1}$$

Respect to angle deviations (vectorial values), it is proceeded to obtain, under normalized values among [0,1] the angle deviation for each a channel in an independent way. These values are computed using a methodology developed only for this algorithm where it has to omit two of the three channels that compose a color frame. It is obtained for the “SE” direction, the “basic” and “related” vectorial values described as: $\theta_{(1,1)}^{\beta}x(0,0)=\theta_{SE(basic)}^{\beta}$, $\theta_{(0,2)}^{\beta}x(i-1,j+1)=$

$\theta^{\beta}_{SE(rel1)}$, $\theta^{\beta}_{(2,0)}x(i+1,j-1)=\theta^{\beta}_{SE(rel2)}$, $\theta^{\beta}_{(0,0)}x(i-1,j+1)=\theta^{\beta}_{SE(rel3)}$, and $\theta^{\beta}_{(0,0)}x(i+1,j-1)=\theta^{\beta}_{SE(rel4)}$. Here, the angle deviation "θ" is computed as:

$$\theta^{\beta}_{\gamma=SE(basic)}=\cos^{-1}\left\{\frac{\varepsilon+x^{\beta}_{(0,0)}\cdot x^{\beta}_{(1,1)}}{\sqrt{\varepsilon+\left(x^{\beta}_{(0,0)}\right)^2}\cdot\sqrt{\varepsilon+\left(x^{\beta}_{(1,1)}\right)^2}}\right\}, \tag{2}$$

where $\varepsilon=2\cdot(255)^2$. Figure 1 illustrates the directions involved as so as the *basic* and *related* components.

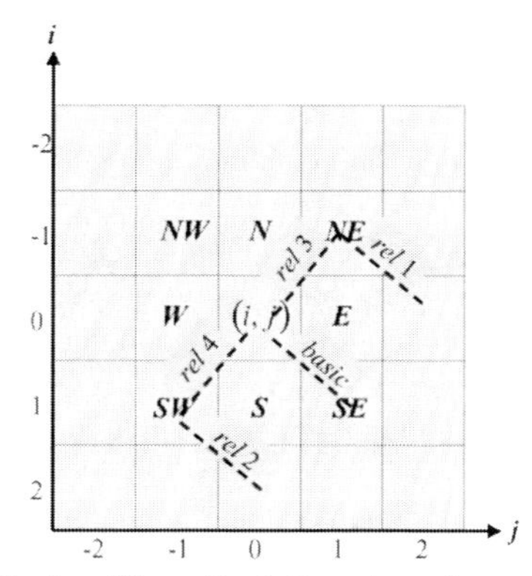

Figure 1. "Basic" and "Related" cardinal directions for vectorial and gradient values.

2.2 Fuzzy Sets and Membership Functions

Let define two fuzzy sets, BIG and SMALL, to characterize the noise level presence in the central pixel of the sample 5×5 denoted in Fig. 1. A big membership degree (≈ 1) in SMALL fuzzy set represents a free noise central pixel; by the contrary, if a big membership degree is present in BIG fuzzy set, it represents a noisy central pixel for sure. To compute membership degrees it is proposed to use a Gaussian membership function. This function is used to compute fuzzy gradient values using gradient values:

$$\mu_{\nabla^{\beta}_{\gamma=SE(basic)}(SMALL,BIG)}=\begin{cases}1, & if\left(\nabla^{\beta}_{\gamma}<med2,\nabla^{\beta}_{\gamma}>med1\right)\\ \left(\exp\left(-\left\{\frac{\left(\nabla^{\beta}_{\gamma}-med2\right)^2}{2\sigma_1^2}\right\}\right),\exp\left(-\left\{\frac{\left(\nabla^{\beta}_{\gamma}-med1\right)^2}{2\sigma_1^2}\right\}\right)\right), & otherwise\end{cases}, \tag{3}$$

where $med1=60, med2=10,$ and $\sigma_1^2=1000$.

Let proceed also to computing the fuzzy vectorial values:

$$\mu_{\theta^{\beta}_{\gamma}(SMALL,BIG)}=\begin{cases}1, & if\left(\theta^{\beta}_{\gamma}<med3,\theta^{\beta}_{\gamma}>med4\right)\\ \left(\exp\left(-\left\{\frac{\left(\theta^{\beta}_{\gamma}-med3\right)^2}{2\sigma_2^2}\right\}\right),\exp\left(-\left\{\frac{\left(\theta^{\beta}_{\gamma}-med4\right)^2}{2\sigma_2^2}\right\}\right)\right), & otherwise\end{cases}, \tag{4}$$

where $med3=0.8, med4=0.1,$ and $\sigma_2^2=0.8$, values obtained experimentally in agree to the best PSNR and MAE criteria.

2.3 Fuzzy Rules and Denoising Scheme

The corruption level is estimated using two fuzzy rules applied for gradient and vectorial values for each a channel in independent way.

1) **Fuzzy Rule 1** defines the membership level of $x^{\beta}_{(i,j)}$ in fuzzy set BIG for each γ direction, that is: IF (∇^{α}_{γ} is BIG AND $\nabla^{\beta}_{\gamma(rel1)}$ is SMALL AND $\nabla^{\beta}_{\gamma(rel2)}$ is SMALL AND $\nabla^{\beta}_{\gamma(rel3)}$ is BIG AND $\nabla^{\beta}_{\gamma(rel4)}$ is BIG) AND (θ^{β}_{γ} is BIG AND $\theta^{\beta}_{\gamma(rel1)}$ is SMALL AND $\theta^{\beta}_{\gamma(rel2)}$ is SMALL AND $\theta^{\beta}_{\gamma(rel3)}$ is BIG AND $\theta^{\beta}_{\gamma(rel4)}$ is BIG) THEN $\nabla^{\beta F}_{\gamma}\theta^{\beta F}_{\gamma}$ (*gradient-vectorial*

values) is BIG. The AND operator outside of the parenthesis is defined as $\min(A,B)$, and inside of the parenthesis is defined as $A \text{ AND } B = A * B$.

2) **Fuzzy Rule 2** if we have the eight fuzzy vector-directional values computed for each cardinal direction, we proceed to calculate the noisy factor, that is: IF $\nabla_N^{\beta F}\theta_N^{\beta F}$ is BIG OR $\nabla_S^{\beta F}\theta_S^{\beta F}$ is BIG OR $\nabla_E^{\beta F}\theta_E^{\beta F}$ is BIG OR $\nabla_W^{\beta F}\theta_W^{\beta F}$ is BIG OR $\nabla_{SW}^{\beta F}\theta_{SW}^{\beta F}$ is BIG OR $\nabla_{NE}^{\beta F}\theta_{NE}^{\beta F}$ is BIG OR $\nabla_{NW}^{\beta F}\theta_{NW}^{\beta F}$ is BIG OR $\nabla_{SE}^{\beta F}\theta_{SE}^{\beta F}$ is BIG THEN r^β is BIG.

In this way we have computed the *fuzzy gradient-vectorial values* and the noisy factor. The noisy factor is used as a threshold to distinguish among a noisy or a free noise pixel. The OR operation is defined as a triangular co-norm and is equal to $\max(A,B)$. We use Fig. 2 to perform the denoising process agree the noisy factor r^β computed in **Fuzzy Rule 2**.

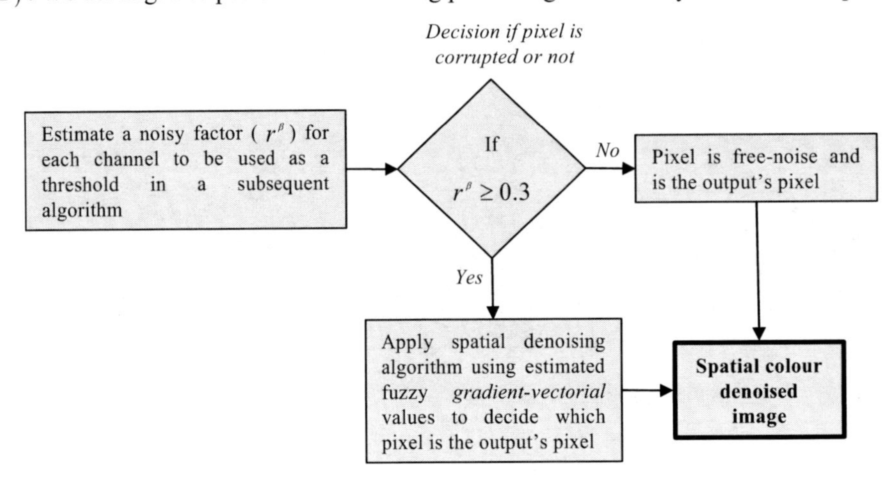

Figure 2. Processing Diagram of spatial algorithm proposed **FTSCF-2D**.

Following Fig. 2, we use r^β as a threshold to define if the central pixel is corrupted or not. If $r^\beta \geq 0.3$, it is realized a robust filtering using fuzzy *gradient-vectorial* values as the weights, in other case, the output's filtered is $y_{out}^\beta = x_{(i,j)}^\beta$.

In case of $r^\beta \geq 0.3$, the weight fuzzy values are used in a standard negator ($\varsigma(x) = 1 - x$, with $x \in [0,1]$) defined as $\rho(\nabla_\gamma^{\beta F}\theta_\gamma^{\beta F}) = 1 - \nabla_\gamma^{\beta F}\theta_\gamma^{\beta F}$, where $\nabla_\gamma^{\beta F}\theta_\gamma^{\beta F} \in [0,1]$, this value origins the fuzzy membership value in a new fuzzy set defined as "NO BIG" (*free noise* values) where their membership values indicate the free noise degree. Let propose a weight fuzzy *gradient-vectorial* value for the central pixel in NO BIG fuzzy set as $\rho(\nabla_{(0,0)}^{\beta F}\theta_{(0,0)}^{\beta F}) = 3 \cdot \sqrt{1 - r^\beta}$. We proceed to perform an ordering scheme to enhance the robust capabilities of the spatial denoising algorithm, the ordering permits to remove some values more outlying from the central pixel. Ordering scheme is defined as follows: $x_{\dot\gamma}^\beta = \{x_{SW}^\beta, \ldots, x_{(i,j)}^\beta, \ldots, x_{NE}^\beta\}$, $x_{\dot\gamma}^{\beta(1)} \leq x_{\dot\gamma}^{\beta(2)} \leq \cdots \leq x_{\dot\gamma}^{\beta(9)} \overset{implies}{\Rightarrow} \rho(\nabla_{\dot\gamma}^{\beta F}\theta_{\dot\gamma}^{\beta F})^{(1)} \leq \rho(\nabla_{\dot\gamma}^{\beta F}\theta_{\dot\gamma}^{\beta F})^{(2)} \leq \cdots \leq \rho(\nabla_{\dot\gamma}^{\beta F}\theta_{\dot\gamma}^{\beta F})^{(9)}$, where $\dot\gamma = ($ *N, E, S, W,* (i,j), *NW, NE, SE, SW*). Following procedure in Fig. 3, we can obtain the output's filtered value under fuzzy *gradient-vectorial* values and to perform ordering properties to enhance robust capabilities of our proposal.

The output's filtering is found selecting one of the neighbor or central components that avoids the smoothing of the image. $j \leq 2$ is chosen to avoid the selection of the farther pixels in the ordering scheme. If $j \leq 2$ condition is satisfied, it should be upgraded the total weight. It is selected the jth ordered value as the output's filtered value.

3. SPATIO-TEMPORAL METHOD DESCRIPTION

3.1 Fuzzy Two Step Color Filter Three Dimension (FTSCF-3D)

Three dimensional processing is achieved using FTSCF-2D as a first step in denoising the first frame of a video sequence (spatial denoising). After temporal denoising, FTSCF-2D will be used again to suppress non-stationary impulsive noise left in the temporal algorithm. There is only used the past and present frames in a video color sequence.

The last can decrease the computational charge and hardware requirements in comparison with three frames algorithms. This is illustrated in Fig. 4.

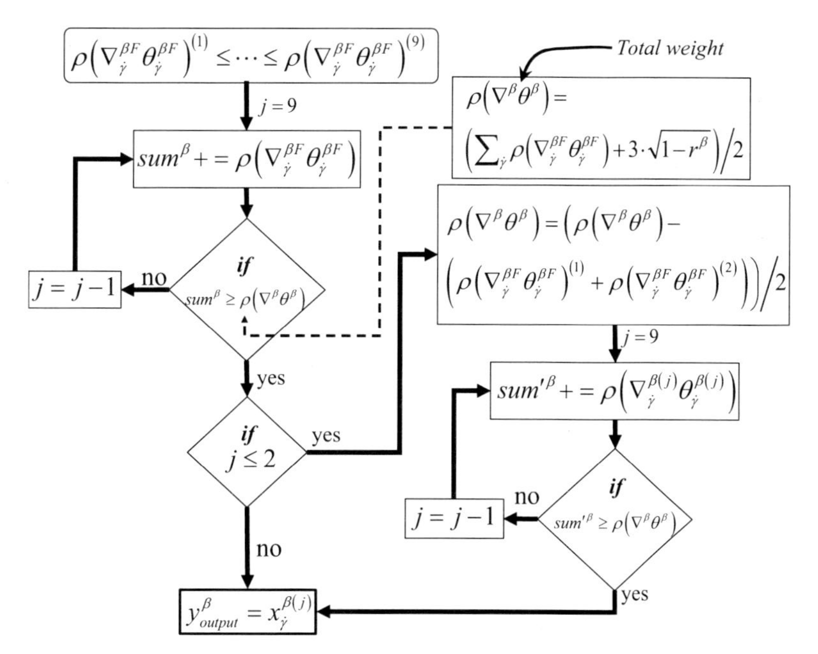

Figure 3. Spatial denoising scheme used in **FTSCF-2D** filter.

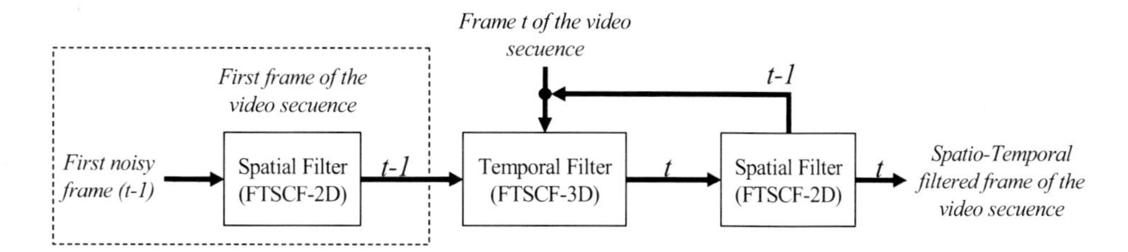

Figure 4. Generalized Block diagram of Temporal-Spatial Algorithm.

Using only two frames of a video sequence is possible to compute movement and noise levels of the central pixel, basically they are used the same procedures discussed in Section 2. The only difference here is in the pixel values, so, in the temporal algorithm we use a 5x5x2 sliding window forming by *past* and *present* frames, and compute a difference values among these frames using Eq. (5):

$$\lambda^{\beta}_{(k,l)}\left(A(i,j),B(i,j)\right)=\left|A(i+k,j+l)-B(i+k,j+l)\right|, \tag{5}$$

where $A(i,j)$ represents pixels in *t-1* frame, and $B(i,j)$ represents pixels in *t* frame, with $(k,l)\in\{-2,-1,0,1,2\}$. Fig. 5 illustrates the procedure to compute difference values among two frames defined as *gradient* values.

Using $\lambda^{\beta}_{(k,l)}$ (Fig. 5a) and Eq. (5)) denominated as *gradient difference value*, we obtain an *error frame* (Fig. 5b) and 5c)). We proceed to calculate the *absolute difference gradient values* of central pixel in respect to its neighbors now for a 5x5x1 window processing. Performing Eq. (6) the *absolute difference gradient values* are computed. Procedure described in Eq. (6) is shown only for the *SE*(*basic*) direction, the same procedure realized in Section 2.1 (for ∇^{β}_{γ}) is achieved for all the other *basic* and *related* values in all cardinal directions (operations in $\nabla^{\beta}_{SE(basic)}$ are symmetrical to the achieved by $\nabla'^{\beta}_{SE(basic)}$):

$$\nabla'^{\beta}_{(1,1)}\lambda(0,0)=\nabla'^{\beta}_{SE(basic)}, \text{ where } \nabla'^{\beta}_{SE(basic)}=\left|\lambda^{\beta}_{(0,0)}-\lambda^{\beta}_{(1,1)}\right|, \tag{6}$$

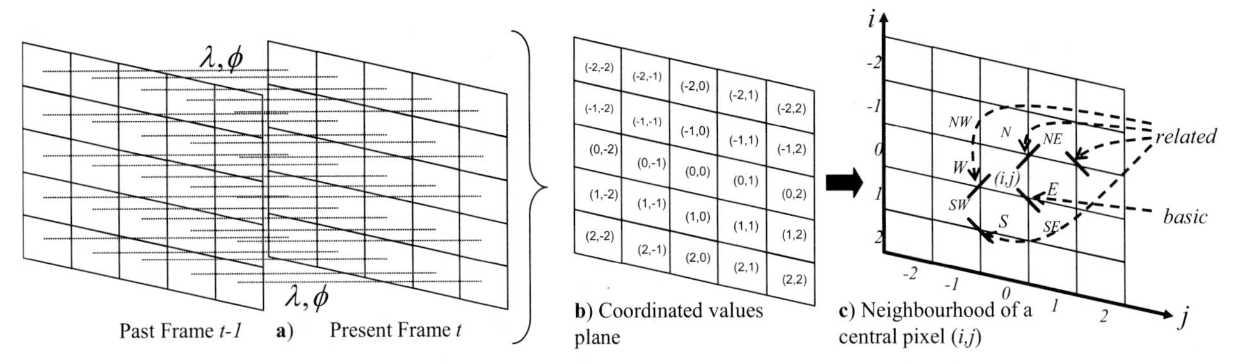

Figura 5. a) Processed frames; b) Coordinated plane; c) Neighborhood of central pixel.

The next step is computing the *absolute difference vectorial values* of central pixel in respect to its neighbors. First of all, it is performed an *angle difference value* among *t-1* and *t* frames, this obey the next expression:

$$\phi^{\beta}_{(k,l)}\left(A(i,j),B(i,j)\right)=\cos^{-1}\left\{\frac{\varepsilon+A(i+k,j+l)\cdot B(i+k,j+l)}{\sqrt{\varepsilon+\left(A(i+k,j+l)\right)^2}\cdot\sqrt{\varepsilon+\left(B(i+k,j+l)\right)^2}}\right\}, \tag{7}$$

Using $\phi^{\beta}_{(k,l)}$, denominated as *angle difference value* (denoting angle deviation), in the next Equation, we can obtain the *absolute difference vectorial values*:

$$\nabla''^{\beta}_{(1,1)}\phi(0,0)=\nabla''^{\beta}_{SE(basic)},\text{ where }\nabla''^{\beta}_{SE(basic)}=\left|\phi^{\beta}_{(0,0)}-\phi^{\beta}_{(1,1)}\right|, \tag{8}$$

The same reasoning done by $\nabla'^{\beta}_{SE(basic)}$ in respect to $\nabla^{\beta}_{SE(basic)}$ is realized by $\nabla''^{\beta}_{SE(basic)}$. The membership functions used to carry out fuzzy values are the same applied in Eq. (3) and Eq. (4). In this way we can compute the *fuzzy gradient-vectorial difference values*. The only difference is in the change of parameter values found experimentally in Eq. (4), these changes are in $med3=0.1$, $med4=0.01$, and $\sigma_2^2=0.1$, these values are proposed agree to the best experimentally PSNR, and MAE criteria results.

3.2 Fuzzy Rules and Denoising Scheme

Fuzzy rules used to characterize movement and noise levels in central pixel components are:

Fuzzy Rule 1: is defined the *FIRST fuzzy gradient-vectorial difference value* as $\left(\nabla'^{\beta F}_{\gamma}\nabla''^{\beta F}_{\gamma}\right)_{FIRST}$:

IF $\left(\nabla'^{\beta}_{\gamma}\right.$ is BIG AND $\nabla'^{\beta}_{\gamma(rel1)}$ is SMALL AND $\nabla'^{\beta}_{\gamma(rel2)}$ is SMALL AND $\nabla'^{\beta}_{\gamma(rel3)}$ is BIG AND $\nabla'^{\beta}_{\gamma(rel4)}$ is BIG$\left.\right)$ AND $\left(\nabla''^{\beta}_{\gamma}\right.$ is BIG AND $\nabla''^{\beta}_{\gamma(rel1)}$ is SMALL AND $\nabla''^{\beta}_{\gamma(rel2)}$ is SMALL AND $\nabla''^{\beta}_{\gamma(rel3)}$ is BIG AND $\nabla''^{\beta}_{\gamma(rel4)}$ is BIG$\left.\right)$ THEN $\left(\nabla'^{\beta F}_{\gamma}\nabla''^{\beta F}_{\gamma}\right)_{FIRST}$ is BIG. This fuzzy rule defines *movement and noise confidence* in the central pixel component including in the fuzzy rule the neighbor fuzzy values computed in determined γ direction. "AND" outside of parenthesis is defined as $\min(A,B)$.

Fuzzy Rule 2: it is defined the *SECOND fuzzy gradient-vectorial difference value* as $\left(\nabla'^{\beta F}_{\gamma}\nabla''^{\beta F}_{\gamma}\right)_{SECOND}$:

IF $\left(\nabla'^{\beta}_{\gamma}\right.$ is SMALL AND $\nabla'^{\beta}_{\gamma(rel1)}$ is SMALL AND $\nabla'^{\beta}_{\gamma(rel2)}$ is SMALL$\left.\right)$ OR $\left(\nabla''^{\beta}_{\gamma}\right.$ is SMALL AND $\nabla''^{\beta}_{\gamma(rel1)}$ is SMALL AND $\nabla''^{\beta}_{\gamma(rel2)}$ is SMALL$\left.\right)$ THEN $\left(\nabla'^{\beta F}_{\gamma}\nabla''^{\beta F}_{\gamma}\right)_{SECOND}$ is SMALL. This fuzzy rule defines *no movement confidence* in the central pixel component in determined γ direction, in this way we can determine if the central pixel component is a uniform region, an edge or a fine detail. "OR" is defined as $\max(A,B)$, and in fuzzy rules 1 and 2 "AND" inside of parenthesis is defined as $A*B$.

Fuzzy Rule 3: it is defined the *fuzzy noisy factor* r^{β}:

IF $\left(\left(\nabla'^{\beta F}_{SE}\nabla''^{\beta F}_{SE}\right)_{FIRST}\right.$ is BIG OR $\left(\nabla'^{\beta F}_{S}\nabla''^{\beta F}_{S}\right)_{FIRST}$ is BIG OR $\cdots$ OR $\left(\nabla'^{\beta F}_{N}\nabla''^{\beta F}_{N}\right)_{FIRST}$ is BIG$\left.\right)$ THEN r^{β} is BIG. In this way, using Fuzzy Rule 1 to compute the *fuzzy noisy factor* we can estimate the movement and noise level in central component using all fuzzy values determined for all cardinal directions.

Fuzzy Rule 4: it is defined the *no movement confidence factor* η^{β} :

IF $\left(\left(\nabla'^{\beta F}_{SE}\nabla''^{\beta F}_{SE}\right)_{SECOND}\right.$ is SMALL OR $\left(\nabla'^{\beta F}_{S}\nabla''^{\beta F}_{S}\right)_{SECOND}$ is SMALL OR $\cdots$ OR $\left(\nabla'^{\beta F}_{N}\nabla''^{\beta F}_{N}\right)_{SECOND}$ is SMALL$\left.\right)$ THEN η^{β} is SMALL. The same reasoning done by the Fuzzy Rule 3 is done for the *no movement confidence factor*.

Using r^{β} and η^{β} parameters effectively in the decision if the central pixel component is noisy, in movement or free of both of them; we have designed the next denoising algorithm using these fuzzy parameters in the process of decision. Fig. 6 illustrates this scheme.

As Fig. 6 exposes, it should be chosen the *j*-ths component pixel, which satisfies the conditions proposed. Such a methodology guarantees that edges and fine details are preserved due to ordering criterion applied and the selection of the nearest pixels in respect to the central one in *t-1* and *t* frames.

To enhance denoising capabilities of the filter, the final step the FTSCF-2D to suppress non-stationary noise left by the temporal filter is employed. The FTSCF-2D applied after the FTSCF-3D is the same algorithm proposed in Section 2, the only modifications are in their parameters, and these modifications are due to the processing of the *temporal filtered* image that means only in non-stationary noise processing:

1) Condition $r^{\beta} \geq 0.3$ changes to $r^{\beta} \geq 0.5$.
2) Total weight $\rho\left(\nabla^{\beta}\theta^{\beta}\right)=\left(\sum_{\dot{\gamma}}\rho\left(\nabla^{\beta F}_{\dot{\gamma}}\theta^{\beta F}_{\dot{\gamma}}\right)+3\cdot\sqrt{1-r^{\beta}}\right)/2$ is rearranged as $\rho\left(\nabla^{\beta}\theta^{\beta}\right)=\left(\sum_{\dot{\gamma}}\rho\left(\nabla^{\beta F}_{\dot{\gamma}}\theta^{\beta F}_{\dot{\gamma}}\right)+5\cdot\sqrt{1-r^{\beta}}\right)/2$.
3) The central weight $\rho\left(\nabla^{\beta F}_{(0,0)}\theta^{\beta F}_{(0,0)}\right)=3\cdot\sqrt{1-r^{\beta}}$, is modified as $\rho\left(\nabla^{\beta F}_{(0,0)}\theta^{\beta F}_{(0,0)}\right)=5\cdot\sqrt{1-r^{\beta}}$.
4) If condition $sum^{\beta} \geq \rho\left(\nabla^{\beta}\theta^{\beta}\right)$ until $\rho\left(\nabla^{\beta F}_{\dot{\gamma}}\theta^{\beta F}_{\dot{\gamma}}\right)^{(2)}$ is not satisfied, *total weight* is updated in such a way $\rho\left(\nabla^{\beta}\theta^{\beta}\right)=\left(\rho\left(\nabla^{\beta}\theta^{\beta}\right)-\rho\left(\nabla^{\beta F}_{\dot{\gamma}}\theta^{\beta F}_{\dot{\gamma}}\right)^{(1)}\right)/2$.

These changes are necessary to provide to guarantee the best performance in agree to the PSNR and MAE criteria results during simulation.

4. NUMERICAL RESULTS

Performance of the filters is measured under different commonly used criteria such as PSNR, MAE [15], and NCD [16-17]. Color images and Video sequences were contaminated artificially by different impulsive noise percentages in independent way for each a channel. QCIF format was used in these video sequences.

Real-time implementations were achieved on Digital Signal Processor (EVM DM642) manufactured by Texas Instruments (TI). This device provides the necessary tools in software and hardware requirements to implement rapidly the designs in its architecture and to evaluate processing capabilities for filters used. The designs were based on the Reference Framework RF5 [18] developed by TI specifically for multichannel applications, it permits to process multiple tasks in a parallel way outperforming time requirement to produce a result.

Table 1 presents the performance in noise suppression for 2D proposal algorithm against other existed filters exposing PSNR values for low and middle noise percentages. One can see the best performance by proposal until 15% of corruption for Lena color image.

Table 1 exposes the operation of novel 2D filter for Baboon color image giving the best results until 20% of corruption. This is an important result because this image is highly detailed and presents high variability in its chromaticity properties and texture. For Peppers color image, the best processing results are obtained for proposed algorithm until 10% of impulsive noise only. Finally, we can notice that our filter operates efficiently in low levels of corruption decreasing in its performance in agree noise levels increase.

We note in Table 2 that practically in all levels of impulsive noise for all tested color images, the best performance in preserving edges and fine details is obtained, that is because selecting a neighbor pixel as an output's pixel in the algorithm does not present any processing of pixels, this avoids smooth of data present in images.

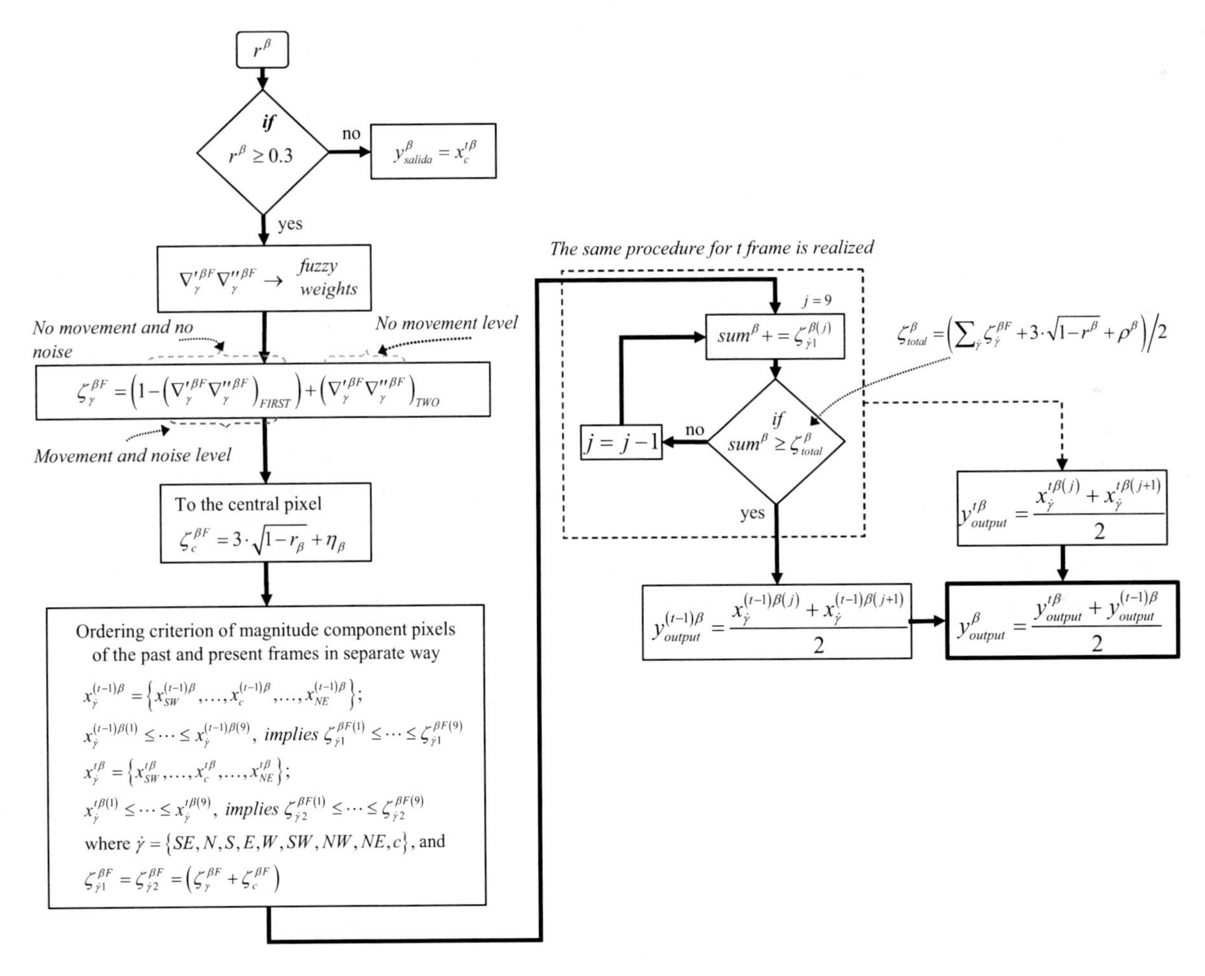

Figure 6. Denoising scheme applied in FTSCF-3D filter..

Table 1. 2D PSNR Numerical Results for Lena, Baboon, and Peppers Images

(%)Noise	FTSCF-2D	AVMF	MF	VMF	AMNF3	CWVDF	VMF_FAS	INR
				Lena				
0	**37,124**	31,578	31,344	30,471	29,45	33,047	***36,461***	31,714
2	**35,546**	31,314	31,031	30,3	29,34	32,151	***34,877***	31,577
5	**33,988**	30,953	30,615	30,072	29,211	31,237	***31,853***	31,452
10	**31,502**	30,098	29,685	29,458	28,934	29,042	28,803	***30,926***
15	***29,204***	29,063	28,642	28,641	28,59	26,606	26,276	**30,243**
20	26,857	27,832	27,203	27,58	***28,175***	24,3	24,799	**29,029**
30	22,712	24,885	24,344	24,831	**27,039**	20,318	22,189	***26,404***
				Baboon				
0	***29,185***	24,437	25,765	24,149	23,761	24,963	**30,143**	27,436
2	***28,641***	24,389	25,568	24,111	23,71	24,669	**29,267**	27,248
5	**27,851**	24,269	25,224	24,016	23,636	24,158	***27,216***	27,006
10	**26,602**	23,974	24,583	23,777	23,482	23,144	25,294	***26,389***
15	***25,239***	23,48	23,906	23,347	23,295	21,946	23,679	**25,693**
20	***23,724***	22,881	23,057	22,792	23,085	20,668	22,473	**24,949**
30	20,688	21,209	21,144	21,179	***22,481***	18,009	20,113	**23,028**
				Peppers				
0	**38,062**	31,546	31,91	30,884	29,465	32,872	***35,49***	32,563
2	**35,612**	31,284	31,569	30,681	29,249	31,47	***33,39***	32,095
5	**33,635**	30,808	31,018	30,304	29,023	29,745	31,194	***31,535***
10	**31,094**	29,796	29,995	29,438	28,705	27,339	29,009	***30,814***
15	28,363	***28,662***	28,63	28,438	28,316	24,463	25,627	**29,382**
20	26,065	27,299	26,991	27,193	***27,818***	22,124	24,454	**28,413**
30	21,563	24,015	23,453	23,995	**26,417**	18,257	21,531	***25,378***

Table 2. 2D MAE Numerical Results for Lena, Baboon, and Peppers Images

(%)Noise	FTSCF-2D	AVMF	MF	VMF	AMNF3	CWVDF	VMF_FAS	INR
				Lena				
0	***0,41***	1,891	3,555	4,078	4,847	2,626	**0,256**	4,345
2	***0,619***	2,093	3,687	4,166	4,922	2,828	**0,541**	4,374
5	**0,911**	2,392	3,869	4,29	5,034	3,099	***1,194***	4,413
10	**1,477**	2,97	4,25	4,57	5,224	3,817	***2,35***	4,529
15	**2,174**	***3,627***	4,73	4,922	5,459	4,873	3,699	4,7
20	**3,106**	***4,404***	5,394	5,414	5,743	6,375	5,004	4,984
30	**5,953**	6,53	7,318	7,035	6,687	10,823	8,091	***5,977***
				Baboon				
0	***2,136***	6,974	5,896	8,547	10,457	5,415	**0,942**	7,544
2	***2,417***	7,1	6,153	8,591	10,567	5,815	**1,4**	7,632
5	***2,867***	7,355	6,573	8,708	10,73	6,424	**2,542**	7,77
10	**3,669**	7,867	7,365	8,959	11,015	7,649	***4,062***	8,089
15	**4,628**	8,594	8,242	9,427	11,357	9,167	***5,828***	8,502
20	**5,832**	9,489	9,332	10,111	11,745	10,998	***7,693***	8,989
30	**9,284**	11,996	12,195	12,294	12,879	15,978	11,839	***10,52***
				Peppers				
0	***0,305***	1,505	2,321	2,861	4,165	1,315	**0,199**	3,622
2	**0,515**	1,677	2,461	2,957	4,267	1,633	***0,56***	3,676
5	**0,815**	1,969	2,684	3,136	4,417	2,15	***1,139***	3,751
10	**1,38**	2,488	3,115	3,494	4,657	3,196	***2,071***	3,898
15	**2,155**	***3,128***	3,691	3,945	4,921	4,818	3,703	4,177
20	**3,129**	***3,919***	4,456	4,529	5,259	6,92	4,84	4,469
30	6,538	***6,269***	7,001	6,579	6,484	12,976	8,287	**5,729**

Observing the capabilities of the algorithm in 3D processing; one can see in Table 3 the best numerical results in 3D processing for the proposed one in MAE criterion. In PSNR results, the best performances for the proposal are found for low, middle, and in some cases, in high noise levels, resulting in the best performance for all levels of impulsive noise. These results were obtained averaging the results of one hundred of the Flowers video sequence.

Table 3. 3D Averaging Criteria Results for Flowers video color sequence

(%) Noise	FTSCF-3D		MF_3F		VVMF		VVDKNNVMF		VGVDF		VAVDATM		VATM		KNNF	
	MAE	PSNR	MAE	PSNR	MAE	PSNR	MAE	PSNR	MAE	PSNR	MAE	PSNR	MAE	PSNR	MAE	PSNR
0	**1,576**	***30,53***	6,444	27,04	6,472	26,96	7,020	26,15	7,352	25,61	5,071	27,53	6,609	27,06	***3,198***	**33,13**
5	**2,134**	***29,52***	6,650	26,83	6,639	26,78	7,449	25,77	7,442	25,56	5,446	27,25	6,806	26,85	***3,979***	**31,30**
10	**2,726**	***28,61***	6,902	26,54	6,865	26,52	7,816	25,45	7,550	25,46	5,843	26,93	7,054	26,58	***5,155***	**28,95**
15	**3,377**	**27,76**	7,194	26,22	7,140	26,20	8,203	25,11	7,717	25,29	***6,278***	26,57	7,353	26,27	6,787	26,63
20	**4,112**	**26,93**	7,550	25,83	7,476	25,80	8,697	24,57	8,121	24,84	***6,771***	26,14	7,722	25,87	8,928	24,49
30	**6,083**	**25,04**	8,590	24,77	8,496	24,69	10,03	23,17	9,741	23,24	***8,132***	24,99	8,880	24,79	14,62	20,86
40	**9,154**	22,82	10,86	***22,82***	***10,67***	22,70	12,42	21,17	11,42	22,08	10,95	**22,99**	11,56	22,85	21,68	18,02

NCD criterion results in Table 4 characterizes chromaticity properties preservation, so, for designed filter we can observe the best numerical results for all levels of corruption in Flowers video sequences. These video sequences have variable texture and a high diversity in colors distributions.

Table 4. 3D Averaging NCD Criterion Results for Flowers video color sequence

(%)Noise	FTSCF_3D	MF_3F	VVMF	VVDKNNVMF	VGVDF	VAVDATM	VATM	KNNF
0	**0,003**	0,014	0,014	0,015	0,016	0,011	0,014	***0,006***
5	**0,004**	0,014	0,014	0,016	0,016	0,012	0,014	***0,008***
10	**0,006**	0,015	0,015	0,017	0,016	0,012	0,015	***0,010***
15	**0,007**	0,015	0,015	0,017	0,016	***0,013***	0,015	***0,013***
20	**0,009**	0,016	0,016	0,018	0,017	***0,014***	0,016	0,017
25	**0,010**	0,017	0,017	0,019	0,018	***0,015***	0,017	0,022
30	**0,012**	0,018	0,018	0,020	0,020	***0,017***	0,018	0,027

In less detailed video sequences (rarely presented in real environment), such as Miss America, designed filter produces the best filtering performance for low and middle noise levels showed in Table 5.

Table 5. 3D Averaging Criteria Results for Miss America video color sequence

(%)	FTSCF-3D		MF_3F		VVMF		VVDKNNVMF		VGVDF		VAVDATM		VATM		KNNF	
Noise	MAE	PSNR	MAE	PSNR	MAE	PSNR	MAE	PSNR	MAE	PSNR	MAE	PSNR	MAE	PSNR	MAE	PSNR
0	**0,037**	**48,544**	2,457	35,324	2,500	35,033	2,905	34,218	2,993	33,491	***0,821***	37,746	2,516	35,434	1,487	***40,412***
5	**0,372**	**39,591**	2,514	35,121	2,544	34,861	3,106	33,484	2,905	33,762	***1,112***	36,971	2,569	35,221	1,909	***37,217***
10	**0,738**	**36,616**	2,591	34,809	2,611	34,587	3,277	32,912	2,866	33,792	***1,410***	36,182	2,650	***34,884***	2,674	33,348
15	**1,180**	34,334	2,701	34,365	2,708	34,184	3,428	32,378	2,847	33,707	***1,714***	**35,383**	2,767	***34,421***	3,838	30,093
20	**1,757**	32,234	2,855	33,726	2,847	33,580	3,598	31,603	2,891	33,312	***2,044***	**34,479**	2,924	***33,794***	5,470	27,385
30	3,578	28,262	3,354	31,948	3,314	31,815	4,392	28,483	***3,283***	31,611	**2,852**	**32,334**	3,439	***32,034***	10,106	23,167
40	6,910	24,522	4,418	**29,151**	**4,326**	28,846	6,775	23,904	4,719	27,707	***4,351***	***29,142***	4,725	29,039	16,221	20,016

In Table 6, NCD criterion presents the best results for low and middle percentages of impulsive noise for novel filter. Miss America video color image is structured with low texture and low chromaticity properties, phenomenon rarely found in nature.

Table 6. 3D Averaging NCD Criterion Results for Miss America video color sequence

(%)Noise	FTSCF-3D	MF_3F	VVMF	VVDKNNVMF	VGVDF	VAVDATM	VATM	KNNF
0	**0,000**	0,009	0,009	0,011	0,011	***0,003***	0,009	0,006
5	**0,002**	0,009	0,009	0,011	0,011	***0,004***	0,009	0,007
10	**0,003**	0,010	0,009	0,012	0,010	***0,005***	0,010	0,010
15	**0,005**	0,010	0,010	0,012	0,010	***0,006***	0,010	0,014
20	***0,008***	0,010	0,010	0,012	0,010	**0,007**	0,010	0,020
30	0,016	***0,012***	***0,012***	0,015	***0,012***	**0,010**	***0,012***	0,038

To provide better evaluation capabilities, we propose to implement some of the algorithms on DSP to a Real-Time evaluation with real video sequences; this is to provide reliability of the proposed filter against some other robust algorithms found in scientific literature. Table 7 shows video processing times in 2D and 3D algorithms.

Table 7. 2D and 3D Time processing in a DSP

Algorithm	Frames	Total Time	Maximum Time (s)	Average Time (s)
		2D Algorithms		
MF	20	0.062	0.0037	0.0031
VMF	20	0.552	0.0283	0.0278
AMNF3	20	85.917	7.216	4.295
VMF_FAS	20	41.116	2.093	2.055
AVMF	20	1.591	0.0805	0.0796
GVDF	20	117.382	5.887	5.869
CWVDF	20	58.18	5.806	2.909
FTSCF-2D	**20**	**24.822**	**1.243**	**1.241**
		3D Algorithms		
MF_3F	20	0.114	0.0065	0.0057
VVMF	20	1.496	0.075	0.075
VATM	20	2.681	0.1347	0.134
KNNF	20	2.04	0.103	0.102
VGVDF	20	512.02	28.52	25.6
VAVDATM	20	497.356	25.551	24.867
FTSCF-3D	**20**	**148.806**	**7.533**	**7.440**

Figure 7. Original frames of Flowers and Miss America video color sequences and zoomed regions analyzed.

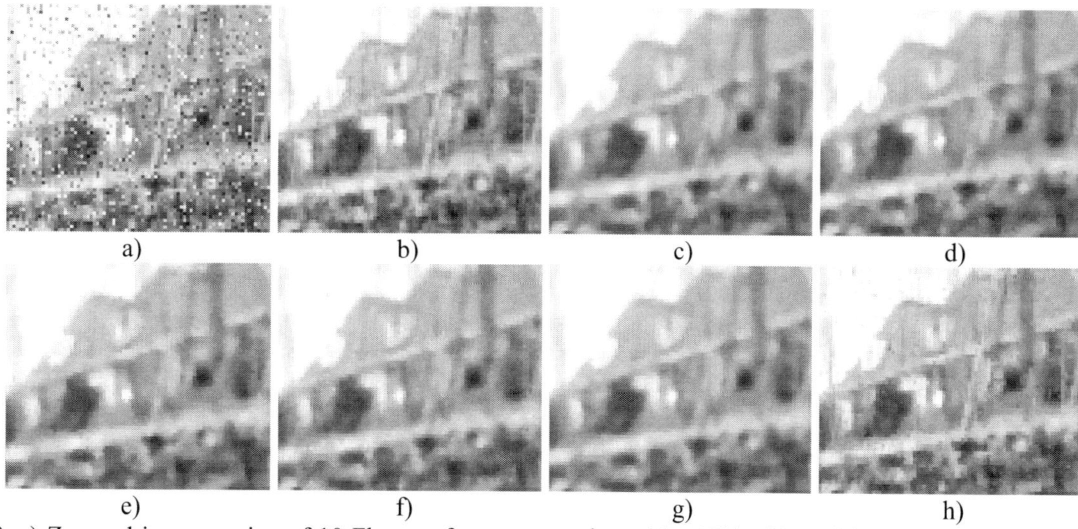

Figure 8. a) Zoomed image region of 10 Flowers frame contaminated by 15% of impulsive noise b) Proposal **FTSCF-3D**, c) MF_3F; d) VVMF, e) VGVDF, f) VAVDATM; g) VATM; h) KNNF.

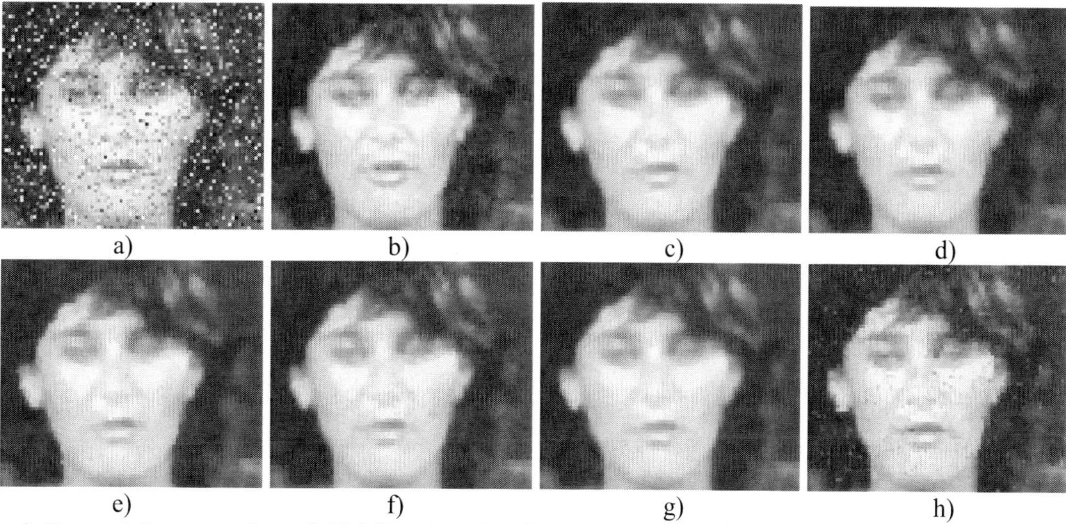

Figure 9. a) Zoomed image region of 10 Miss America frame contaminated by 15% of impulsive noise b) Proposal **FTSCF-3D**, c) MF_3F; d) VVMF, e) VGVDF, f) VAVDATM; g) VATM; h) KNNF.

In Fig. 8, we perceived in the zoomed filtered Flowers frame where novel filter preserves better the edges, fine details, and chromaticity properties against the other ones. The same comments are made for Fig. 9, where in Miss America frame one can observe the best preservation capabilities and noise suppression for novel 3D algorithm.

5. CONCLUSIONS

Novel FTSCF-2D filter has demonstrated a better performance in denoising color and multispectral images against existed robust techniques. FTSCF-3D outperforms rendering of 2D version as numerous simulation results indicate. Fuzzy and directional techniques are used in the design of the filter. Directional processing techniques takes the pixels as vectors providing angular deviations. It is demonstrated good capabilities achieved for proposed robust filters (FTSCF-2D and FTSCF-3D) under subjective and objective criteria. Real-Time evaluation of novel filter was efficiently implemented under RF5 Reference Framework Structure.

ACKNOWLEDGEMENTS

The authors would thank National Polytechnic Institute of Mexico and CONACYT for their support to realize this work.

REFERENCES

[1] Schulte S., Morillas S., Gregori V., Kerre E., "A New Fuzzy Color Correlated Impulse Noise Reduction Method," Trans. on Image Proc., 16(10), 2565-2575(2007).

[2] Plataniotis K. N., Androutsos D., Vinayagamoorthy S., Venetsanopoulos A. N., "Color Image Processing Using Adaptive Multichannel Filters," IEEE Transactions on Image Processing, 6(7), 933-949(1997).

[3] Ponomaryov VI, Gallegos-Funes FJ, Rosales-Silva A, "Real-time color imaging based on RM-Filters for impulsive noise reduction," Journal of Imaging Science and Technology, 49(3), 205-219(2005).

[4] Trahanias P. E., Venetsanopoulos A. N., "Vector Directional Filters. "A new class of multichannel image processing Filters," IEEE Trans. on Image Processing, Vol.2, 528-534(1993).

[5] Lukac R., Smolka B., Plataniotis K. N., Venetsanopoulos A. N., "Selection Weighted Vector Directional Filters," Comput. Vision and Image Understanding, Vol. 94, 140-167(2004).

[6] Smolka B., Lukac R., Chydzinski A., Plataniotis K. N., Wojciechowski W., "Fast Adaptive Similarity Based Impulsive Noise Reduction Filter," Real-Time Imaging, 9(4), 261-276(2003).

[7] Ponomaryov V, Rosales A., Gallegos F., Loboda I., "Adaptive Vector Directional Filters to Process Multichannel Images," IEICE Trans. on Fundamentals of Electronics Communications and Computer Sciences, E90-B (2), 429-430(2007).

[8] Ponomaryov V., Gallegos-Funes F., Rosales-Silva A., Loboda I., "3D Vector Directional Filters to Process Video Sequences," Lecture Notes in Computer Science, Springer Berlin, Heidelberg Vol. 4225/2006, ISBN 978-3-540-46556-0(2006).

[9] Zlokolica V., Philips W., Van De Ville D., "A new non-linear filter for video processing," in Proceedings of the third IEEE Benelux Signal Processing Symposium (SPS-2002), 221-224(2002).

[10] Ponomaryov V., Rosales A., Gallegos-Funes F., "3D Order Statistics Filters in Processing of Video Sequences," Signal Processing, Patterns Recognition, and Applications, Austria, 45-50(2007).

[11] Reference Frameworks for eXpressDSP Software: RF5, An Extensive, High Density System (SPRA795A).

[12] A Multi-Channel Motion Detection System Using eXpressDSP RF5 NVDK Adaptation (SPRA904).

[13] Schulte S., De Witte V. Nachtegael M., Van del Weken D., Kerre E., "Fuzzy two-step filter for impulse noise reduction from color images," IEEE Trans. Image Processing, 15(11), 3567-3578(2006).

[14] Zlokolica V., Schulte S., Pizurica A., Philips W., and Kerre E., "Fuzzy logic recursive motion detection and denoising of video sequences," Journal of Electronic Imaging, 15(2), 023008(2006).

[15] A. Bovik, [Handbook of Image and Video Processing], Academic Press, San Diego CA,(2000).

[16] Prat W. K., [Digital Image Processing], Wiley, 2nd Edition, New Cork, 1991.

[17] Poynton C. A., [Poynton's Color FAQ], Electronic Preprint(1995).

[18] Adapting the SPRA904 Motion Detection Application Report to the DM642 EVM.

Quality Based Approach for Adaptive Face Recognition

Ali J. Abboud, Harin Sellahewa and Sabah A. Jassim
Department of Applied Computing
The University of Buckingham, Buckingham, MK18 1EG, UK
{ali.abboud, harin.sellahewa, sabah.jassim}@buckingham.ac.uk

ABSTRACT

Recent advances in biometric technology have pushed towards more robust and reliable systems. We aim to build systems that have low recognition errors and are less affected by variation in recording conditions. Recognition errors are often attributed to the usage of low quality biometric samples. Hence, there is a need to develop new intelligent techniques and strategies to automatically measure/quantify the quality of biometric image samples and if necessary restore image quality according to the need of the intended application. In this paper, we present no-reference image quality measures in the spatial domain that have impact on face recognition. The first is called symmetrical adaptive local quality index (SALQI) and the second is called middle halve (MH). Also, an adaptive strategy has been developed to select the best way to restore the image quality, called symmetrical adaptive histogram equalization (SAHE). The main benefits of using quality measures for adaptive strategy are: (1) avoidance of excessive unnecessary enhancement procedures that may cause undesired artifacts, and (2) reduced computational complexity which is essential for real time applications. We test the success of the proposed measures and adaptive approach for a wavelet-based face recognition system that uses the nearest neighborhood classifier. We shall demonstrate noticeable improvements in the performance of adaptive face recognition system over the corresponding non-adaptive scheme.

Key words: Image Quality Measures, Face Recognition, Adaptive Quality Restoration, Wavelet Transform.

1. INTRODUCTION

Face recognition systems are designed to recognize persons based on physiological or behavioral characteristics. It is an important and challenging process that is affected by interclass variations. The face recognition has many applications such as security access control, personal ID verification, e-commerce, and video surveillance[1]. Face recognition systems need to be robust against changes in recording conditions due to variation in illumination, occlusion, expression, shadowing, background, low-resolution, image noise, and pose [2]. Therefore, there is a need to build reliable systems that work under different recording conditions.

Many algorithms have been developed and proposed to address the problem of variation in illumination conditions. These methods can be categorized into appearance-based, feature-based and hybrid approaches [3]. In feature based methods, a face is represented by geometrical measurements and relationships between automatically located significant facial feature points (such as eyes, mouth, nose and chin) and similarities between two face images are determined by holistic similarity measures. The success of these approaches relies heavily on accurate features location in the presence of extreme variation in lighting conditions. In appearance–based approaches, face images are represented by vectors in a high dimensional space, without consideration for the facial features that are usually subjected to dimension reduction techniques and/or some image pre-processing procedures. Methods within this approach include PCA, LDA, ICA, and wavelet-based schemes that are known to keep intra-class variations due to illumination and pose [4]. In the hybrid approach, the whole face modal and face features are used for recognition. These various approaches perform well under normal recording conditions but are affected by external factors such as extreme variations in illumination, expression, and occlusion. To deal with varying recording conditions, these schemes adopt normalization procedures that are applied irrespective of the recording conditions. Such strategies are known to improve accuracy in adverse conditions at the expense of deteriorated performance in somewhat normal recording conditions, and thereby yielding little or no improved overall accuracy. Therefore, there is a need to develop adaptive approaches to deal with such variations. In this paper, we investigate and develop quantitative quality measures for use in adaptive face recognition systems in the presence of extreme variation in illumination. The rest of the paper is organized as follows: section2 gives essential information about quality assessment measures. In section 3, explains new quality measures (SALQI, MH). Section 4

Mobile Multimedia/Image Processing, Security, and Applications 2009, edited by Sos S. Agaian, Sabah A. Jassim,
Proc. of SPIE Vol. 7351, 73510N · © 2009 SPIE · CCC code: 0277-786X/09/$18 · doi: 10.1117/12.817922

describes experimental database and algorithms. In section 5 experimental results are presented and section 6 is devoted for conclusions and future work.

2. QUALITY ASESSMENT MEASURES

Quality measures play an important role in improving the performance of biometric systems. The quality of biometric sample can be considered from three different views: (1) character: an indicator of inherent physical features, (2) fidelity: a measure of the degree of similarity with a reference signal, (3) utility: a reference to the impact on system performance [21]. In order to predict the utility of biometric sample for identification, we need some measures to calculate its quality [5, 6]. Quality measures can be classified as modality-dependent and modality-independent. Modality dependent measures (such as pose or expression) can be used for face biometric only, while modality-independent measures such as (contrast and sharpness) can be used for any modality because they do not need any knowledge about the specific biometric[7]. Quality measures can also be classified in terms of the availability of reference information: full reference, reduced reference, and no reference quality assessment approaches [8, 9].The biometric classification errors can be classified into: systematic, presentation-dependent, and user-dependent. Systematic errors are related to the design of pattern classification system. Presentation-dependent are those caused because of variations in the biometric signal. User-dependent happens only when some users do not fit within the population[7].

There is increasing interest by researchers in using quality information to make more robust and reliable recognition systems. Automatic face image quality assessment is an important step for other stages in the face recognition system. Face image quality measures must reflect some or all aspects variation from a "norm" in terms of lighting, expression, pose, contrast, eyes location, mouth location, ears location, red-eye, near/far, centered image, blur and so forth [10,11,12]. A variety of quantitative quality measures have been defined for different biometric modalities. Azeddine Beghdadi[13] developed new measures based on wavelet analysis tools for different biometrics. Also, Li Ma, Tieniu Tan[14] developed a new approach based on the wavelet transform which is used to select best image from a sequence of images and then used some classification method to arrange images in four classes, clear, defocused, blurred and occluded. Yi Chen, Sarat C. Dass, and Anil K. Jain[15] designed a new approach for measuring the quality of different regions of iris biometric and then used these measures as weights for the Hamming distance similarity measure. The usage of multibiometric can improve the accuracy of recognition systems because it is more robust against variations in the biometric quality and also it has more confidence pieces about the individual modalities. Furthermore, we can use the various biometric sample's qualities and accordingly adaptive weighting associated with the matching scores produced by their individual matchers[16, 25]. Quality measures also used to develop an adaptive multibiometric authentication score–level fusion algorithms[17]. The quality information can be exploited in many ways such as invoking different normalization algorithms, use different matchers, change threshold, score normalization and fusion [18]. For example, in [19] quality measures are used to develop an adaptive threshold scheme based on measuring the variation in the quality between enrolled and tested images. In this paper, we are mainly concerned with image quality measures that reflect image distortion as a result of variation in lighting conditions.

2.1 UNIVERSAL IMAGE QUALITY INDEX

Illumination image quality measures need to either reflect luminance distortion of a given image in comparison to a known reference image or regional variations in the image itself. The *universal image quality index* (*Q*) proposed by Wand and Bovik [22] does incorporate a number of image quality components fro which one can extract the necessary ingredients an illumination image quality measure that fits the above requirements. Let $X=\{x_i \mid i=1,2,.....N\}$ and $Y=\{y_i \mid i=1,2,..N\}$ be the original and the test image signals, respectively .The proposed quality index is defined as:

$$Q = \frac{4\,\sigma_{xy}\,\overline{X}\,\overline{Y}}{(\sigma^2_x + \sigma^2_y)[(\overline{x})^2 + (\overline{y})^2]} \quad \textbf{(1)}$$

Where,

$$\bar{x} = \frac{1}{N}\sum_{i=1}^{N} x_i \quad , \bar{y} = \frac{1}{N}\sum_{i=1}^{N} y_i \, , \sigma_x^{\,2} = \frac{1}{N-1}\sum_{i=1}^{N}(x_i - \bar{x})^2 \, , \sigma_{xy} = \frac{1}{N-1}\sum_{i=1}^{N}(x_i - \bar{x})(y_i - \bar{y})$$

This measure models any distortion as a combination of three components: loss of correlation, luminance distortion, and contrast distortion. In fact, Q is the product of three quality measures reflecting these components, respectively:

$$Q = \frac{\sigma_{xy}}{\sigma_x \sigma_y} \bullet \frac{2\,\overline{xy}}{(\bar{x})^2 + (\bar{y})^2} \bullet \frac{2\,\sigma_x \sigma_y}{\sigma_x^2 + \sigma_y^2} \quad \textbf{(2)}$$

In this paper, we only the luminance distortion component, i.e.:

$$\text{LQI} = \frac{2\,\bar{x}\,\bar{y}}{(\bar{x})^2 + (\bar{y})^2} \quad \textbf{(3)}$$

In practice, the LQI of an image with respect to another reference image is calculated for each window of size 8x8 pixels in the two images, and the average of all these blocks defines the LQI of the entire image.

In[25], Sellahewa and Jassim investigate the use of LQI for adaptive face recognition schemes with improved performance over the non-adaptive schemes. They proposed to use the LQI of face images in comparison to a fixed reference image that has a perceptually good illumination quality, as a pre-processing procedure prior to single-stream and all multi-streams wavelet-based face recognition schemes. In the case of multi-streams schemes, the LQI values are also used to adapt, the fusion weights. This paper continues the work towards adaptive face recognition by investigating other illumination quality measures. In particular, we shall propose a new approach for adaptive illumination normalization without using any reference image.

3. SALQI and MH QUALITY MEASURES

The adaptive normalization approach presented in[25] relies on the use of a good quality reference face image. However, in many instances, such a reference image may not be available or very hard to get [26]. For this reason, we shall present two novel illumination quality measures that do not need for any reference information. The first measure is based on the universal image quality index and the second is based on the idea of histogram partitioning. The quality scores of these measures are used to develop an adaptive approach for face illumination normalization that will be described in section 5. The two proposed quality measures are introduced in the reminder of this section.

3.1 SALQI QUALITY MEASURE

The direction of light source with respect to the camera axis has a significant effect on the appearance of face biometric samples. The images in the extended Yale B database can be divided into 3 groups: in the first group light is evenly distributed in some images (i.e. good quality); in the second group there is shadowing on either the left or the right side of images; and in the third group shadowing is distributed on the entire image (i.e. bad quality). This is a consequence of non-symmetric illumination distribution between left and right halves of face images. Therefore, we have developed a new quality measure based on the universal image quality index to measure the left-right lighting symmetry of a face image. For given face image (I), we define the symmetrical adaptive local quality index (SALQI) as follows.:

1. Divide I into left and right half subimages, I_L and I_R respectively.
2. Let I_{FR} be the horizontal flipping of I_R .
3. Starting from the top left corner, use equation (3) to compute LQI of the 8x8 windows in I_{FR} with respect to the corresponding windows in I_L.
4. After calculating the quality map $\{m_i = LQI_i: i=1,\dots,N\}$, we use a pooling strategy as indicated in equations (4) and (5) to calculate the final quality-score of the image (I) as a weighted average of the m_i's:

$$Q = \frac{\sum_{i=1}^{N} m_i * w_i}{\sum_{i=1}^{N} w_i} \quad (4)$$

Where, $w_i = g(x_i, y_i)$, **and**

$$g(x, y) = \sigma_x^2 + \sigma_y^2 + C \quad (5)$$

Here, $x_i = I_{L,i}$ and $y_i = I_{FR,i}$, where $I_{FR,i}$ is the mirrored block of $I_{L,i}$ of a row. The σ_x , σ_y are the standard deviations of X and Y, respectively, and C is a constant representing a baseline minimal weight.

The idea behind equation (4) is that regions of high quality have high energy (high variance) and are likely to contain more information. Since the image has non-uniform quality distribution [23] , different weights are assigned to different regions in the image according to their importance in the image space. Other weighting strategies can be used.

Fig.1 summarizes how the proposed SALQI approach is applied to a given face image to compute its illumination quality. The value range of SALQI is [0, 1] and its equals 1 if and only if the quality of left and right regions is equal..,

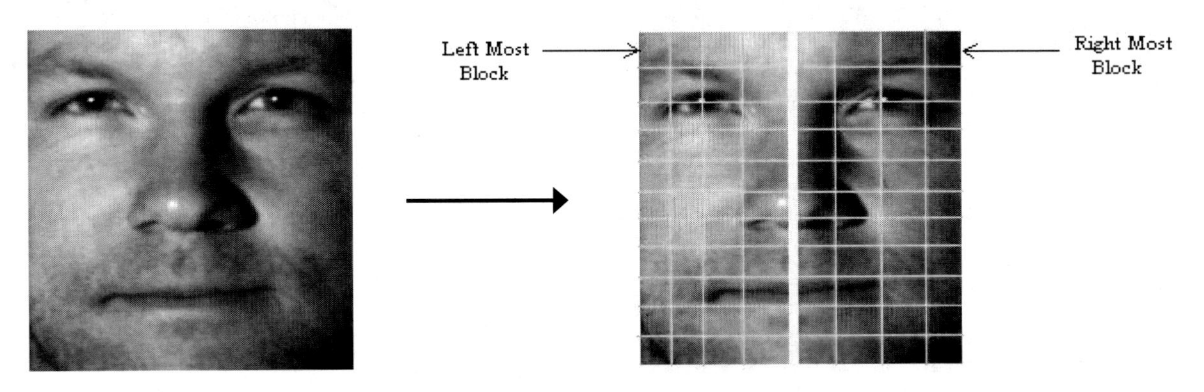

Fig.1 Face Image is divided into left and right regions.
Each region is divided into a set of blocks of size (8X8 pixels).

3.2 MH QUALITY MEASURE

The quality score of SALQI provides an indication of how symmetrical the light is distributed between the left and right regions of given image. However, this measure on its own could not effectively distinguish between well-lit face images, where shadowing artifacts are evenly distributed on both side of the face, SALQI produces high quality scores for such images. To overcome this problem, we developed a second quality measure the middle half (MH).

The histogram of good quality face images (e.g. Fig .3 (a) & (b) below), pixel intensities are mostly distributed in the middle part of their histogram (see Fig 2.A,B) and a reasonable amount of intensities are distributed on the left and right sides of the histogram.. On other hand, the histogram of poor quality images (see Fig. 3 (c),(d) & (e)) show only a small number of pixels with mid-range intensity values. Based on this observation, we propose the following measure:

$$MH = \frac{Middle}{Bright + Dark} \quad (6)$$

Where,

Middle= number of pixels in the middle region of the histogram between a (Lower bound LB) to Upper bound UB.
Bright = number of pixels in the bright region of the histogram greater than UB
Dark = number of pixels in the dark region of the histogram less than LB

The best choices for the values of (LB) and (UB) can be determined empirically, and throughout our experiments we use 63 and 243, respectively. The value of MH ranges from 0 to Max = (M/2), where M is the size of the image. The larger the value of MH is, the better s the quality. Normally, the maximum value of MH depends on the used dataset, and in the Extended Yale B database it was estimated to be 36.6.

The quality scores based on (SALQI), (MH) and (LQI) for the example images in Fig. 3 are shown in table.2.

4. EXPERIMENTAL DATA AND ALGORITHMS

4.1 Database

We used the cropped face images of extended Yale B database [20] for the face recognition experiments to test the use of the proposed illumination quality measures as the basis for an adaptive approach to illumination normalization. The database has 38 subjects and each one, in frontal pose, has 64 images captured under different illumination conditions. Hence, the total number of images in the database is 2414 images. In addition to these frontal pose images, an ambient illumination image was captured but it is not used in our experiments. The images in the database are divided into five subsets according to the direction of the light-source from the camera axis as shown in Table.1. Samples of images taken from the Extended Yale B database are shown in the Fig.3.

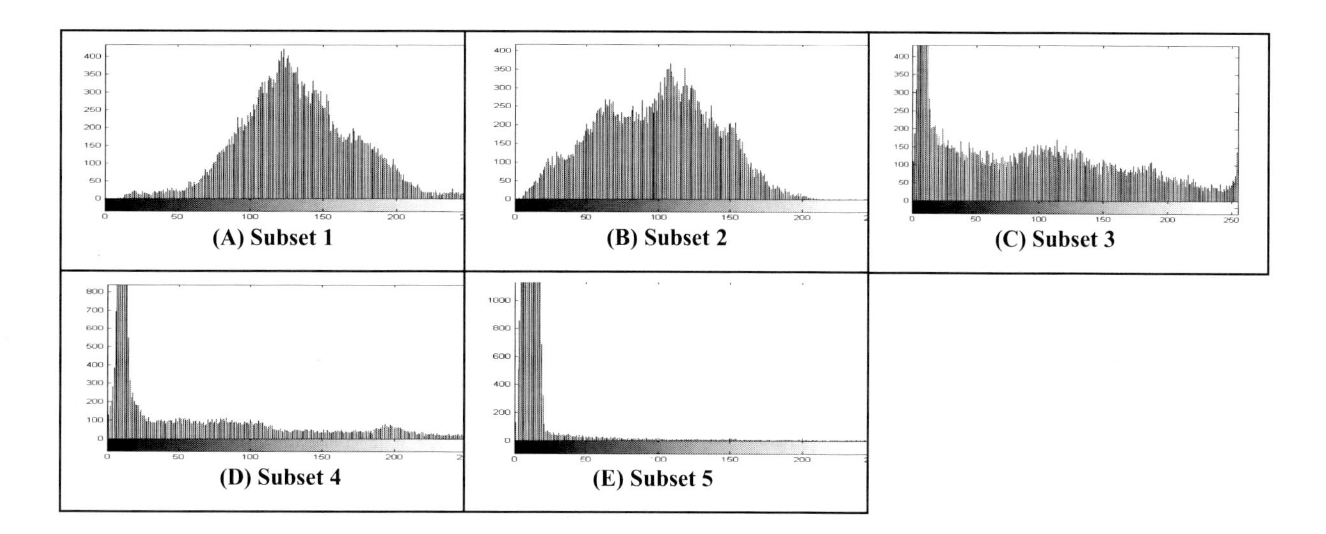

Fig. 2 Sample of image histograms in the spatial domain for different illumination subsets

Subsets	Angles	Image Numbers
1	$\theta < 12$	263
2	$20 < \theta < 25$	456
3	$35 < \theta < 50$	455
4	$60 < \theta < 77$	526
5	$85 < \theta < 130$	714

Table .1 Different illumination sets in the extended Yale B database

4.2 Face Recognition System

A wavelet-based face recognition system is used in our experiments [4,24]. It is based on the idea of using different subbands of a wavelet transformed face image as a face feature representation. During recognition, feature vectors of gallery and probe images are compared by calculating the distance between the two feature vectors (e.g. using City Block distance) and the identification of the probe image is based on nearest-. We followed the same procedure for feature representation and classification as used by Sellahewa and Jassim[4].

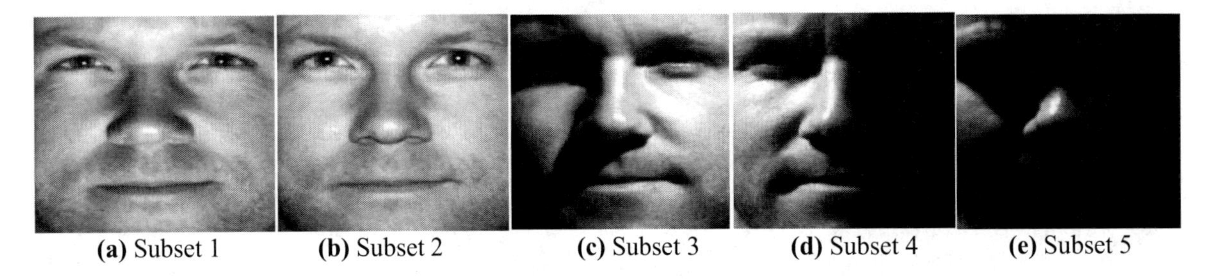

(a) Subset 1 **(b)** Subset 2 **(c)** Subset 3 **(d)** Subset 4 **(e)** Subset 5

Fig. 3 Sample images from different illumination subsets in the Extended Yale B

Quality Measure	Subset1	Subset2	Subset3	Subset4	Subset5
LQI	1	0.9838	0.8090	0.4306	0.2213
SALQI	0.9919	0.9690	0.6193	0.2975	0.4671
MH	24.344	25.992	5.649	0.4346	0.1142

Table.2 Quality scores of sample images

5. ADAPTIVE ILLUMINATION NORMALIZATION FOR FACE RECOGNITION

Having developed the new SALQI quality measure has been developed based on the LQI measure that does not need any reference image, we initially used the value of SALQI to decide adaptively to apply lighting normalization procedures on the whole image or not. But the results were almost similar to the study conducted in[25]. This led to development of the MH quality measure to complement the use of SALQI. Together these two measures are used to design a new adaptive approach to the use of the histogram equalization (HE) on whole image, on its left half, on its right half, or nowhere. This new approach, called symmetrical adaptive histogram equalization (SAHE), is designed specifically as the adaptive illumination normalization preprocessing that is usually needed for face recognition. The pseudo code for SAHE is given in Fig.4, below, and . Fig.5a shows the results when applied to the images of extended Yale B database. The SAHE performs better than all the previously developed approaches for adaptive normalization. Furthermore, we can see also that contrast has improved for some subsets of extended Yale B database as shown in the Fig5.b. Hence, we can draw two main advantages of using quality measures (1) avoidance of excessive unnecessary enhancement procedures that may cause undesired artifacts, and (2) reduced computational complexity, which is essential for real time applications. Furthermore, we observed that SALQI measure can separate the subset1 and subset2 of the extended Yale B database that do not need for any enhancement from other subsets. The MH measure separates subset 5 from all other subsets.

Finally, we designed a SAHE-adaptive wavelet-based recognition scheme and tested its performance in comparison to the corresponding versions with no normalization and with adaptive LQI. The latter wavelet-based face recognition system was in [25]. In our experiments, two different wavelets are used: Daubechie-1, and Daubechie-2 at three decomposition levels. For testing we used the Extended Yale B database, which includes face image recordings for 38 subjects. Each subject has 65 images, but the ambient image is excluded, therefore, the total number images per subject are 64. The dataset are divided into two groups: training set and testing set. The training set has (38) images, one image per subject, which is chosen to be (P00A+000E+00). The testing set consists of all the remaining (2394) images, i.e. 63 images per subject. In our experiments, different values of SALQI and MH quality measures are used as thresholds for

SAHE approach. The recognition results, displayed in Fig.6, has shown that the LH2 subband gives the best results under varying illumination and the error rate for (the SAHE with SLQI <0.6) is about 0.30% less than what was achieved before with AHE and LQI <0.8 threshold [25]. However, SAHE resulted in slightly increased error rates for subset 3 images while it reduced the errors of subset 4 and subset 5 and the results for LL2 features are significantly better, although these error rates are much higher than the errors with LH2. However, the improved results for LL2 may be used improve the score level fusion. Overall, we can say that there is noticeable improvement in the performance of face recognition system in terms of recognition accuracy.

1. Calculate the quality scores for the image (I) using (SALQI) and (MH)

2. If (SALQI < Thershold1) and (MH < Threshold 2) Then

IF (MH < Thershold3) Then

{Apply normalization algorithm on the whole image (I)}

Else if (MH >= Thershold3) Then

a. Apply HE on the left region of image (I) and compute SALQI

b. Apply HE on the right region of image (I) and compute SALQI

c. Apply HE on left and right regions of the image (I) and compute SALQI

Select the case that has higher SALQI value

End if

3. Else if (SALQI > Thershold1) and (MH >= Thershold2) Then

{Do not apply histogram normalization algorithm on image (I)}

4. End if

Feed image (I) to the face recognition system

Fig.4 Symmetrical Adaptive Histogram Equalization Algorithm

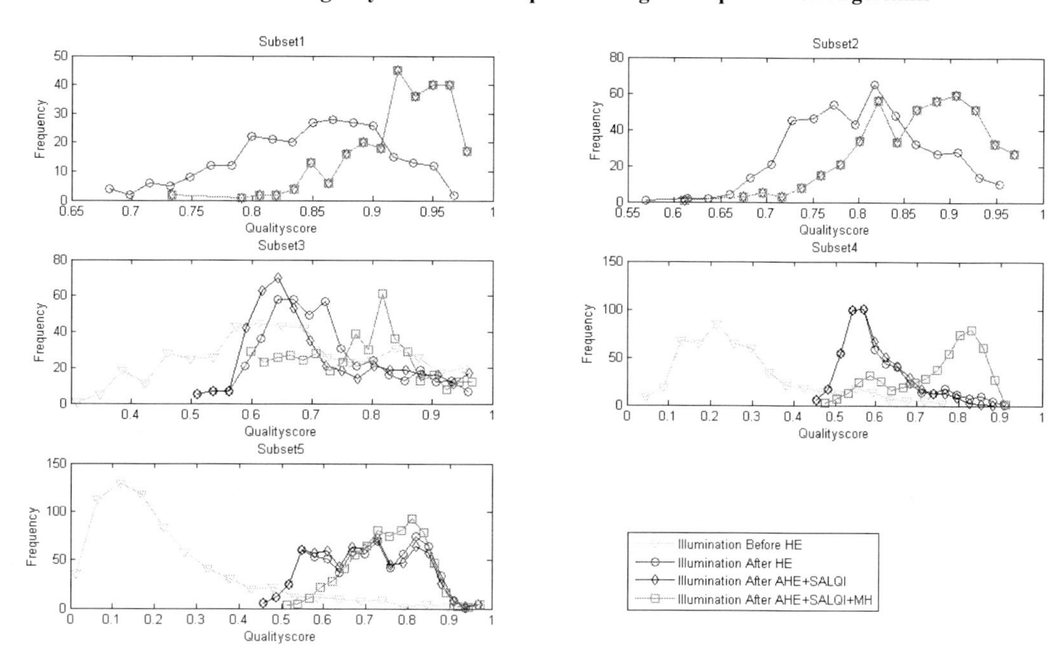

(a) Illumination

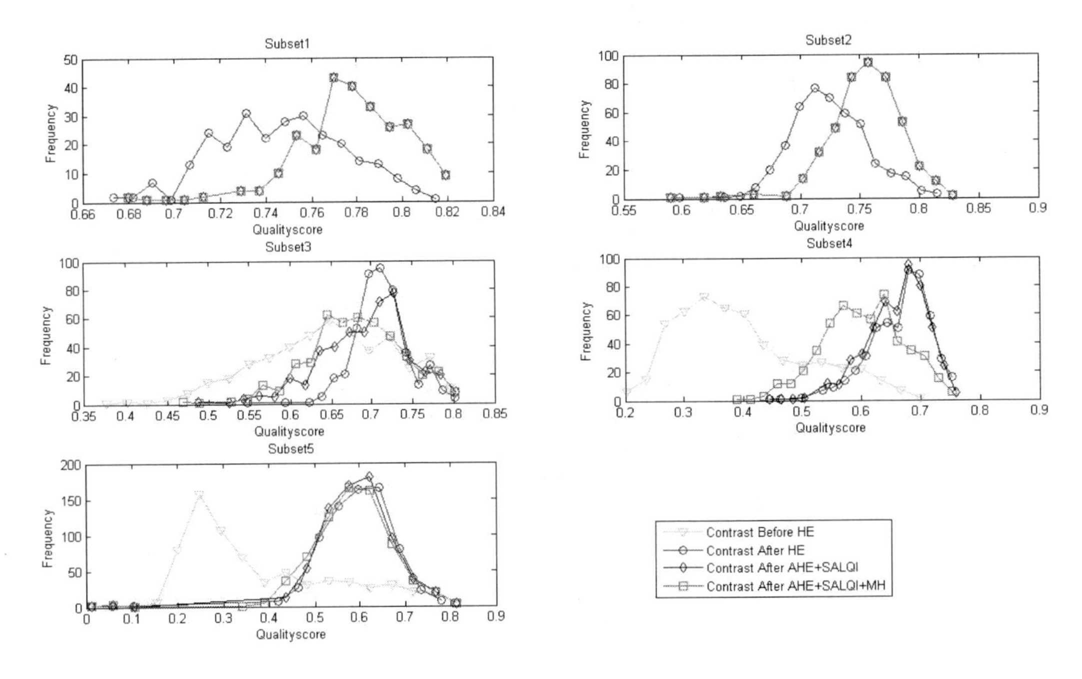

(b) Contrast

Fig .5 Illumination and contrast of extended Yale B database

	Set1	Set2	Set3	Set4	Set5	All
No pre-process	8.89	18.20	83.30	95.82	97.20	70.71
HE, ZN	3.11	25.88	70.99	90.11	85.57	64.52
AHE, LQI < 0.80	2.67	22.81	69.01	90.11	84.03	63.05
SAHE, SALQI < 0.60	2.67	7.89	37.8	73.76	76.61	48.36
SAHE, SALQI < 0.70	2.67	7.89	38.02	73.76	76.47	48.36
SAHE, SALQI < 0.80	2.67	20.83	40	73.76	76.47	51.22
SAHE, SALQI < 0.90	2.67	7.89	38.24	75.1	76.05	48.57

(a) Wavelet db1, suband: LL2

	Set1	Set2	Set3	Set4	Set5	All
No pre-process	8.00	0	30.5	71.1	95.2	50.97
HE, ZN	7.56	0.4	17.5	26.6	14.1	14.31
AHE, LQI < 0.80	7.11	0	11.6	20.3	11.7	10.94
SAHE, SALQI < 0.60	7.11	0	12.9	18.2	11.3	10.61
SAHE, SALQI < 0.70	7.11	0	12.9	18.6	11.4	10.73
SAHE, SALQI < 0.80	7.11	0	12.5	18.4	12.3	10.86
SAHE, SALQI < 0.90	7.11	0	12.7	18.6	11.3	10.65

(b) Wavelet db1, suband: LH2

	Set1	Set2	Set3	Set4	Set5	All
No pre-process	8.44	14.25	80.66	95.63	97.20	69.36
HE, ZN	1.78	20.83	67.47	90.30	85.71	62.84
AHE, LQI < 0.80	0.89	17.54	64.84	90.87	84.45	61.36
SAHE, SALQI < 0.60	0.89	4.61	30.99	72.05	77.03	46
SAHE, SALQI < 0.70	0.89	4.61	31.21	71.86	76.89	45.96
SAHE, SALQI < 0.80	0.89	15.79	33.19	72.05	77.03	48.57
SAHE, SALQI < 0.90	0.89	4.61	31.43	73.38	76.47	46.21

(c) Wavelet db2, suband: LL2

	Set1	Set2	Set3	Set4	Set5	All
No pre-process	14.6	0	35.6	66.3	89.6	49.83
HE, ZN	13.3	0	24.8	28.3	18.3	17.80
AHE, LQI < 0.80	13.3	0	21.7	22.2	16.1	15.19
SAHE, SALQI < 0.60	13.3	0	20.2	21.4	15.8	14.65
SAHE, SALQI < 0.70	13.3	0	20.2	21.4	15.8	14.65
SAHE, SALQI < 0.80	13.3	0	20.6	21.4	16.3	14.90
SAHE, SALQI < 0.90	13.3	0	20.2	21.2	15.6	14.56

(d) Wavelet db2, suband: LH2

Fig.6 Identification error rates of wavelet-based face recognition system

6. CONCLUSIONS AND FUTURE WORK

A current trend in biometric recognition is to augment biometric signal quality information in different stages of the biometric recognition system. In this paper, we investigated and developed new image quality measures that reflect the distribution of illumination distortions such as shadows that would have adverse effect on face recognition. These measures were used as the bases of a new approach for adaptive illumination normalization. The first quality measure is based on the universal quality measure but instead of relying on the use of a good quality reference image it exploit the symmetry of face to select one half of the face image as a reference. The second quality measure is based on the nature of histogram distribution. We tested the performance of these adaptive normalization procedures, and demonstrated much improved results. We also tested the performance of an adaptive wavelet-based face recognition scheme, which uses the symmetric adaptive histogram equalization and demonstrated noticeable improvement in recognition accuracy rate. Our future work will focus on developing new ideas about usage of the quality scores provided by quality measures in different ways to improve the accuracy of face recognition systems. Such actions may use quality scores to change the weight that is given for different modalities, match score normalization, selection of best feature extraction algorithms, features fusion, and decision fusion. We will try also to exploit these quality scores to adapt the fusion of face recognition algorithms such as (PCA, LDA, LFA, ICA, and DCTmode2) in what is called monomodality multialgorithm fusion and see which algorithms are better at specific quality conditions.

ACKNOWLEDGMENT

The first author would like to thank the ministry of higher education and scientific research in Iraq for funding this work as part of his PhD studies.

7. REFERENCES

[1] Dmitry O. Gorodnichy,"Face Databases and Evaluation", (Online at http://www.videorecognition.com/doc), 5-10(2008).

[2] Enrique, "Quality-Based Score Normalization and Frame Selection for Video–based Person Authentication", Proc. Biometric and identity management 5372, 1-9 (2008).

[3] Yang, Lai, Chang, "Robust Face Image Matching under Illumination Variations", EURASIP journal on Applied Signal Processing 2004(16), 2533–2543 (2004).

[4] Harin Sellahewa, Sahah A. Jassim, "Illumination and Expression Invariant Face Recognition: Toward Sample Quality –based Adaptive Fusion", in Biometrics: Theory, Applications and Systems, Proc. Second IEEE Int'l Conference 10414693, 1-6 (2008).

[5] Andy Adler, Tatyana Dembinsky, "Human VS. Automatic Measurement of Biometric Sample Quality ", Proc. Conference of Electronic Computer Engineering in Ottawa, 7-10 (2006).

[6] H. Fronthaler, K. Kollreider, J. Bigun, "Automatic Image Quality Assessment with Application in Biometrics", Proc. Conference on Computer Vision and Pattern recognition Workshop, 30-35. (2006).

[7] Jonas Richiardi, Krzysztof Kryszczuk, Andrzej Drygajlo, "Quality Measures in Unimodal and Multimodal Biometric Verification ", Proc. 15th European Conference on Signal Processing EUSIPCO,2007.

[8] Z wang, Hamid R. Sheikh and Alan C. Bovik, [Objective Video Quality Assessment], CRC Press publishers, Furht and O. Marqure, 1041-1078 (2003).

[9] Hanghang Tong, Mingjing Li, Hong-Jiang Zhang, Changshui Zhang, Jingrui He. "Learning No-Reference Quality Metric by Examples", Proc. The 11th International Multi-Media Modelling Conferene 1550, 247- 254 (2005)

[10] Oriana Yuridia Gonzalez Castillo, "Survey about Facial Image Quality", Fraunhofer Institute for Computer Graphics Research, 10-15 (2005).

[11] Stefan Winkler, "Vision Models and Quality Metrics for Image Processing Applications", Ph.D. Thesis, (2000).

[12] Robert Yen, "A New Approach for Measuring Facial Image Quality", Biometric Quality Workshop II, Online Proc. National Institute of Standards and Technology, 7-8, (2007).

[13] Azeddine Beghdadi, "A New Image Distortion Measure Based on Wavelet Decomposition ", invited paper, Proc. of IEEE ISSPA2003, 1-4 (2003).

[14] Li Ma, Tieniu Tan," Personal Identification Based on Iris Texture Analysis", IEEE Transactions on Pattern Analysis and Machine Intelligence, 25(12), 20-25 (2003).

[15] Yi Chen, Sarat C. Dass, and Anil K. Jain, " Localized Iris Image Quality using 2-D Wavelets ", Proc. Advances in Biometrics International Conference, ICB 2006, 5-7 (2006).

[16] Sarat C. Dass, Anil K. Jain, "Quality-based Score Level Fusion in Multibiometric Systems", Proc. of the 18th International Conference on Pattern Recognition 4(1), 473 - 476 (2006).

[17] Julian, Joaquin, Josef Bigun,"Discriminative Multimodal Biometric Authentication Based on Quality Measures", Proc. Biometric Quality Workshop on Pattern Recognition, 38(5),777–779 (2005).

[18] International Technology Standards, "Working Draft 29794-1, Biometric Sample Quality – part1: Framework ", ISO/IEC JTC 1/SC 37 Biometrics, 5-10 (2006).

[19] Krzysztof Kryszczuk, Andrzej, "Gradient–based Image Segmentation for Face Recognition Robust to Directional Illumination", Proc. SPIE, 2005.

[20] Georghiades, A .S. and Belhumeur, P.N. and Kriegman, D. J, "From Few to Many: Illumination Cone Models for Face Recognition under Variable Lighting and Pose", IEEE Transactions on Pattern Analysis and Machine Intelligence"23(6), 643-660 (2001).

[21] Fernando, Julian,"A Comparative Study of Fingerprint Image-Quality Estimation Methods", IEEE Transactions on Information Forensics and Security,2(4), 734-743 (2007).

[22] Alan C. Bovik, Zhou Wang",A Universal Image Quality Index", IEEE Signal Processing Letters.9(.3), 81-84 (2002).

[23] Alan C. Bovik, Hamid Rahim Sheikh, Gustavo,"An Information Fidelity Criterion for Image Quality Assessment Using Natural Scene Statistics", IEEE Transactions on the Image Processing,14(12), 2117-2128 (2005).

[24] H.Sellahewa,"Wavelet–based Automatic Face Recognition for Constrained Devices", Ph.D. Thesis, University Of Buckingham, 10-20 (2006).

[25] H. Sellahewa, S. A. Jassim, "Image Quality-based Adaptive Illumination Normalization for Face Recognition", 2-3 (2008).

A New Approach for Direct Image Registration

Guy Brodetzki, Alexander Notik, Dan Azaria, Yaacov Krips*

Elisra Electronic Systems Ltd. Israel

Abstract

We present a novel algorithm for direct image registration, based on a generic description of the geometric transformation. The direct image registration algorithm consists of minimizing the intensity discrepancy between images. We propose the Gauss – Newton algorithm for the solution of this minimization problem. The method solves the optical flow equation iteratively to reduce the cost function. Registration was successfully performed on images taken from both aerial and ground platforms.

Introduction

Image registration is an important tool for many applications (for example, super-resolution, change detection, target tracking). Aerial platforms introduce difficulties for image registration due to high velocity, strong maneuvers and vibrations.

Today, it is common to perform image registration by using navigation data (6 degrees of freedom) to compensate for the platform maneuver. Using navigation data is limited to navigation errors, sensor vibrations and occasionally lack of navigation (for example, by GPS jamming).We propose an algorithm for image registration based only on the optical data.

Method

Using the optical flow principle, we propose an intensity model that assumes the total intensity of a pixel in unchanged from one frame to another as described in Equation 1.

---

* yaacovk@elisra.com

Mobile Multimedia/Image Processing, Security, and Applications 2009, edited by Sos S. Agaian, Sabah A. Jassim,
Proc. of SPIE Vol. 7351, 73510O · © 2009 SPIE · CCC code: 0277-786X/09/$18 · doi: 10.1117/12.818352

$$I(x,y) = J(x+\delta x, y+\delta y) + noise$$

Equation 1

Where I is the intensity image at time t_0 and J is the intensity image at time t_1. x and y are the pixel coordinates and δx and δy are the displacements of the pixels (velocity field).

We assume that the noise is a thermal noise (white noise with an average of zero)

Our goal is to estimate the velocity field between the two images. The estimation is done in order to minimize the log likelihood error function as described in Equation 2.

$$[\delta x, \delta y] = \arg\min\left(\sum_x \sum_y \left(I(x,y) - J(x+\delta x, y+\delta y)\right)^2\right)$$

Equation 2

In order to find δx and δy parameterization of the problem is needed. We propose a polynomial function to model to displacement and distortion between the two images.

Parameterization of the problem was first suggested by Bruss and Horn [1], they showed that the velocity field is compiled of two components, translation and rotation. The rotation component can be described by a 2nd degree polynom of x and y. The translation component can be described by a linear function of x and y, scaled by the distance from the sensor to the pixel's footprint (which depends on the topography of the observed area).

Assuming the topography of the observed area is approximately flat, it is possible to approximate the translation component to a polynomial function as well. Taking all assumption into account, we suggest using a polynomial model for the velocity field:

$$\delta x = \sum_{i=0}^{k} \sum_{j=0}^{l} C_{ij} x^i y^j, \delta y = \sum_{i=0}^{k} \sum_{j=0}^{l} D_{ij} x^i y^j$$

Equation 3

Now it is possible to write Equation 2 like this:

$$[C,D] = \arg\min\left(\sum_x \sum_y \left(I(x,y) - J\left(x + \sum_{i=0}^{n} \sum_{j=0}^{m} C_{ij} x^i y^j , y + \sum_{i=0}^{n} \sum_{j=0}^{m} D_{ij} x^i y^j \right) \right)^2 \right)$$

Equation 4

Where C and D are the coefficients matrices of the polynoms from Equation 3.

C, and D can be estimated using the Gauss-Newton method. Basically, the Gauss-Newton method replaces a non-linear least square optimization problem with an iterated linear least square optimization problem:

$$X^{n+1} = X^n + \Delta^n$$

$$\text{where } \mathrm{X}^{\mathrm{n}} = \begin{bmatrix} \mathrm{C}^{\mathrm{n}} \\ \mathrm{D}^{\mathrm{n}} \end{bmatrix}$$

$$\mathrm{X}^1 \text{ is a zero vector (zero - motion assupmtion)}$$

$$\Delta^n = \begin{bmatrix} \Delta\mathrm{C}^{\mathrm{n}} \\ \Delta\mathrm{D}^{\mathrm{n}} \end{bmatrix}$$

Looking at Equation 3**,** one can see that the sum is a linear function of coefficients. Using this feature, it is possible to separate the polynom into two polynoms, one of the coefficients from the nth iteration C^n, and one of the current iteration increment:

$$\begin{bmatrix} \delta x^n \\ \delta y^n \end{bmatrix} = \sum_{i=0}^{k} \sum_{j=0}^{l} X^n{}_{ij} x^i y^j + \sum_{i=0}^{k} \sum_{j=0}^{l} \Delta^n{}_{ij} x^i y^j$$

Equation 5

From the linearization of Equation 2, and using Equation 5, we get the following equations[2]:

$$\frac{\partial I(x^n, y^n)}{\partial x} \delta x^n + \frac{\partial I(x^n, y^n)}{\partial y} \delta y^n = -\left(I(x^n, y^n) - J(x^m, y^m) \right)$$

Equation 6

Equation 6 is similar to the standard optical flow equation and can be solved using various techniques[3].

We developed a direct image registration that solves the above equations and tested it on images taken from aerial platforms.

Results

Figure 1 and Figure 2 shows two images taken at different times from an aerial platform flying at 10 m/sec at 930 m above ground level while performing a 4.7 deg/sec maneuver. The scenery in the images is industrial and contains sharp edges and strong contrasts, mainly around roads and factories.

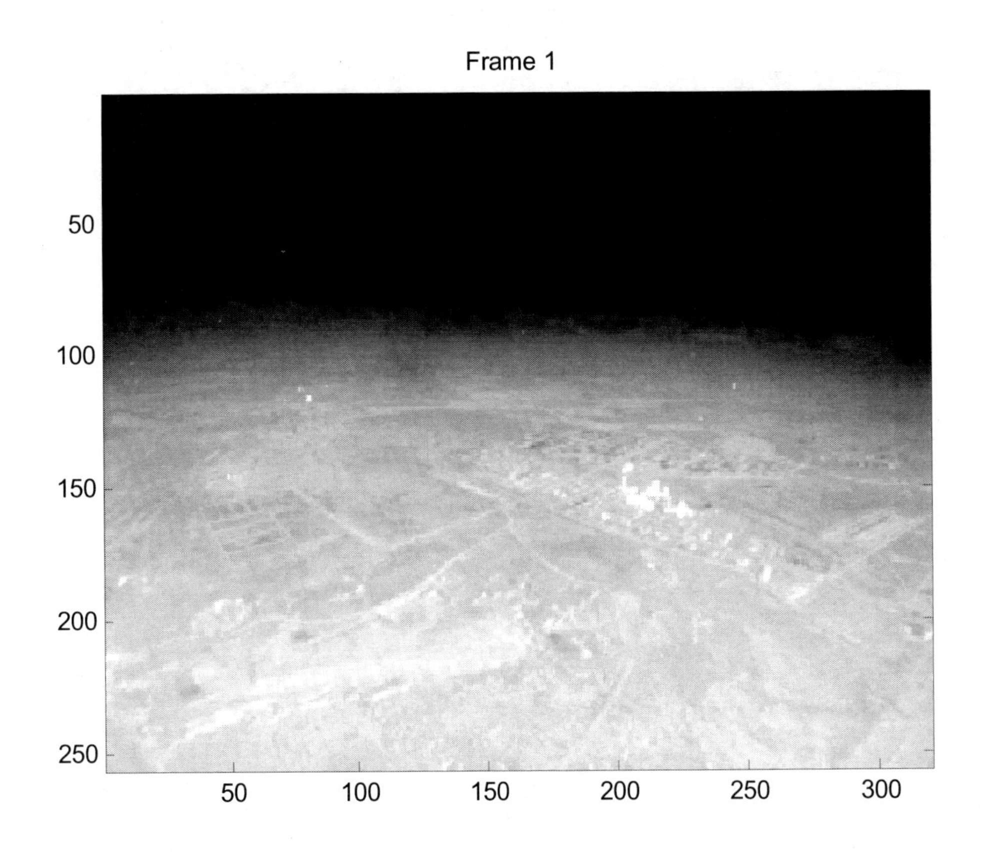

Figure 1. Image taken at time t_0

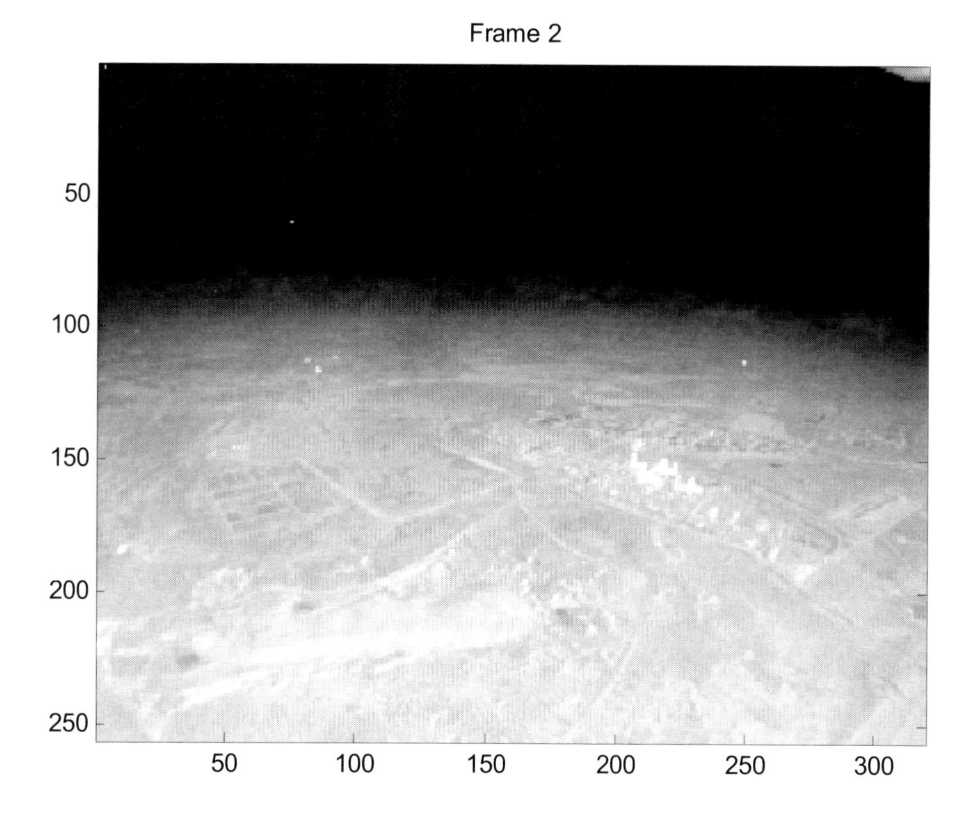

Figure 2. Image taken at time $t_1 = t_0 + \Delta t$ sec

The following figure shows the residual image obtained by subtracting the two images with no registration.

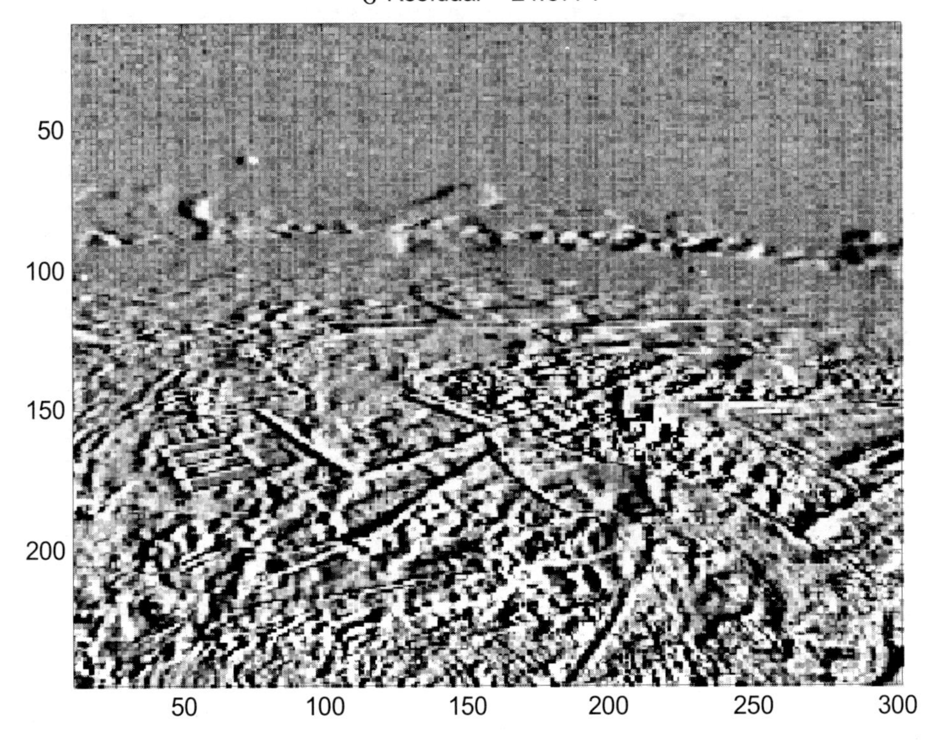

Figure 3. Residual image obtaind be simple subtraction of images 1 and 2

It is seen in the figure that the strong contrasts in the original images transform into strong borderlines after image subtraction due to the translation and maneuver of the platform. The standard deviation of the residual image serves as a figure of merit to evaluate the registration performance and is 24.9 gray levels in this case.

The following figure shows the residual image obtained by subtracting the two images with registration by navigation data. This registration uses the navigation data and a rigid-body transformation matrix to rotate & translate frame 1 to the orientation of frame 2.

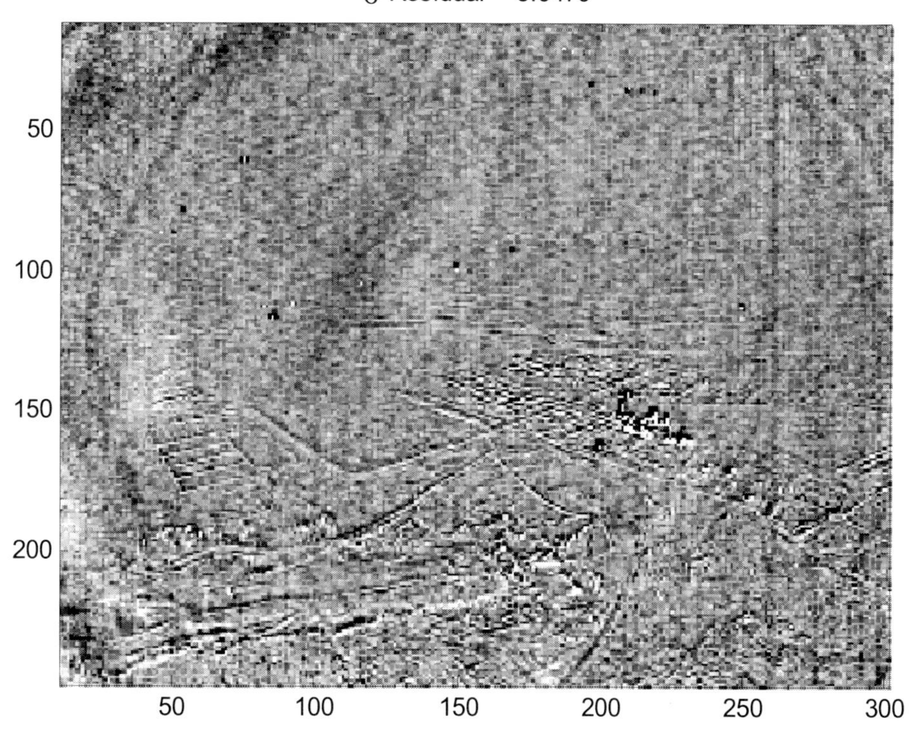

Figure 4. Residual images obtained by subtration of frame 2 and frame 1 after registation by navigation

It is seen in the figure that the strong contrasts in the original images are less obvious in the residual image thanks to the registration. The registration by navigation compensates for some of the displacement in location and orientation of the two images. The standard deviation of the residual image is 5.95 gray levels in this case.

The following figure shows the residual image obtained by subtracting the two images with direct image registration. This registration uses only data obtained from the 2 images.

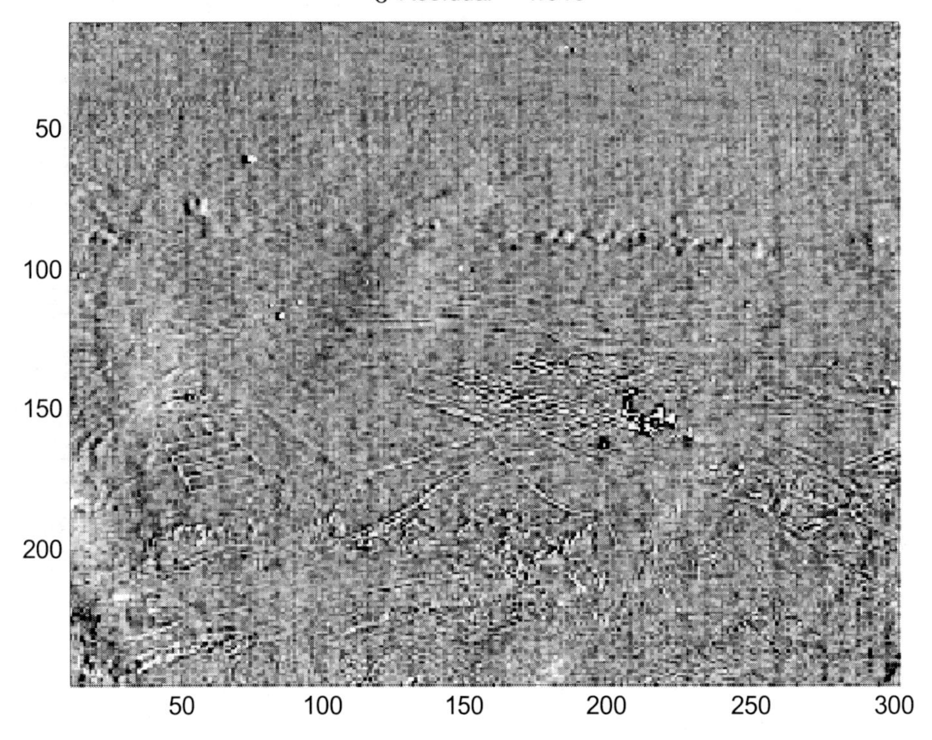

Figure 5. Residual image obtained by subtracting frame 2 and frame 1 after direct registration

It is obvious from the figure that the residual image obtained by direct registration is better than that of the registration by navigation. The direct registration managed to create a relatively smooth residual image, looking like white Gaussian noise (WGN) at vast areas of the image, proving it is able to compensate the platform's motion between the two images. The standard deviation of the residual image is 4.6 gray levels in this case which is lower by 25% than the standard deviation of the previous case (registration by navigation).

Conclusions

We have shown that direct image registration can be used to registrate images taken by aerial platforms. The parameters of the algorithm can be calculated in a pseudo-linear method which is efficient and relatively fast. The performance of the algorithm is similar to, and sometimes better than, the registration by navigation algorithm and is independent of data other than the images.

Future Work

Investigating functions other polynomial function, to improve the registration. Employing the results of the direct registration to calculate the navigation data – this can be useful to improve the navigation system errors and might be used to generate navigation data in a GPS blocked arena.

---

[1] Bruss, A., R., Horn, B., K., P., "Passive Navigation", Computer Vision Graphics and Image Processing. 21(1),3-20(1982).

[2] Bard, Y.,[Nonlinear parameter estimation],ACADEMIC PRESS, New York & London, 1-200,(1974).

[3] Lucas B., Kanade T., "An Iterative image registration technique with an application to stereo vision", Proceedings of the International Joint Conference on Artificial Intelligence, 674-679(1981).

A Lightweight approach for biometric template protection

Hisham Al-Assam, Harin Sellahewa, & Sabah Jassim

University of Buckingham, Buckingham MK18 1EG, U.K.
{hisham.al-assam , harin.sellahewa, sabah.jassim}@buckingham.ac.uk

ABSTRACT

Privacy and security are vital concerns for practical biometric systems. The concept of cancelable or revocable biometrics has been proposed as a solution for biometric template security. Revocable biometric means that biometric templates are no longer fixed over time and could be revoked in the same way as lost or stolen credit cards are. In this paper, we describe a novel and an efficient approach to biometric template protection that meets the revocability property. This scheme can be incorporated into any biometric verification scheme while maintaining, if not improving, the accuracy of the original biometric system. However, we shall demonstrate the result of applying such transforms on face biometric templates and compare the efficiency of our approach with that of the well-known random projection techniques. We shall also present the results of experimental work on recognition accuracy before and after applying the proposed transform on feature vectors that are generated by wavelet transforms. These results are based on experiments conducted on a number of well-known face image databases, e.g. Yale and ORL databases.

Keywords: Cancelable biometrics, revocable biometrics, random projections, template protection, biometric security.

1. INTRODUCTION

The growing use of biometric systems in many real life applications raises some privacy and security concerns. In the past, the focus of biometrics research has been on accuracy, speed, cost, and robustness challenges but only recently some attentions have been paid to security and privacy issues of biometric systems [1]. Public acceptance of biometric systems has a crucial impact on their success due to potential misuse of collected biometric data. Therefore, questions such as "What if my biometric data has been stolen or misused?" becomes an urgent question which needs satisfactory answers. The concept of cancelable or revocable biometrics has been proposed by Ratha et al. [2] as a solution for biometric template security. Revocability means that biometric templates are no longer fixed over time and could be revoked in the same way as lost or stolen credit cards and forgotten or compromised passwords are. This paper is primarily concerned with biometric template protection through proposing a new lightweight approach that meets revocability requirements.

Generally speaking, any biometric system consists of four main parts: acquisition and preprocessing, feature extractor, matcher, and the decision module. Firstly, a biometric data sample (e.g. a face image) is captured and is preprocessed (i.e. to normalize variations in scale, rotation and/or remove noise). The feature extractor extracts specific information from the acquired biometric data and represents it as a feature vector, which is smaller than the acquired biometric data. The feature vector(s) is stored as a *template(s)* in a database or on a secure token during the enrollment stage, while it is referred to as a *query or input* when a new biometric sample is presented for matching during the recognition stage.

Mobile Multimedia/Image Processing, Security, and Applications 2009, edited by Sos S. Agaian, Sabah A. Jassim,
Proc. of SPIE Vol. 7351, 73510P · © 2009 SPIE · CCC code: 0277-786X/09/$18 · doi: 10.1117/12.818291

Although biometric-based authentication is known to be more reliable than traditional authentication schemes, the security of biometric systems can be undermined in a number of ways. A biometric template can be replaced by an imposer's template in a system database or it might be stolen and replayed [3]. As a result, the imposter will gain unauthorized access to a place or a system. Moreover, researchers have shown that it is possible to create a physical spoof starting from biometric templates [3, 4]. Andy Adler proposed a "hill climbing attack" [4] on a biometric system. This attack is claimed to be able, after a finite number of iterations, to generate a good approximation of the target template. Also a method has been proposed in [6] to reconstruct fingerprint images from standard templates which might fool fingerprint recognition systems. Hence, securing the biometric template plays a vital role in maintaining the security of biometric systems.

An ideal biometric template protection scheme must satisfy four properties[7]: (i) *Diversity*: templates cannot be used for cross-matching across different databases in which users can be tracked without their permissions. (ii) *Revocability*: templates can be revoked and new one can be issued whenever needed. (iii) *Security*: it is computationally infeasible to reconstruct the original template starting from the transformed one. (iv) *Performance*: the recognition accuracy must not degrade significantly when the protection scheme is applied.

In this paper, we shall focus on using random projections as secure transformations and propose an efficient approach to biometric template protection that meets the afore mentioned properties. This approach can be incorporated into any biometric recognition scheme while maintaining the accuracy of the original biometric system.

The rest of this paper is organized as follows: Section II reviews existing template protection schemes and focuses on the random projection method. In section III, we shall describe our proposed approach in detail. Accuracy analysis for the proposed approach will be in section IV while its security will be discussed in section V. Finally, our conclusions and future work are presented in section VI.

2. EXISTING TEMPLATE PROTECTION SCHEMES

Modern encryption algorithms whether symmetric (e.g. AES) or public key (e.g. RSA) have been used by most commercial biometric systems to protect biometric templates, even though protecting biometric template in this way is inappropriate for several reasons [3]. Firstly, applying the matching process in the encrypted domain results in significantly reduced accuracy since distances/similarities between two encrypted feature vectors should not give any indication on the distances between original vectors. Secondly, decrypting the stored templates for every authentication attempt will make the templates vulnerable to eavesdropping. Finally, the security of the system depends totally on the security of the cryptographic key. Once the key is compromised and the template is stolen, the template cannot be revoked at all in such systems. Steganography and watermarking methods for the protection of biometric templates such as using have also been described in [9]. However, revocability is not catered for in such schemes. Moreover, in the absence of an established robust watermarking/steganographic scheme, designing such schemes remains an open challenge.

Template protection schemes which meet the revocability property in the literature can be classified into two categories [7]: feature transformation and biometric cryptosystem. The basic idea behind feature transformation is to use a function $\mathcal{F}$ to transform the original biometric template to a secure domain. The function $\mathcal{F}$ typically depends on a key (the key might be a user-based or a system-based depending on usage scenario and/or application type). The transformed template will be used instead of the original template. At the matching stage, the same transformation is applied the freshly input biometric data and the matching process will take place in the secure transformed domain. Whenever necessary, the transformed template could be revoked and one transformation is selected and applied to raw biometric template(s).

Depending on the characteristics of $\mathcal{F}$, feature transformation can be farther categorized as salting and non-invertible transforms. In salting, $\mathcal{F}$ is invertible i.e. if the key is known, the original template or a good approximation of it can be recovered. However, it is computationally infeasible to reconstruct the original template using the secured template and the key in the non-invertible transform.

Biometric cryptosystems [8] on the other hand combine biometrics with cryptography to produce what is known as biometric-based key. They were originally proposed for protecting cryptographic keys or even generating them from biometric data. However, biometric cryptosystems can be used as biometric template protection schemes through the generation of biometric-based keys that can be used as revocable representations of biometric templates.

Each of the above template protection schemes has its own advantages and limitations in terms of template security, computational cost, storage requirements, applicability to different kinds of biometric representations and ability to handle inter-class variations in biometric data i.e. maintaining the accuracy [3].

In this paper we shall be concerned with the implementation of, and the consequences of using, a recently developed feature transformation protection scheme namely the random projection scheme. Although a random projection is similar to salting in that it is invertible, it can be developed into a non-invertible transforms and/or used as a biometric cryptosystem.

2.1. Random projection

Random projection (RP) is a technique that uses random orthonomal matrices to project data point into other spaces where the distances among the data points before and after the transform are preserved. RP has been proposed as a secure transform for biometric templates and it was used in the literature [5, 10, 11, 12, 13, 14, 15, 16] to meet the revocability property. RP was proposed as standalone template protection scheme [10, 16] which can be considered as a salting approach. However, it was applied to generate a cancelable template for fingerprint data [15] and face image data [5, 11] and then a quantization step was added to make the transform non-invertible. Also RP has been used as a step for generating biometric based key from biometric data [12, 13, 14] to guarantee the revocability then the biometric-based key is used as a cancelable template in the recognition process.

The basic idea of random projection is to generate m orthonomal vectors that have the same dimensionality of the original template feature n (usually $n=m$ otherwise if $m<n$ RP will function as dimensionality reduction method [10]). To generate those m orthonomal vectors, m pseudo random vectors are generated first and then an orthogonalization process such as Gram-Schmidt process is applied to transform the random vectors into orthonomal ones. In practice, the output of Gram-Schmidt algorithm is a set of orthonomal vectors if and only if the input vectors are linearly independent [17]. When the size of the generated random vectors n is large, the chance of these vectors being linearly independent is high but this cannot be guaranteed. In the literature, RP transformed templates are created in three stages:

1. Generate m pseudo random vectors $\in \Re^n$ from a user-based key or token.

2. Apply Gram-Schmidt on the previous set of random vectors to get an orthonomal matrix. A matrix A is called an orthonomal matrix if it is orthogonal and each column/row vector has a unit norm, equivalently $AA^t = I$, where A^t is the transpose of A and I is the identity matrix of the same size of A.

3. Transform the original template feature x to a secure domain using matrix product: $y=Ax$.

2.1.1. Gram-Schmidt process:

Let $\{v_1, v_2, .., v_n\}$ be a set of linearly independent vectors. The Gram-Schmidt algorithm can be summarized as follows [16]:

$$u_1 = v_1 \ , \quad \mathrm{u}_1 = \frac{u_1}{\|u_1\|}$$

$$u_2 = v_2 - proj_{u_1} v_2 \ , \quad \mathrm{u}_2 = \frac{u_2}{\|u_2\|}$$

$$u_3 = v_3 - proj_{u_1} v_3 - proj_{u_2} v_3 \ , \quad \mathrm{u}_3 = \frac{u_3}{\|u_3\|}$$

$$u_i = v_i - \sum_{k=1}^{i-1} proj_{u_k} v_i \ , \quad \mathrm{u}_i = \frac{u_i}{\|u_i\|}$$

Where $proj_u v = \langle u, v \rangle u$, $\langle u, v \rangle$ is the inner product operation, and $\|v\|$ is the norm of a vector v.

Input vectors for the Gram-Shmidt process must be linearly independent. However, this property is not always guaranteed when a set of random vectors are generated. Moreover, generating large number of random vectors of large size n starting from a token then applying a complex process such as Gram-Schmidt is a computationally demanded process which might becomes a critical problem when applied on computationally constraint devices such as PDA and handheld devices. In the next section, we shall propose a novel lightweight approach for generating an orthonomal matrix to be used for transforming original templates to a secure domain. The proposed approach can be considered as a simple random projection and it might function as a standalone template protection or a step for meeting revocability property.

3. THE PROPOSED APPROACH

The main idea of the proposed approach is based on the fact that small size orthonomal matrices can be generated without a need for Gram-Schmitd procedure. For example, for any θ the 2x2 matrix

$$\begin{bmatrix} \cos\theta & \sin\theta \\ -\sin\theta & \cos\theta \end{bmatrix}$$

is an orthonomal matrix(it might be seen as a result of rotating 2x2 identity matrix by an angle θ). Here we shall describe our procedure in terms of 2x2 matrices, but it can easily be generalized for other small size matrices. Suppose we want to generate a random orthonomal matrix of size $2n \times 2n$. The proposed approach works as follows:

1. Select a set of n random values $\{\theta_1, \theta_1, ..,\theta_n\}$ of real numbers in the range $[0..2\pi]$ according to a user-based key or token and create the corresponding 2x2 orthonomal matrices. Now the matrix A whose diagonal is a set of the n 2x2 orthonomal matrices is an $2n \times 2n$ orthonomal matrix, where the 2x2 matrices are placed on its 3-band diagonal and all other entries of A are zeros.
2. Transform the original biometric templates, as well as any fresh biometric sample to a secure domain, as follows:

$$y = Ax \tag{1}$$

The resulting matrix A is a random sparse $2n \times 2n$ orthonomal matrix whose entries away from a diagonal band are zeros. Therefore, the computation of the product matrix Ax to generate the cancelable templates will be much faster compared to that obtained using the Gram-Schmidt procedure, and so is the case at the matching stage.

In practice, we replace the matrix A by AP, where P is a secret permutation 2nx2n matrix. For simplicity, in the rest of the paper we do not apply the permutation matrix P. For example, in order to increase the security of the process we add a blinding vector ***b*** of size 2n and equation (1) above is replaced with:

$$y = Ax + \boldsymbol{b} \tag{1´}$$

Whenever the security of the templates is undermined, the transformed templates can be revoked and new ones can be constructed easily by choosing a new set of thetas $\{\theta_1, \theta_2, .., \theta_n\}$ and ***b***.

If ***b*** is the zero vector, then the fact that A (or AP) is orthonormal implies that equation (1) preserves the Euclidean distances and hence the outcome of matching in transformed domain is equivalent to that of matching in the original domain. We shall now determine the effect of using a nonzero blinding vector ***b*** on the process of matching for 3 different distance functions including the Euclidean distance.

2.1. Matching using Euclidean distance:

Let x_1, x_2 be two original template vectors of size *2n*; the Euclidian distance between the two vectors is given by

$$\|x_1 - x_2\|^2 = \sum_{i=1}^{2n} (x_{1i} - x_{2i})^2 = (x_1 - x_2)^T (x_1 - x_2)$$

Let y_1, y_2 be the cancelable version of x_1, x_2 that have been created by (1).

$$\begin{aligned}
&\|y_1 - y_2\|^2 = (y_1 - y_2)^T (y_1 - y_2) \\
&= (Ax_1 + b - Ax_2 - b)^T (Ax_1 + b - Ax_2 - b) \\
&= (x_1 - x_2)^T A^T A (x_1 - x_2) \\
&= (x_1 - x_2)^T I_{2n} (x_1 - x_2) = (x_1 - x_2)^T (x_1 - x_2) = \|x_1 - x_2\|^2
\end{aligned}$$

As a result, the Euclidean distances will be exactly the same before and after applying the proposed RP transform in case of using same key (***A***,***b***) for all users (identification scenario).

3.2. Matching using Cosine function:

The Cosine similarity function between the two vectors is given by:

$$\cos(x_1, x_2) = \frac{x_1 . x_2}{\|x_1\| \|x_2\|} = \frac{x_1^T x_2}{\sqrt{x_2^T x_2} \sqrt{x_2^T x_2}}$$

Let y_1, y_2 be the cancelable version of x_1, x_2 that have been created by (1). The accuracy of biometric systems when cosine similarity function is in use is preserved only when the random vector ***b*** is set to zeros.

$$\cos(y_1, y_2) = \frac{y_1 . y_2}{\|y_1\| \|y_2\|}$$
$$= \frac{x_1^T A^T A x_2}{\sqrt{x_1^T A^T A x_1} \sqrt{x_2^T A^T A x_2}} = \frac{x_1^T I x_2}{\sqrt{x_1^T I x_1} \sqrt{x_2^T I x_2}}$$
$$= \cos(x_1, x_2))$$

3.3. Matching using CityBlock distance function (CB):

The CityBlock distance, also called the Manhattan distance function, between the two transformed vectors y_1, y_2 is given by:

$$CB(y_1, y_2) = \sum_{i=1}^{2n} |y_{1i} - y_{2i}| = \sum_{i=1}^{2n} |x_1 A_i - x_2 A_i|$$

where A_i is the column i from the transformation matrix *A*. One can see that orthonomal matrices do not preserve the CB distances. However, our experiments show that the accuracy will have only small fluctuation when our approach is applied.

4. EXPERIMENT RESULT & DISCUSSION

To evaluate our proposed approach, we shall compare the biometric system accuracy before and after applying it in two cases: authentication and identification using number of matching function. We shall also analyze the efficiency of the proposed approach and compare it to that of random projection which has been used in the literature.

For testing purposes we use two different databases, the Yale database and the ORL database. For face recognition we shall use a wavelet-based recognition scheme, whereby the feature vector representation of a face is the coefficients in a specific frequency subband of wavelet decomposed image of the face 20. The wavelet transform is a mathematical tool for short time analysis of quasi-stationary signals, such as speech and image signals. It decomposes such signals into its different frequency subbands at different scales, providing a multi-resolution view of the signal. In particular, wavelet transforms are capable of representing smooth patterns as well anomalies (e.g. edges and sharp corners) in images. The most commonly used scheme is the pyramid scheme, which we have adopted. At a resolution depth of k, the pyramidal scheme decomposes an image I into 3k+1 frequency subbands (LL_k, HL_k, LH_k, HH_k,..., HL_1, LH_1, HH_1), each of which can be used to create a feature vector representation of the input image. The lowest-pass subbands LL_k represent the k−level resolution approximation of the image I, while the other high frequency subbands highlight the significant image features at different scales. In all experiments, the lowest-pass subband LL3 is selected as a representative of the face image.

4.1. Accuracy analysis

4.1.1 Experimental data

Yale Face database [18] consists of 15 subjects with 11 different 320x240 grey scale images per person. These images include five different expressions (normal, happy, sad, sleepy, surprised, and wink), two face details (with glasses and without glasses), and three types of illuminations (center-light, left-light and right-light). The original images are manually cropped and resized to *80x96* pixels.

ORL database [19] has been established by Cambridge University. It contains 40 distinct persons; each person has ten different images, taken at different times. There are variations in facial expressions such as open/closed eyes, smiling/non-smiling, and facial details such as glasses/no-glasses. All the images were taken against a dark background with the subjects in an up-right, frontal position, with tolerance for a modest side movements.

4.1.2. Testing protocol

The $2n \times 2n$ orthonormal matrix in all experiments is generated using n different angles $\{\theta_1, .., \theta_n\}$. Experiments can be divided into two categories: authentication and identification. In authentication, each user is assigned a different key whereas in identification, the same transformation key is used for all users.

In Yale, one reference image is used (no-glasses image of each person) for the enrolment stage and the other 10 images for testing i.e. 150 images are used for testing. For ORL, the first image is used as a reference and the other 9 images are for testing i.e. 360 images are used for testing. Nearest neighbor classification algorithm is used for the matching stage. The three different distance functions, described above) have been used in our experiments, (i.e. Euclidean, Cosine, and CityBlock)

4.1.3. Authentication Tests

In this experiment, we are calculating the False Acceptance Rate (FAR) and the False Rejection Rate (FRR) in two cases: (a) no transformation is applied, (b) one key per a user.

Figure 1 and figure 2 show the recognition accuracy results in term of FAR and FRR using nearest neighbor classification method applied on Yale and ORL respectively where only one image per person is chosen for enrolment and the Euclidean distance is used as a distance function. The random projection, for these experiments were generated by our procedure.

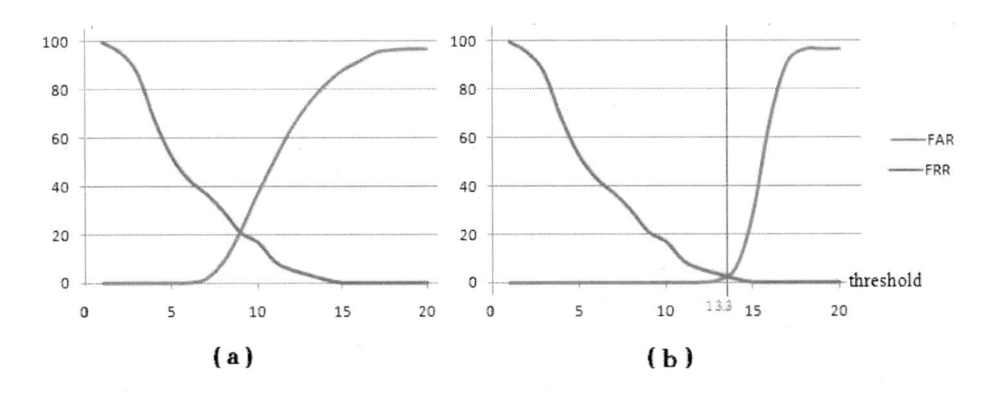

Fig1: Yale database: FAR and FRR (a) no transform (b) after applying RP

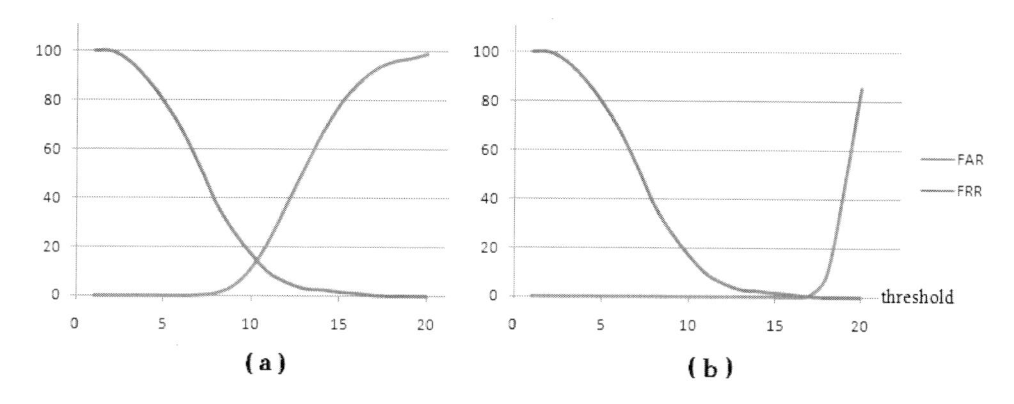

Fig2: ORL database: FAR and FRR (a) no transform (b) after applying RP

It can be clearly seen that the accuracy has improved significantly when RP is applied (EER has fallen from 21% to 2% and from 17% to 0.2% in Yale and ORL respectively).

4.1.3. Identification

In the following experiments, we shall apply our approach according to the formula (1) in two cases: (a) the random vector b is non-zero vector and (b) the vector b is a zero vector. To evaluate the result of using different matching functions (Euclidean, Cosine, and CityBlock), the experiments are repeated 20 times. A different transformation key is used for all users each time.

Figure 3 and 4 show the identification error rate (%) using 20 different transformation keys and three distance functions: Euclidean, Cosine, and CityBlock. The results support our mathematical explanation presented in section 3. It is clear that accuracies are preserved when the Euclidean distance function is used whether the random vector is non-zero or zero vector whereas accuracies computed by cosine function are preserved only when the random vector is zeros. On the other hand, using CityBlock distance function will result in accuracy fluctuation in both cases. As a result, one can conclude that applying the proposed approach to secure the biometric template has a very slight effect on the identification accuracy in some cases where in most cases the identification accuracy is preserved.

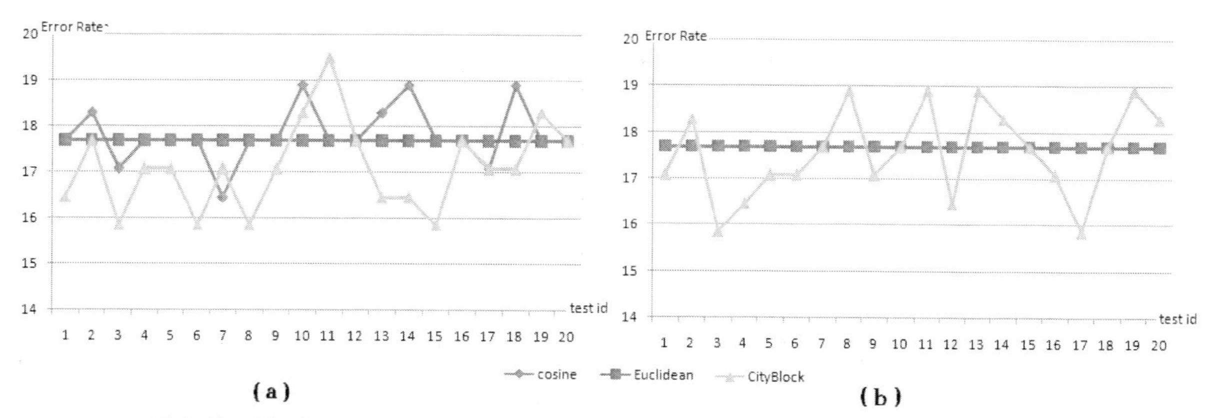

Fig3: identification error rate (%) using 20 different transformation keys and three distance functions for the Yale face database

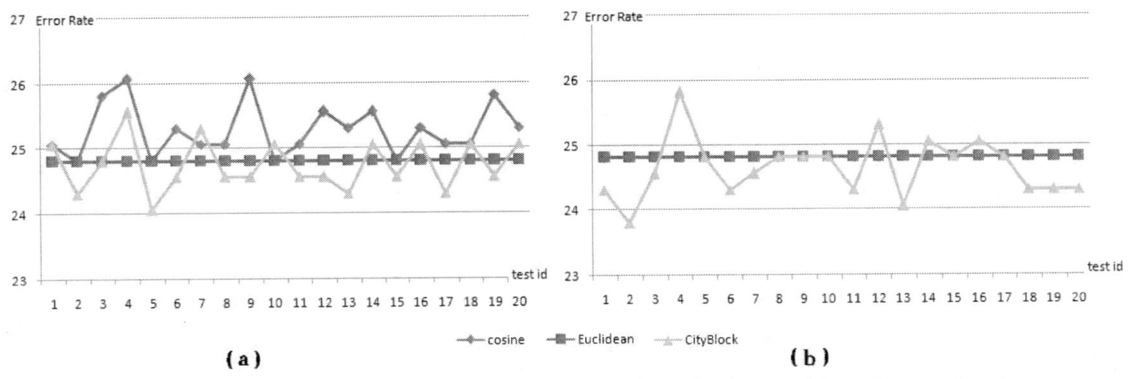

Fig4: identification error rate (%) using 20 different transformation keys and three distance functions for the ORL face database

4.2. Efficiency analysis:

Let *2n* be the size of original templates; a $2n \times 2n$ orthonomal matrix *A* is needed to transform the template to a secure domain. Table 1 shows the complexity of our approach compared to traditional Grahm-Schmidt approach which we denote by RP. One can clearly see from the table that our approach is much more efficient than RP.

No. of random numbers to be generated		Generating $2n \times 2n$ orthonomal matrix A		Applying the transform		
Our app	RP	Our app	RP		Our app	RP
n	$(2n)^2$	Up to n rotation for the 2x2 matrix	Gram-Schmidt	Multiplication	$4n$	$(2n)^2$
				Summation	$2n$	$2n(2n-1)^2$

Table1: complexity of our approach compared to RP

5. SECURITY ANALYSIS OF THE PROPOSED APPROACH

The security analysis of the proposed approach can be accomplished by discussing two properties mentioned in [6] namely: security and diversity. Security property is satisfied when it is computationally infeasible to reconstruct the original template starting from the transformed one. Since the proposed approach is reversible, its security depends mainly on the security of the used key, i.e. we need to answer the question: is the key robust against brute force attack?. On the other hand, diversity property addresses three security issues. i) Is it possible to track users without their permissions by cross-matching across different databases? i.e. how easy is it to know that two cancelable templates belong to the same user ii) If an adversary has captured the biometric data, is it possible for her to get access to the system without knowing the key? iii) If a transformed template has been stolen from one database, is it possible to apply replay attack to gain access to another system which uses another cancelable version of templates? Diversity property can be tested by generating two different cancelable templates for each user and using one of them as a reference template and the other as an input or a query.

5.1. Security :Robustness against brute force attack

If the size of original feature vector is $2n$, then the number of different θ values needed to generate $2n \times 2n$ orthonomal matrix is n. In theory, each of the angle values can assume any number in the interval $[0,2\pi[$. However, for practical reason, we might limit our choice to a fixed number of quantized values in that range. If we to use k>10 quantized values then even when $\boldsymbol{b}$ is the zero vector, the transformation key space will be bigger than 10^n. In the Yale database, the size of the original feature vector that has been used is 120. As a result the transformation key space will be bigger than $10^{60} \approx 2^{200}$ which is more than enough to be robust against brute force attacks.

5.2. Diversity :The effect of using two different keys for the same user on the recognition accuracy

Suppose a user has enrolled in two different databases where each template is secured by a different transformation key. Figure 5 compares the results of two cases: (a) the same key is used to transform the template and the query image, (b) One key is used to transform the stored template and the other is used to transform the presented biometric. One can see that FRR at the same threshold of 13.3 has risen from 2% to 99.33% i.e. it is very difficult to track users without their permissions using cross-matching. Moreover, it is very difficult for an adversary to get access to the system without knowing the key even if he/she captures the biometric of a genuine user.

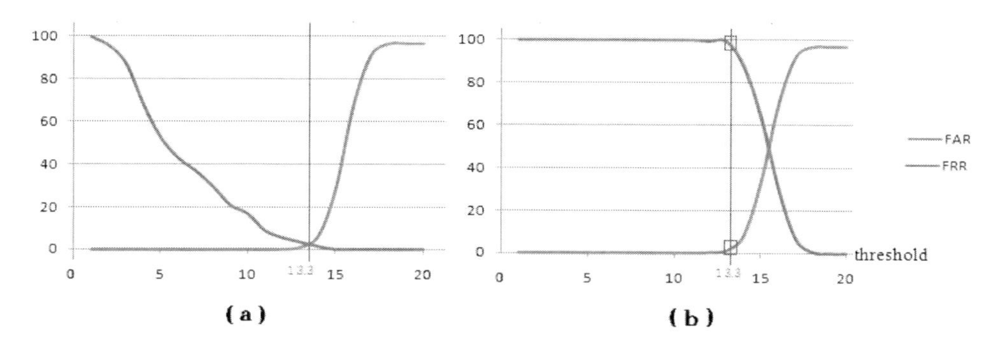

Fig4: FAR and FRR (%) for Yale database when (a) same transformation key is used for both template and query (b) one key is used to secure the template and another is used to secure the query

6. CONCLUSION & FUTURE WORK

In this paper we proposed a new lightweight approach for securing biometric template, based on a simple efficient and stable procedure to generate random projections. The proposed approach can be incorporated into any biometric verification schemes. We have demonstrated that the proposed scheme meets the four properties: diversity, revocability, security, and performance. Moreover, the accuracy of the biometric system will be preserved in identification scenario whereas it will improve significantly in authentication scenario where each user is assign to a different key.

The proposed approach and all current template protection techniques in general do not solve the problem of the replay attack. Therefore, our future work will focus on combining cancelable biometric with secret sharing concepts by adding some blinding factors to address the problem of replay attack, and can also be used between two untrusting parties in a way they could share the biometric verification in the cancelable domain.

7. ACKNOWLEDGMENTS

The first author would like to thank the Ministry of Higher Education in Syria (Al-Baath University) for sponsoring his PhD study.

REFERENCES

[1] U. Uludag, S. Pankanti, S. Prabhakar and AK Jain, "Biometric Cryptosystems: Issues and. Challenges", Proc. of the IEEE. Special Issue on Multimedia Security for Digital Rights, (2004).

[2] N. K. Ratha, J. H. Connell, and R. M. Bolle, "Enhancing security and privacy in biometrics-based authentication systems", IBM Systems Journal, 40(1), 614–634 (2004).

[3] K. Nandakumar, "Multibiometric Systems: Fusion Strategies and Template Security", PhD thesis. Michigan State University, (2008).

[4] A. Adler, "Vulnerabilities in biometric encryption systems", Proc. of the 5th International Conference on Audio and Video-Based Biometric Person Authentication, (2005).

[5] Y C Feng, Pong C Yuen and Anil K Jain, "A Hybrid Approach for Face Template Protection", Proceedings of the SPIE, 6944(11), (2008).

[6] R. Cappelli, A. Lumini, D. Maio, and D. Maltoni, "Fingerprint Image Reconstruction from Standard Templates", IEEE Transactions on Pattern Analysis and Machine Intelligence, 29(9), 1489-1503 (2007).

[7] Anil K. Jain, Karthik Nandakumar and Abhishek Nagar," Biometric Template Security", Advances in Signal Processing. Special Issue on Biometrics, (2008).

[8] U. Uludag, S. Pankanti, S. Prabhakar, and A. K.Jain, "Biometric Cryptosystems: Issues and Challenges", Proceedings of the IEEE. Special Issue on Enabling Security Technologies for Digital Rights Management, 92(6), (2004).

[9] A. K. Jain, A. Ross, and U. Uludag, "Biometric Template Security: Challenges and Solutions," Proc. of European Signal Processing Conference (EUSIPCO), Antalya, Turkey, (2005).

[10] N. Goal, G. Bebis, and A. Nefian, "Face recognition experiments with random projection", Proc. of SPIE 5779, 426–437 (2005).

[11] Yongjin Wang Plataniotis, K.N, "Face Based Biometric Authentication with Changeable and Privacy Preservable Templates", Biometrics Symposium, 1-6 (2007).

[12] ABJ Teoh, DCL Ngo, A Goh, "Personalized cryptographic key generation based on FaceHashing", Computers & Security, 23(7), 606-614 (2004).

[13] Alwyn Goh and David C.L. Ngo, "Computation of Cryptographic Keys from Face Biometrics", Lecture notes in computer science, 1-13 (2003).

[14] DCL Ngo, ABJ Teoh, and A Goh "Biometric hash: high-confidence face recognition" Circuits and Systems for Video Technology, IEEE Transactions, 771- 775 (2006).

[15] Andrew Teoh Beng Jin, David Ngo Chek Ling, and Alwyn Goh, "Biohashing: two factor authentication featuring fingerprint data and tokenised random number", Pattern Recognition, 37, 2245–2255 (2004).

[16] O Koval, S Voloshynovskiy, T Pun," Privacy-preserving multimodal person and object identification", Proceedings of the 10th ACM workshop on Multimedia and security, 77-184 (2008).

[17] Stephen Andrilli , David Hecker, "Elementary Linear Algebra", 298-309 (1993).

[18] A. Georghiades, D. Kriegman, and P. Belhumeur, "From Few to Many: Generative Models for Recognition under Variable Pose and Illumination," IEEE Trans. Pattern Analysis and Machine Intelligence, 40, 643-660 (2001).

[19] F.S. Samaria and A.C. Harter, "Parameterisation of a stochastic model for human face identification", Proceedings of the 2nd IEEE workshop on Applications of Computer Vision, Sarasota, Florida, (1994).

[20] H. Sellahewa and S. Jassim, "Wavelet-based face verification for constrained platforms", Proc. SPIE Biometric Technology for Human Identification II, 5779, 173–183 (2005).

Using Artificial Neural Networks to Statistically Fuse Current Iris Segmentation Techniques to Improve Limbic Boundary Localization

Randy P. Broussard
Systems Engineering Department, U.S. Naval Academy, Annapolis, MD 21402, USA

and

Robert W. Ives
Electrical and Computer Engineering Department, U.S. Naval Academy, Annapolis, MD 21402, USA

***Abstract*— One of the basic challenges to robust iris recognition is iris segmentation. This paper proposes the use of an artificial neural network and a feature saliency algorithm to better localize boundary pixels of the iris. No circular boundary assumption is made. A neural network is used to near-optimally combine current iris segmentation methods to more accurate localize the iris boundary. A feature saliency technique is performed to determine which features contain the greatest discriminatory information. Both visual inspection and automated testing showed greater than 98 percent accuracy in determining which pixels in an image of the eye were iris pixels when compared to human determined boundaries.**

Keywords: Iris, neural networks, information fusion, segmentation

I. INTRODUCTION

The majority of current iris identification techniques use a rectangular-to-polar coordinate transform as

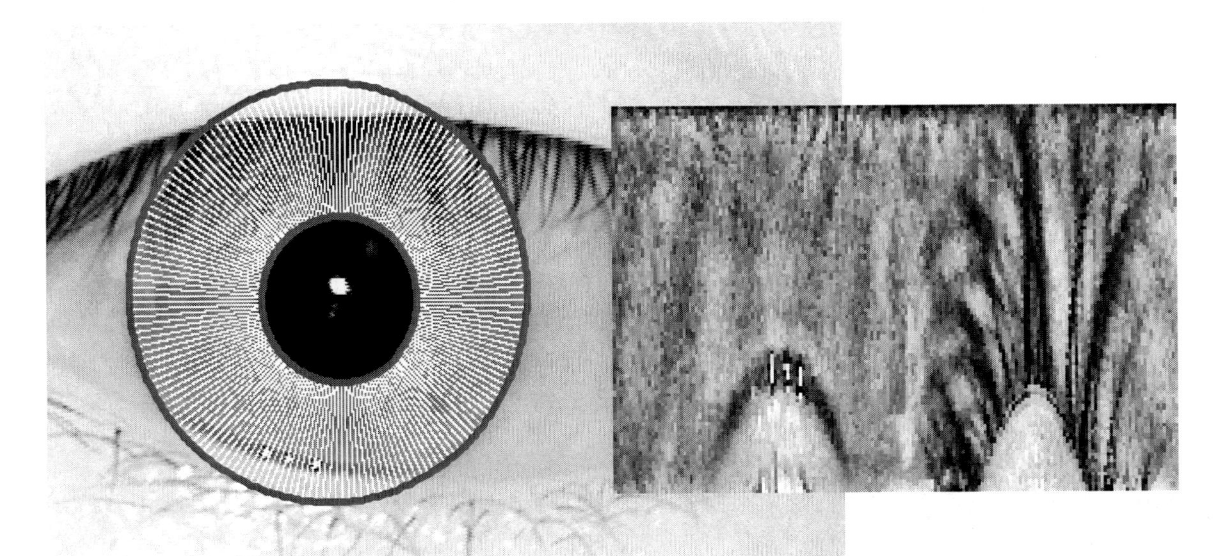

Fig. 1. Example of an image that is transformed from rectangular to polar coordinates. The left image is in rectangular coordinates. The polar measurement pattern used to form the polar image is overlaid on the rectangular image.

Manuscript received March 15, 2009.
Randy P. Broussard is with the Weapons and Systems Engineering Department, United States Naval Academy, Annapolis, MD 21402 USA (e-mail: broussar@ usna.edu).
Robert W. Ives is with the Electrical and Computer Engineering Department, United States Naval Academy, Annapolis, MD 21402 USA.

Mobile Multimedia/Image Processing, Security, and Applications 2009, edited by Sos S. Agaian, Sabah A. Jassim,
Proc. of SPIE Vol. 7351, 73510Q · © 2009 SPIE · CCC code: 0277-786X/09/$18 · doi: 10.1117/12.820247

the first step in the process of creating an iris code. This iris code typically contains structural phase information measured using Gabor wavelets. In the identification process, this structural phase is compared at each location and a match determination is made. The transformation to polar coordinates allows eye rotation to be compensated for by simply shifting the data left or right. Since positional content is being measured and compared, the accuracy of the polar transformation process is imperative. In this transformation process, radial points located between the pupil and limbic boundary are used to create a square image whose axes are polar coordinates. Figure 1 shows an example of the measurement pattern and the resulting polar image. Since radial points relative to the pupil and limbic boundary are used to create this polar image, the positional accuracy within the polar image is dependent on the accuracy of the iris boundaries. Many techniques for localizing iris the boundary have been published. The pupil boundary is typically considered easier to find than the limbic boundary. The limbic boundary may have such a gradual transition that the gradient and edge based measurements fail to clearly identify the boundary. Often, a circular or elliptical shape assumption is made to improve boundary detection. Occlusion by the eyelid and specular reflections can cause the visible boundary to have an irregular shape. Irregular shaped boundaries can cause a segmentation system, based on a circular/elliptical assumption, to produce inaccurate results. In this research, we use an artificial neural network to combine many of the standard methods used to localize and identify the iris boundary. The goal is to fuse the strengths of multiple segmentation techniques, to produce a more accurate pixel-level iris boundary. Many existing segmentation methods use a circular Hough transform to localize the iris boundary. This imposes an implicit circular assumption on the resulting boundary. Many irises are slightly elliptical, and the pupil and iris centers are typically not co-located. More advanced boundary representations such as ellipses and active contours have been used to overcome some of these issues, but the iris boundary must still be accurately identified if these methods are to operate accurately. In contrast to the circular Hough transform, the neural network identifies iris pixels on a pixel-by-pixel basis. Since the majority of the input features used by the neural network do not contain a circular boundary assumption, the resulting boundary is free to take non-circular patterns. This resulting pixel-level boundary can then be used by any of the boundary representation methods to concisely store the boundary location.

II. BACKGROUND

Much research has been produced on iris segmentation in recent years. It has been demonstrated that very high identification accuracy rates can be achieved on good images when accurate iris segmentation is present [1]. Non-ideal iris databases often pose a segmentation challenge and the resulting identification accuracies achieved are often orders of magnitudes less than that achieved on pristine images [2-6]. Many of the accuracy improvements recently published were the result of improved segmentation accuracy.

The majority of fielded iris identification systems in use today make the assumption that the pupil and iris were circular [7], though this is quickly changing as segmentation research advances. Current segmentation techniques can be coarsely grouped in one of the following categories; circular gradient maximization, edge detection followed by Hough transforms, and statistical pattern searches. The most cited segmentation methodology was published by Daugman in 1993. This method has been the basis for much of the segmentation research performed in recent years. The method operates in the image domain using integrodifferental operators to identify the maximum partial derivative with respect to increasing radius from a particular center point in the image. The center point is varied and the radius and center point that contains the maximum gradient are selected as the iris boundary.

Many variations to the Daugman method have been proposed to improve segmentation accuracy. Wildes proposed using a gradient based binarized image (an edge detected image) followed by a circular Hough transform to locate the center and radius of the maximum gradient [8]. Currently, the most commonly used segmentation methodologies in the literature are based on a variant of this method [5, 9-14]. Du et. al. proposed using a Sobel edge detector and looking for a straight line in the rectangular-to-polar image as an approximation to the circular Hough transform [15]. Other methodologies based on morphology [16], statistical distributions [17], and local image statistics [18] have also been proposed, but the most commonly adopted method is edge detection followed by a Hough transform. Typically an image is blurred with a Gaussian function to reduce noise. Next, the image is edge detected using the Canny edge detector which is the method of choice. Last, the iris boundary is located using a circular or elliptical Hough transform. Since an elliptical Hough transform has a very large parameter search space, the search space is typically reduced by assuming the outer iris boundary has the same ellipticity as the pupil boundary.

III. METHODOLOGY

In this research, we use an artificial neural network to statistically classify each pixel of a polar iris image as either an iris pixel or not an iris pixel. The pupil boundary and center are assumed to be known, thus the goal of the segmentation is to locate the outer iris boundary (limbic boundary). The artificial neural network is used as a classifier to determine the statistical boundaries that can separate the pixel of the iris, from the rest of the eye image. Each pixel based measurements supplied to the neural network, as part of an input vector, either directly represents a common segmentation methodology, or is a close approximation of the method. Both real valued and binary measurements are used. The real valued measurements typically represent local statistical measurements such as gradient or kurtosis. The binary measurements represent edge detected images that have been thresholded. To simulate Daugman's circular search or the Hough transform, the maximum of each measurement is found through a shaped search (circular). To avoid inserting a circular assumption into the data measurements, the circle was divided into 2, 4, 8, and 16 sectors (arcs). These arc measurements were added to the feature set along with the full circular measurement. Local neighborhood statistical measurements were also included in the feature set. The neighborhood size was varied, and these measurements contained no circular assumption. In all, gradient measurements, brightness measurements, local image statistics (mean, standard deviation, skewness, and kurtosis) and binary edge maps were included in the feature set. When computing each measurement, multiple neighborhood sizes and multiple threshold levels were applied when an empirical value was required in the process.

To determine which segmentation measurements to use as input to the neural network, many measurements were generated and a feed-forward feature saliency technique (described in [19]) was performed to determine which combination of features contains the most discriminatory information. A total of 323 measurements were passed through a feature selection process. A more detailed description of the feature set and the results of the feature selection process can be found in [19]. The saliency process identifies a subset of features that jointly contains the most discriminatory information for the problem at hand.

Using the subset of most salient features found in the feature saliency process, a neural network was trained to near-optimally segment the image pixels to statistically best approximate the ground truth data. A major goal of this research is that the neural network produces a boundary that slightly deviates from the ground truth boundary to produce a more statistically accurate boundary. This goal is based on the premise that a pixel that statistically belongs to the iris or non-iris class should be classified as such even if its location slightly deviates from the highly structured, circular based ground truth boundary. This premise allows the neural network to form oddly shaped boundaries even when circular or elliptical ground truth is used.

IV. IMAGE DATABASE AND TRUTH DATA

A subset of the publically available ICE 1 database [2] was used to train the neural network. This database is composed of 2953 total images from 132 subjects. Of these, 1425 were right eyes from 124 individuals and 1528 were left eyes from 120 individuals. Testing was performed on a different subset of the database. Ground truth was generated using an in-house algorithm, which used local statistics and a circular iris boundary assumption [18]. Upper and lower eyelid detection was included in the binary ground truth image. Figure 2 shows an example of a ground truth polar image. To increase accuracy, the ground truth was visually adjusted by hand to an edge detected version of the original rectangular image.

V. ARTIFICIAL NEURAL NETWORK

The neural network is used as a multidimensional statistical classifier to near-optimally combine the segmentation measurements into a statistically determined iris boundary estimation. In this research, a multi-layer perceptron (MLP) feed-forward artificial neural network was used for classification. The training method used was the text book standard, gradient decent based error back-propagation training method. A simple neural network and training method were purposely chosen to direct this research away from classifier research and towards iris segmentation research. A more complex network or training algorithm may train faster or produce improved results, but the selected network/algorithm have been shown to approach the accuracy of a statistical Bayes optimal solution [20] and therefore can near-optimally weight segmentation measurements, based on the discriminatory content of the individual measurements, to form a near-optimal combination of those measurements. To deter data memorization, the number of nodes within the hidden layer of the neural network was kept low. This also decreased computational run time. The number of hidden nodes was varied for thoroughness, but ten hidden nodes provided acceptable accuracy.

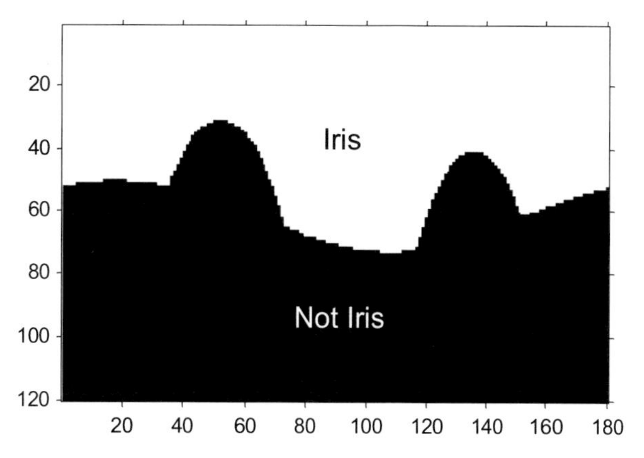

Fig. 2. An example of the ground truth image in polar coordinates.

The error back-propagation training algorithm traverses the gradients of an error surface to minimize classification error [21]. Using error back-propagation training, the neural network adjusts mathematical hyper-plane boundaries to form a near-optimal discriminate between statistical class distributions in a multidimensional space. This training methodology can offer a performance advantage over other statistical classifier for those problems where the feature distributions do not match a simple Bayesian Probability Density Function (PDF) assumption [20] or when the feature PDFs are unknown .

Fusing the real valued input data based on statistical distributions also has advantages. Most current segmentation algorithms contain some empirically determined internal threshold to create an edge map or decision criteria. The neural network offers the advantage that all decision boundaries (thresholds) are multidimensional, near-optimal, and statistically based on the properties of the training data. In this process, no PDF assumption is made. The neural network learns the statistical distributions of each class during training and weights each feature according to its discriminatory contribution. Because of this weighting, measurements that statistically have greater discriminatory information content will contribute more to the final boundary decision than measurements that statistically have lower discriminatory information content.

VI. FEATURE SET

The features used as input to the neural network consist of several of the segmentation measurements commonly used in iris research and fielded systems. This work was performed in [19]. Taken from [19], "the feature set was composed of both binary (edge detected) and real valued measurements. Image gray scale value, mean, standard deviation, skewness, kurtosis, horizontal gradient and vertical gradient measurements composed the core of the real valued features. These measurements were taken from the original polar image and various processed forms of that image. The processed forms of the polar image included Gaussian smoothing and adaptive histogram equalization." Figure 3 shows examples of several features. By using a rectangular to polar conversion, the maximization or summation of circular regions simple consist of summing the rows. Shifting the center used in the rectangular to polar conversion and selecting the maximum row sum across all shifts, approximates a circular Hough transform. Various measurements were maximized in way and the location and value of the maximum was included in the feature set.

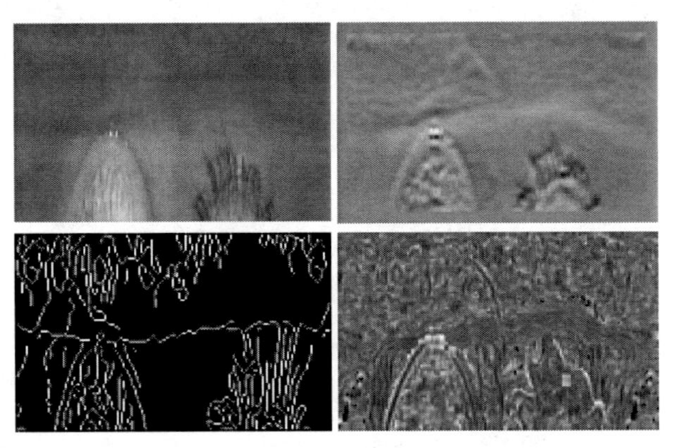

Fig. 3. Examples of polar images used to create features for the neural network. Clockwise from the top left are: original image, vertical gradient image, local kurtosis image and Canny edge detected image.

A total of 323 features were evaluated for saliency. The incorporation of this large number of features is intended to represent each major aspect of the common iris segmentation methodologies. Only the subset of features selected in the feature saliency process was used in this portion of the research.

1) 11 pixel wide Hough transform of right half of Canny image
2) Presence of eyelid at this pixel
3) 3 pixel wide Hough transform of left fourth of Canny image
4) 9 row sum of Canny polar image corresponding to eighth located at 10 o'clock
5) 7 row sum of Canny polar image corresponding to eighth located at 6 o'clock
6) Original image pixel value
7) 1x5 kurtosis of Hist Equalized polar image
8) 9 row sum of Canny polar image corresponding to eighth located at 8 o'clock

Fig 4. The eight most salient features found in the feature saliency process

VII. FEATURE SALIENCY

Feature saliency was performed to reduce the original 323 iris measurements to a subset of features that jointly contain the greatest discriminatory information that distinguishes the iris from the rest of the eye. To find this feature subset, a feed-forward, MLP classifier based feature saliency technique was used. The feature saliency process and results are described in detail in [19]. Figure 4 shows the eight most salient features found in the saliency process. Using fewer features as input to the neural network has the benefit of increasing computation speed and improving neural network statistical generalization.

VIII. SEGMENTATION AND POST PROCESSING

The neural network segmentation process starts with the assumption that the location and size of the pupil has been determined, and an approximate radius of the limbic boundary is known. The approximate radius of the limbic boundary is measured from the pupil center. The pupil and limbic centers are seldom collocated, and knowing the true location of the limbic boundary center is not required for this segmentation process. The role of the neural network is to better determine the iris boundary and compensate for the possible difference in location of the pupil and limbic boundary centers. First a polar image was constructed such that the approximate circular boundary of the iris was in the vertical center of the image. The polar measurements extended half the distance to the pupil boundary, and the same distance on the outer portion of the limbic boundary. This configuration allowed for the number of iris and non-iris pixels to be roughly equal over the entire training set. As stated before, the segmentation process does not know the true location of the center of the limbic boundary. All rectangular to polar conversions were based around the pupil center.

The MLP was trained on 120x180 pixel polar images. The training set consisted of 500 polar images which contained a total of 10.8 million pixels. The 8 features, selected in the feature saliency process (Fig 4), were used as input to the MLP. To remove any effects caused by the random neural network internal weight settings at the start of training, the neural network was trained 30 times and the weights from the best training run were retained for use in the testing phase.

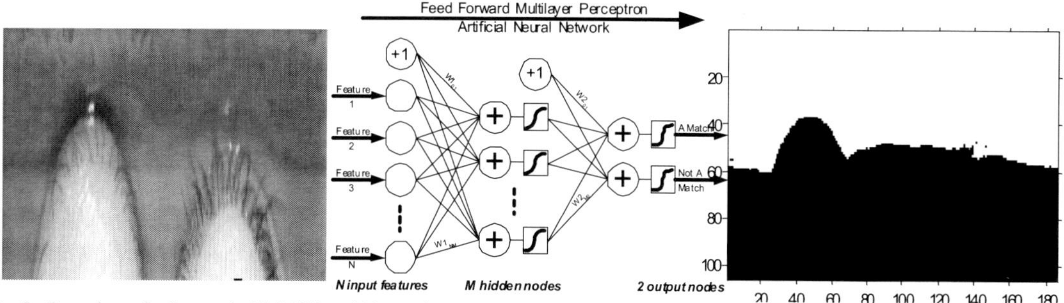

Fig. 5. Overview of using an Artificial Neural Network to segment the iris.

Figure 5 shows the neural network configuration used to segment the iris. For each pixel in the iris image, the selected features were computed and presented to the neural network for classification. Post processing was used to increase the accuracy and quality of the iris mask. A number of the incorrectly classified pixels were not contiguously connected to the largest grouping of pixels in the mask. Locating and retaining only the largest group of contiguous pixels increased the iris mask accuracy to 98.2 percent. To smooth the resulting boundary, a second-order median filter, with a width of five pixels, was applied to the vertical location of the boundary pixels. Additional accuracy could have been pursued by using basic morphological operations, such as open or close, but no attempt was performed.

IX. RESULTS

The neural network was trained on the 500 image training set, and then tested on a separate 200 image test set. The test set was composed of left and right eyes of 20 individuals. No training images were contained in the test set. The artificial neural network achieved 97.32 percent classification accuracy on the test set when compared to the truth data. Though this accuracy number is relative to the polar image size, it does represent a quality of fit metric for the classification process. Figure 6 shows an example polar image created using the neural network output. The pixel colors represent the neural network class determination, and the red line represents the median filtered boundary limbic boundary. Note the neural network produced boundary at horizontal locations between 70 and 100. This was a non-circular region of the limbic boundary.

In many images, the location of the iris boundaries was subjective due to the gradual color transition of the limbic boundary. This subjective nature of this boundary called into question the accuracy of the truth masks used to measure classification correctness. For many images, visual inspection often provided greater insight to correctness than numerical accuracy.

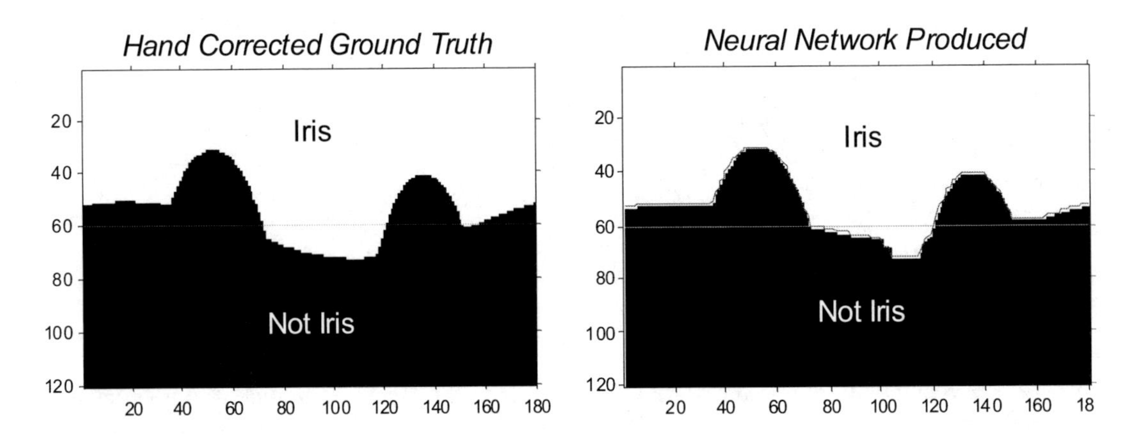

Fig. 6. A neural network produced iris boundary as compared to ground truth image. The light (green) line denotes the estimated iris radius, relative to the pupil center, that was used in the rectangular-to-polar conversion.

Figure 7 shows this estimated limbic boundary overlaid on the original iris image. The neural network produced boundary is shown in white and the ground truth data is shown in red. Figure 7a shows the boundary estimate the neural network produced using only one feature (11 pixel wide Hough transform of right half of Canny polar image) as input. This estimate is equivalent to a segmentation process that uses a circular boundary assumption. Figures 7b and 7c shows the non-circular boundary developing as additional features are added. Figure 7c was created using all eight features. Note the boundary deformations in the lower left and lower right quadrants of the eye. This is the neural networks attempt to form a non-circular boundary that best fits the statistical distributions of the iris features. The boundary building process performed by the neural network gives some insight into a technique that can be adapted to other segmentation methodologies. The neural network first established a circular boundary, and then used successively smaller sized measurements to adjust that boundary. Part of this is due to the feature ordering

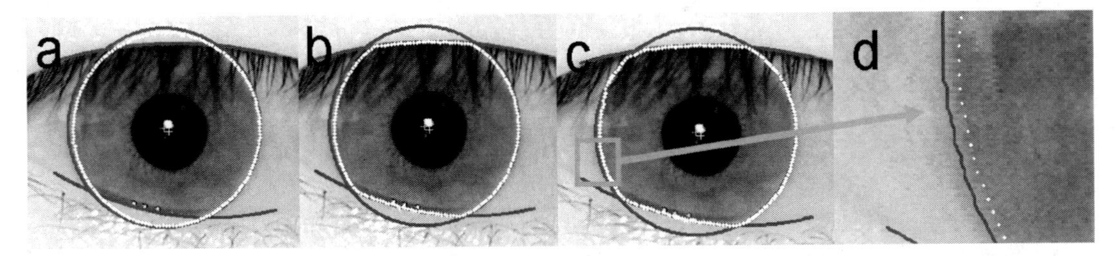

Fig. 7. The neural network produced boundary overlaid on the original image. White represents the neural network boundary and dark (red) represents the truth data. Each image shows the result as additional features are provided to the neural network.

which was a function of the feed-forward feature selection process [19]. A half circle Hough transform was used to establish the initial boundary. Next, a one-fourth circle Hough transform was used to adjust the radius of the opposite side. Successive one-eighth circle Hough transforms were then used to adjust for the elliptical shape of the boundary. Last, neighborhood and pixel measurements were used to refine the boundary.

Figure 7 is an image of an Asian eye which is often more difficult to segment than eyes of other races. Eyelashes also occlude this iris. No attempt was made to remove eyelashes from the truth masks, thus the neural network also failed to remove them.

The image in Fig 8 represents a typical result from the test set. Note this iris is slightly elliptical. The limbic boundary is slightly wider than the circular ground truth data. The neural network boundary is not contiguous because the missing portion was not contained with the polar image used to form the boundary. Space precludes including more of the test set images in this paper.

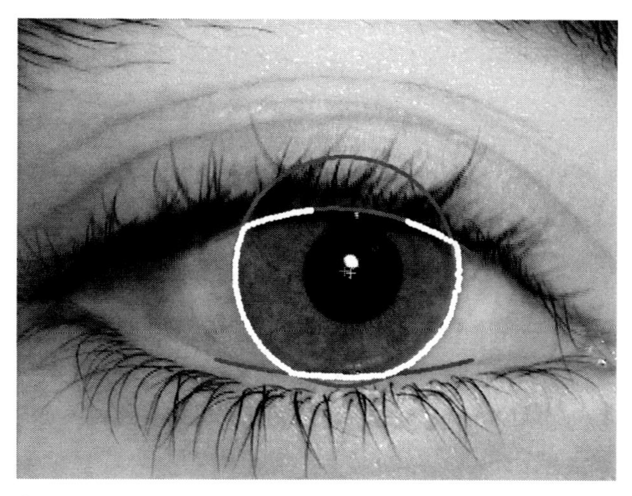

Fig. 8. A typical segmentation result on an image from the test set. White represents the neural network boundary and dark (red) represents the truth data.

X. CONCLUSION

A neural network can effectively be used to fuse multiple segmentation techniques. Feature selection can be used to narrow the list of features, used as input to the neural network, to those that contain the most discriminatory information. In conjunction, the process shows that identifying a non circular boundary can be achieved by starting with a circular boundary and then using smaller scoped local measurements to adjust the boundary. The neural network improved segmentation accuracy over the circular assumption based ground truth data. Visually, the neural network segmentation results matched or exceeded the perceived accuracy of the manually created truth masks.

REFERENCES

[1] J. Daugman, "Probing the Uniqueness and Randomness of IrisCodes: Results From 200 Billion Iris Pair Comparisons", *Proceedings of the IEEE*, Vol. 94, No. 11, pages 1927-1935, November 2006.

[2] 12. Iris Challenge Evaluation (ICE) database, available at http://iris.nist.gov/ice/, accessed June 12, 2008.

[3] A. Ross, S. Shah, "Segmenting Non-Ideal Irises Using Geodesic Active Contours", *Biometrics Consortium Conference, 2006 Biometrics Symposium.*

[4] J. Zuo, N.D. Kalka, N.A. Schmid, "A Robust Iris Segmentation Procedure for Unconstrained Subject Presentation", *Biometrics Consortium Conference, 2006 Biometrics Symposium.*

[5] H. Proenca, L.A. Alexandre, "Iris segmentation methodology for non-cooperative recognition", IEEE Proceeding on Vision, Image and Signal Processing, Volume 153, Issue2, Pages 199-205, April 2006.

[6] Proenca, H., and Alexandre, L.A., 'UBIRIS: a noisy iris image database'. ICIAP 2005, 13th Int. Conf. on Image Analysis and Processing, Cagliari, Italy, 6–8 September 2005, Lect. Notes Comput. Sci., 3617, pp. 970–977, ISBN 3-540-28869-4. http//iris.di.ubi.pt

[7] Daugman, J.G., 'High confidence visual recognition of persons by a test of statistical independence', IEEE Trans. Pattern Anal. Mach. Intell., 1993, 15, (11), pp. 1148–1161

[8] Wildes, R.P., 'Iris recognition: an emerging biometric technology', Proc. IEEE, 1997, 85, (9), pp. 1348–1363

[9] Cui, J., Wang, Y., Tan, T., Ma, L., and Sun, Z., 'A fast and robust irislocalization method based on texture segmentation', 2004. http://nlprweb.ia.ac.cn/english/irds/papers/cuijl/SPIE.pdf

[10] Huang, J., Wang, Y., Tan, T., and Cui, J., 'A new iris segmentation method for recognition'. Proc. 17th Int. Conf. on Pattern Recognition (ICPR04), Cambridge 2004, Vol. 3, pp. 554–557

[11] Ma, L., Wang, Y., and Zhang, D., 'Efficient iris recognition by characterizing key local variations', IEEE Trans. Image Process., 2004, 13, (6), pp. 739–750

[12] Ma, L., Tan, T., Wang, Y., and Zhang, D., 'Personal identification based on iris texture analysis', IEEE Trans. Pattern Anal. Mach. Intell., 2003, 25, (12), pp. 1519–1533
[13] Ma, L., Wang, Y., and Tan, T., 'Iris recognition based on multichannel Gabor filtering'. ACCV2002: 5th Asian Conf. on Computer Vision, Melbourne, Australia, January 2002, Vol. 1, pp. 279–283
[14] Kong, W.K., and Zhang, D., 'Accurate iris segmentation method based on novel reflection and eyelash detection model'. Proc. 2001 Int. Symp. on Intelligent Multimedia, Video and Speech Processing, Hong Kong, May 2001
[15] Y. Du, et. al., 'A new approach to iris pattern recognition'. SPIE European Symposium on Optics/Photonics in Defence and Security, London, UK, October 2004
[16] Kim, J., Cho, S., and Choi, J.: 'Iris recognition using wavelet features', J. VLSI Signal Process., 2004, 38, (2), pp. 147–156
[17] Dempster, A.P., Laird, N., and Rubin, D.: 'Maximum likelihood from incomplete data via the EM algorithm', J. R. Stat. Soc., 1977, 39, pp. 1–38
[18] Lauren R. Kennell, Robert W. Ives, Ruth M. Gaunt, "Binary morphology and local statistics applied to iris segmentation for recognition," *Proceedings of the 13th Annual International Conference on Image Processing*, Oct. 2006.
[19] R.P. Broussard, R.W. Ives, "Using Artificial Neural Networks and Feature Saliency to Identify Iris Measurements that Contain the Most Discriminatory Information for Iris Segmentation", *2009 IEEE Workshop on Computational Intelligence in Biometrics: Theory, Algorithms, and Applications*, March 2009
[20] Ruck, D. W., et al., "The multiplayer perceptron as an approximation to a Bayes optimal discriminant function", IEEE Transactions on Neural Networks, Vol 4, pp. 296-298, 1990
[21] Cybenko, G. "Approximation by Superpositions of Sigmodial Functions", Mathematics of Control, Signals, and Systems, 1982
[22]
[23] John Daugman, "Biometric personal identification system based on iris analysis," United States Patent, Patent Number 5,291,560, 1994.
[24] R.P. Wildes, J.C. Asmuth, G.L. Green, S.C. Hsu, R.J. Kolczynski, J.R. Matey, and S.E. McBride, "A Machine Vision System for Iris Recognition", Mach. Vision Application, Vol. 9, pp.1-8, 1996.
[25] W.W. Boles and B. Boashash, "A Human Identification Technique Using Images of the Iris and Wavelet Transform", *IEEE Transactions on Signal Processing*, Vol. 46, No. 4, pp. 1185-1188, 1998.
[26] Y. Zhu, T. Tan, and Y. Wang, "Biometric Personal Identification Based on Iris Patterns", *15th International Conference on Pattern Recognition*, Vol. 2, pp. 801–804, 2000.
[27] R.C. Gonzales and R.E. Woods, *Digital Image Processing*, Second Edition, Prentice Hall, Upper Saddle River, New Jersey, 2001.
[28] Masek L., Kovesi P., MATLAB Source Code for a Biometric Identification System Based on Iris Patterns. The School of Computer Science and Software Engineering, The University of Western Australia. 2003.

Rate-Adaptive Video Compression (RAVC)
Universal Video Stick (UVS)

David Hench
Air Force Research Laboratory/Information Directorate/Information Grid Division

ABSTRACT

The H.264 video compression standard, aka MPEG 4 Part 10 aka Advanced Video Coding (AVC) allows new flexibility in the use of video in the battlefield. This standard necessitates encoder chips to effectively utilize the increased capabilities. Such chips are designed to cover the full range of the standard with designers of individual products given the capability of selecting the parameters that differentiate a broadcast system from a video conferencing system. The SmartCapture commercial product and the Universal Video Stick (UVS) military versions are about the size of a thumb drive with analog video input and USB (Universal Serial Bus) output and allow the user to select the parameters of imaging to the. Thereby, allowing the user to select video bandwidth (and video quality) using four dimensions of quality, on the fly, without stopping video transmission. The four dimensions are: 1) spatial, change from 720 pixel x 480 pixel to 320 pixel x 360 pixel to 160 pixel x 180 pixel, 2) temporal, change from 30 frames/ sec to 5 frames/sec, 3) transform quality with a 5 to 1 range, 4) and Group of Pictures (GOP) that affects noise immunity. The host processor simply wraps the H.264 network abstraction layer packets into the appropriate network packets. We also discuss the recently adopted scalable amendment to H.264 that will allow limit RAVC at any point in the communication chain by throwing away preselected packets.

Keywords:

RAVC, Rate Adaptive Video Coding, SmartCapture™, UVS™, Universal Video Stick, H.264, AVC, Military Video

1. Introduction

RAVC[1,2,3] (Rate Adaptive Video Coding) utilizes the universality of the H.264/AVC[4] (Advanced Video Coding) standard and makes practical a new type of video encoding that is believed to have multiple Department of Defense (DoD) specific applications.

The H.264/AVC video encoding standard, aka MPEG-4 Part 10, aka (AVC) is designed to cover all applications in video coding and is rapidly being implemented in many areas of video coding. Because of the convergence of multiple previous standards as an unwanted byproduct, this standard gets referred to by at least six different names H.264, H.26L, ISO/IEC 14496-10, JVT, MPEG-4 AVC and MPEG-4 Part 10. This paper will follow the lead of Sullivan[5] and refer to the standard as H.264/AVC to reflect the difference between the names in the cooperating agencies.

DoD is now afforded the opportunity to bypass legacy commercial digital video solutions since H.264/AVC is expected to be with us for long time[6] and replace the highly successful legacy MPEG-2 standard with the more flexible and higher performing new standard. It makes sense to be an early implementer of the new technology instead of implementing MPEG-2 in new applications.

FastVDO, LLC has implemented the RAVC solution in a commercial solution, SmartCapture, with a proposed military solution, UVS, partly under an AFRL/RIGC contract[7]. These devices comply with Motion Imagery Standards Board (MISB) standards[8].

Approved for Public Release; distribution unlimited:88ABW-2009-1129, 20MAR09

Mobile Multimedia/Image Processing, Security, and Applications 2009, edited by Sos S. Agaian, Sabah A. Jassim, Proc. of SPIE Vol. 7351, 73510S · © 2009 SPIE · CCC code: 0277-786X/09/$18 · doi: 10.1117/12.817764

In order to understand RAVC, consider the two major differences between traditional commercial video encoding applications such as picture phone, video conferencing, high definition broadcast, video archiving, web video, and computer video. These differences are image quality in its multiple dimensions and the resulting data rate. RAVC provides a "dial a bandwidth" capability that allows trading off quality and bandwidth without stopping the video encoding. DoD certainly has requirements for all of these traditional commercial applications. In addition, we are seeing a proliferation of video sensors on the battlefield and a slow growth of communication bandwidth which can be aided with the new approach. Indeed, many military communications bands are being lost to commercial use and the spectrum is crowded. Improper implementation of video is and has been a bandwidth hog.

Consider a typical targeting application. Most of the operational time of a targeting scenario does not require high bandwidths and the application should operate more as a picture phone. When a potential target is in view, more bandwidth should be given to the video in an appropriate manner to aid in targeting and the video coding application should become more like commercial broadcast. These are two extremes of RAVC operation

Rate adaptivity can be performed in H.264/AVC by scaling: 1) in the spatial domain, 2) in the temporal domain, 3) in the encoder fidelity domain, and 4) in the group of pictures (GOP) domain. It will be shown that useful video can be produced from 32 kbps to 4 Mbps using a combination of these scalings. In the spatial domain, image resolutions of 720x480 pixels, 640x480, 352x288, 320 X 240 and 160 x 120 can produced. In the temporal domain, frame rates of 30, 15, 10, and 5 can be used. In the fidelity domain at standard resolution typical rates vary from 2.5 Mbps to 500 kbps with appreciable transform artifacts at lower rates. A typical GOP at 2.5 Mbps may be 30 frames; lower GOP values produce more noise immunity and higher data rates.

RAVC operation is practical by making available to the user the full flexibility of the standard that ASIC designers provide to software developers to allow these developers to design many products from one chip.

RAVC SmartCapture™ technology, designed and developed by FastVDO, LLC partially under AFRL/RI contract[7], is based on H.264/RAVC and makes operation such as the above currently practical in an attractive packaging footprint. A currently commercial available implementation of much of this technology is the size of a thumb drive. This commercial implementation is considered to serve as evaluation platform for the technology.

1.1. Overview of H.264/AVC Standard

H.264/AVC is currently the most powerful and state-of-the-art video decoding standard as recently reviewed by Sullivan, Topiwala, and Luthra[5] (as in other video standards only the decoder is specified and the encoder is up to the implementer). This powerful standard was developed by a Joint Video Team (JVT) consisting of experts from ITU-T's Video Coding Experts Group (VCEG) and ISO/IEC's Moving Picture Experts Group (MPEG).

H.264/AVC improves coding efficiency by a factor of at least about two (on average) over MPEG-2 (the most widely used video coding standard today) in the applications for which MPEG-2 was designed. (Advanced profiles provide at least a factor of three improvements.) H.264/AVC has kept implementation costs within an acceptable range considering advances in ASIC technology.

With the wide breadth of applications considered by the two organizations, the application focus for the work was correspondingly broad – from videoconferencing to entertainment (including broadcasting over cable, satellite, terrestrial, cable modem, DSL, etc.; storage on DVDs and hard disks; video on demand etc.) to streaming video, surveillance and military applications, and digital cinema. Three basic feature sets called profiles were established to address these application domains:

For DoD use, the United States Department of Defense Motion Imagery Standards Board (MISB) recommends[8] MPEG-2. for applications over 1 Mbps and recognizes the factor of two improvement in coding efficiency for H.264/AVC. Only H.264/AVC is recommended for applications under 1 Mbps. The older MPEG-4 part 2 standard is not recommended for use.

Table 1: History of Video Coding Standards

NAME	Date	org	Title	Major Use
H.261[9]	v1: Nov 1990, v2: Mar.1993.	ITU-T	codec for audiovisual services at px64 kbits/s	Video conferencing
MPEG-1[10]	Nov. 1993	ISO/ IEC	Coding of moving pictures and associated audio for digital storage media at up to about 1.5 Mbit/s – Part 2: Video	Computer storage
MPEG-2[11]	Nov. 1994 (with several subsequent amendments and corrigenda)	ISO/IEC & ITU-T	Generic coding of moving pictures and associated audio information – Part 2: Video	Broadcast
H.2631[12]	v1: Nov. 1995, v2: Jan. 1998, v3: Nov. 2000	ITU-T	Video coding for low bit rate communication	Video Conferencing
MPEG-4 Part 2[13]	1999 (with several subsequent amendments and corrigenda).	ISO/IEC	Coding of audio-visual objects – Part 2: Visual	Internet video
H.264/ AVC[14]	Version 1: May 2003, Version 2: May 2004, Version 3: Mar. 2005, Version 4: Sept. 2005, Version 5 and Version 6: June 2006, Version 7: Apr. 2007, Version 8 (including SVC extension): Consented in July 2007.	ISO/IEC & ITU-T	International Standard of Joint Video Specification	All

In order to fit a variety of delivery frameworks (e.g., broadcast, wireless, storage media), the H.264/AVC data is divided into a Network Abstraction Layer (NAL) and a Video Coding Layer (VCL). The VCL data and the highest levels of NAL data can be sent together as part of one single bit stream or can be sent separately.

Table 1 show a history of major digital video standards. Two major threads of development are shown coming together in the H.264/AVC standard. The ITU T H.261 and H.263 series was designed for low bit rate applications such as video conferencing and picture phone. The ISO/ IEC series is MPEG-1, MPEG-2 and MPEG-4 part 2. MPEG-1 was designed for digital storage media. MPEG-2 was designed for the broadcast industry. This standard has been highly successful, has been adopted by ITU-T as a recommendation, and will leave many legacy installations. MPEG-4 Part 2 was designed as an universal video coding standard. This standard has many modes such as embedded video objects that have not been implemented. A common approach to MPEG 4 Part 2 implementation is to use a modified H.263 encoder. H.264/AVC represents the coming together of the best of all these standards with AVC and MPEG 4 Part 10 being the ISO/IEC designation.

A scalable video extension to H.264/AVC has recently been added. This extension is currently under evaluation by industry and will be discussed later in this paper.

2. RAVC Operation

Typical RAVC operation is illustrated in Table 2. In this typical operation the bit rate varies from 2.5 Mbps to 32 kbps by varying image quality in fidelity, spatial, and temporal domains. The designations V, Q, and S refer to spatial resolutions of respectively 720 pixels x 480 pixels, 320 pixels x 240 pixels, and 160 x 120 pixels. A GOP of 12 has been used throughout this section.

The OldTownCross[14] test sequence by Sveriges Television AB (SVT), Sweden was used in this study. The original frame size was 1280 pixels x 720 pixels. It was resized to 480 pixels high using the Lanczos-3 algorithm in MATLAB.

Fig. 1 shows three frames at the different spatial resolutions from the movie produced for this study. This movie is to be shown at the conference.

Table 2: RAVC Transfer Points

RATE (KBPS)	SPATIAL {V,Q,S}	FRAMES/ SEC
2500	V	30
1500	V	30
1000	V	30
750	V	30
500	Q	30
300	Q	30
200	Q	15
100	Q	15
64	S	30
50	S	15
32	S	8

3. SmartCapture™and UVS™ RAVC H.264/AVC Technology

The RAVC SmartCapture™ technology from FastVDO, LLC[15] provides an integrated hardware/firmware/ software solution available for situational awareness, military laptops, surveillance, unmanned systems, reconnaissance and robotics. This technology is low power and light weight. It weighs less than an ounce and takes less than 1

(a) 720 pixels x 480 pixels (V)

(b) 360 pixels x 240 pixels (Q)

(c) 180 pixels x 120 pixels

Figure 1: Three frames from RAVC movie at the three spatial resolutions

Watt to operate. It is standards compliant and has metadata support. Video is compliant to H.264/AVC with on-the-fly RAVC capability from 32 kbps to 4 Mbps. Audio is compliant to AAC. Metadata support includes CE608 (Line 21 analog) and SMPTE 336 (KLV metadata). A custom API (Application Programmer Interface) is also available.

High definition will be available in quarter 4 of 2009. Video conferencing and live streaming are currently available.

Three embodiments of this technology are shown in Fig.'s 2, 3, and 4. Fig. 2 shows the commercially available unit. As is shown this unit is the size of a thumb drive and is inserted into the USB port of a laptop. Standard analog video (NTSC or PAL) connects to one side and H.264/AVC NAL (Network Abstraction Layer) packets are sent to the laptop leaving very little work for the laptop to create streaming or file storage data. The decoder is of much less complexity then the encoder and can be implemented in software using readily available software packages.

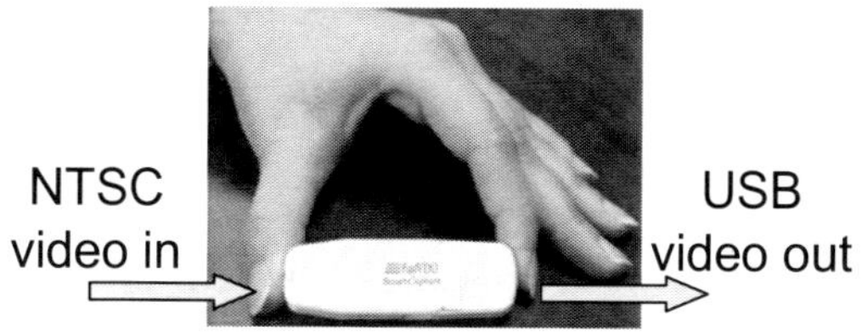

NTSC
video in
USB
video out

Figure 2: Commercially Available Embodiment of RAVC SmartCapture Technology

The SmartCast embodiment is shown in Fig. 3 and uses an ARM processor to make a net enabled unit for video and audio streaming.

A prototype of a militarized UVS version is shown in Fig. 4. A pigtail is supplied with this version supplying video and RS232 metadata. This pigtail can be trimmed to connect to military connectors and a mounting screw allows the unit to be mounted in a variety of vehicles or carrying cases. This version is now being implemented under Quick Reaction Funds.

Figure 3: SmartCast embodiment of SmartCapture technology

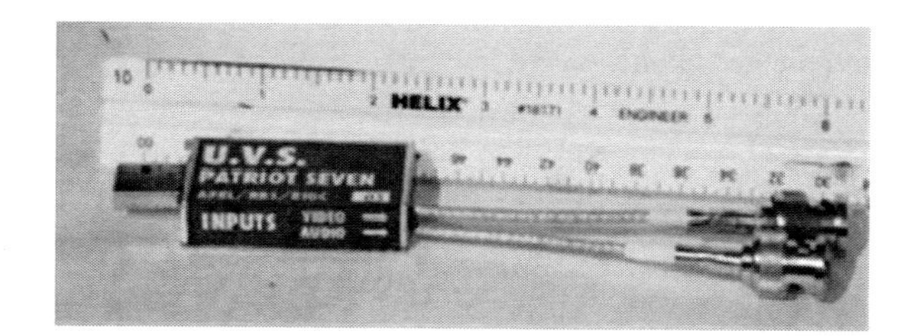

Figure 4: Prototype Universal Video Stick

For example the USB connection is plugged into a field laptop and the video BNC connector is plugged into a field video source. The audio connect is available for a microphone.
The additions that are being made in the military UVS over the commercial SmartCapture are listed below:

- Ruggedize as in Fig. 4
- Stream to PDA (Personal Digital Assistant)
- Develop secure video server
- Provide continuous adjustment of spatial resolution
- Provide a player with a ROVER look and feel
- Develop MISB compliant MPEG 2 Transport Streams
- Develop KLV for metadata
- Provide training materials
- Test at Patriot Seven

4. Scalable Video Extensions

Scalable video coding (SVC) is a form of RAVC that has been pursued with many of the referenced standards. SVC has never been found to be practical with previous standards. JVT has made an attempt to add a practical SVC extension to H.264/AVC. This extension was the subject of a special issue of IEEE Transactions on Systems and Circuits for Video and was well reviewed by Schwarz, Marpe, and Wiegard.[17]

5. Summary

RAVC technology allows a "dial a bandwidth capability" to adjust video bandwidth with a reduction in picture quality. Quality can be changed in spatial, temporal, transform quality, and Group of Pictures size. RAVC allows adjustment from sub picture phone to commercial quality "on the fly".

A commercial embodiment exists as a thumb drive sized USB devise with analog video input. Military applicability has be demonstrated and militarized version is expected within four months.

The host computer has been replaced by an ARM computer to allow a small net ready device.

References

[1] W. Dai, S. Patil, P. Topiwala, and D. Hench, [Rate Adaptive Live Video Communications Over IEEE 802.11 Wireless Networks], Proc. SPIE 6696, (2007).
[2] P. Kota, K. Kannan, Z. Xiong, P. Topiwala, and D. Hench, [Rate-Adaptive H.264 Video for Information for Global Reach (IFGR)", Proc. SPIE 5909, (2005).
[3] D. Hench, P. Topiwala, and Z. Xiong, [Channel Adaptive Video Compression for Unmanned Aerial Vehicles (UAVs)], Proc. SPIE 5558, (2004).
[4] [Advanced Video Coding for Generic Audiovisual Services, ITU-T Rec. H.264 and ISO/IEC 14496-10 (MPEG-4 AVC), ITU-T and ISO/IEC JTC 1], Version 1: May (2003), Version 2: May (2004), Version 3: Mar. (2005), Version 4: Sept.(2005), Version 5 and Version 6: June (2006), Version 7: Apr.(2007), Version 8 (including SVC extension): Consented in July (2007).
[5] G. J. Sullivan, P. Topiwala, and A. Luthra, [The H.264/AVC Advanced Video Coding Standard: Overview and Introduction to the Fidelity Range Extensions], SPIE Annual Conference on Applications of Digital Image Processing XXVII, Special Session on Advances in the New Emerging Standard H.264/AVC, Aug., pp. 454-474, (2004).
[6] G.J. Sullivan,[The H.264/MPEG-4 AVC video coding standard and its deployment status], Proceedings of SPIE -- Volume 5960, Visual Communications and Image Processing, (2005).
[7] FastVDO LLC, [Final Technical Report], AFRL/ RI contract FA8750-06-C-0190, [Rate-Adaptive Communications], (2008)
[8] Motion Imagery Standards Board, [Motion Imagery Standards Profile], Version 4.4 13 December (2007).
[9] ITU-T[Video Codec for Audiovisual Services at px64 kbits/s], ITU-T Rec. H.261 (v1: Nov 1990), (v2: Mar. 1993).

[10] ISO/IEC JTC 1,[Coding of Moving Pictures and Associated Audio for Digital Storage Media at up to about 1.5 Mbit/s – Part 2: Video," ISO/IEC 11172 (MPEG-1), (Nov. 1993).
[11] ITU-T and ISO/IEC JTC 1, [Generic Coding of Moving Pictures and Associated Audio Information – Part 2: Video], ITU-T Rec. H.262 and ISO/IEC 13818-2 (MPEG-2), (Nov. 1994) (with several subsequent amendments).
[12]. ITU-T, [Video Coding for Low Bit Rate Communication], ITU-T Rec. H.263;V1:(Nov. 1995), V2:(Jan. 1998), V3:(Nov. 2000).
[13] ISO/IEC JTC 1[Coding of audio-visual objects – Part 2: Visual,] ISO/IEC 14496-2 (MPEG-4 Part 2), (Jan. 1999) (with several subsequent amendments and corrigenda).
[14] [The SVT High Definition Multi Format Test Set Version 1.0] (Feb.2006), http://www.ebu.ch/CMSimages/en/tec_svt_multiformat_v10_tcm6-43174.pdf .
[15] [www.FastVDO.com/SmartCapture]
[16] [Introduction to the Special Issue on Special Issue --Scalable Video Coding — Standardization and Beyond IEEE Transactions on Circuits and Systems for Video Technology],Volume 17, Issue 9, (Sept. 2007).
[17] H. Schwarz, D. Marpe,T; Wiegand, [Overview of the Scalable Video Coding Extension of the H.264/AVC Standard], IEEE Transactions on Circuits and Systems for Video Technology, Volume 17, Issue 9, Page(s):1103 – 1120,(Sept. 2007).

WiMAX-WiFi Convergence with OFDM Bridge

Ali Al-Sherbaz, Chris Adams & Sabah Jassim*
Applied Computing Department, University of Buckingham
UK – MK181EG

ABSTRACT

Nowadays, Wireless and mobile communications technologies are the most important areas, which are rapidly expanding either in horizontal or vertical directions. WiMAX is trying to compete with WiFi in coverage and data rate, while the inexpensive WiFi still very popular in both personal and business use. Efficient bandwidth usage, Multi-Standard convergence and Wireless Mesh Networks (WMN) are the main vertical tends in the wireless world. WiMAX-WiFi convergence as an ideal technology that provides the best of both worlds: WiMAX new features and the low cost of the WiFi. In order to create a heterogeneous network environment, differences between the two technologies have been investigated and resolved. In the Multi-Carrier WiMAX-WiFi Convergence, the mismatch between the fixed WiMAX-OFDM (N_{fft}=256) and the WiFi-OFDM (N_{fft}=64) has been confirmed as a physical layer issue that will never be solved as MAC layer problem; therefore the current proposal is how to build what we called the "*Convergence-Bridge*". This bridge is like an extra thin layer, which is responsible for harmonizing the mismatch. For the WiFi-OFDM physical layer, the paper has selected the IEEE 802.11n OFDM standard while it is being developed. The proposal does not suggest changing the standard itself but modifying some functions to be configurable. The IEEE 802.11 standard has fixed the configurations for WiFi mode only, while our proposal is to set up these functions for WiFi and WiMAX modes.

Keyword list: Convergence, WiMAX, WiFi, Bridge, OFDM, IEEE 802.11, IEEE 802.16

1. INTRODUCTION

The Convergence-Bridge is a smart modification in the WiFi OFDM Physical layer to enable WiFi devices to join the WiMAX-OFDM wireless network. In this paper the WiMAX-Fixed (OFDM-256) and the WiFi-OFDM-64 have been selected to achieve the multi-carrier convergence. The convergence idea is initiated from the similarities between the WiMAX and the WiFi, however the dissimilarities are still real obstacles to enable them to communicate with each other. Dissimilarities between wireless standards are usually in the lower layers so that the investigations are focused on the PHY and MAC layers. In the standard investigations, it has been discovered that the convergence in WiMAX-WiFi multi-carrier OFDM is a physical layer issue, [1]. The proposal does not suggest changing the standard itself but modifying some functions to be configurable. The IEEE 802.11 standard has fixed the configurations for WiFi mode only, while our proposal is to set up these functions for WiFi and WiMAX modes

World Interoperability for Microwave Access (WiMAX) is the trade name of the IEEE 802.16 standard and is expected to dominate wireless networking technology for decades to come. It is designed to meet the requirements of the last-mile applications of wireless technology for broadband access with mobility, high bit rate, security and long distance coverage. The 802.16 is a set of evolving IEEE standards that are applicable to a vast array of the spectrum ranging from 2GHz to 66 GHz, which presently include both licensed and unlicensed (licence exempt) bands, [2]. The 802.16 is the enabling technology standard that is intended to provide Wireless Metropolitan Area Network (WMAN) access to locations, usually buildings, by the use of exterior illumination typically from a centralized base station (BS), [3].

In 2001 the IEEE 802.16 standard was released, whereas the groups continued to modify it to work on NLOS (Non Line-of-Sight) deployments. These modifications have covered the licensed and licensed-exempt bands between 2GHz-11GHz. In 2003 the IEEE 802.16a released with an extending OFDM techniques added for supporting the multi-path propagation problem. Meanwhile, the IEEE 802.11n standard group has also developed the OFDM as a part of the

* {ali.al-sherbaz, chris.adams , sabah.jassim}@buckingham.ac.uk

Mobile Multimedia/Image Processing, Security, and Applications 2009, edited by Sos S. Agaian, Sabah A. Jassim,
Proc. of SPIE Vol. 7351, 73510T · © 2009 SPIE · CCC code: 0277-786X/09/$18 · doi: 10.1117/12.818472

physical layer of the WiFi. Besides the OFDM physical layers, the 802.16a established an optional MAC-Layer function that includes support for Orthogonal Frequency Division Multiple Access (OFDMA),[4].

In 2004, revisions to IEEE 802.16a were made which were called IEEE 802.16-2004. It replaces 802.16, 802.16a and 802.16c with a single standard. Moreover, this revised standard has also been adopted as the basis for HIPERMAN (High-Performance Metropolitan Area Network) by ETSI (European Telecommunication Standards Institute). In 2005, 802.16e-2005 was completed, a further MAC-PHY layers modification was formulated by using a scalable OFDM to accommodate high-speed mobility,[5].

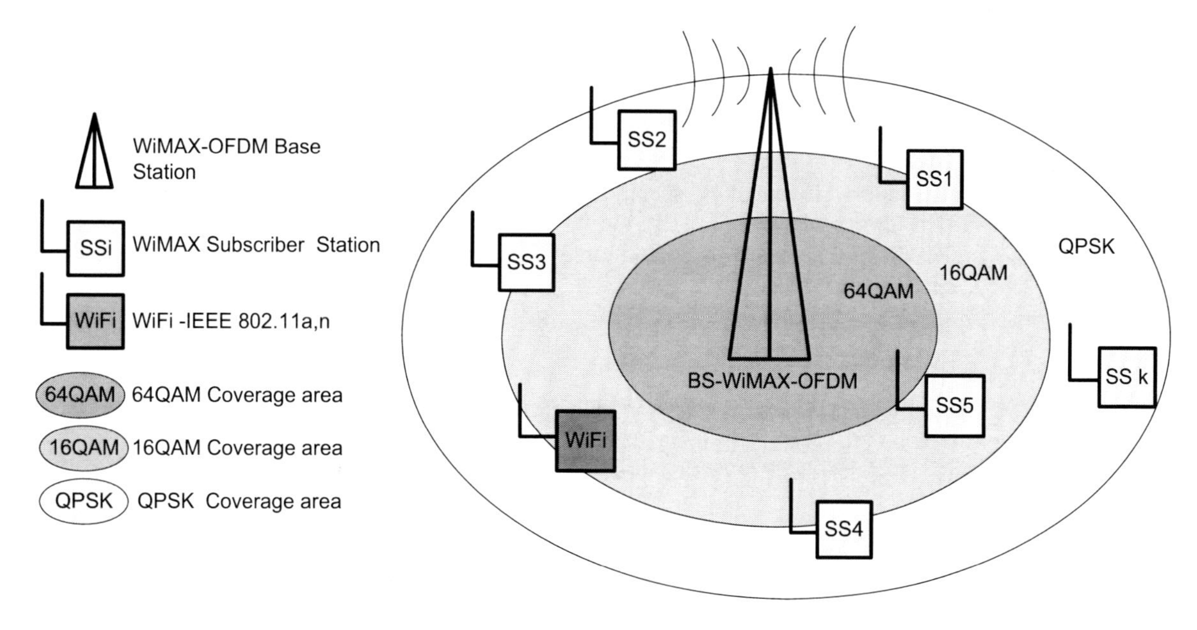

Figure (1) WiMAX Point to Multipoint (PMP) Architecture Diagram

In addition to the Point-to-Point (PTP) and Point-to-Multi Point (PMP) topologies, the 802.16a introduces the WiMAX-mesh topology as a possible alternative to the cellular networks. This alternative (Mesh) applies to wireless network in which nodes cooperatively route packets. Mesh could act as an Ad-Hoc wireless network or as edge networks to a larger network infrastructure such as the Internet, [6]. It gains flexibility, reliability and nomadic network architecture based on multi-hop model. Adding the mesh concept to the 802.16 enlarge the geographical area of a network. As mentioned in the previous section, the WiFi has also taken over the Mesh topology and released the IEEE 802.11s draft standard, [7].

A new vision of the convergence has turned up when two different standards have created the same topology (Mesh) as vertical developments. While the Mesh standards are under development, the paper has studied the potentials of the WiMAX-WiFi convergence in the Mesh topology also. Apparently, it would appear that the WiMAX-WiFi convergence has a good chance to be involved in the mesh topologies, because of the many common parameters are present there (e.g. OFDM).

WiFi is the predecessor technology of WiMAX that has been widely deployed. It is cheap, available, applicable, and has multi-vendors. It has many advantages over WiMAX, while WiMAX fills many gaps that have been found in WiFi. The WiMAX technology in its current form will complement the 802.11 or WiFi standard. The deployment and adoption of the 802.16e standard could decrease the number of WiFi users while increasing WiMAX users and WiMAX "hot spots." The 802.16a standard will help corporations and Internet service providers by expanding their services to rural markets or the "last mile", [8].

In spite of the vast development in the WiMAX, WiFi is still the dominant wireless technology at the present time. In fact, tri mode WiFi (IEEE 802.11 a/b/g) is already built in the laptop machines, and day-by-day PDAs and iPhones contain 802.11. However, it is my contention that while there are many advantages to be had from using WiFi, the early

WiFi standards have less security and poor reliability with low data rate than WiMAX. WiFi standard developers and vendors have tried to overcome these problems by releasing IEEE 802.11i for security and IEEE 802.11e for QoS (Quality of Service) as vertical improvements for the IEEE 802.11. Ultimately, the IEEE 802.11n has been released as a new WiFi standard, which has covered (standard's claim) the previous problems by using the MIMO-OFDM mechanism. 802.11n has the ability to be peer-to-peer with WiMAX due to the higher data throughput and longer wireless distance. The increased performance from 802.11n WLAN should eliminate the last bottleneck in the enterprise-wide WLAN deployment, [4]. The paper has focused on adding the smart modification while the IEEE 802.11n is in the process of being developed.

The security improvement (802.11i) and the MIMO-OFDM mechanism (802.11n) can extensively enhance the WiFi. This enhancement enthused the task group (TGs) to define the Extended Service Set (ESS) Mesh Networking Standards. Presently, the WiFi mesh draft standard has been released as IEEE 802.11s. A lot of challenges against the 802.11s have to be harmonized to efficiently provide a large bandwidth over a large coverage area, [9].

IEEE 802.16 standard proposes four different PHY specifications .Following the work in the WiMAX Single Carrier Access that has been solved in the previous paper [10], the methodology WiMAX-OFDM has been selected for this paper while the WiMAX-OFDMA will be investigated in future work. The WiMAX-SC PHY (10-66 GHz) has not been selected due to frequency band incompatibility

The rest of the paper is organized as follows. Section-2 illustrates in detail, the Fixed WiMAX PHY layer (WirelessMAN-OFDM PHY) and the related mathematical background for the WiMAX-WiFi OFDM RF signals and also gives a full description of the WiMAX-OFDM-TDD frame structure. Section-3 describes WiFi-OFDM PHY layer specifications and compromise with the WiMAX-OFDM. Section-4 is the proposed convergence bridge with explanation of the smart modification in the physical layers functionalities.

2. THE PHYSICAL LAYER FOR WIMAX-OFDM

2.1 WiMAX-IEEE 802.16 PHY (2-11 GHz)

Both licence spectrum and unlicensed spectrum are found in the range of (2-11 GHz), and both are discussed in detail in the IEEE 802.16 standard. The overall design of the (2-11 GHz) physical layer is based on the clear demand for a non-line-of-sight (NLOS). This came about because of a realization on the part of standards developers that residence application of 802.16 would encounter multipath propagation issues because of trees and other signal obstacles. Three air interface specifications for 802.16 are described in the original standards, [3]. They are WirelessMAN-SCa, which relies on a single carrier modulation schema; WirelessMAN-OFDM (Fixed WiMAX), which relies on OFDM with a 256-point transform scheme with TDMA access, which is mandatory for unlicensed bands; and WirelessMAN-OFDMA (Mobile WiMAX), which uses OFDMA with 2048-point transform. Here, multiple access is provided by targeting subset of various sub-carriers to individual receiving devices. In this paper, WirelessMAN-OFDM physical layer has been selected for the WiMAX-WiFi convergence (mixed standards). The first stage is to convert the carrier frequency of the WiFi-OFDM signal, which is either 2.4GHz or 5 GHz, via a straightforward frequency converter to be the same as WiMAX Wireless MAN-OFDM Base-station carrier frequency as shown in figure (2).

2.2 WiMAX-OFDM

This paper suggests the Multi-Carrier aspects of WiMAX-WiFi Convergence: the WiMAX-OFDM (N_{fft}=256) and the WiFi-OFDM (N_{fft}=64), as shown in figure (2). The mismatch in the number of samples (N_{fft}) cannot be resolved at the MAC layer, and we deal with it as a physical layer issue, by creating a *Convergence-Bridge* . This bridge is like an extra thin layer, which is responsible to harmonize the mismatch. Generally, any OFDM signal, *s(t),* could be produced from equation (1) whether it is WiMAX or WiFi,[11]. This equation underpins the design of the proposed Convergence-Bridge.

$$S(t) = Re\left\{e^{j2\pi f_c t} . \sum_{\substack{k=-N_{used}/2 \\ k\neq 0}}^{N_{used}/2} C_k . e^{j2\pi k\Delta f(t-T_g)}\right\} \dots\dots\dots\dots\dots\dots\dots\dots. (1)$$

- N_{used}, Number of used subcarriers (N_{used}=200, for the WiMAX OFDM)
- C_k is a complex number that represent the Data,
- Δf is the subcarriers frequency spacing, (Δf = 15.625 KHz)
- f_c is the central frequency ,(f_c= 2-11 GHz)
- T_g is the Guard Time

Mathematically, equation (1) has been structured from three main parts:

Part1: Carrier Frequency part: $e^{j2\pi f_c t}$
Part2: Data part: C_k
Part3: Subcarriers part: $e^{j2\pi k\Delta f(t-T_g)}$

In part1, the carrier frequency (f_c) is the main factor in deciding which technology is being used, e.g. the WiFi carrier frequency is 2.4 GHz or 5 GHz while the WiMAX carrier frequency bands are (2 GHz-11 GHz). In part2, the C_k is the data to be transmitted on the subcarriers whose frequency offset index is k, during an OFDM symbol. Part3 is the summation of the N_{FFT} samples of the orthogonal subcarriers; therefore N_{FFT} is another factor which differs between the WiFi and WiMAX, ([12], [13]).

Fig (2) illustrates the frequency domain of the WiFi-OFDM signal (spectrum) which is either IEEE802.11a (5 GHz carrier frequency) or IEEE802.11n (2.4 or 5 GHz carrier frequency) .The WiMAX WirelessMAN-OFDM signal (spectrum) has a 256 sub-carrier frequencies within (2-11 GHz) band, as pointed out in fig (2). After the frequency conversion both WiFi-OFDM and WiMAX WirelessMAN-OFDM signals have been held at the same centre frequency ([12], [14]).

The IEEE 802.11a,n and the IEEE 802.16e standards specify the number of the subcarriers and the bandwidth of the related OFDM signals. The WiFi-OFDM signal has got 64 subcarriers spread over 20MHz bandwidth, while the WiMAX-OFDM signal has got 256 subcarriers spread over 3.5MHz bandwidth. These differences complicate the convergence between the two technologies. Figure (2) illustrates the RF conversion of the WiFi-OFDM signal's central frequency to be the same as the WiMAX-OFDM signal's central frequency. Even though the WiFi signal can exist in the WiMAX band but they cannot interoperate due to the differences outlined above.

2.3 WiMAX-OFDM-TDD Frame Structure

The IEEE 802.16 standard subcategorised the WiMAX-OFDM PHY in terms of the duplexing method. These duplexing methods are the TDD (Time Davison Duplexing) and the FDD (Frequency Division Duplexing). In the Point to Multipoint (PMP) WiMAX topology, the duplexing method is either TDD or FDD for the licence bands, while in the licence exempt bands the duplexing method shall be TDD (5). In this paper, the WiMAX-OFDM-TDD has been selected to satisfy the WiMAX-WiFi convergence, while WiMAX-OFDM-FDD will be investigated in the future.

A frame-based transmission has been used in the WiMAX-OFDM. The frame duration is a central periodic time based on the frame start preambles, which has been selected by the BS (base station), and it should not be changed. The specific frame durations that are allowed are (2.5ms, 4ms, 5ms, 8ms, 10ms, 12.5ms and 20ms). A frame consists of a downlink subframe and an uplink subframe. A downlink subframe consists of only one downlink PHY PDU,[11]. An uplink subframe consists of contention intervals scheduled for initial ranging and bandwidth request purposes and one or multiple uplinks PHY PDUs, each transmitting from a different SS (Subscriber Station). Moreover, according to the OFDM frequency domain, equ (1) shows the OFDM symbol is made up from N_{FFT} subcarriers that are classified into three subcarriers types: Data subcarriers, Pilot subcarriers and Null subcarriers. In the WiMAX-OFDM the data subcarriers (N_{used}) are dedicated for data transmission and it is 200 out of 256 subcarriers, whereas the Pilot subcarriers are dedicated for various estimation purposes and it is only 8 subcarriers. The Null subcarriers are not for transition at all, such as the lower 28 frequency guard subcarriers, the higher 27 frequency guard subcarriers and the one and only one DC subcarrier. Figure (3) shows the subcarriers (sub channel frequency index) consist of the three categories of the subcarriers. The data subcarrier index range is $-100 < k < +100$, exempt from that the pilot subcarrier indices (-88, -63, -38, -13, 13, 38, 63, 88), while the DC subcarrier index is when k=0. The other two null subcarrier ranges are the *lower* frequency guard subcarrier band ($-128 < k < -101$) and the *higher* frequency guard subcarrier band ($+101 < k < +127$), (Hints: these guard bands do not appear in figure-3). The basic structure of a WiMAX-OFDM frequency domain has 256 subcarriers (N_{FFT}) spread over 3.5MHz bandwidth with 15.625KHz subcarrier spacing as determined by the IEEE 802.16 WirelessMAN-OFDM, ([11], [15]).

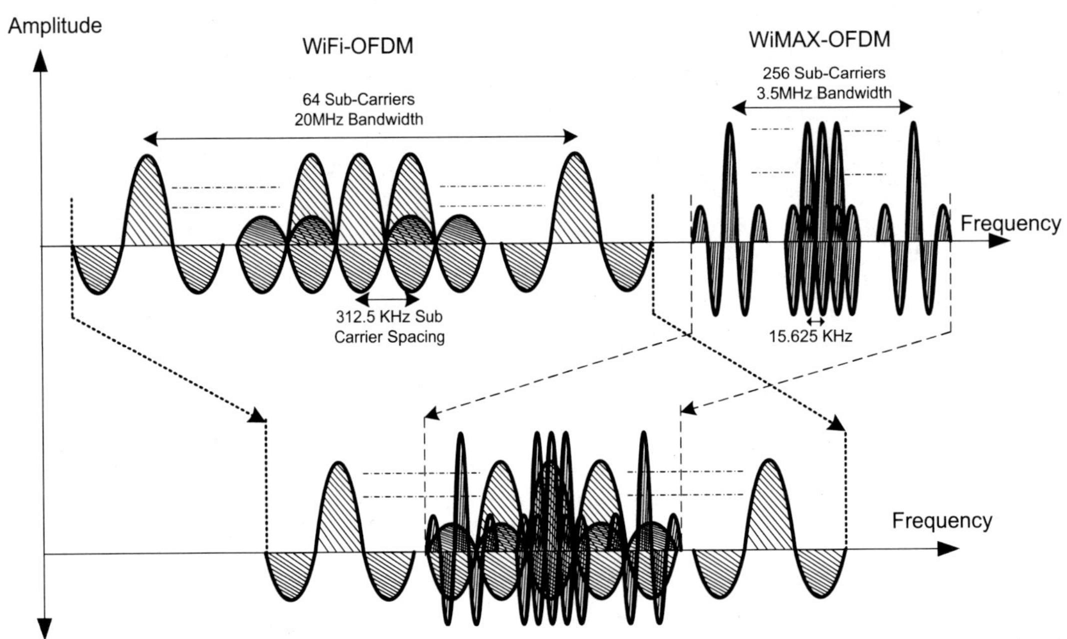

Figure (2) WiMAX-OFDM, WiFi-OFDM spectrums and the frequency conversion stage for WiFi-OFDM Signal

3. THE PHYSICAL LAYER FOR WIFI-OFDM

3.1 WiFi-IEEE 802.11a,n PHY (2.4 or 5 GHz)

The current IEEE 802.11b standard and IEEE 802.11g standard operate in the 2.4GHz radio frequency band, while the IEEE 802.11a operates in the 5GHz radio band. The IEEE 802.11n standard has been established as a new WiFi standard that provides backward compatibility with the IEEE 802.11a at 5GHz and with the IEEE 802.11b, g at 2.4GHz. The IEEE 802.11n is being developed to overcome (as the standard's claims) previous problems by using three primary modern technologies to satisfy higher data throughput, longer wireless range and reliability of the wireless LAN. The amendment features are not only allowing the IEEE 802.11n standard to achieve an approximate fivefold increase performance over current 802.11a/b/g network, but also be peer to peer with the WiMAX, ([4], [5]).

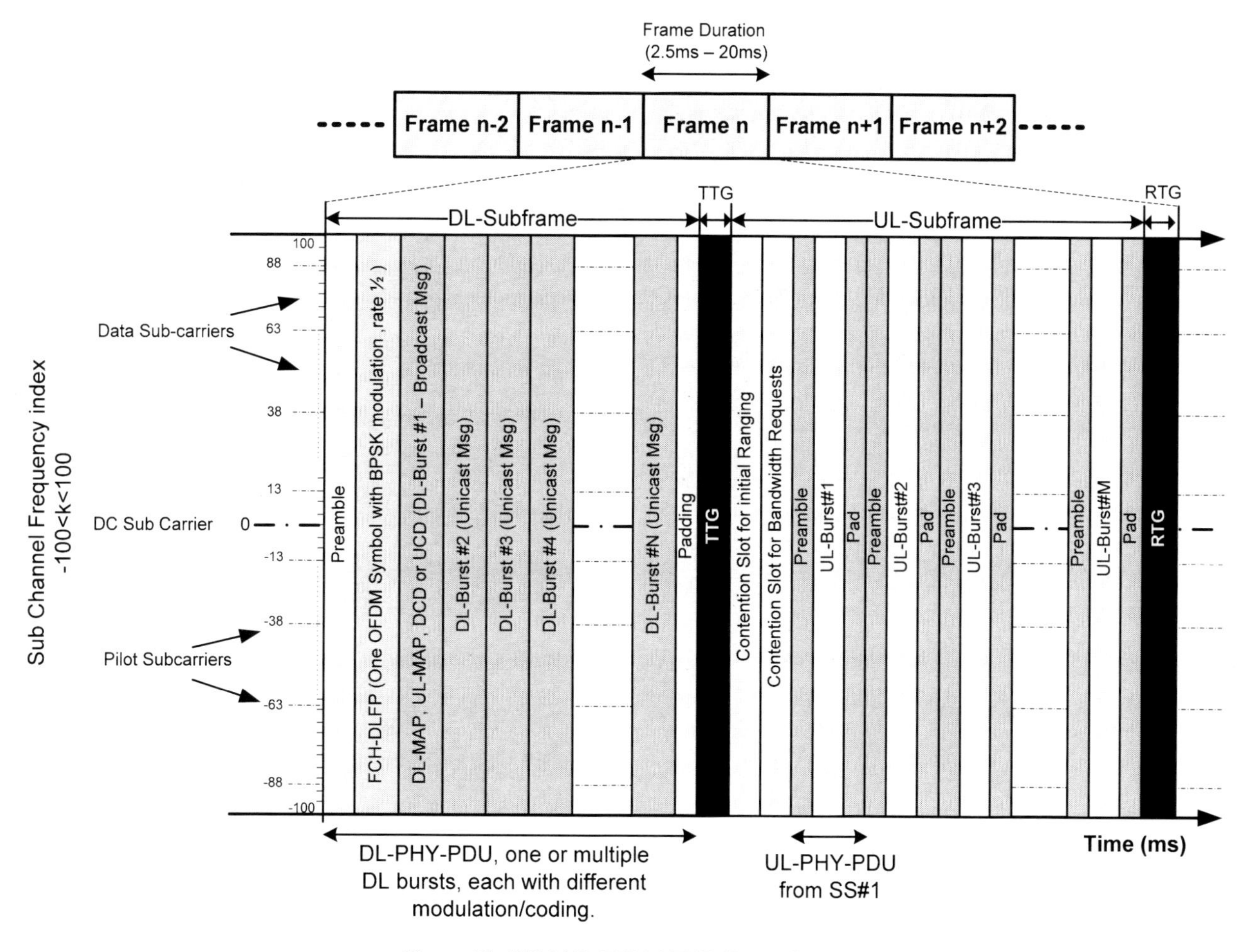

Figure (3), WiMAX-OFDM-TDD Frame Structure

3.2 WiFi-OFDM vs. WiMAX-OFDM

Exploring the similarities and dissimilarities among the wireless standards is the initial step towards the convergence. The WiFi is the predecessor version of the WiMAX; thus a lot of the similarities have been underlined. A major similarity is the OFDM, which will be the common ground between them to initiate the convergence, but resolving the dissimilarities would be our main challenge. Consequently, it has been discovered that the possibilities of the convergence in the single carrier could be resolved in the MAC layer; while in the multi carrier we deal with it as a physical layer issue by creating a '*Convergence-Bridge*' to harmonize the mismatch. In the Multi-Carrier WiMAX-WiFi Convergence, the mismatch between the fixed WiMAX-OFDM (N_{fft}=256) and the WiFi-OFDM (N_{fft}=64) has been confirmed as a physical layer issue that will never be solved as MAC layer problem; therefore the current proposal is how to build what we called the "*Convergence-Bridge*". This bridge is a smart modification in the IEEE 802.11a,n standard physical layer, which is responsible for harmonizing these mismatches,[16].

The OFDM technique is the common ground among the multi-carriers wireless technologies. However, the IEEE 802.16 and IEEE 802.11 standard developers were keen to use the OFDM as a major change for the new standards. The OFDM has been shown to be effective over the ISI (Inter- Symbol Interference) and the channel fading, therefore; the use of OFDM has been relied on, because of its elegant handling of multi path interference. On the other hand, the OFDM requires a costly and a power inefficient transmitter front end, especially a problem in the uplink where the transmitter is a battery driven terminal,[17]. It has been argued that using the single carrier technique is better than the OFDM in terms of data rate and the packet error rate (PER), so this argument goes, the single carrier is a very good candidate for portable data rate terminal in the indoor environment. However, the new wireless standards are being developed under the OFDM techniques because, from cost/performance point of view, it is still seen as the more attractive solution, which has the ability to overcome the multi path propagation problems,[18]. Another argument about the use of the OFDM, [19] had proposed to use a mixed OFDM downlink and single carrier uplink for the IEEE 802.16 which, as they claimed, will ensure to fully benefit from the features of each technology to make cost effective CPE with NLOS operation capability. Ultimately, the final draft of the IEEE 802.16 has not approved [19] approach to avoid the dissimilarities between the downlink and the uplink. However, it is my contention that there are many advantages to be had from using OFDM or single carrier techniques, which depends on the application itself.

4. THE PROPOSED CONVERGENCE-BRIDGE

The Convergence-Bridge is a smart modification in the WiFi OFDM Physical layer to enable the WiFi devices to join the WiMAX-OFDM wireless network. In this paper the WiMAX-Fixed (OFDM-256) and the WiFi-OFDM-64 have been selected to achieve the multi-carrier convergence. The convergence idea was initiated from the similarities between the WiMAX and the WiFi, however the dissimilarities are still real obstacles to enable them to communicate with each other,[20]. Dissimilarities between wireless standards are usually in the lower layers so that the investigations are focused on the PHY and MAC layers. As mentioned, due to the standards investigations, it has been discovered that the convergence in WiMAX-WiFi multi-carrier OFDM is a physical layer issue. RF carrier, Bandwidth, Guard Time, FFT samples and the OFDM duration are the main issues for those physical layer differences that should be harmonized, as shown in figure (4). In this convergence, several WiFi and WiMAX PHY specifications could be involved. However choosing a specific WiMAX-PHY is vital because these diverse PHY specifications will generate different MAC level management messages. As mentioned above in section 1, IEEE 802.16 standard proposes four different PHY specifications. Following the work in the WiMAX Single Carrier Access that has been solved in the previous paper [10], the methodology WiMAX-OFDM has been selected for this paper while the WiMAX-OFDMA will be investigated in future work. The WiMAX-SC PHY (10-66 GHz) has not been selected due to frequency band incompatibility. For the WiFi-OFDM physical layer, the paper has selected the IEEE 802.11n OFDM standard while it is being developed. Figure (4) illustrates the IEEE 802.11 OFDM Physical layer and the proposed modification that will satisfy the convergence. The proposal does not suggest changing the standard itself but modifying some functions to be configurable. These configurable functions have been highlighted as Grey areas in figure (4). The IEEE 802.11 standard has fixed the configurations for WiFi mode only, while our proposal is to set up these functions for WiFi and WiMAX modes.

4.1 Unified the WiFi-WiMAX Frequency Bands

The RF Central Frequency is the main factor in deciding which technology is being used and it also represents the identification factor of a certain frequency band. Figure (2) illustrates two different OFDM spectrums in different frequency bands. The IEEE 802.11a,n standard is being carried on 2.4GHz or 5GHz, while the IEEE 802.16 OFDM – TDD standard is being carried on 3.5GHz. The first step of the convergence is to unify the two spectrums in a one band. Converting the WiFi RF- signal to the WiMAX frequency band has been illustrated in figure (2). Even though the WiFi signal can exist in the WiMAX band but they cannot interoperate due to the differences outlined above.
As mentioned in section 2.2, generally by changing the equation parameters, any OFDM signal could be produced from equation (1) whether it is WiMAX-OFDM or WiFi-OFDM. This equation underpins the design of the proposed *Convergence-Bridge*. Basically, the convergence enables the WiFi device to receive a signal *S(t)* from a WiMAX Base Station and transmit a signal *S(t)* to a WiMAX Base Station,[21]. The following steps are the mathematical implantation of these signals:

For the WiFi-OFDM,

$$S_1(t) = Re\left\{ e^{j2\pi f_{c1} t} \cdot \sum_{\substack{k=-26 \\ k\neq 0}}^{+26} C_k \cdot e^{j2\pi k \Delta f_1 (t-T_{g1})} \right\} \ldots\ldots\ldots\ldots\ldots\ldots\ldots\ldots\ldots\ldots (2)$$

Whereas, $S_1(t)$ is the time domain equation for the WiFi-OFDM-64, f_{c1} is the central frequency which is either 2.4GHz or 5GHz, k is the frequency index (52 subcarrier indices) which is $-26 \leq k \leq +26$, N_{used} is 52 subcarriers, 48 data subcarriers + 4 pilot subcarriers. There are also 14 frequency guard subcarriers (7 *lower* frequency guard subcarriers band + 7 *higher* frequency guard subcarriers band), which have not appeared in the equation. In total 64 subcarriers (48 data subcarrier + 4 pilot subcarriers+ 14 frequency guard subcarriers) are there in the WiFi-OFDM. Δf_1 is the subcarrier frequency spacing (Δf_1=BW/N_{fft}=20MHz/64= 312.5 KHz). T_{g1} is the guard time (T_{g1}=0.8 μs).

For the WiMAX-OFDM,

$$S_2(t) = Re\left\{ e^{j2\pi f_{c2}t} . \sum_{\substack{k=-100 \\ k\neq 0}}^{+100} C_k . e^{j2\pi k\Delta f_2(t-T_{g2})} \right\} \ldots\ldots\ldots\ldots\ldots\ldots\ldots\ldots\ldots (3)$$

Whereas, $S_2(t)$ is the time domain equation for the WiMAX-OFDM-256, f_{c2} is the central frequency which is 3.5 GHz, k is the frequency index (200 subcarrier indices) which is $-100 \leq k \leq +100$, N_{used} is 200 subcarriers, 192 data subcarriers + 8 pilot subcarriers. There are also 55 frequency guard subcarriers (28 *lower* frequency guard subcarriers band + 27 *higher* frequency guard subcarriers band), which have not appeared in the equation. In total 256 subcarriers (192 data subcarrier + 8 pilot subcarriers+ 55 frequency guard subcarriers +1 DC Subcarrier) are there in the WiMAX-OFDM. Δf_2 is the subcarrier frequency spacing (Δf_2= 15.625 KHz). T_{g2} is the guard time (T_{g2}=18.24 μs).

Despite the fact that WiFi-OFDM-64 and the WiMAX-OFDM-256 signals have been generated from the same equation, the above dissimilarities of the equation's parameters have denied them to communicate with each other. The current proposal in this paper suggests enabling specific functions in the WiFi PHY-layer designed to be configurable. These functions have been highlighted as grey areas in figure (4). The first part (function) is the RF-oscillator; which participates alternatively in the down/up conversions of the RF signals in the receiver/transmitter tasks. There is only one RF-OSC device, which is being used by both the transmissions and receiving processes. Currently, the WiFi PHY layer can detect the WiMAX signal but it cannot convert the WiMAX signal in the RF level because the RF oscillator of the WiFi does not have the 3.5 GHz carrier frequency. The suggestion is to enable the RF-OSC to have the 3.5 GHz carrier frequency in addition to the original RF carrier frequencies 2.4 GHz and 5GHz. As shown in figure (2), having this facility could only exist WiFi and WiMAX signals in the same RF bands but they cannot interoperate due to the other differences. The second part (device) is the Automatic Frequency Control (AFC) clock recovery, which detects/maintains the drift of the clock and also participates in separating the IQ signals. While the AFC device has been designed to work over 20MHz bandwidth (WiFi-OFDM), the paper suggests enabling the AFC device to also work over 3.5 MHz bandwidth (WiMAX-OFDM). The third part is how to remove the guard time length; part of the proposal is to configure this stage to be able to remove not only the o.8 μs guard time (WiFi), but also 18.24 μs for the WiMAX. The fourth part is the bottleneck part, the FFT block of the WiFi-PHY layer transforms only 64 samples simultaneously, while the receiving signal $S_2(t)$ has 256 samples. Having a 256/64 FFT block instead is our proposal to overcome this bottleneck.

4.2 WiFi-WiMAX Convergence Steps

As mentioned in the previous section, modifying those functions (grey blocks) to be configurable is our suggestion to satisfy the *convergence bridge* requirements. Figure (4) represents the WiFi PHY layer; the top part is the receiver part of the PHY layer while the bottom part is the transmitter. Most of the transmitter / receiver stages (functions) are reversible functions. The middle part in the figure (4) is the WiFi/WiMAX configuration parameters table that illustrates the two different modes; mode-1 is the original configuration parameters of the WiFi device, while mode-2 is the additional configuration parameters that will enable the WiFi PHY layer to behave like a WiMAX device. It is not possible to activate the two modes simultaneously, because they are using the same physical layer blocks in different configurations.

The following steps explain how a WiMAX signal $S_2(t)$ could be converted and processed in the WiFi PHY layer. Figure (4) shows number of test points that will be used to track the signal through the PHY layer stages:

1. The WiMAX-OFDM-256 signal, $S_2(t)$, is being carried on 3.5GHz carrier frequency with 256 OFDM samples.
2. The WiFi antenna detects between 2.4 and 5GHz carrier frequencies, therefore; 3.5 GHz is within the antenna's detection range.
3. At the first test point (T_1), $S_2(t)$ went through the LNA (Low Noise Amplifier) form noise elimination and signal amplifications.

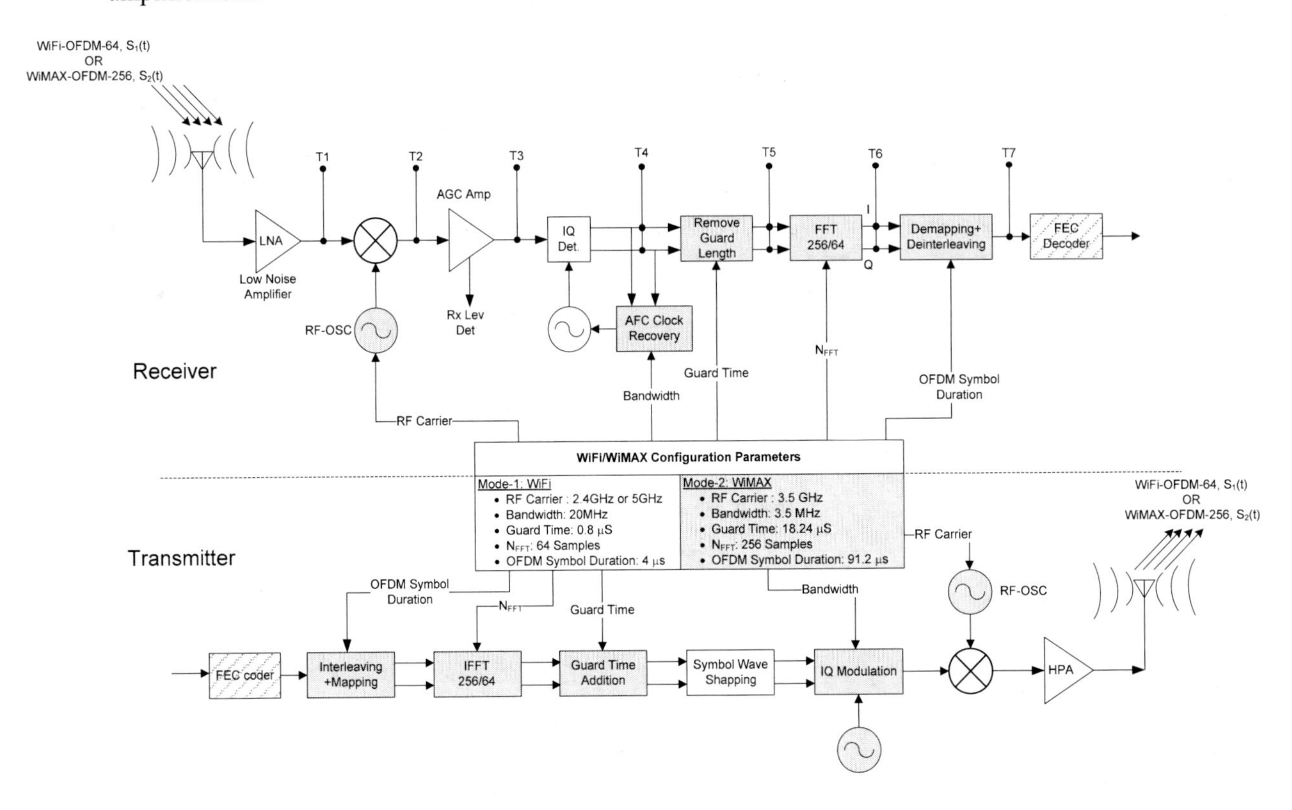

Figure (4), the WiFi-OFDM Physical layer
(The Grey area is the modified blocks)

4. At T_2, the signal will be down-converted through the RF-conversation stage.
The RF-OSC generates a sinusoidal signal, $\mathrm{Cos}(2\pi f_{c2}t) = \frac{1}{2}(e^{-j2\pi f_{c2}t} + e^{j2\pi f_{c2}t})$, that will be multiplied in the time domain by an OFDM symbol $S_2(t)|_{T1}$, ($S_2(t)$ at T_1):

$S_2(t)|_{T2} = S_2(t)|_{T1} \times \mathrm{Cos}(2\pi f_{c2}t)$

$$= \frac{1}{2} Re \left\{ e^{-j2\pi f_{c2}t} . e^{j2\pi f_{c2}t} . \sum_{\substack{k=-100 \\ k\neq 0}}^{+100} C_k . e^{j2\pi k\Delta f_2(t-T_{g2})} + e^{j2\pi f_{c2}t} . e^{j2\pi f_{c2}t} . \sum_{\substack{k=-100 \\ k\neq 0}}^{+100} C_k . e^{j2\pi k\Delta f_2(t-T_{g2})} \right\}$$

$$S_2(t)|_{T2} = \frac{1}{2} Re \left\{ \sum_{\substack{k=-100 \\ k\neq 0}}^{+100} C_k . e^{j2\pi k\Delta f_2(t-T_{g2})} + e^{j4\pi f_{c2}t} . \sum_{\substack{k=-100 \\ k\neq 0}}^{+100} C_k . e^{j2\pi k\Delta f_2(t-T_{g2})} \right\}$$

The second part of the above equation is a by-product signal, which represents the second harmonic of the carrier frequency. It has been generated as a result of the multiplication of the positive frequency part ($e^{j2\pi f_{c2}t}$) of the RS-OSC sinusoidal signal ($\mathrm{Cos}(2\pi f_{c2}t)$). Any harmonic signal is being eliminated by a Low Pass Filter. Therefore a WiMAX-OFDM Symbol $S_2(t)$ at T_2 will be defined by the following equation:

$$S_2(t)|_{T2} = \frac{1}{2}\sum_{\substack{k=-100 \\ k\neq 0}}^{+100} C_k\,.\,e^{j2\pi k\Delta f_2(t-T_{g2})} \ldots\ldots\ldots (4)$$

5. At T_3, the **Automatic gain control** (AGC) is an adaptive device to adjust the average output signal level to an appropriate level for a range of input signal levels.

$$S_2(t)|_{T3} = \sum_{\substack{k=-100 \\ k\neq 0}}^{+100} C_k\,.\,e^{j2\pi k\Delta f_2(t-T_{g2})} \ldots\ldots\ldots (5)$$

6. At T_4, the IQ time domain separation happens at this sage. The WiMAX-OFDM Symbol, $S_2(t)$signal, separates into two parts; real and imaginary parts. As mentioned in section 2.2, C_k is the data to be transmitted on the subcarriers whose frequency offset index is k, during an OFDM symbol:

$$\boldsymbol{C_k}.\,e^{j2\pi k\Delta f_2(t-T_{g2})} = \boldsymbol{I_k}.\,Cos(j2\pi k\Delta f_2(t-T_{g2})) + j.\,\boldsymbol{Q_k}.\,Sin(j2\pi k\Delta f_2(t-T_{g2}))$$

$S_2(t)|_{T4}$ =I(t)+j.Q(t), where I(t) and Q(t) are:

$$\mathrm{I(t)} = \sum_{\substack{k=-100 \\ k\neq 0}}^{+100} \boldsymbol{I_k}.\,Cos(j2\pi k\Delta f_2(t-T_{g2})) \ldots\ldots\ldots (6)$$

$$\mathrm{Q(t)} = \sum_{\substack{k=-100 \\ k\neq 0}}^{+100} \boldsymbol{Q_k}.\,Sin(j2\pi k\Delta f_2(t-T_{g2})) \ldots\ldots\ldots (7)$$

7. At T_5 the guard time length is removed from the signals I(t) and Q(t).The guard time is one of the modified configuration parameters that has been highlighted in figure(4). For the WiMAX-OFDM-256 signal the guard time is (T_{g2}=18.24 μs). This stage prepares the IQ signals (an OFDM Symbol) to be transformed from time domain to frequency domain using the Fast Fourier Transform stage . The WiFi PHY layer has been designed to transform only 64 samples in the FFT. But, part of the *convergence bridge* proposal is to modify the FFT to transform 256 samples also. The IQ signals equations (an OFDM symbol) will be:

$$\mathrm{I(t)} = \sum_{\substack{k=-100 \\ k\neq 0}}^{+100} \boldsymbol{I_k}.\,Cos(j2\pi k\Delta f_2(t)) \ldots\ldots\ldots (8)$$

$$\mathrm{Q(t)} = \sum_{\substack{k=-100 \\ k\neq 0}}^{+100} \boldsymbol{Q_k}.\,Sin(j2\pi k\Delta f_2(t)) \ldots\ldots\ldots (9)$$

8. At T_6, the FFT function transforms the I(t) and the Q(t) signals to the frequency domain .It generates two vectors : I-vector and Q-vector with 256 length each. The combination of I and Q vectors represents a single OFDM symbol. At this point the IQ-vectors (data) appears will contain real numbers. $I=[I_1,I_2,I_3,\ldots,I_{256}]$ and $Q=[Q_1,Q_2,Q_3,\ldots,Q_{256}]$, where I_k and Q_k are real numbers, and $C_k= I_k + j\,Q_k$.

9. At T_7, each IQ symbol is converted to a binary number. The number of bit per symbol could be determined by knowing the modulation type that has been used in the current OFDM symbol. The number of bit per symbol is equal to 2,4 or 6 bits per symbol if the modulation type is BPSK, 16QAM or 64QAM respectively. For instance, if the current OFDM symbol has been sent using 16QAM modulation type, each C_k (whereas $C_k = I_k + j.Q_k$) is converted to 4 bits binary number. Therefore, a full IQ-vector (one OFDM symbol) generates (4x256) 1024 bits as an input vector to the FEC (Forward Error Correction) block.
10. As shown in figure (4), the FEC block has been highlighted in a different pattern to indicate a future investigation of the encoder/decoder of WiFi PHY layer. The implementation of Read Salmon block code and Vertabi convolution code are in principal simple bit polynomial manipulations,[22].

5. CONCLUSION AND FUTURE WORKS

The effort of this paper is to produce multi-carrier convergence between Fixed WiMAX-TDD (OFDM-256) and WiFi-OFDM-64, which is an ideal technology that provides the best of both worlds: the new features of the WiMAX and the low cost of the WiFi. In order to create a heterogeneous network environment, differences between the two technologies have been investigated and resolved. Dissimilarities among wireless standards are usually in the lower layers so that the investigations are focused on the PHY and MAC layers. In the standards investigations, it has been discovered that the convergence in WiMAX-WiFi multi-carrier OFDM is a physical layer issue. RF carrier, Bandwidth, Guard Time, FFT samples and the OFDM duration are the main issues for those physical layer differences that should be harmonized. For the WiFi-OFDM physical layer, the paper has selected the IEEE 802.11n OFDM standard while it is being developed. The proposal does not suggest changing the standard itself but modifying some functions to be configurable. The IEEE 802.11 standard has fixed the configurations for WiFi mode only, while our proposal is to set up these functions dynamically for WiFi and WiMAX modes. Following the work in the WiMAX Single Carrier Access that has been solved in the previous paper (10), the methodology WiMAX-OFDM (256) has been selected for this paper while the Mobile WiMAX-OFDMA (512, 1024, and 2048) will be investigated in the future. In this paper only the WiMAX-TDD (OFDM-256) has been involved in the convergence investigations, but it is also recommended as a future work to involve the WiMAX-FDD (OFDM-256) in these convergence investigations. Future improvement would include the implementation of simple Read Salmon block code and Vertabi convolution code.

A new vision of the convergence has emerged when two different wireless standards have created the MESH topology as vertical developments. As a future work, while the Wireless Mesh Network standards are under development, the paper suggests studying the potentials of the WiMAX-WiFi convergence in the Mesh topology also. Apparently, the WiMAX-WiFi convergence has a good chance to be involved in the mesh topology applications.

ACKNOWLEDGMENT

- The first author thanks the Kurdistan Regional Government of Iraq for funding this work as part of his PhD studies.
- This work was supported by the COST Action IC0803 "RF/Microwave Communication Subsystems for Emerging Wireless Technologies", (RFCSET).

REFERENCES

[1] Al-Sherbaz, Ali; Jassim, Sabah; Adams, Chris. *Convergence in wireless transmission technology promises best of both worlds.* Florida : SPIE Opt electronics & Optical Communications newsroom, Nov-2008.

[2] Intel. *Understanding Wi-Fi and WiMAX as Metro-Access Solutions.* s.l. : Intel, 2004. White Paper.

[3] Clint, Smith and Meye, John. *3G Wireless with WiMAX and WiFi.* New York : McGRAW-HALL, 2004.

[4] SIEMENS. *Practical Considerations for Deploying 802.11n.* Munchen, Germany : Siemins Enterprise Communications, 2008.

[5] *Shorter wireless technologies, wireless fidelity (WiFi) & Worldwide interoperability for microwave access (WiMAX).* Bradley K., Patton, Dr. Richard, Aukerman and Dr. Jack, D. 2, s.l. : Issues in Information Systems Volume, 2005, Vol. 4.

[6] Madden, Samuel; Levis, Philip;. Mesh Networking: Research and Technology for Multihop Wireless Networks. *IEEE Computer Society.* July/Augest 2008, pp. 9-11.

[7] *Towards Guaranteed QoS in Mesh Networks: EmulatingWiMAX Mesh over WiFi Hardware* . Petar, Djukic and Shahrokh, Valaee. 2007. 27th International Conference on Distributed Computing Systems Workshops (ICDCSW'07) . pp. 15-20.
[8] DeBeasi, Paul. *8011.n:Enterprise Deployment Considerations.* Utah : Burton Group, 2008. White Paper.
[9] *A Survey on Wirless Mesh Networks.* Akyildiz, Ian F. and Wang, Xudong. September 2005, IEEE Radio Communications, pp. 23-30.
[10] *Private synchronization technique for heterogeneous wireless network (WiFi and WiMAX).* Al-Sherbaz, Ali, Adams, Chris and Jassim, Sabah. Florida-USA : SPIE--The International Society for Optical Engineering, 2008. Mobile Multimedia/Image Processing, Security, and Applications 2008. pp. 3 -11.
[11] IEEE Standard for Local Metropolitan Area Network. *Part16: Air Interface for Fixed Broadband Wireless Access System.* s.l. : IEEE, 2004.
[12] Shepard, Steven. *WiMAX Crash Course.* New York : McGraw Hill, 2006.
[13] Altera. *An OFDM FFT Kernel for WiMAX.* San Jose, CA : Altera Corporation, 2007. Application Note.
[14] Stallings, William. *Wireless Communication and Network.* New Jersey : Prince Hall, 2002.
[15] *Bridging solutions for a heterogeneous WiMAX-WiFi scenario.* Roman, Fantacci and Daniele, Tarchi. 4, Dec 2006, Journal of Communications and Networks, Vol. 8, pp. 1-9.
[16] *Mobility Using IEEE 802.21 in A Hetrogeneous IEEE 802.16/802.11-Based, IMT-Advanced (4G) Network.* Eastwood, Les, et al. April 2008, IEEE Wireless Communications, pp. 26 - 34.
[17] *From WiFi to WiMAX: Techniques for High-Level IP Reuse across Different OFDM Protocols.* Ng, Man Cheuk, et al. Nice, France : s.n., 2007. Formal Methods and Models for Codesign, 2007. MEMOCODE 2007. 5th IEEE/ACM International Conference on. pp. 71-80.
[18] *WiMAX Networks: From Access to Service Platform.* Lu, Kejie, et al. May 2008, IEEE Network, pp. 38 - 45.
[19] Ran, Moshe. *A mixed OFDM downlink and single carrier uplink for the 2-11 GHz licensed bands.* s.l. : IEEE 802.16a, 2002.
[20] *Impact of Wireless (WiFi,WiMAX) on 3G and Next Generation-An intial Assessment.* Behmann, Fawzi. Lincoln,NE : IEEE Xplore, 2005. Electro Information Technology,2005 IEEE International Confrence. pp. 1-6.
[21] 802.11, IEEE Std. *Part 11: Wireless LAN Medium Access Control (MAC) and Physical Layer (PHY) Specification.* New York : IEEE Comuter Society, 2007.
[22] Altera. *Accelerating WiMAX System Design with FPGAs.* San Jose, CA : Altera Corporation, 2004.

On A Nascent Mathematical-Physical Latency-Information Theory, Part I: The Revelation Of Powerful And Fast Knowledge-Unaided Power-Centroid Radar

Erlan H. Feria
Department of Engineering Science and Physics
The College of Staten Island of the City University of New York
E-mail feria@mail.csi.cuny.edu Web site http://feria.csi.cuny.edu

ABSTRACT

In this first part of the latest latency-information theory (LIT) and applications paper series powerful and fast 'knowledge-unaided' power-centroid (F-KUPC) radar is revealed. More specifically, it is found that for real-world airborne moving target indicator radar subjected to severely taxing environmental conditions F-KUPC radar approximates the signal to interference plus noise ratio (SINR) radar performance derived with more complex knowledge-aided power-centroid (KAPC) radar. KAPC radar was discovered earlier as part of DARPA's 2001-2005 knowledge-aided sensor signal processing expert reasoning (KASSPER) Program and outperforms standard prior-knowledge radar schemes by several orders of magnitude in both the compression of sourced intelligence-space of prior-knowledge, in the form of SAR imagery, and the compression of processing intelligence-time of the associated clutter covariance processor, while also yielding an average SINR radar performance that is approximately 1dB away from the optimum. In this paper, it is shown that the average SINR performance of significantly simpler F-KUPC radar emulates that of KAPC radar and, like KAPC radar, outperforms a conventional knowledge-unaided sample covariance matrix inverse radar algorithm by several dBs. The matlab simulation programs that were used to derive these results will become available in the author's Web site.

Index Terms— *Latency, Information, Intel-Space Compression, Intel-Time Compression, Knowledge-Aided, Knowledge-Unaided, Adaptive Radar, Sample Covariance Matrix Inverse*

1. INTRODUCTION

A straight forward approach to adaptive airborne moving target indicator (AMTI) radar, see Fig. 1, that does not use clutter prior-knowledge is the sample covariance matrix inversion (SCMI) scheme that is used to approximate the optimum Wiener-Hopf weighting vector that arises from maximizing target signal to interference plus noise ratio (SINR) for an investigated range-bin [1]. The construction of the optimum Wiener-Hopf algorithm requires knowledge of the interference plus noise covariance of the range-bin. The inverted range-bin interference plus noise covariance is then multiplied by the steering vector of the assumed target to yield a complex weighting vector of dimension NM where N is the number of antenna elements and M is the number of its transmitted pulses during a coherent pulse interval (CPI). This weighting vector is then multiplied by an NM dimensional complex vector that is measured by the radar receiver and reflects range-bin target, clutter and additional radar interferences and noise. The result of this multiplication is a complex scalar variable that is used by the AMTI to determine if a target appears on the range-bin or not. Since the actual interference plus noise covariance is not available in a real-world scenario, the SCMI scheme addresses this issue by using one or more measurements from range-bins that are adjacent to the range-bin in question to construct a sample covariance matrix (SCM). Unfortunately, however, in a real-world non-stationary environment this very simple scheme only crudely approximates the exact range-bin covariance, which often manifests itself in an unsatisfactory SINR radar performance. To address this problem clutter prior-knowledge in the form of SAR imagery has been advanced, such as was done by DARPA in its 2001-2005 knowledge-aided sensor signal processing expert reasoning (KASSPER) Program [2]. The use of these SAR images as prior-knowledge achieves SINR performances that significantly outperform the SCMI algorithm as well as approximate the optimum results derived from the optimum Wiener-Hopf scheme. On the down side, unfortunately, the use of SAR imagery prior-knowledge has the drawback of increasing the complexity of the

This work was supported in part by the Defense Advanced Research Projects Agency (DARPA) under Grant No. FA8750-04-1-004 and the PSC-CUNY Research Awards PSCREG-37-913, 38-247, 39-834, 40-1013

Mobile Multimedia/Image Processing, Security, and Applications 2009, edited by Sos S. Agaian, Sabah A. Jassim,
Proc. of SPIE Vol. 7351, 73510U · © 2009 SPIE · CCC code: 0277-786X/09/$18 · doi: 10.1117/12.819040

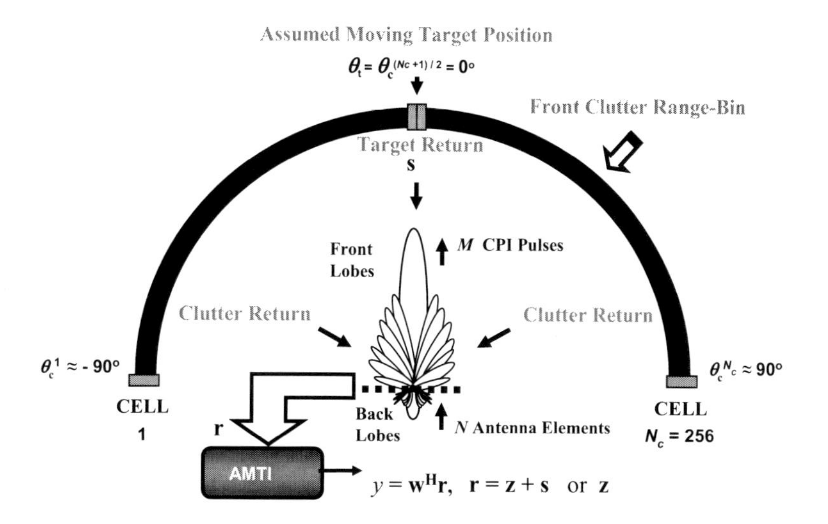

Fig. 1 Airborne Moving Target Indicator Radar System

radar system due to the storage space needs for the 'sourced intelligence-space (or intel-space in short)', i.e. the SAR imagery, and the 'processing intelligence time (or intel-time in short)' needs for the on-line evaluation of the clutter covariance from SAR imagery. To address this problem a novel latency-information theory (LIT) methodology has been advanced [3] whose catalyst was a DARPA KASSPER grant [4] for the compression of SAR imagery via minimum mean squared error (MMSE) predictive-transform (PT) source coding [5], as well as previous Ph.D. work in control theory [6]. In particular, the advanced control work formulated a fundamental parallel processing approach to quantized control using as motivation the discovery in 1978 of an uncertainty-communication/certainty-control duality, unexplored until then, that exists between an uncertainty 'digital' communication problem, i.e. Matched Filters, and a certainty 'quantized' control problem, given the name Matched-Processors from a duality perspective. Thus not surprisingly LIT was formulated by the author as a unified approach to the compression of intel-space—in binary digit or bit units of a passing of time uncertainty nature—and the compression of intel-time—in binary operator or bor units of a configuration of space certainty nature. Using this novel LIT conceptualization a knowledge-aided power-centroid (KAPC) radar algorithm was found that yielded orders of magnitude improvements in both intel-space and intel-time savings over standard techniques. The fundamental idea behind this low storage and very fast clutter prior-knowledge scheme was the use of off-line designed predicted clutter covariances (PCCs). The PCCs were developed using a mathematical model of the physically produced antenna pattern that pointed in the direction of the range-bin power-centroid rather than the target. The objective of this model was to serve as a compensating antenna pattern (CAP) for the information lost from SAR images that had been compressed by a factor of 8,172 in a highly lossy manner. The intel-space compression of these images was also done using a radar blind scheme that made it possible for their use in any type of radar system. The main purpose of this paper, is to show that in fact the aforementioned powerful and fast KAPC scheme can be replaced with an even more powerful and fast 'knowledge-unaided' power-centroid (F-KUPC) scheme. The main virtue of F-KUPC radar is that it does not need the use of clutter prior-knowledge in the form of SAR imagery for its evaluation of the range-bin power-centroid as can be easily verified using matlab simulations of standard radar systems. This is the case since with a very small number of on-line scannings of each range-bin in question a very simple on-line algorithm can be used yielding an outstanding on-line estimation of the prerequisite power-centroid.

The rest of the paper is organized as follows. In Section 2 the optimum AMTI radar defining equations are stated. In Section 3 the sample covariance matrix inverse scheme is summarized. In Section 4 the previously offered KAPC radar algorithm is summarized. In Section 5 the newly revealed F-KUPC radar scheme is advanced. In Section 6 simulations for real-world AMTI radar under severe environmental disturbances are given and then conclusions are drawn.

2. OPTIMUM AIRBORNE MOVING TARGET INDICATOR RADAR

The main defining equations for AMTI radar are summarized next for ease of reference.

2.1 The AMTI Radar System

In Fig. 1 the AMTI radar system is displayed. It consists of: a) N antenna elements that emit M pulses during a coherent pulse interval (*CPI*); b) an antenna pattern with its mainbeam pointing to the assumed target location; and c) a front clutter range-bin mathematically modeled as N_C clutter patches or cells that radiate to the AMTI receiving antenna reflections of the M pulses transmitted by the antenna elements as well as the investigated target steering vector times its power. In our simulations the target power times its steering vector will be normalized to one and the target boresight angle θ_t with respect to the moving AMTI antenna is of zero degrees, i.e. $\theta_t = 0^0$. In addition, the assumed N_C number of cells is even where the boundary line between cells $N_C/2$ and $(N_C+2)/2$ is investigated to determine if a moving target appears there. When N_C has a large value the first and last clutter cells have a clutter boresight angle θ_C^i that is slightly greater than -90° for $i=1$ and slightly less than 90° for $i=N_C$, respectively. In Table 1 the N_C, N and M values used in our simulations are summarized along with other radar parameters.

2.2 The Analytical Antenna Pattern Model

Regarding the antenna pattern of Fig. 1, it is assumed that the ith clutter cell receives from this pattern the ith antenna gain which is derived from the following analytical expression

$$g_i(\theta_t) = K^f \left| \frac{\sin\left\{ N\pi \frac{d}{\lambda}(\sin(\theta_C^i) - \sin(\theta_t)) \right\}}{\sin\left\{ \pi \frac{d}{\lambda}(\sin(\theta_C^i) - \sin(\theta_t)) \right\}} \right|^2 / N^2 \tag{1}$$

where i=1,..,N_C, θ_t is the target boresight angle, d is the antenna inter-element spacing, λ is the operating-wavelength, and K^f is the front antenna gain constant.

2.3 The AMTI Scalar Complex Output

The scalar complex output of the AMTI system is given by the expression

$$y = \mathbf{w}^H(\mathbf{z}+\mathbf{s}) \tag{2}$$

where: a) **s** is an NM dimensional complex normalized steering vector representing the target; b) **z** is an NM dimensional complex interference plus noise complex vector; c) 'H' denotes transpose and complex conjugation; and d) **w** is an NM dimensional complex weighting vector that is designed with the view of maximizing SINR.

2.4 The Signal to Interference Plus Noise Ratio (SINR)

The SINR expression that is maximized is given by

$$\text{SINR} = \mathbf{w}^H\mathbf{s}\mathbf{s}^H\mathbf{w} / \mathbf{w}^H C\mathbf{w}. \tag{3}$$

where $\mathbf{w}^H\mathbf{s}\mathbf{s}^H\mathbf{w}$ is the signal power in (2) and $\mathbf{w}^H C\mathbf{w}$ is the interference plus noise power in (2) with C being the covariance of the interference plus noise, i.e. $C=E[\mathbf{z}^H\mathbf{z}]$.

2.5 The Target Steering Vector s

The normalized steering vector for the target is given by the expressions

$$\mathbf{s} = [\underline{\mathbf{s}}_1(\theta_t)\ \underline{\mathbf{s}}_2(\theta_t) \ldots \underline{\mathbf{s}}_M(\theta_t)]^T / \sqrt{NM} \tag{4}$$

$$\underline{\mathbf{s}}_k(\theta_t) = e^{j2\pi(k-1)\overline{f}^t_D}\, \underline{\mathbf{s}}_1(\theta_t) \quad for\ k = 1,\ldots,M \tag{5}$$

$$\underline{\mathbf{s}}_1(\theta_t) = [\mathrm{s}_{1,1}(\theta_t)\ \mathrm{s}_{2,1}(\theta_t)\ \ldots\ \mathrm{s}_{N,1}(\theta_t)] \tag{6}$$

$$s_{k,1}(\theta_t) = e^{j2\pi(k-1)\bar{\theta}_t} \quad for\ k = 1,\ldots,N \tag{7}$$

$$\bar{f}^t_D = f^t_D / f_r \tag{8}$$

$$f^t_D = 2v_p / \lambda = 2(v_p / c) f_c \tag{9}$$

$$f_r = 1 / T_r \tag{10}$$

$$\bar{\theta}_t = (d / \lambda)\sin(\theta_t) \tag{11}$$

where: a) θ_t is the boresight position of the target which is of 0° for the case displayed in Fig. 1 as well as in the simulations; b) f_c is the carrier (or operating) frequency; c) $\bar{\theta}_t$ is the normalized θ_t; d) T_r is the pulse repetition interval (PRI); e) f_r is the pulse repetition frequency (PRF); f) v_p is the target radial velocity; g) c is the speed of light; h) f^t_D is the target Doppler; and i) $\bar{f}^t_D$ is the normalized Doppler.

2.6 The Interference Plus Noise Covariance C=E[$\mathbf{z}^H\mathbf{z}$]

The interference plus noise covariance C is given by the following covariance matrix tapers (CMTs) structure [7]

$$C = \{ (C^f_c + C^b_c) \circ (C_{RW} + C_{ICM} + C_{CM})\} + \{C_J \circ C_{CM}\} + C_n \tag{12}$$

where $\mathbf{C}_n$, C^f_c, C^b_c, $\mathbf{C}_J$, $\mathbf{C}_{RW}$, $\mathbf{C}_{ICM}$ and $\mathbf{C}_{CM}$ are covariance matrices of dimension *NM x NM* and the symbol 'O' denotes a Hadamard product or element by element multiplication. These covariances correspond to: $\mathbf{C}_n$ to thermal white noise; C^f_c to front clutter; C^b_c to back clutter; $\mathbf{C}_J$ to jammer; $\mathbf{C}_{RW}$ to range walk; $\mathbf{C}_{ICM}$ to internal clutter motion; and $\mathbf{C}_{CM}$ to channel mismatch. The defining expressions for each of these covariances are given next starting with the front clutter covariance C^f_c which in knowledge-aided radar is evaluated making use of prior-knowledge such as SAR imagery (2). Note that the back clutter covariance C^b_c will be assumed to offer a negligible contribution to C since we will be assuming in our simulations that the back antenna gain constant K^b is of 10^{-4}, i.e. – 40 dBs.

2.6.1 The Front Clutter Covariance C^f_c

$$C^f_c = \sum_{i=1}^{N_C} x_i g_i(\theta_t)\mathbf{c}_i(\theta_{AAM})\mathbf{c}^H_i(\theta_{AAM}) \tag{13}$$

$$\mathbf{c}_i(\theta_{AAM}) = [{}_f\underline{c}_1(\theta^i_c,\theta_{AAM})\ \ {}_f\underline{c}_2(\theta^i_c,\theta_{AAM})\ \cdots\ {}_f\underline{c}_M(\theta^i_c,\theta_{AAM})]^T \tag{14}$$

$${}_f\underline{c}_k(\theta^i_c,\theta_{AAM}) = e^{j2\pi(k-1)\bar{f}^{c_f}_D(\theta^i_c,\theta_{AAM})}\,\underline{c}_1(\theta^i_c) \quad for\ k = 1,\ldots,M \tag{15}$$

$$\underline{c}_1(\theta^i_c) = [c_{1,1}(\theta^i_c)\ \ c_{2,1}(\theta^i_c)\ \ \ldots.\ \ c_{N,1}(\theta^i_c)] \tag{16}$$

$$c_{k,1}(\theta^i_c) = e^{j2\pi(k-1)\bar{\theta}^i_c} \quad for\ k = 1,\ldots,N \tag{17}$$

$$\bar{f}^{c_f}_D(\theta^i_c,\theta_{AAM}) = \beta\bar{\theta}^i_C \tag{18}$$

$$\beta = (v_p T_r)/(d/2) \tag{19}$$

$$\bar{\theta}^i_C = (d/\lambda)\sin(\theta^i_C + \theta_{AAM}) \tag{20}$$

where: a) the index i refers to the ith front clutter cell on the front range bin section shown in Fig. 1; b) θ^i_c is the boresight angle of the ith clutter cell; c) θ_{AAM} is the antenna array misalignment angle; d) x_i is the ith front clutter source cell power; e) θ_t is the target boresight angle; f) $g_i(\theta_t)$ is the antenna gain linked to the ith front clutter cell; g) $\mathbf{c}_i(\theta_{AAM})$ is the front *NM* x 1 dimensional and complex i^{th} clutter cell steering vector; h) v_p is the radar platform speed; i) T_r is the PRI; j) f_r is the PRF; k) $\bar{\theta}^i_c$ is the normalized θ^i_c; *l*) d is the antenna inter-element spacing; m) λ is the operating wavelength; and n) β is the ratio of the distance traversed by the radar platform during the PRI, i.e. $v_p T_r$, to the half antenna inter-element spacing, $d/2$. Finally, the first element of the NM by NM matrix C_c^f divided by the noise variance is the front clutter to noise ratio given by

$$\text{CNR}^{\text{f}} = C_c^f(1,1)/\sigma_n^2 = \sum_{i=1}^{N_c} x_i g_i(\theta_t)/\sigma_n^2 \tag{21}$$

where σ_n^2 is the variance of the thermal white noise whose value is assumed to be of one in our simulations, i.e., $\sigma_n^2=1$.

2.6.2. Thermal White Noise Covariance C_n

$$C_n = \sigma_n^2 I_{NM} \tag{22}$$

where σ_n^2 is the average power of thermal white noise and I_{NM} is an identity matrix of dimension NM by NM.

2.6.3. Jammer Covariance C_J

$$C_J = \sum_{i=1}^{N_J} p_i g_i(\theta_t)(I_M \otimes 1_{N\times N}) \text{O}(\text{j}(\theta_J^i) \bullet \text{j}(\theta_J^i)^H) \tag{23}$$

$$\mathbf{j}(\theta_J^i) = [\mathbf{j}_1(\theta_J^i)\ \mathbf{j}_2(\theta_J^i) .... \mathbf{j}_M(\theta_J^i)]^T \tag{24}$$

$$\mathbf{j}_k(\theta_J^i) = \mathbf{j}_1(\theta_J^i) \quad for\ k = 1,..., M \tag{25}$$

$$\mathbf{j}_1(\theta_J^i) = [\text{j}_{1,1}(\theta_J^i)\ \text{j}_{2,1}(\theta_J^i)\ ....\ \text{j}_{N,1}(\theta_J^i)] \tag{26}$$

$$j_{k,1}(\theta_J^i) = e^{j2\pi(k-1)\overline{\theta}_J^i} \quad for\ k = 1,..., N \tag{27}$$

$$\overline{\theta}_J^i = \frac{d}{\lambda}\sin(\theta_J^i) \tag{28}$$

where: a) the index i refers to the ith jammer on the range bin; b) N_J is the total number of jammers; c) θ_J^i is the boresight angle of the ith jammer ; d) $\otimes$ is the Kronecker (or tensor) product; e) I_M is an identity matrix of dimension M by M; f) 1_{NxN} is a unity matrix of dimension N by N; g) p_i is the ith jammer power; and h) $\mathbf{j}(\theta_J^i)$ is the NM x 1 dimensional and complex i^{th} jammer steering vector that is noted from (23)-(28) to be Doppler independent.

The first element of the NM by NM matrix C_J defines the jammer to noise ratio

$$\text{JNR} = C_J(1,1)/\sigma_n^2 = \sum_{i=1}^{N_J} p_i g_i(\theta_t)/\sigma_n^2 \tag{29}$$

2.6.4. Range Walk Covariance CMT C_{RW}

$$C_{RW} = C_{RW}^{time} \otimes C_{RW}^{space} \tag{30}$$

$$[C_{RW}^{time}]_{i,k} = \rho^{|i-k|} \tag{31}$$

$$C_{RW}^{space} = 1_{NxN} \tag{32}$$

$$\rho = \Delta A / A = \Delta A / \{\Delta R \Delta\theta\} = \Delta A / \{(c/B)\Delta\theta\} \tag{33}$$

where: a) c is the velocity of light; b) B is the bandwidth of the compressed pulse; c) ΔR is the range-bin radial width; d) Δθ is the mainbeam width; e) A is the area of coverage on the range bin associated with Δθ at the beginning of the range walk; f) ΔA is the remnants of area A after the range bin migrates during a CPI; and g) ρ is the fractional part of A that remains after the range walk.

2.6.5. Internal Clutter Motion Covariance CMT C_{ICM}

$$C_{ICM} = C_{ICM}^{time} \otimes C_{ICM}^{space} \tag{34}$$

$$[\mathbf{C}_{ICM}^{time}]_{i,k} = \frac{r}{r+1} + \frac{1}{r+1}\frac{(b\lambda)^2}{(b\lambda)^2 + (4\pi\,|\,k\text{-}i\,|\,T_r)^2} \quad (35)$$

$$\mathrm{C}_{ICM}^{space} = 1_{NxN} \quad (36)$$

$$10\log_{10} r = -15.5\log_{10}\omega - 12.1\log_{10} f_c + 63.2 \quad (37)$$

where: a) f_c is the carrier frequency in megahertz; b) ω is the wind speed in miles per hour; c) r is the ratio between the dc and ac terms of the clutter Doppler power spectral density; d) b is a shape factor that has been tabulated; e) c is the speed of light; and f) T_r is the pulse repetition interval.

2.6.6. Channel Mismatch Covariance CMT C_{CM}

$$\mathrm{C}_{CM} = \mathrm{C}_{NB} \circ \mathrm{C}_{FB} \circ \mathrm{C}_{AD} \quad (38)$$

where C_{NB}, C_{FB} and C_{AD} are composite *CMT*s that are defined next.

2.6.7. Finite Bandwidth: C_{FB} is a finite (nonzero) bandwidth (FB) channel mismatch CMT

$$\mathrm{C}_{FB} = \mathrm{C}_{FB}^{time} \otimes \mathrm{C}_{FB}^{space} \quad (39)$$

$$\mathrm{C}_{FB}^{time} = 1_{MxM} \quad (40)$$

$$[\mathrm{C}_{FB}^{space}]_{i,k} = (1-\Delta\varepsilon/2)^2 \,\mathrm{sin\,c}^2(\Delta\phi/2) \quad for \quad i \neq k \quad (41)$$

$$[\mathrm{C}_{FB}^{space}]_{i,i} = 1 - \Delta\varepsilon + \frac{1}{3}\Delta\varepsilon^2 \quad for \quad i = 1,..,N \quad (42)$$

where $\Delta\varepsilon$ and $\Delta\phi$ denote the peak deviations of decorrelating random amplitude and phase channel mismatch, respectively. The square term in (42) corrects an error in the derivation of equation (4.21) in [7].

2.6.8. Angle Dependent: C_{AD} is a reasonably approximate angle-independent CMT for angle-dependent (AD) channel mismatch [7] given by

$$\mathrm{C}_{AD} = \mathrm{C}_{AD}^{time} \otimes \mathrm{C}_{AD}^{space} \quad (43)$$

$$\mathrm{C}_{AD}^{time} = 1_{MxM} \quad (44)$$

$$[\mathrm{C}_{AD}^{space}]_{i,k} = \mathrm{sinc}(B\left|k-i\right|\frac{\mathrm{d}}{\lambda f_{\mathrm{c}}}\sin(\Delta\theta)) \; for \;\; i \neq k \quad (45)$$

$$[\mathrm{C}_{AD}^{space}]_{i,i} = 1 \quad (46)$$

where B is the bandwidth of an ideal bandpass filter and $\Delta\theta$ is a suitable measure of mainbeam width.

2.6.9. Angle Independent Narrowband: C_{NB} is an angle-independent narrowband or NB channel mismatch CMT

$$\mathrm{C}_{NB} = \mathbf{q}\mathbf{q}^H \quad (47)$$

$$\mathrm{q} = \left[\underline{\mathrm{q}}_1 \; \underline{\mathrm{q}}_2 \cdots \underline{\mathrm{q}}_M\right]^T \quad (48)$$

$$\underline{\mathrm{q}}_k = \underline{\mathrm{q}}_1 \quad for \; k = 1,..,M \quad (49)$$

$$\underline{\mathrm{q}}_1 = \left[\varepsilon_1 e^{j\gamma_1} \;\; \varepsilon_2 e^{j\gamma_2} \; \ldots \; \varepsilon_N e^{j\gamma_N}\right] \quad (50)$$

where $\Delta\varepsilon_1,\ldots,\Delta\varepsilon_N$ and $\Delta\gamma_1,\ldots,\Delta\gamma_N$ denote amplitude and phase errors, respectively.

2.7 The Weighing Vector w

The weighting vector **w** that maximizes the SINR (3) is derived via the use of Schwarz's inequality which yields the Wiener-Hopf equation

$$\mathbf{w} = \mathrm{C}^{-1}\mathbf{s} \tag{51}$$

where C is the interference plus noise covariance (12) and **s** is the normalized steering vector of the target.

2.8 The Optimum SINR Performance

The optimum SINR performance, $\mathrm{SINR}_{\mathrm{Opt}}$, is derived from the substitution of the weighting vector (51) in (3) to yield

$$\mathrm{SINR}_{\mathrm{Opt}} = \mathbf{s}^{H}\mathrm{C}^{-1}\mathbf{s}. \tag{52}$$

3. THE SAMPLE COVARIANCE MATRIX INVERSE

The sample covariance matrix inverse (SCMI) algorithm is stated next.

The SCMI Weighting Vector $\mathbf{w}_{\mathrm{SCMI}}$

The SCMI counterpart of the optimum SINR Wiener-Hopf weighting vector (51) is as follows

$$\mathbf{w}_{\mathrm{SCMI}} = {}^{SCM}\mathrm{C}^{-1}\mathbf{s} \tag{53}$$

$${}^{SCM}\mathrm{C} = \frac{1}{L_{SCM}}\sum_{i=1}^{L_{SCM}} \mathbf{Z}_i\mathbf{Z}_i^{H} + \sigma^2_{diag} I \tag{54}$$

where: a) ${}^{SCM}\mathrm{C}$ denotes the sample covariance matrix (SCM); b) $\sigma^2_{diag}I$ is a diagonal loading term where $\sigma^2_{diag}=10$ is used in our simulations to address numerical problems linked with the ${}^{SCM}\mathrm{C}$ inversion: and c) $\{\mathbf{Z}_i, i=1,..,L_{SCM}\}$ denotes L_{SCM} samples from L_{RB} range-bins. Thus

$$L_{SCM} = L_{Scan}L_{RB} \tag{55}$$

where L_{Scan} represents the number of scans of L_{RB} range-bins. In our simulations we will use $L_{SCM}=256$ where the number of range bins L_{RB} is 64 range-bins for a 1024 by 256 SAR image to be discussed in Section 6 and the number of scans L_{Scan} of this image will be 4.

3.2 The Simulation Scheme

To derive the set of range-bin measurements $\{\mathbf{Z}_i, i=1,..,L_{SCM}\}$ the following simulation algorithm will be used

$$\mathbf{Z}_i = \mathrm{C}_i^{1/2}\mathbf{n}_i \tag{56}$$

where $\mathbf{n}_i$ is a zero mean, unity variance, NM dimensional complex random draw and C_i is the interference plus noise covariance (12) associated with the ith range-bin. All the radar and environmental conditions that are used to evaluate expression (12) for each range-bin are summarized in Table 1, inclusive of jammer assumptions.

4. THE KNOWLEDGE-AIDED POWER-CENTROID ALGORITHM

In this section the knowledge-aided power-centroid (KAPC) algorithm is stated which derives the prerequisite range-bin power-centroid from the stored SAR imagery.

4.1 The Knowledge-Aided Power-Centroid (KAPC) Algorithm

The knowledge-aided power-centroid (KAPC) algorithm is given by the following weighting vector expressions

$$\mathbf{w}_{KAPC} = [{}^{KAPC}\mathrm{C}]^{-1}\mathbf{s} \tag{57}$$

$${}^{KAPC}\mathrm{C} = \{ ({}^{KAPC}\mathrm{C}_c^{f} + \mathrm{C}_c^{b}) \,\mathrm{O}\, (\mathrm{C}_{RW} + \mathrm{C}_{ICM} + \mathrm{C}_{CM})\} + \{\mathrm{C}_J \,\mathrm{O}\, \mathrm{C}_{CM}\} + \mathrm{C}_n \tag{58}$$

where the covariances C_c^b, $\mathbf{C}_{RW}$, $\mathbf{C}_{ICM}$, $\mathbf{C}_{CM}$, $\mathbf{C}_J$, and $\mathbf{C}_n$ are the same as those for the interference plus noise covariance $\mathbf{C}$ given by expression (12), and ${}^{KAPC}\mathrm{C}_c^f$ predicts C_c^f of (12).

4.1.1 The *KAPC* Predicted Front Clutter Cell Covariance ${}^{KAPC}\mathrm{C}_c^f$

The defining expressions for ${}^{KAPC}\mathrm{C}_c^f$ are given by

$$ {}^{KAPC}\mathrm{C}_c^f = \sum_{i=1}^{N_C} \overline{\mathrm{g}}_i(\theta_C^{\mathbf{C(xOg)}})\mathbf{c}_i(\theta_{AAM})\mathbf{c}_i^H(\theta_{AAM}) \tag{59} $$

$$ \mathbf{C(xOg)} = \sum_{i=1}^{N_C} i x_i g_i(\theta_t) / \sum_{i=1}^{N_C} x_i g_i(\theta_t) \tag{60} $$

$$ \theta_C^{\mathbf{C(xOg)}} = \mathbf{C(xOg)}\pi / N_C - \pi/2 \tag{61} $$

$$ \mathbf{xOg} = [x_1 g_1 \;\; x_2 g_2 \;\; .... \;\; x_{N_C} g_{N_C}] \tag{62} $$

where: a) $\mathbf{c}_i(\theta_{AAM}), x_i$ and $g_i(\theta_t)$ are as defined for C_c^f (13); b) $\mathbf{xOg}$ is the Hadamard product or element by element product of the power emitted by the investigated range-bin $\mathbf{x} = [x_1 \; x_2 \; .... \; x_{N_C}]$ and the front antenna gain $\mathbf{g} = [g_1 \; g_2 \; .... \; g_{N_C}]$; c) $\mathbf{C(xOg)}$ is the power-centroid which can be any real number from 1 to N_C; d) $\theta_C^{\mathbf{C(xOg)}}$ is the power-centroid boresight angle in radians; and e) $\overline{\mathrm{g}}_i(\theta_C^{\mathbf{C(xOg)}})$ is an appropriately normalized antenna pattern version of (1) that is shifted to the power-centroid and serves as a compensating antenna pattern (CAP) for the lack of either partial or total prior knowledge about the clutter.

4.1.2 The Compensating Antenna Pattern $\overline{\mathrm{g}}_i(\theta_C^{\mathbf{C(xOg)}})$

The CAP predicts the clutter-antenna-gain product set $\{x_i g_i(\theta_t) : \mathrm{i} = 1,...,\mathrm{N_C}\}$ of C_c^f (13) and is defined by the following antenna pattern expression

$$ \overline{g}_i(\theta_C^{\mathbf{C(xOg)}}) = K \left| \frac{\sin\left\{ N\pi \frac{d}{\lambda}(\sin(\theta_C^i) - \sin(\theta_C^{\mathbf{C(xOg)}}))\right\}}{\sin\left\{ \pi \frac{d}{\lambda}(\sin(\theta_C^i) - \sin(\theta_C^{\mathbf{C(xOg)}}))\right\}} \right|^2 \tag{63} $$

$$ {}^{KAPC}\mathrm{C}_c^f(1,1) = \mathrm{C}_c^f(1,1) \tag{64} $$

where K is a normalizing gain that results in the matching of the average clutter power of the front clutter covariance C_c^f and its prediction ${}^{KAPC}\mathrm{C}_c^f$. This scheme which is generally suboptimum since it only approximates the true expression for C_c^f will be found in our simulations to be approximately 1 dB away from the optimum Wiener-Hopf algorithm due to its evaluation of the clutter power-centroid from the true clutter (60) which is available as prior knowledge. However, it must also be noted that earlier in [3]-[4] it had been shown that this scheme also produces outstanding results when the power-centroid of (60) is derived using a SAR image that is a reconstructed version of an original SAR image that is highly lossy compressed.

5. THE KNOWLEDGE-UNAIDED POWER-CENTROID ALGORITHMS

Two power-centroid schemes that do not use clutter prior-knowledge are stated next. The first is a knowledge-unaided power-centroid (KUPC) algorithm that evaluates the power-centroid from the same on-line range-bin measurements $\{\mathbf{Z}_i, i=1,..,L_{SCM}\}$ that are used to construct the SCMI algorithm. This KUPC scheme requires a predicted clutter covariance (PCC) matrix to be derived on-line and has at its disposal full or partial knowledge of other radar interferences and noise covariances. The second scheme is a fast KUPC (F-KUPC) algorithm that quantizes the power-centroid derived from $\{\mathbf{Z}_i, i=1,..,L_{SCM}\}$. The quantized power-centroid is then used to extract from a memory a PCC that had been previously designed off-line using a CAP directed towards the quantized power-centroid.

5.1 The Knowledge-Unaided Power-Centroid (KUPC) Algorithm

The KUPC algorithm is given by the following weighting vector expressions

$$\mathbf{w}_{KUPC} = [{}^{KUPC}\mathrm{C}]^{-1}\mathbf{s} \tag{65}$$

$${}^{KUPC}\mathrm{C} = \{ ({}^{KUPC}\mathrm{C}_c^f + \mathrm{C}_c^b) \mathrm{O} (\mathrm{C}_{RW} + \mathrm{C}_{ICM} + \mathrm{C}_{CM})\} + \{\mathrm{C}_J \mathrm{O}\, \mathrm{C}_{CM}\} + \mathrm{C}_n \tag{66}$$

where the covariances C_c^b, $\mathbf{C}_{RW}$, $\mathbf{C}_{ICM}$, $\mathbf{C}_{CM}$, $\mathbf{C}_J$, and $\mathbf{C}_n$ are the same as those for the interference plus noise covariance $\mathbf{C}$ given by expression (12) and are either assumed to be zero or known, and ${}^{KUPC}\mathrm{C}_c^f$ predicts the front clutter cell covariance for C_c^f in (12).

5.1.1 The *KUPC* Predicted Front Clutter Cell Covariance ${}^{KUPC}\mathrm{C}_c^f$

The defining expressions for ${}^{KUPC}\mathrm{C}_c^f$ are given by

$${}^{KUPC}\mathrm{C}_c^f = \sum_{i=1}^{N_C} \hat{g}_i(\theta_C^{\overline{C}})\mathbf{c}_i(\theta_{AAM})\mathbf{c}_i^H(\theta_{AAM}) \tag{67}$$

$$\overline{C} = \begin{cases} 1, & \overline{\mathbf{M}} < -\frac{N_C-1}{2} \\ \frac{N_C+1}{2} + \overline{\mathbf{M}}, & -\frac{N_C-1}{2} \le \overline{\mathbf{M}} \le \frac{N_C-1}{2} \\ N_C, & \frac{N_C-1}{2} < \overline{\mathbf{M}} \end{cases} \tag{68}$$

$$\overline{\mathbf{M}} = \sum_{i=2}^{N+M-1} k_i \mathrm{Imag}[m_i] \Big/ m_1 \tag{69}$$

$$\theta_C^{\overline{C}} = \frac{\pi}{N_C}\left(\overline{C} - \frac{N_C+1}{2}\right) \tag{70}$$

$$-\frac{N_C-1}{N_C}\frac{\pi}{2} \le \theta_C^{\overline{C}} \le \frac{N_C-1}{N_C}\frac{\pi}{2} \tag{71}$$

$$\hat{g}_i(\theta_C^{\overline{\mathbf{C}}}) = \hat{K}\left| \frac{\sin\left\{ N\pi\frac{d}{\lambda}(\sin(\theta_C^i) - \sin(\theta_C^{\overline{\mathbf{C}}}))\right\}}{\sin\left\{ \pi\frac{d}{\lambda}(\sin(\theta_C^i) - \sin(\theta_C^{\overline{\mathbf{C}}}))\right\}} \right|^2 \tag{72}$$

$${}^{KUPC}\mathbf{C}_c^f(1,1) = m_1 \tag{73}$$

where: a) $\mathbf{c}_i(\theta_{AAM})$ is as defined for C_c^f (13); b) $\overline{C}$ is the estimated power-centroid of the range-bin using the sample covariance matrix in (54), and inclusive of any type of disturbance and noise; c) $\overline{\mathrm{M}}$ is a real scalar quantity which is a function of an appropriately determined set of constants $\{k_i\}$, the real power of the on-line derived measurements $\{\mathbf{Z}_i\}$ or m_1, and the imaginary part or Imag[.] of each element of a set of N+M-2 distinct and complex correlations $\{m_i\}$ selected from the first row of the NM x NM dimensional sample covariance matrix (SCM) in (54), e.g. when $M=N=2$ the three correlation elements $\{m_i: i=1,2,3\}$ are the first, second, and fourth elements of the SCM's first row as seen below

$$\frac{1}{L_{SCM}}\sum_{i=1}^{L_{SCM}} \mathbf{Z}_i\mathbf{Z}_i^H = \begin{bmatrix} m_1 & m_2 & X & m_3 \\ X & X & X & X \\ X & X & X & X \\ X & X & X & X \end{bmatrix} \tag{74}$$

and when $M=N=3$ the five correlation elements $\{m_i: i=1,2,3,4,5\}$ are as seen below

$$\frac{1}{L_{SCM}}\sum_{i=1}^{L_{SCM}} \mathbf{Z}_i\mathbf{Z}_i^H = \begin{bmatrix} m_1 & m_2 & m_3 & X & X & m_4 & X & X & m_5 \\ X & X & X & X & X & X & X & X & X \\ X & X & X & X & X & X & X & X & X \\ X & X & X & X & X & X & X & X & X \\ X & X & X & X & X & X & X & X & X \\ X & X & X & X & X & X & X & X & X \\ X & X & X & X & X & X & X & X & X \\ X & X & X & X & X & X & X & X & X \\ X & X & X & X & X & X & X & X & X \end{bmatrix}, \tag{75}$$

and so on for higher dimensions; and d) $\theta_C^{\overline{C}}$ is the boresight angle in radians associated with $\overline{C}$; and d) the values of the constant gains $\{k_i\}$ used in our simulations are found from the following expression

$$k_i = -K_{\overline{\mathbf{M}}}(-1)^i \frac{1}{2^{i-2}} \tag{76}$$

where the value of $K_{\overline{\mathbf{M}}}$ is 60.

5.1.2 Justification for KUPC expressions (68), (69) and (76)

The basic idea behind the power-centroid expressions of (68), (69) and (76) is explained next using the optimum $M=N=2$ case depicted in Fig. 2 for $\beta=1$ as motivation. This figure shows a clutter range-bin made of $N_C=4$ clutter cells which are symmetrically spaced with respect to the target that is being investigated at the boresight angle of $\theta_t=0°$. Thus we have that the four clutter cell locations $\{\theta_1, \theta_2, \theta_3, \theta_4\}$ send to the two elements of the receiving antenna the clutter-antenna-gain modulated steering expressions $\left\{\sqrt{x_1g_1}\mathbf{v}_1, \sqrt{x_2g_2}\mathbf{v}_2, \sqrt{x_3g_3}\mathbf{v}_3, \sqrt{x_4g_4}\mathbf{v}_4\right\}$, respectively. Furthermore, the two antenna elements in this example produce during a CPI two different measurements. All of these measurements are represented in Fig. 3 with the matrix

$$\begin{bmatrix} z(s_1,t_1) & z(s_2,t_1) \\ z(s_1,t_2) & z(s_2,t_2) \end{bmatrix} \tag{77}$$

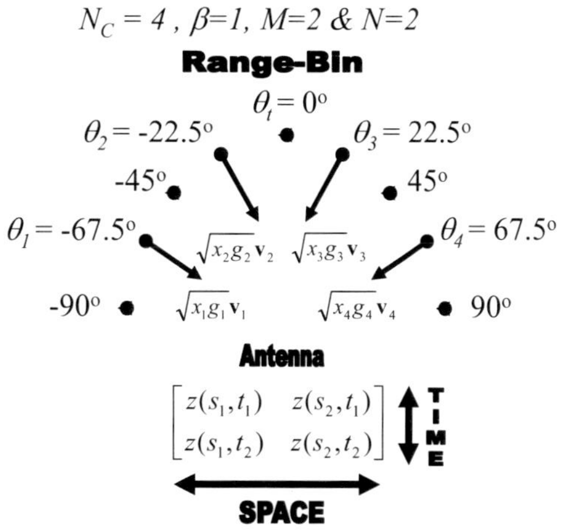

$N_C = 4$, $\beta=1$, $M=2$ & $N=2$

Range-Bin

$\theta_t = 0°$

$\theta_2 = -22.5°$ $\theta_3 = 22.5°$

-45° 45°

$\theta_1 = -67.5°$ $\sqrt{x_2g_2}\mathbf{v}_2$ $\sqrt{x_3g_3}\mathbf{v}_3$ $\theta_4 = 67.5°$

-90° $\sqrt{x_1g_1}\mathbf{v}_1$ $\sqrt{x_4g_4}\mathbf{v}_4$ 90°

Antenna

$$\begin{bmatrix} z(s_1,t_1) & z(s_2,t_1) \\ z(s_1,t_2) & z(s_2,t_2) \end{bmatrix}$$

TIME

SPACE

Fig. 2 Space-Time Geometry For An Optimum Power- Centroid Algorithm Example

where the left element of each (s_i,t_j) pair, i.e. s_i, indicates the ith antenna element and the right element t_j denotes the jth received pulse during a CPI, while the four measurement values depicted on the matrix, i.e. $\{z(s_i,t_j)\}$, are a function of the received modulated steering vectors as shown below

$$\mathbf{z} = \begin{bmatrix} z(s_1,t_1) & z(s_1,t_2) & z(s_2,t_1) & z(s_2,t_2) \end{bmatrix}^T = \sum_{i=1}^{4} \sqrt{x_i g_i}\,\mathbf{v}_i \tag{78}$$

with the four clutter steering vectors given by the expression

$$\mathbf{v}_i = \begin{bmatrix} v_i(s_1,t_1) & v_i(s_2,t_1) & v_i(s_1,t_2) & v_i(s_2,t_2) \end{bmatrix}^T = \begin{bmatrix} 1 & e^{j\frac{2\pi d}{\lambda}\sin\theta_i} & e^{j\frac{2\pi d}{\lambda}\sin\theta_i} & e^{j\frac{4\pi d}{\lambda}\sin\theta_i} \end{bmatrix}^T, \quad i=1,2,3,4 \tag{79}$$

Next an expression is found for the correlation matrix $E[\mathbf{xx}^H]$ under the assumption that each clutter return is uncorrelated from each other, i.e. it is assumed that $E[\sqrt{x_i g_i}\sqrt{x_j g_j}]=0$ for $i \neq j$ and $E[\sqrt{x_i g_i}\sqrt{x_j g_j}]=x_i g_i$ for $i=j$. Thus it is found via straight forward algebraic manipulations and the symmetry condition $\theta_3=-\theta_2=22.5^\circ$ and $\theta_4=-\theta_1=67.5^\circ$ deduced from Fig. 3 that

$$E[\mathbf{xx}^H] = \begin{bmatrix} \mathbf{M}(t_1,t_1) & \mathbf{M}(t_1,t_2) \\ \mathbf{M}(t_2,t_1) & \mathbf{M}(t_2,t_2) \end{bmatrix} = \begin{bmatrix} \mathbf{M}(t_1,t_1) & \mathbf{M}(t_1,t_2) \\ \mathbf{M}^*(t_1,t_2) & \mathbf{M}(t_1,t_1) \end{bmatrix} \tag{80}$$

$$\mathbf{M}(t_1,t_1) = \begin{bmatrix} m_1 & m_2 \\ m_2^* & m_1 \end{bmatrix}, \qquad \mathbf{M}(t_1,t_2) = \begin{bmatrix} m_2 & m_3 \\ m_1 & m_2 \end{bmatrix} \tag{81}$$

where the three correlation elements in (81) are found from the following three expressions

$$m_1 = x_1 g_1 + x_2 g_2 + x_3 g_3 + x_4 g_4 \tag{82}$$

$$m_2 = x_1 g_1 e^{j\frac{2\pi d}{\lambda}\sin\theta_4} + x_2 g_2 e^{j\frac{2\pi d}{\lambda}\sin\theta_3} + x_3 g_3 e^{-j\frac{2\pi d}{\lambda}\sin\theta_3} + x_4 g_4 e^{-j\frac{2\pi d}{\lambda}\sin\theta_4} \tag{83}$$

and

$$m_3 = x_1 g_1 e^{j\frac{4\pi d}{\lambda}\sin\theta_4} + x_2 g_2 e^{j\frac{4\pi d}{\lambda}\sin\theta_3} + x_3 g_3 e^{-j\frac{4\pi d}{\lambda}\sin\theta_3} + x_4 g_4 e^{-j\frac{4\pi d}{\lambda}\sin\theta_4}. \tag{84}$$

Next the moment expressions (82)-(84) are related to the desired evaluation of C(**xOg**) (60) for the clutter range-bin of Fig. 2 which is for this case as follows:

$$C(\mathbf{xOg}) = \frac{x_1 g_1.1 + x_2 g_2.2 + x_3 g_3.3 + x_4 g_4.4}{x_1 g_1 + x_2 g_2 + x_3 g_3 + x_4 g_4}. \tag{85}$$

An algebraic manipulation of this expression then yields

$$C(\mathbf{xOg}) = \frac{4+1}{2} + \frac{\begin{bmatrix} 1 & 3 \end{bmatrix}}{2m_1} \begin{bmatrix} x_3 g_3 - x_2 g_2 \\ x_4 g_4 - x_1 g_1 \end{bmatrix}. \tag{86}$$

where the denominator of (85) now appears as m_1 (82) in (86) and the first term to the right of (86) is given by 2.5 which is the assumed boresight position of the target of 0° as well as the position of the power-centroid when either a symmetrical condition for the clutter, i.e. $x_3g_3=x_2g_2$ and $x_4g_4=x_1g_1$ exits or the clutter difference closest to the target, i.e. x_3g_3-x_2g_2, is equal to the negative of three times the clutter difference away form the target, i.e. $-3(x_4g_4-x_1g_1)$. We next use expressions (82)-(84) to derive the following relationships between the real and imaginary parts of m_2 and m_3

$$\begin{bmatrix} \mathrm{Real}[m_1] \\ \mathrm{Real}[m_2] \end{bmatrix} = \begin{bmatrix} 1 & 1 \\ \cos\left(\frac{2\pi d}{\lambda}\sin\theta_3\right) & \cos\left(\frac{2\pi d}{\lambda}\sin\theta_4\right) \end{bmatrix} \begin{bmatrix} x_3 g_3 + x_2 g_2 \\ x_4 g_4 + x_1 g_1 \end{bmatrix} \tag{87}$$

$$\begin{bmatrix} \mathrm{Imag}[m_2] \\ \mathrm{Imag}[m_3] \end{bmatrix} = -\begin{bmatrix} \sin\left(\frac{2\pi d}{\lambda}\sin\theta_3\right) & \sin\left(\frac{2\pi d}{\lambda}\sin\theta_4\right) \\ \sin\left(\frac{4\pi d}{\lambda}\sin\theta_3\right) & \sin\left(\frac{4\pi d}{\lambda}\sin\theta_4\right) \end{bmatrix} \begin{bmatrix} x_3 g_3 - x_2 g_2 \\ x_4 g_4 - x_1 g_1 \end{bmatrix} \tag{88}$$

Solving the linear system of equations (88) for the clutter differences vector under the constraint that the matrix must be invertible, and then substituting this result in (86) yields the desired optimum expression for the range-bin power-centroid C(**xOg**) in term of the three correlation elements of $E[\mathbf{xx}^H]$, i.e. m_1, m_2 and m_3, as follows

$$C(\mathbf{xOg}) = \frac{4+1}{2} - \frac{[1 \quad 3]}{2m_1}\begin{bmatrix} \sin\left(\frac{2\pi d}{\lambda}\sin\theta_3\right) & \sin\left(\frac{2\pi d}{\lambda}\sin\theta_4\right) \\ \sin\left(\frac{4\pi d}{\lambda}\sin\theta_3\right) & \sin\left(\frac{4\pi d}{\lambda}\sin\theta_4\right)\end{bmatrix}^{-1}\begin{bmatrix}\text{Imag}[m_2] \\ \text{Imag}[m_3]\end{bmatrix} \tag{89}$$

To get an idea of the values derived for this simple and optimum case we evaluate (89) for the assumed symmetrical conditions of Fig. 2 and under the assumption that $d/\lambda=0.5$ (also used in our simulations) to yield

$$C(\mathbf{xOg}) = \frac{5}{2} - [2.10405 \quad -2.17625]\begin{bmatrix}\text{Imag}[m_2] \\ \text{Imag}[m_3]\end{bmatrix}\Big/ m_1 \tag{90}$$

This expression can then be rewritten as follows

$$C(xOg) = \frac{5}{2} + \sum_{i=2}^{3} k_i \text{Imag}[m_i]\Big/ m_1, \quad k_2 = -2.10405(-1)^2 \ \& \ k_3 = -2.17625(-1)^3 \tag{91}$$

where a comparison of expressions (68), (69) and (76) with (91) should help explain why the forms for expressions (68), (69) and (76) are being used for the real-world radar problem simulated in this paper. It should also be emphasized that the selected simulation values of N=16, M=16 and N_C=256 do not allow us to generalize (89) to include this case. Moreover even if this was possible it is likely that numerical stability problems will surface when dealing with the inversion of high dimensionality matrices. This is why the simple and well behaved expressions (68), (69) and (76) are used by us.

5.2 The Fast Knowledge-Unaided Power-Centroid Algorithm

The fast KUPC or F-KUPC algorithm is characterized by the following weighting vector expressions

$$\mathbf{w}_{F\text{-}KUPC} = [{}^{F\text{-}KUPC}\mathbf{C}]^{-1}\mathbf{s} \tag{92}$$

$${}^{F\text{-}KUPC}\mathbf{C} = \{ ({}^{F\text{-}KUPC}\mathbf{C}_c^f + \mathbf{C}_c^b) \circ (\mathbf{C}_{RW} + \mathbf{C}_{ICM} + \mathbf{C}_{CM})\} + \{\mathbf{C}_J \circ \mathbf{C}_{CM}\} + \mathbf{C}_n \tag{93}$$

where the covariances $\mathbf{C}_c^b$, $\mathbf{C}_{RW}$, $\mathbf{C}_{ICM}$, $\mathbf{C}_{CM}$, $\mathbf{C}_J$, and $\mathbf{C}_n$ are the same as those for the interference plus noise covariance $\mathbf{C}$ given by expression (12), and ${}^{F\text{-}KUPC}\mathbf{C}_c^f$ predicts the front clutter cell covariance for $\mathbf{C}_c^f$ in (12).

The defining expressions for ${}^{F\text{-}KUPC}\mathbf{C}_c^f$ are similar to those for ${}^{KUPC}\mathbf{C}_c^f$ except that the centroid value derived via (68), (69) and (76) are now quantized. Thus

$${}^{F\text{-}KUPC}\mathbf{C}_c^f = \sum_{i=1}^{N_C} \bar{g}_i(\theta_C^{\hat{C}})\mathbf{c}_i(\theta_{AAM})\mathbf{c}_i^H(\theta_{AAM}) = {}^{KUPC}\mathbf{C}_c^f\big|_{C(\mathbf{xOg})=\hat{C}(\mathbf{xOg})} \tag{94}$$

where the quantized power-centroid $\hat{C}$ is defined as follows:

$$\hat{C} = \min_{\hat{C}_i}\left|\bar{C} - \hat{C}_i\right| \tag{95}$$

$$\hat{C}_i \in \left\{\frac{N_C-1}{L}, \frac{2(N_C-1)}{L}, \frac{3(N_C-1)}{L}, \ldots, \frac{(L-1)(N_C-1)}{L}\right\} \tag{96}$$

$$\bar{g}_i(\theta_C^{\hat{C}}) = \hat{K}\left|\frac{\sin\left\{N\pi\frac{d}{\lambda}(\sin(\theta_C^i) - \sin(\theta_C^{\hat{C}}))\right\}}{\sin\left\{\pi\frac{d}{\lambda}(\sin(\theta_C^i) - \sin(\theta_C^{\hat{C}}))\right\}}\right|^2 \tag{97}$$

$${}^{F\text{-}KUPC}\mathbf{C}_c^f(1,1) = m_1 \tag{98}$$

with L denoting the finite number of CAPs that may be used in (94) to predict the $\mathbf{C}_c^f$, and $\bar{C}$ is the unquantized centroid of (68). Clearly this algorithm lends itself to an extremely fast implementation where sets of L predicted clutter covariances (PCCs) are designed off-line for as many antenna array misalignment angles $\{\theta_{AAM}\}$ as desired. In our simulations, that we discuss next, we will consider two cases for L (or equivalently the number of CAPs). These cases are L=3 and L=11 where in addition N=M=16 and N_C=256, and L_{SCM}=256.

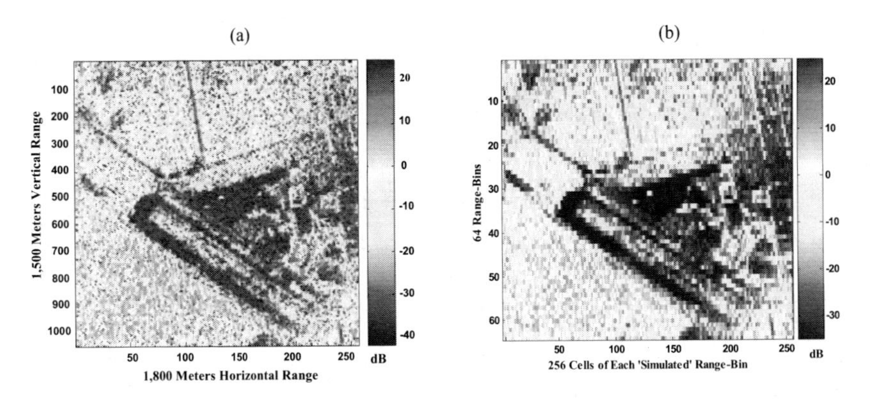

Fig. 3 Synthetic Aperture Radar (SAR) Image of Mojave Airport in California

6. SIMULATION RESULTS

In this section several simulations results are presented that use the 64 by 256 SAR image seen in Fig. 3. The original 4 megabytes SAR image of Fig. 3a is made up of 1,024 by 256 pixel elements representing 1,800 by 1,500 meters of the Mojave Airport in California where each pixel denotes clutter power. Sixteen consecutive rows of Fig. 3a are then averaged to form the 64 x 256 image of Fig. 3b. In our simulation each row of Fig. 3b is a range-bin with 256 clutter cells for each. In addition, the simulation radar parameters, disturbance and noise values used are those of Table 1. The SAR image of Fig. 3 as well as the matlab program used in our simulations will become available in the author's Web site [8] in the near future.

The results are summarized in four figures, i.e. Figs. 4-7. The basic difference between the four figure cases is in the use of jammers and the number of CAPs used. Figs. 4-5 present cases where three jammers are used at boresight angles of -60°, -30 ° and 45 ° with corresponding JNR values of 52, 55, and 66 dBs, respectively, while Figs. 6-7 show simulation results with no jammers used. On the other hand, Figs. 4 and 6 present results for eleven CAPs and Figs. 5 and 7 for three CAPs. Each figure case has seven displays. First Figs. 4a, 5a, 6a and 7a show the SINR error in dBs as a function of range-bin where note is made of the average SINR error over all range-bins for the three cases which are the SCMI scheme of (53)-(56), the KAPC scheme of (57)-(64), and the F-KUPC scheme of (92)-(98). For example, in Fig. 4a KAPC is noted to yield an average SINR error of 0.87 dBs, F-KUPC of 1.32 dBs, and SCMI of 7.59 dBs. These results are satisfactory since more than 6 dBs improvements are derived over the SCMI with both the KAPC and F-KUPC schemes, while also yielding in both cases an average SINR radar performance close to the optimum SINR radar performance. In Figs. 4b-7b the power of the KAPC and F-KUPC schemes, i.e. (64) and (98), respectively, is depicted where the fluctuations from range-bin to range-bin found with the F-KUPC case reflects its derivation from the average power m_l that is derived from the use of the sample covariance matrix (54) without the use of the loading factor of course. In Figs. 4c-7c the range-bin power-centroids corresponding to the KAPC algorithm (60), the KUPC algorithm without centroid quantizations (68), and the F-KUPC algorithm with centroid quantizations (95) are shown. Figs. 4d-g, 5d-g, 6d-g and 7d-g specifically display the radar system performance for range-bin number one. In Figs. 4d-7d the actual range-bin clutter, KAPC predicted clutter power {(59),(63),(64)} and F-KUPC predicted clutter power {(94),(97),(98)} is displayed as a function of clutter cell number. In Fig. 4e-7e the optimum, KAPC, F-KUPC and SCMI SINRs are displayed as a function of normalized Doppler. In Fig. 4f-7f the optimum, KAPC, F-KUPC and SCMI adapted patterns [7] are shown as a function of clutter cell number. Finally in Fig. 4g-7g the optimum, KAPC, F-KUPC and SCMI interference plus noise covariance eigenvalues [7] are plotted for each case.

The results presented in Figs. 4-7 are typical results for the compared schemes as researchers should be able to confirm using their own matlab simulations or those that will become available in [8]. It is thus concluded that F-KUPC radar offers a major improvement over KAPC radar. This result also brings further support to the mathematical uncertainty-information/certainty-latency duality that has resulted in both KAPC and F-KUPC radar. In the second part of this two parts paper series the recently discovered [3] physical dual for our mathematical uncertainty-information/certainty-latency duality is presented in some detail and illustrated with simple physical examples in a more general unification setting.

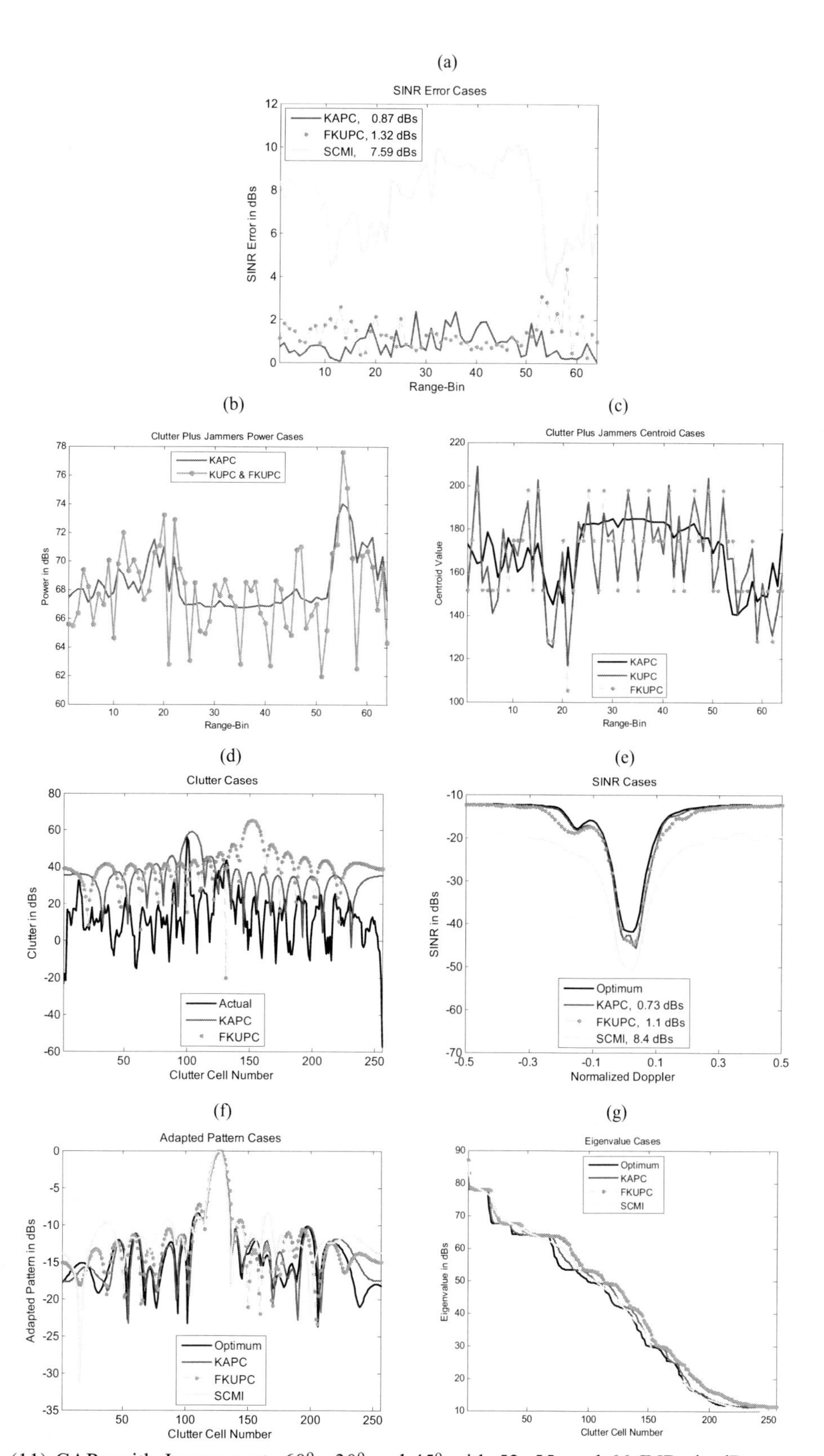

Fig. 4. Eleven (11) CAPs with Jammers at -60°, -30° and 45° with 52, 55, and 66 JNRs in dBs, respectively. (a) SINR Error. (b) Range-Bin Power of Clutter Plus Jammer. (c) Range-Bin Centroid of Clutter Plus Jammer. (d) Range-Bin #1 Clutter and its Predictions. (e) Range-Bin #1 SINRs. (f) Range-Bin #1 Adapted Patterns. (g) Range-Bin #1 Eigenvalues

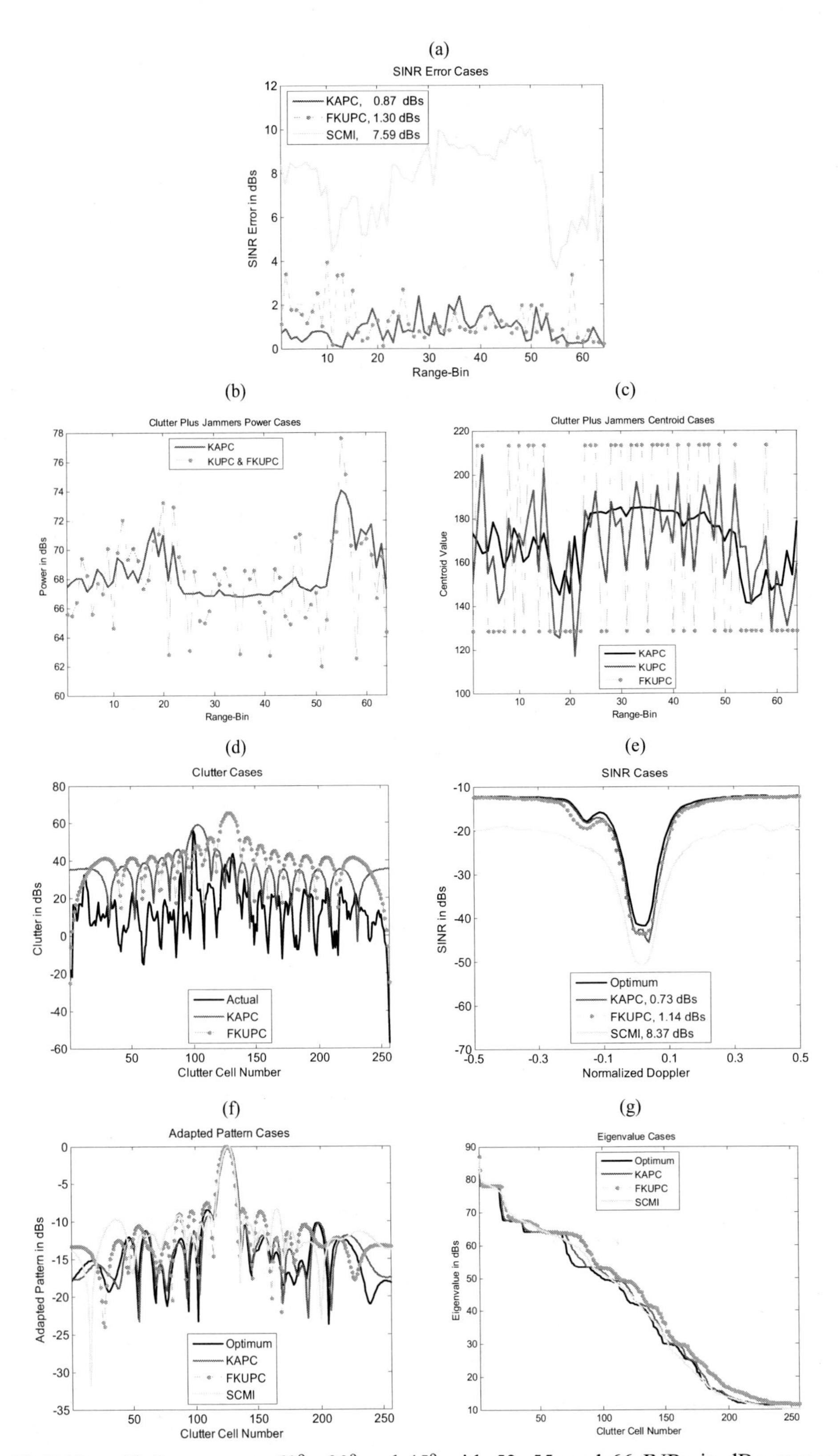

Fig. 5. Three (3) CAPs with Jammers at -60°, -30° and 45° with 52, 55, and 66 JNRs in dBs, respectively. (a) SINR Error. (b) Range-Bin Power of Clutter Plus Jammer. (c) Range-Bin Centroid of Clutter Plus Jammer. (d) Range-Bin #1 Clutter and its Predictions. (e) Range-Bin #1 SINRs. (f) Range-Bin #1 Adapted Patterns. (g) Range-Bin #1 Eigenvalues

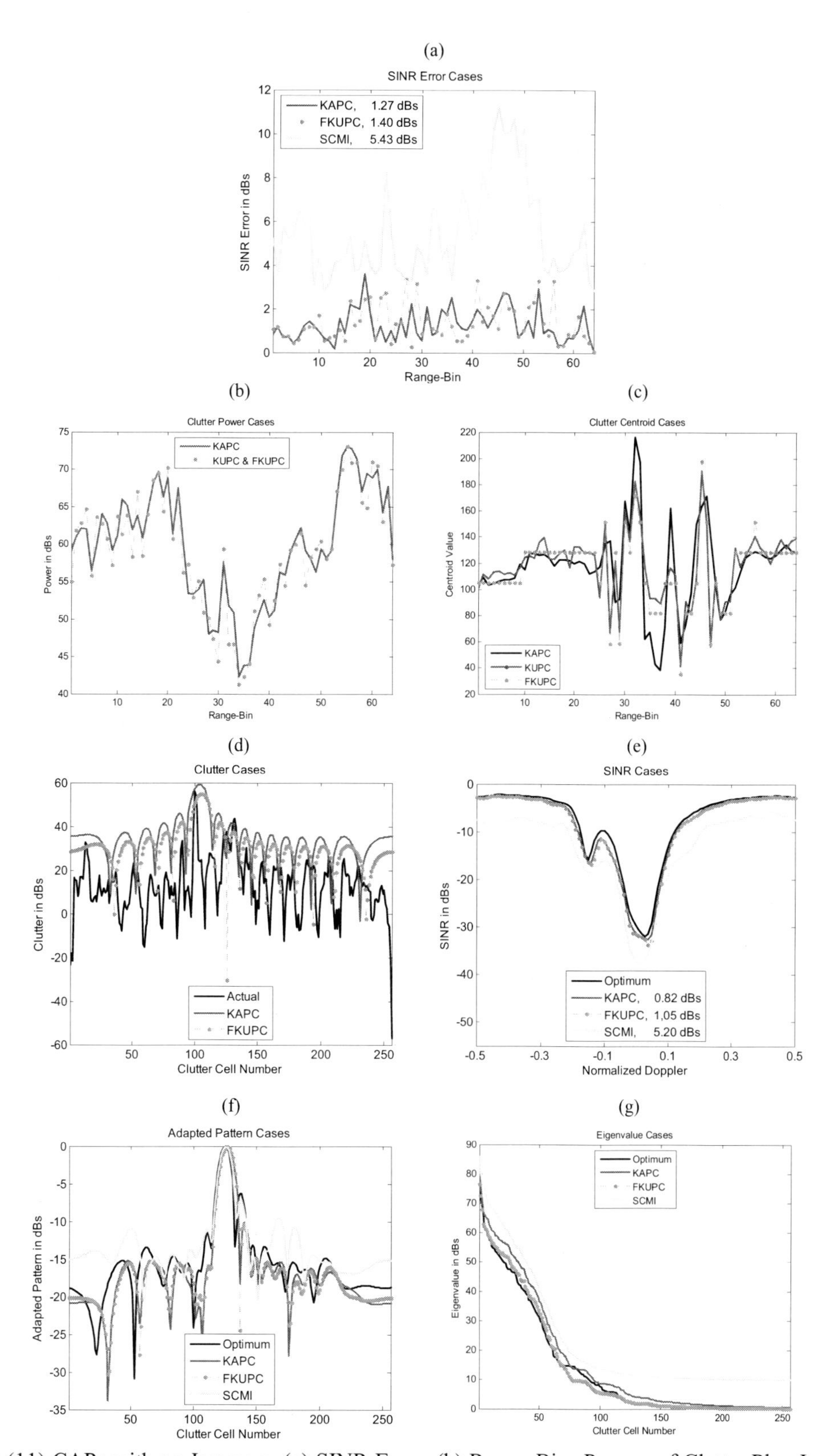

Fig. 6. Eleven (11) CAPs with no Jammers. (a) SINR Error. (b) Range-Bins Powers of Clutter Plus Jammer. (c) Range-Bins Centroids with Clutter Plus Jammer. (d) Range-Bin #1 Clutter and its Predictions. (e) Range-Bin #1 SINRs. (f) Range-Bin #1 Adapted Patterns. (g) Range-Bin #1 Eigenvalues

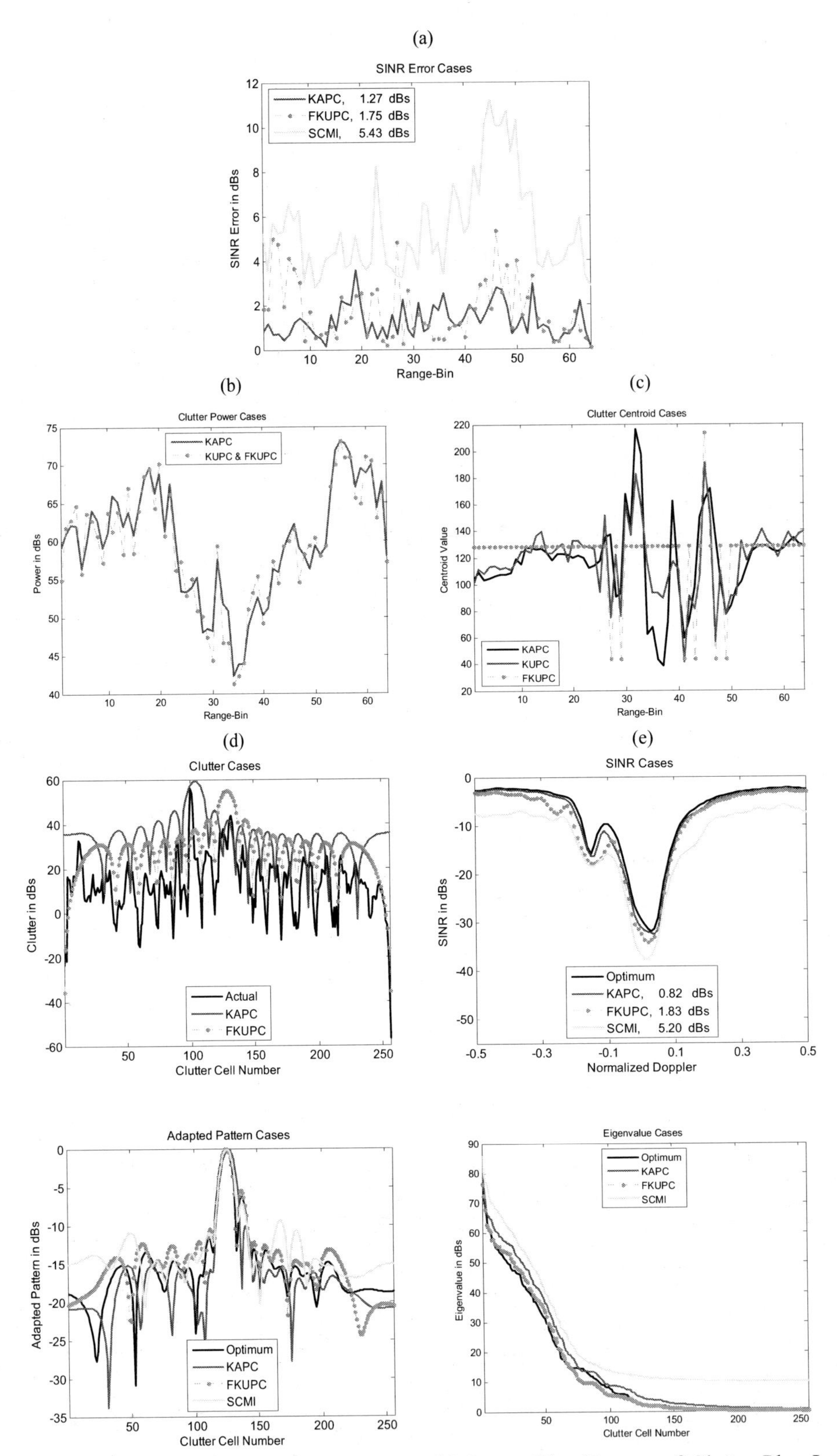

Fig. 7. Three (3) CAPs with no Jammers. (a) SINR Error. (b) Range-Bins Powers of Clutter Plus Jammer. (c) Range-Bins Centroids with Clutter Plus Jammer. (d) Range-Bin #1 Clutter and its Predictions. (e) Range-Bin #1 SINRs. (f) Range-Bin #1 Adapted Patterns. (g) Range-Bin #1 Eigenvalues

Table 1. Radar Simulation Parameters

a.	Antenna	N=16, M=16, d/λ=½, $\sigma_n^2 = 1$ Front antenna gain constant, (1): K^f = 56 dBs Back antenna gain constant : K^b = -40 dBs Carrier frequency, (9): $f_c = 10^9$ Hz Pulse repetition frequency, (10): $f_r = 10^3$ Hz Antenna array misalignment, (13): $\theta_{AAM} = 2^o$
b.	Clutter	$N_c = 256$ Radar's *ratio* β, (19): $\beta = 1$
c.	Jammers	Jammers were used at -60°, -30° and 45° with 52, 55 and 66 JNRs in dBs, respectively, inclusive of antenna gains.
d.	Range Walk	Fraction of remaining area after range walk, (33): ρ=0.999999.
e.	Internal Clutter Motion	Shape factor, (35): $b = 5.7$ Wind-speed, (37): $\omega = 15$ mph
f.	Channel Mismatch: Narrowband	Amplitude error, (50): $\Delta\varepsilon_i = 0$ for all i, Phase-error, (50): $\Delta\gamma_i$ fluctuates with a 5° rms for all i
h.	Channel Mismatch: Finite-Bandwidth	Amplitude peak deviation, (41): $\Delta\varepsilon = 0.001$, Phase peak deviation, (41): $\Delta\phi = 0.1^o$
i.	Channel Mismatch : Angle-Dependent	Bandwidth, (45): $B = 10^8$ Hz Mainbeam width, (45): $\Delta\theta = 28.6^o$

REFERENCES

[1] Guerci, J. R. and Feria, E. H., "Application of a least squares predictive-transform modeling methodology to space-time adaptive array processing," *IEEE Trans. On Signal Processing,* vol. 44, no. 7, pp.1825-1833, July 1996.

[2] Guerci, J. R. and Baranoski, E., " Knowledge-aided adaptive radar at DARPA", *IEEE Signal Processing Magazine*, vol. 23, no. 1, pp. 41-50, January 2006

[3] Feria, E. H., "Latency-information theory: A novel latency theory revealed as time-dual of information theory", *Proceedings of IEEE Signal Processing Society:* DSP Workshop and SPE Workshop, Marco Island, Florida, Jan. 2009

[4] ---------, "A predictive-transform compression architecture and methodology for KASSPER," *Final Technical Report*, DARPA Grant FA8750-04-1-0047, May 2006.

[5] ---------, "Predictive-transform estimation," *IEEE Transaction On Signal Processing*, vol. **39**, no. 11, pp. 2481-2499, Nov. 1991.

[6] ---------, "Matched processors for optimum control", *City University Of New York (CUNY)*, Ph.D. in E.E., Aug. 1981 and Feria, E. H., "Matched processors for quantized control: A practical parallel processing approach," *International Journal of Controls*, Vol. 42, Issue 3, pp. 695-713, Sept. 1985.

[7] Guerci, J. R. " Space-time adaptive processing for radar", *Artech House*, 2003

[8] Web site, http://feria.csi.cuny.edu

On A Nascent Mathematical-Physical Latency-Information Theory, Part II: The Revelation Of Guidance Theory For Intelligence And Life System Designs

Erlan H. Feria
Department of Engineering Science and Physics
The College of Staten Island of the City University of New York
E-mail feria@mail.csi.cuny.edu Web site http://feria.csi.cuny.edu

Since its introduction more than six decades ago by Claude E. Shannon information theory has guided with two performance bounds, namely source-entropy H and channel capacity C, the design of sourced intelligence-space compressors for communication systems, where the units of intelligence-space are 'mathematical' binary digit (bit) units of a passing of time uncertainty nature. Recently, motivated by both a real-world radar problem treated in the first part of the present paper series, and previous uncertainty/certainty duality studies of digital-communication and quantized-control problems by the author, information theory was discovered to have a 'certainty' time-dual that was named latency theory. Latency theory guides with two performance bounds, i.e. processor-ectropy $\boldsymbol{K}$ and sensor consciousness $\boldsymbol{F}$ the design of processing intelligence-time compressors for recognition systems, where the units of intelligence-time are 'mathematical' binary operator (bor) units of a configuration of space certainty nature. Furthermore, these two theories have been unified to form a mathematical latency-information theory (M-LIT) for the guidance of intelligence system designs, which has been successfully applied to real-world radar. Also recently, M-LIT has been found to have a physical LIT (P-LIT) dual that guides life system designs. This novel physical theory addresses the design of motion life-time and retention life-space compressors for physical signals and also has four performance bounds. Two of these bounds are mover-ectropy $\boldsymbol{A}$ and channel-stay $\boldsymbol{T}$ for the design of motion life-time compressors for communication systems. An example of a motion life-time compressor is a laser system, inclusive of a network router for a certainty, or multi-path life-time channel. The other two bounds are retainer-entropy $\boldsymbol{N}$ and sensor scope $\boldsymbol{I}$ for the design of retention life-space compressors for recognition systems. An example of a retention life-space compressor is a silicon semiconductor crystal, inclusive of a leadless chip carrier for an uncertainty, or noisy life-space sensor. The eight performance bounds of our guidance theory for intelligence and life system designs will be illustrated with practical examples. Moreover, a four quadrants (quadrants I and III for the two physical theories and quadrants II and IV for the two mathematical ones) LIT revolution is advanced that highlights both the discovered dualities and the fundamental properties of signal compressors leading to a unifying communication embedded recognition (CER) system architecture.

Index Terms—Uncertainty/Certainty Duality, Communication/Recognition Duality, Intelligence/Life Duality, Motion Life-Time and Retention Life-Space, Sourced Intel-Space and Processing Intel-Time, Compression, Guidance Theory for Intelligence and Life System Designs, Communication Embedded Recognition System

1. Introduction

Recently information-theory, guiding mathematical communication system designs, was found to have a time-dual that guides mathematical recognition system designs and has been named latency-theory [1]. More specifically, information-theory is characterized by mathematical binary digit (bit) units of a passing of time uncertainty nature that are linked to a signal-source's 'sourced intelligence-space (or intel-space in short)' as well as an uncertainty, or noisy intel-space channel. On the other hand, it has been found that latency-theory is characterized by mathematical binary operator (bor) units of a configuration of space certainty nature that are linked to a signal-processor's 'processing intelligence-time (or intel-time in short)' as well as a certainty, or limited intel-time sensor. This discovered uncertainty-information-space/certainty-latency-time duality is fundamental in nature since every uncertainty intel-space method in mathematical information theory (or M-IT) must then have a certainty intel-time method dual in mathematical latency-theory (or M-LT) or visa-versa. A method in M-IT for which a certainty intel-time method dual has been found in M-LT is *intel-space channel-coding* or 'the mathematical theory of communication' as originally identified by his creator Claude E. Shannon [2] and later by communication experts [3]. Intel-space channel-coding guides the design of intel-space communication

This work was supported in part by the defense advanced research projects agency (DARPA) under grant no. FA8750-04-1-004 and the PSC-CUNY Research Awards PSCREG-37-913, 38-247, 39-834, 40-1013

Mobile Multimedia/Image Processing, Security, and Applications 2009, edited by Sos S. Agaian, Sabah A. Jassim,
Proc. of SPIE Vol. 7351, 73510V · © 2009 SPIE · CCC code: 0277-786X/09/$18 · doi: 10.1117/12.819057

systems using as performance bounds the expected source-information in bits, denoted as the source-entropy H, and the channel-capacity C of a noisy intel-space channel. More specifically, these bounds guide the design of intel-space channel and source integrated (CSI) coders with the integrated intel-space channel-coder advancing overhead knowledge in the form of mathematical signals, e.g. parity bits extracted from bit streams that have the effect of increasing the amount of channeled intel-space. The certainty intel-time dual for this scheme was found to be M-LT's *intel-time sensor-coding* (or 'the mathematical theory of recognition' as identified from a duality perspective) [1]. Intel-time sensor-coding guides the design of intel-time recognition systems using as performance bounds the minmax processor-latency criterion in bor units, denoted as the processor-ectropy K, and the sensor-consciousness F of a limited intel-time sensor. More specifically, these bounds guide the design of intel-time sensor and processor integrated (SPI) coders with the integrated intel-time sensor-coder advancing overhead knowledge in the form of mathematical signals, such as interferences, clutter inclusive, and noise that have the effect of increasing the amount of sensed intel-time. Moreover, the unification of M-LT and M-IT has yielded a mathematical latency-information theory (M-LIT), or mathematical guidance theory for intelligence system designs, that for mathematical intelligence signals addresses in a unified fashion uncertainty communication issues of intel-space and certainty recognition issues of intel-time. Some details and insightful illustrations of the aforementioned mathematical performance bounds are presented in this paper. It is of interest to note that the aforementioned uncertainty-information-space/certainty-latency-time duality revelation had its roots in an earlier 1978 discovery by the author of an existing uncertainty-communication/certainty-control duality between an uncertainty 'digital' communication problem [4] and a certainty 'quantized' control problem [5]. This newly found duality in turn led him to the formulation of a novel and practical parallel processing methodology to quantized control where the controlled processor can be modeled with any certainty linear or nonlinear state-variables representation. This control scheme he named Matched-Processors since it was the 'certainty' control dual of the 'uncertainty' Matched-Filters communication problem. It should be further noticed that this uncertainty/certainty duality perspective for a 'discrete' communication/control problem is also exhibited by Kalman's LQG control formulation of 1960 [6] for the complementary 'continuous' communication/control problem.

Our M-LIT has also been discovered to have a physical dual [1] which has been named physical LIT (P-LIT). Like the aforementioned uncertainty/certainty duality, this newly discovered mathematical-physical duality is fundamental in nature since every mathematical method in M-LIT must then have a corresponding physical method in P-LIT as physical-dual or visa-versa. A mathematical method in M-LIT for which a physical-dual has been found in P-LIT is M-IT's intel-space channel-coding. The physical-dual for this scheme is physical latency-theory (P-LT)'s *'motion life time (or life-time in short)' channel-coding* (or 'the physical theory of communication' as identified from a duality perspective). Life-time channel-coding guides the design of life-time communication systems using as performance bounds the minmax mover-latency criterion in physical time units, denoted as the mover-ectropy A, and the channel-stay T of a multi-path life-time channel. More specifically, these bounds are used to guide the design of life-time channel and mover integrated (*CMI*) coders with the integrated life-time channel-coder advancing overhead knowledge in the form of available physical signals that have the effect of increasing the amount of channeled life-time. An illustration of a mover-coder (encoder/decoder) for physical signals is a laser system, and of a life-time channel-coder (encoder/decoder) for a certainty, or multi-path life-time channel is a network router. Another mathematical method in M-LIT for which a physical-dual has been found in P-LIT is M-LT's intel-time sensor-coding. The physical-dual is physical information-theory (P-IT)'s *'retention life space (or life-space in short)' sensor-coding* (or 'the physical theory of recognition' as identified from a duality perspective). Life-space sensor-coding guides the design of life-space recognition systems using as performance bounds the expected retainer-information in physical space units, denoted as the retainer-entropy N, and the sensor-scope I of a noisy life-space sensor. More specifically, these bounds are used to guide the design of life-space sensor and retainer integrated (*SRI*) coders with the integrated life-space sensor-coder advancing overhead knowledge in the form of available physical signals that have the effect of increasing the amount of sensed life-space. An illustration of a retainer-coder for physical signals is a silicon semiconductor crystal, and of a life-space sensor-coder for an uncertainty, or noisy life-space sensor is a surface mounted leadless chip carrier.

Moreover, as is the case for M-LIT, P-LIT exhibits an uncertainty-information-space/certainty-latency-time duality. Thus like M-LIT this duality is fundamental in nature since for every uncertainty life-space method in P-IT there must be a certainty life-time method dual in M-LT or visa-versa. For example, the certainty speed of light in a vacuum limit of c=2.9979 x 10^8 *m/sec* for energy motion in P-LT, was found in [7] to have an uncertainty intel-space dual in P-IT. This uncertainty intel-space dual is the pace of dark in a black-hole limit of $\not{X}=960\pi c^2/hG=6.1123 \times 10^{63}$ *secs/m*^{3} where h is Plank's constant and G is the Gravitational constant for energy retention [7]. Moreover, while M-LIT is identified to be 'the mathematical guidance theory of intelligence system designs' because for mathematical intelligence signals it addresses in a unified fashion uncertainty communication issues of intel-space and certainty recognition issues of intel-time, P-LIT is identified to be 'the physical guidance theory of life system designs' because for physical signals it addresses in a unified fashion uncertainty recognition issues of life-space and certainty communication issues of life-time. The integration of M-LIT and P-LIT has been found to be conveniently described and remembered in a LIT revolution that summarizes the advanced unification of four fundamental, two mathematical and two physical, methodologies in science. These four quadrants of the LIT revolution, i.e. quadrant I for P-LT, quadrant II for M-IT, quadrant III for P-IT, and quadrant IV for M-LT can be used to highlight the following three quadrant pair dualities. They are: 1) the (II,III)-(I,IV) quadrant pairs exhibiting an uncertainty-information-space/certainty-latency-time duality; b) the (II,IV)-(I,III) quadrant pairs exhibiting a mathematical-intelligence/physical-life duality; and c) the (I,II)-(III,IV) quadrant pairs exhibiting a channel-communication/sensor-recognition duality. The LIT revolution also highlights the main properties of the signal compressors that naturally lead to communication embedded recognition (CER) system architectures first used in a radar application [1]. The eight performance bounds of the LIT revolution will be stated in this paper and illustrated with simple practical examples. In addition, for the rest of the paper the author will use the same sentence constructions to explain each quadrant theory with the idea of facilitating as much as possible the understanding of the advanced duality perspective. Furthermore, in the assignment of names for the symbols, terminology and methods of each theory care has been taken to make them user-friendly, didactic in nature and also vividly reflect their duality roots.

This paper is organized as follows. In Section 2 and 3 the M-LIT guidance of intel-space and intel-time signal compressors is summarized. In Section 4 and 5 the P-LIT guidance of life-time and life-space signal compressors is presented. In Section 6 the unifying LIT revolution is highlighted.

2. Mathematical Information Theory (M-IT)

The principal task of M-IT is to guide the design of intel-space compressors using the performance-bounds of source-entropy H and channel-capacity C for a noisy intel-space channel. The units of intel-space are mathematical bits, thus in this kind of setting information theory offers a mathematical guidance methodology for intel-space compressor designs. The bits of intel-space are sourced through time, and are communicated through space via an uncertainty, or noisy intel-space channel of given capacity. Moreover, the mathematical bit unit of intel-space has a passing of time uncertainty nature since its creation is traced to an uncertainty signal-source. The uncertainty nature of M-IT is modeled by assigning probability distributions to the uncertainty signal-source intel-space that are driven by the passing of time, e.g. a uniform distribution may be associated with the 0/1 outputs of a binary signal-source such as a digital computer.

In Fig. 1 the M-IT's communication system for intel-space compression is shown. It consists of a given signal-source's mathematical signal **X** to be communicated and a given noisy intel-space channel, assumed memoryless, plus a source-coder (encoder/decoder) and an intel-space channel-coder (encoder/decoder) to be designed. Also, when designed together, the two coders form a channel and source integrated (CSI) coder. The signal-source produces a discrete random variable output $\mathbf{X} \in \{x_1, x_2,.., x_\Omega\}$ with Ω possible scalar outcomes $\{x_i\}$.

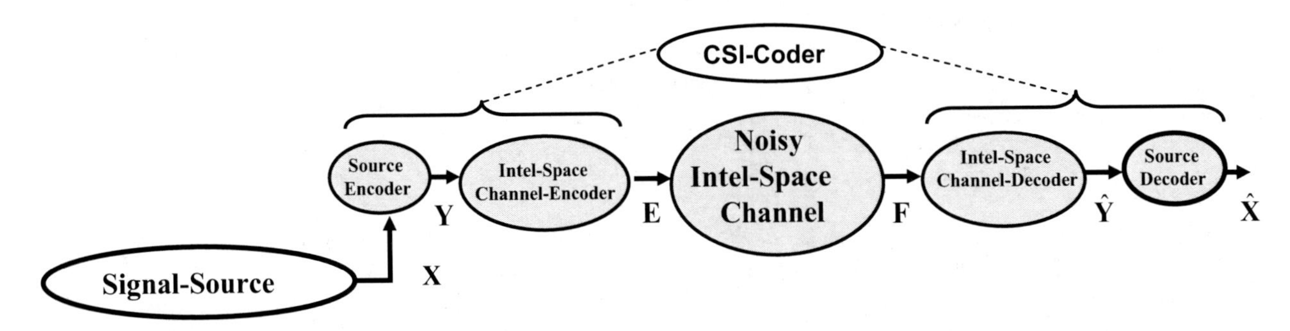

Fig. 1. Mathematical Information Theory's Communication System for Intel-Space Compression

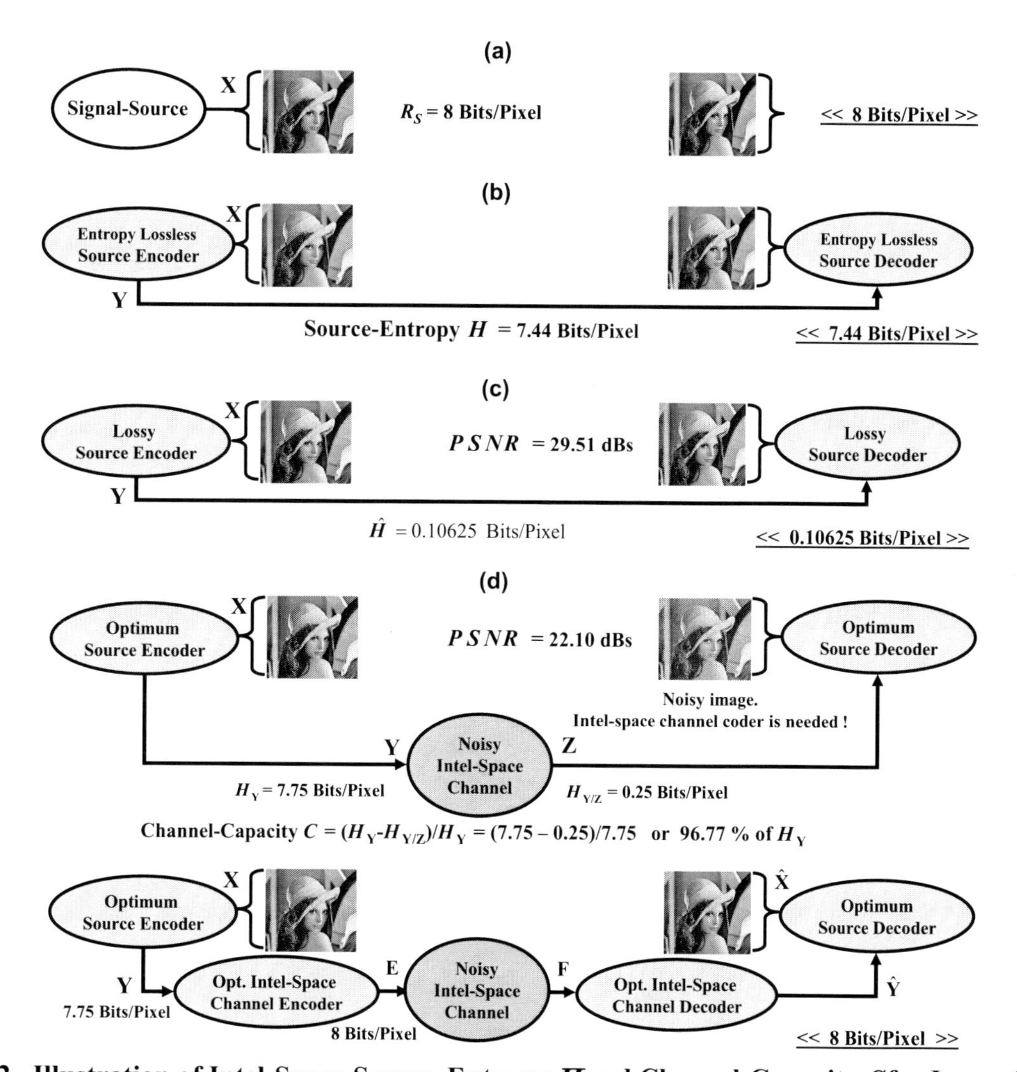

Fig. 2. Illustration of Intel-Space Source-Entropy H and Channel-Capacity C for Image System

The expected number of bits of the signal-source outcomes is called the source-rate R_S. For instance, for the 512x512 Lena image of Fig. 2 [8] each pixel is represented with 8 bits which results in an intel-space of 256 kilo-bytes and R_S= 8 bits/pixel. The uncertainty in the outcome set $\{x_i\}$ is modeled with the source-probability set $\{ P_S[x_i] \}$. These probabilities are then used to derive the amount of source-information in the outcome set $\{x_i\}$ in bit units, i.e.

$$I_S(x_i) = \log_2(1/P_S[x_i]) \text{ for all } i. \tag{1}$$

The expected value of the source-information, called the source-entropy and expressed by

$$H = \sum_{i=1}^{\Omega} P_S[x_i] I_S(x_i), \tag{2}$$

then guides as a lower performance bound source-coder designs. For the Lena image H= 7.44 bits/pixel.

The source-coder of Fig. 1 is characterized by the source-encoder rate R_{SE}. In particular, when the noisy intel-space channel is not present and R_{SE} is greater than the source-entropy H a lossless source-coder, i.e. yielding $\hat{\mathbf{X}} = \mathbf{X}$, can always be designed using algorithms such as Arithmetic, Huffman, and Entropy coding [9]. In Fig. 2b the source encoding and decoding of the Lena image is shown for the case where the source-coder is lossless and its rate is the same as that of the source-entropy, i.e. $R_{SE} = H$. On the other hand, the source-coder is said to be lossy when its rate is less than the source-entropy. For instance, in Fig. 2c the source encoding and decoding of the Lena image is shown for the case where a MMSE predictive-transform source-coder with subbands is used [8]. The lossy source-encoder rate for this system is of R_{SE} = 0.10625 bits/pixel which is significantly less than the image entropy and still yields an image quality that is satisfactory in many practical applications.

Yet when the noisy intel-space channel of Fig. 1 is present an unsatisfactory degradation of the communicated intel-space may occur. Thus when this occurs it becomes necessary to use an intel-space channel-coder that advances overhead knowledge to satisfy the intel-space capacity needs introduced by a noisy intel-space channel. More specifically, for our illustrative case it becomes necessary to use an intel-space channel-coder that introduces sufficient overhead intel-space, such as parity bits from bit streams to elicit satisfactory 'bit corrections' when the sourced bits so require it. This overhead intel-space is an unavoidable channel-induced intel-space penalty. To guide the design of a CSI-coder, which includes the intel-space channel-coder, an upper performance bound is then defined which tell us about the maximum possible percentage of the channeled intel-space that is not overhead intel-space. This upper performance bound is the channel-capacity C that for a memoryless noisy channel with an input $\mathscr{E}$ and output $\mathscr{F}$ denoting n-bit random codewords is defined as

$$C = (H_{\mathbf{E}} - H_{\mathbf{E/F}})/H_{\mathbf{E}} = \max_{\{P_S[e_i]\}} (H_{\mathscr{E}} - H_{\mathscr{E}/\mathscr{F}})/H_{\mathscr{E}} \quad (3)$$

where **E** and **F** are the $\mathscr{E}$ and $\mathscr{F}$ cases with a probability distribution $\{P_S[e_i]\}$ for **E** that maximizes the mutual source-information ratio $(H_{\mathscr{E}} - H_{\mathscr{E}/\mathscr{F}})/H_{\mathscr{E}}$ where $H_{\mathscr{E}/\mathscr{F}}$ is the *channel-induced intel-space penalty*.

In Fig. 2d an illustration is given of the channel-induced intel-space penalty associated with a noisy intel-space channel which results in a noisy image communication when an intel-space channel-coder is not used. The channel-induced intel-space penalty is assumed to be of 0.25 bits/pixel for this case with the channel-capacity C=0.9677 covering 96.77% of the optimum source-entropy $H_{\mathbf{E}}$ = 7.75 bits/pixel. Notice that $H_{\mathbf{E}}$ is also assumed to be larger in value by 0.31 bits/pixel than the source-entropy case displayed in Fig. 2b of 7.44 bits/pixel when the noisy intel-space channel is not used. In the same figure it is also shown how an optimum intel-space channel-coder improves image communications through a noisy intel-space channel. Clearly the improved performance achieved with the CSI-coder yields an added cost. This cost is the need to design and implement an intel-space channel-coder as well as the need to increase the inter-space rate from 7.75 bits/pixel to 8 bits/pixel. To end the discussion it should be noted that the displayed images of Fig. 2d and stated source-compression values are only stated to convey basic signal compression ideas that will be extended to three other major areas of research, and not to give any state-of-art algorithm results such as is done in Figs. 2c for a lossy source-coder that is noisy intel-space channel free [8].

The intel-space mathematical methodology that seeks to achieve the channel-capacity of (3) is called channel-coding or equivalently intel-space channel-coding to differentiate it from life-time channel coding to be discussed later. Intel-space channel-coding is also called 'the mathematical theory of communication' [3] and is a special case of M-IT that may cover other uncertainty intel-space topics.

3. Mathematical Latency Theory (M-LT)

The principal task of M-LT is to guide the design of intel-time compressors using the performance-bounds of processor-ectropy K and sensor-consciousness F for a limited intel-time sensor. The units of intel-time are mathematical bors, thus in this kind of setting latency theory offers a mathematical guidance methodology for intel-time compressor designs. The bors of intel-time process across a space surface area, and are recognized across a time interval via a certainty, or limited intel-time sensor of given consciousness. Moreover, the mathematical bor unit of intel-time has a configuration of space certainty nature since its creation is traced to a certainty signal-processor. The certainty nature of M-LT is modeled by assigning constraints to the certainty signal-processor intel-time that are driven by the configuration of space, e.g. of two-input NAND gates when implementing a binary-adder signal-processor.

In Fig. 3 the M-LT's recognition system for intel-time compression is shown. It consists of a given signal-processor's mathematical signal **y** to be recognized and a given limited intel-time sensor plus a processor-coder and an intel-time sensor-coder to be designed. A signal-source also feeds intel-space to the signal-processor and intel-time sensor-coder. Furthermore, when designed together, the two coders form a sensor and processor integrated (SPI) coder. The signal-processor produces a vector output $\mathbf{y} = [y_1, y_2, .., y_G]$ with G scalar outputs $\{y_i\}$ that may be real or complex. In particular, the following three space/time or uncertainty/certainty dualities between the M-LT and the M-IT

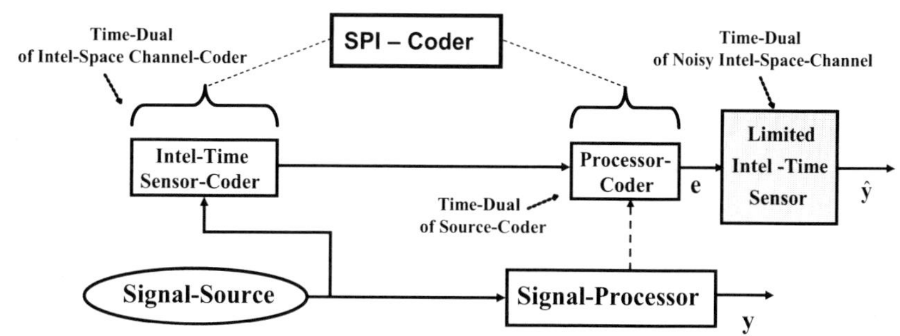

Fig. 3. Mathematical Latency Theory's Recognition System for Intel-Time Compression

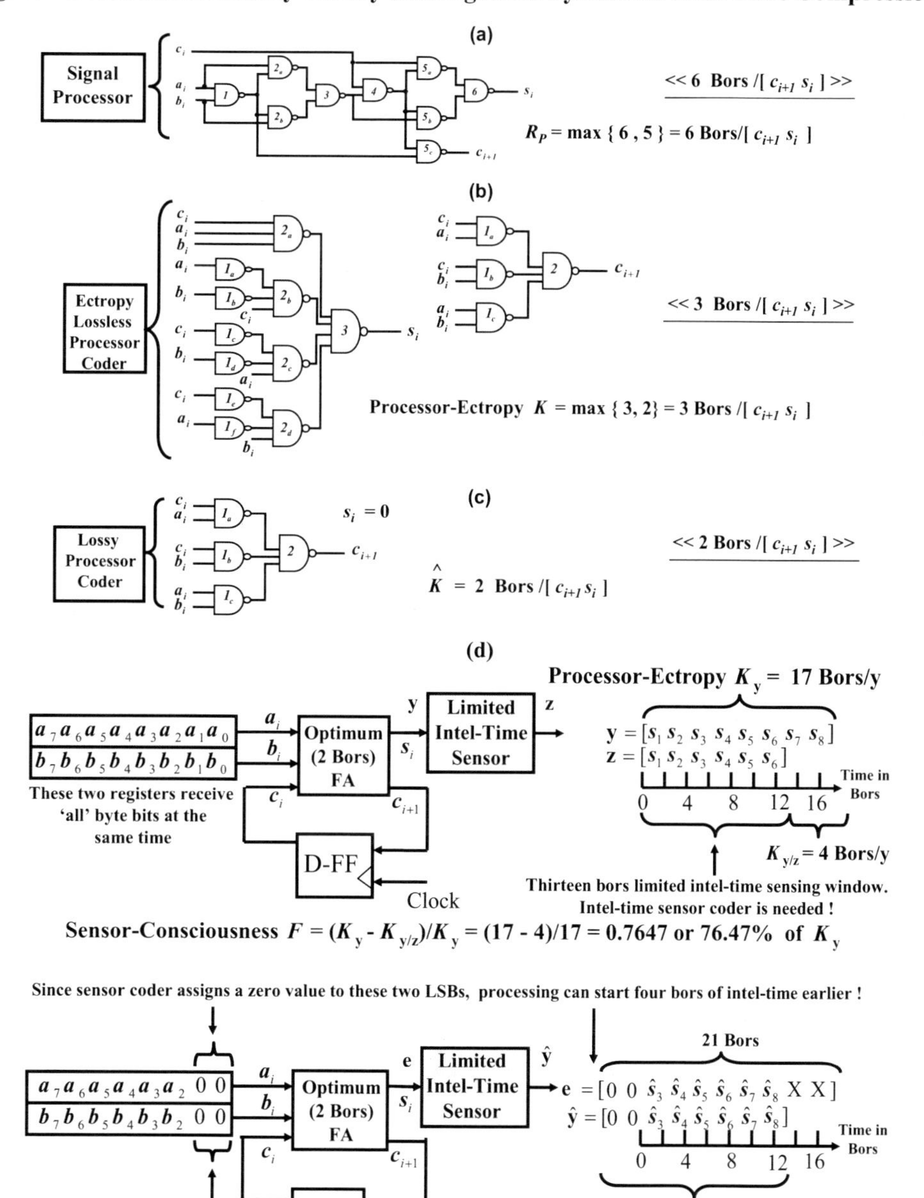

Fig. 4. Illustration of Intel-Time Processor-Ectropy K and Sensor-Consciousness F for Binary-Adder System

subsystems are highlighted: 1) the intel-time sensor-coder is the time (or certainty) dual of the intel-space channel-coder; 2) the processor-coder is the time (or certainty) dual of the source-coder; and 3) the limited intel-time sensor is the time (or certainty) dual of the noisy intel-space channel.

The maximum number of intel-time bors of the signal-processor scalar outputs is called the processor-rate R_P. For instance, for the full adder of Fig. 4a [10] with the two added bit inputs a_i and b_i and the carry in bit c_i, the intel-time of the sum output s_i is of six bors and for the carry out c_{i+1} is of five bors. The maximum of these two intel-times is the processor-rate which is of six bors in this case, i.e. R_P=6 bors/[c_{i+1} s_i]. The certainty in the scalar output set $\{y_i\}$ is modeled with the processor-constraint set $\{C_P[y_i]\}$, e.g. it can be assumed that each of the scalar outputs arises from cascading NAND gates that have an arbitrary number of inputs. These constraints are then used to derive the 'minimum' number of bor levels, denoted as the processor-latency $L_P[y_i]$ in bor units, of each member of the set $\{y_i\}$. For instance, for the full-adder of Fig. 4a, it is noted that the processor-latency of s_i and c_{i+1} is of three and two bors, respectively, since the full-adder can be redesigned as shown in Fig. 4b subject to the constraint of NAND gates with an arbitrary number of inputs which yields

$$L_P[y_1 = s_i] = f_P(C_P[s_i]) = 3 \text{ bors} \quad \text{and} \quad L_P[y_2 = c_{i+1}] = g_P(C_P[c_{i+1}]) = 2 \text{ bors}. \tag{4}$$

The processor-latency with a maximum value, called the processor-ectropy and expressed by

$$K = \max\{L_P[y_1],.., L_P[y_G]\}, \tag{5}$$

then guides as a lower performance bound processor-coder designs. For the full-adder of Fig. 4a the processor-ectropy K is of 3 bors as depicted in Fig. 4b.

The processor-coder of Fig. 3 is characterized by the processor-coder rate R_{PC}. In particular, when the limited intel-time sensor is not present and R_{PC} is greater than or equal to the processor-ectropy K a lossless processor-coder, i.e. yielding $\hat{\mathbf{y}} = \mathbf{y}$, can always be designed, e.g. a fast Fourier transform is such a processor-coder when it replaces a Fourier transform signal-processor. In Fig. 4b the processor-coder is shown for the case where it is lossless and its rate is the same as that of the processor-ectropy, i.e. $R_{PC} = K$. On the other hand, the processor-coder is said to be lossy when its rate is less than the processor-ectropy. For instance, in Fig. 4c the processor-coder is lossy since it only implements the carry out section of the lossless processor-ectropy full adder of Fig. 4b and assigns to s_i the value of zero. The rate for this coder is of 2 bors/[c_{i+1} s_i] which is 66.66 % of the processor-ectropy K= 3 bors/[c_{i+1} s_i].

Yet, when the limited intel-time sensor of Fig. 3 is present an undesirable degradation of the recognized intel-time may occur. Thus when this occurs it becomes necessary to use an intel-time sensor-coder that advances overhead knowledge (also called prior-knowledge) to satisfy the intel-time consciousness needs introduced by the limited intel-time sensor. More specifically, for our illustrative case it becomes necessary to use an intel-time source-coder that introduces sufficient overhead intel-time, such as the bors of the added least significant bits (LSBs) to elicit satisfactory 'bor corrections' when the processing bors so require it. This overhead intel-time is an unavoidable sensor-induced intel-time penalty. To guide the design of a SPI-coder, which includes the intel-time sensor-coder, an upper performance bound is then defined which tell us about the maximum possible percentage of the sensed intel-time that is not overhead intel-time. This upper performance bound is the sensor-consciousness F that for a limited intel-time sensor with a G dimensional vector input $\mathfrak{e}$ and vector output $\mathfrak{f}$ is defined as

$$F = (K_{\mathbf{e}} - K_{\mathbf{e}/\mathbf{f}})/K_{\mathbf{e}} = \max_{\{C_P[e_i]\}} (K_{\mathfrak{e}} - K_{\mathfrak{e}/\mathfrak{f}})/K_{\mathfrak{e}} \tag{6}$$

where $\mathbf{e}$ and $\mathbf{f}$ are the $\mathfrak{e}$ and $\mathfrak{f}$ cases with a constraint distribution $\{C_P[e_i]\}$ for $\mathbf{e}$ that maximizes the mutual processor-latency ratio $(K_{\mathfrak{e}} - K_{\mathfrak{e}/\mathfrak{f}})/K_{\mathfrak{e}}$ where $K_{\mathfrak{e}/\mathfrak{f}}$ is the *sensor-induced intel-time penalty*.

In Fig. 4d an illustration is given of the sensor-induced intel-time penalty associated with a 13 bors limited intel-time sensor which results in the limited sum recognition (only six out of 8 sum bits can be sensed) of an optimum processor-coder (a one byte sequential full adder) output when an intel-time sensor-coder is not used. Notice that the optimum processor-coder has been constrained to have a one byte sequential full adder architecture and thus uses the 2 bors carry-out full-adder of Fig. 4b to yield the smallest possible number of bors. The sensor-induced intel-time penalty is of 4 bors/$\mathbf{y}$ (with $\mathbf{y} = [s_1, s_2, ..., s_8]$ being the sum output) with the sensor-consciousness F=0.7647 covering 76.47% of the optimum processor-entropy $K_{\mathbf{y}}$ = 17 bors/$\mathbf{y}$. In the same figure it is also shown how an optimum intel-time source-coder may be used to improve sum recognitions across a limited intel-time sensor. First it is noticed that to recognize the

17 bors of the 8 bits sum the sequential adder must start at least four bors earlier in time than when the two bytes first become available. This time-dislocation of the processing initiation is achieved in our example by having the sensor-coder advance as prior-knowledge zero bits for the first two least significant bits of the sequential addition. Clearly the improved performance of the SPI-coder is achieved with an added cost. This cost is the need to design and implement an intel-time sensor-coder as well as the need to increase the inter-time rate from 17 bors/**y** to 21 bors/ **y**. Notice that in the count of 21 bors, bors have been included that correspond to the four bors idle time of the processor after the sum ends.

The intel-time mathematical methodology that seeks to achieve the sensor-consciousness of (6) is called intel-time sensor-coding. Furthermore using our duality perspective we have that intel-time sensor-coding is called 'the mathematical theory of recognition', and is a special case of M-LT that may cover other certainty intel-time topics.

4. Physical Latency Theory (P-LT)

The principal task of P-LT is to guide the design of life-time compressors using the performance-bounds of mover-ectropy $\mathcal{A}$ and channel-stay $\mathcal{T}$ for a multi-path life-time channel. The units of life-time are physical time units, thus in this kind of setting latency theory offers a physical guidance methodology for life-time compressor designs. The physical time interval of life-time accompanies energy motion through space (or space-dislocation), and is communicated through a time interval via a certainty, or multi-path life-time channel of given stay. Moreover, the physical time unit of life-time has a configuration of space certainty nature since its origin is traced to a certainty signal-mover for the space-dislocation of physical signals. Furthermore, the life-time required for any desired space-dislocation of a physical signal can never be zero due to an upper bound on energy motion. This upper bound is the 'certainty' speed of light in a vacuum which tells us that the maximum space-dislocation for energy per second is of 2.9979×10^8 meters. The certainty nature of P-LT is modeled by assigning constraints to the certainty signal-mover life-time that are driven by the configuration of space, e.g. of four-wheeled movers for people that are being space-dislocated.

In Fig. 5 the P-LT's communication system for life-time compression is shown. It consists of a given physical-signal **p** to be communicated plus a mover-coder (encoder/decoder) and a life-time channel-coder (encoder/decoder) to be designed. Furthermore, when designed together, the two coders form a channel and mover integrated (CMI) coder. In particular, the following three mathematical-physical dualities between the P-LT and the M-IT subsystems are highlighted: 1) the life-time channel-coder is the physical dual of the intel-space channel-coder; 2) the mover-coder is the physical dual of the source-coder; and 3) and the multi-path life-time channel is the physical dual of the noisy intel-space channel. Also a special case of the mover-coder is any initial mover-coder which is denoted as the signal-mover and is the physical-dual of a signal-source. The signal-mover moves a vector physical signal $\mathbf{p} = [p_1, p_2, .., p_D]$ with D scalar elements $\{p_i\}$.

The maximum amount of life-time physical time due to the signal-mover is called the mover-rate R_M. For instance, consider the persons (or physical signals) signal-mover of Fig. 6a, where the couple is assigned the physical variable p_1 and the single man is assigned the physical variable p_2. It is then noted that it takes the displayed signal-mover, in the form of non-motorized bicycles, 7 hours of life-time for p_1 and 5 ½ hours of life-time for p_2 to move from one location to another in space. Thus the mover-rate for this case is of 7 hours, i.e. $R_M = 7$ hrs/**p**. The certainty in the

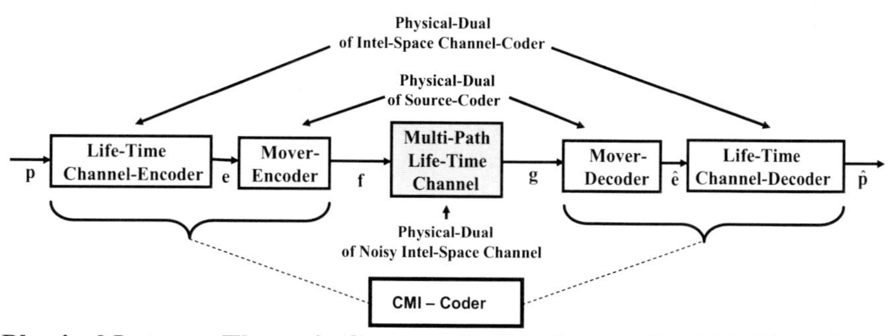

Fig. 5. Physical Latency Theory's Communication System for Life-Time Compression

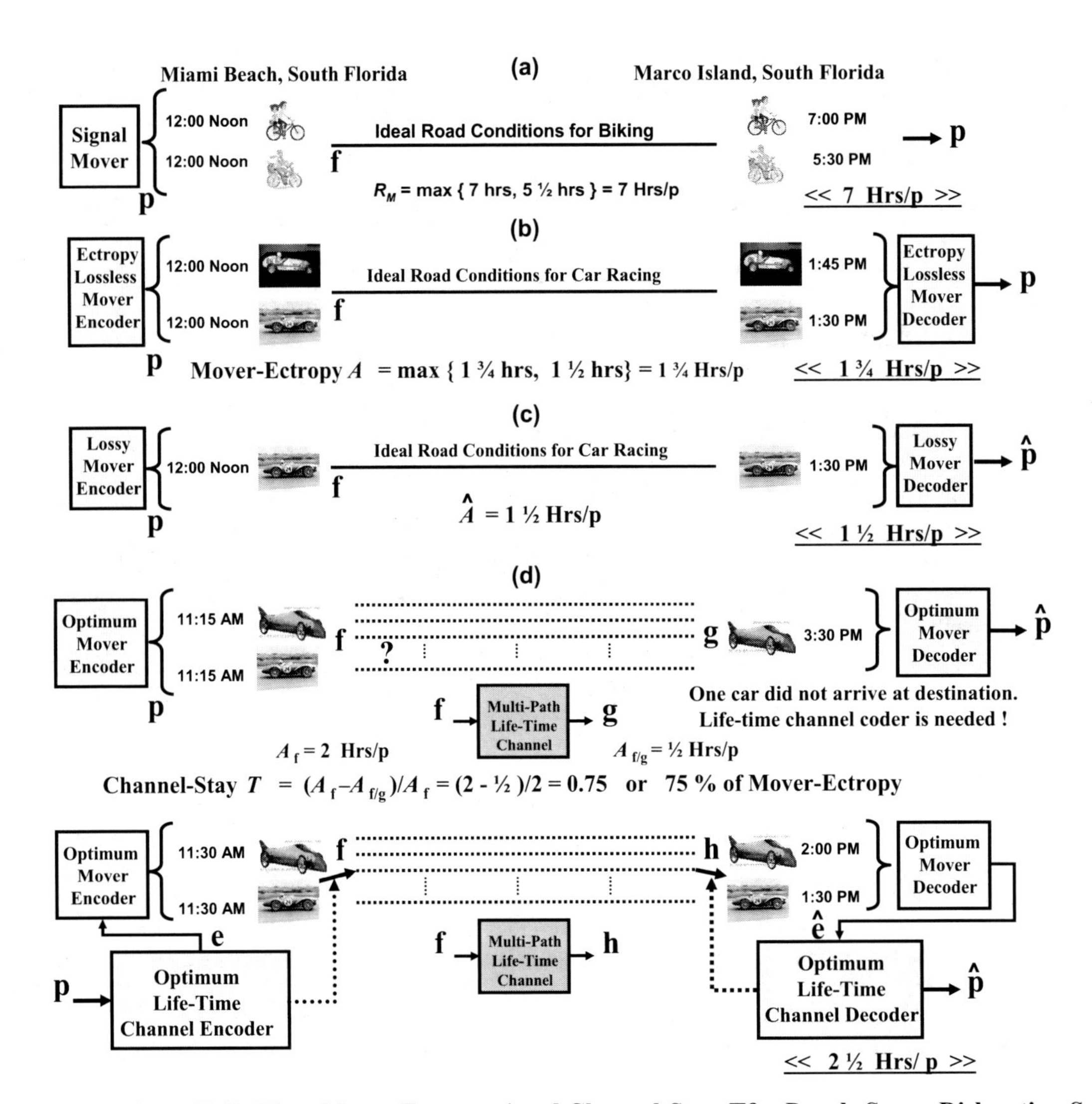

Fig. 6. Illustration of Life-Time Mover-Ectropy *A* and Channel-Stay *T* for People Space-Dislocation System

physical signals $\{p_i\}$ is modeled with the mover-constraint set $\{C_M[p_i]\}$, where these constraints are assumed when seeking physical signals movers, e.g. in the case of Fig. 6a the mover-constraint is a non-motorized two-wheeled mover. These allowed constraints are then used to derive the 'minimum' amount of physical time, denoted as the mover-latency $L_M[p_i]$ in physical time units, of each member of the set $\{p_i\}$. For instance, for the people motion system of Fig. 6a, it is noted that the mover-latency of the illustrated motion system will be of 1¾ hours and 1 ½ hours, respectively, when four-wheeled motorized movers are allowed constraints. More specifically we have for this case the following mover-latencies,

$$L_M[p_1] = f_M(C_M[p_1]) = 1¾ \text{ hours} \quad \text{and} \quad L_M[p_2] = f_M(C_M[p_2]) = 1 ½ \text{ hours}, \tag{7}$$

The mover-latency with a maximum value, called the mover-ectropy and expressed by

$$A = \max\{L_M[p_1], .., L_M[p_D]\}, \tag{8}$$

then guides as a lower performance bound mover-coder designs. For the mover system of Fig. 6a the mover-ectropy A is of 1¾ hours as seen from Fig. 6b.

The mover-coder of Fig. 5 is characterized by the mover-coder rate R_{MC}. In particular, when the life-time channel is not present and R_{ME} is greater than or equal to the mover-ectropy A a lossless mover-coder, i.e. with $\hat{\mathbf{e}} = \mathbf{e}$, can always be designed. In Fig. 6b the mover-coder is shown for the case where it is lossless and its rate is the same as that of the mover-ectropy, i.e. $R_{ME} = \mathit{\Lambda}$. On the other hand, the mover-coder is said to be lossy when its rate is less than the mover-ectropy. For instance, in Fig. 6c the mover-coder is lossy since only the physical signal p_2 is moved. The rate for this coder is of 1 ½ hours which is 85.71 % of the mover-ectropy $\mathit{\Lambda} = 1\frac{3}{4}$ hours.

Yet, when the multi-path intel-time sensor of Fig. 5 is present an undesirable degradation of the communication life-time may occur. Thus when this occurs it becomes necessary to use a life-time channel-coder that advances overhead knowledge to satisfy the life-time stay needs introduced by the multi-path life-time channel. More specifically, for our illustrative case it becomes necessary to use a life-time channel-coder that introduces sufficient overhead life-time such as an earlier departure time to elicit satisfactory 'time corrections' when the moving time so require it. This overhead life-time is an unavoidable channel-induced life-time penalty. To guide the design of a CMI-coder, which includes the life-time channel-coder, an upper performance bound is then defined which tell us about the maximum possible percentage of the channeled life-time that is not overhead life-time. This upper performance bound is the channel-stay T that for a multi-path life-time channel with a D dimensional vector input $\mathfrak{f}$ and vector output $\mathfrak{g}$ is

$$\mathit{T} = (\mathit{\Lambda}_{\mathbf{f}} - \mathit{\Lambda}_{\mathbf{f}/\mathbf{g}})/\mathit{\Lambda}_{\mathbf{f}} = \max_{\{C_M[f_i]\}} (\mathit{\Lambda}_{\mathfrak{f}} - \mathit{\Lambda}_{\mathfrak{f}/\mathfrak{g}})/\mathit{\Lambda}_{\mathfrak{f}} \tag{9}$$

where **f** and **g** are the $\mathfrak{f}$ and $\mathfrak{g}$ cases with a constraint distribution $\{C_M[f_i]\}$ for **f** that maximizes the mutual mover-latency ratio $(\mathit{\Lambda}_{\mathfrak{f}} - \mathit{\Lambda}_{\mathfrak{f}/\mathfrak{g}})/\mathit{\Lambda}_{\mathfrak{f}}$ where $\mathit{\Lambda}_{\mathfrak{f}/\mathfrak{g}}$ is the *channel-induced life-time penalty.*

In Fig. 6d an illustration is given of the channel-induced life-time penalty associated with a multi-path life-time channel which results in a limited people communication when a life-time channel-coder is not used. The channel-induced life-time penalty is assumed to be of ½ hrs/**p** for this case with the channel-stay T of 0.75 covering 75 % of the optimum mover-ectropy $\mathit{\Lambda}_{\mathbf{f}} = 2$ hrs/**p**. Notice that $\mathit{\Lambda}_{\mathbf{f}}$ is also assumed to be larger by ¼ hr/**p** than the mover-ectropy case displayed in Fig. 6b of 1 ¾ hrs/**p** for the case when a multi-path life-time channel is not used. Further notice that the top racing car of Fig. 6b is not available to start the people space-dislocation at the earlier starting time displayed in Fig. 6d. In the same figure it is also shown how an optimum life-time channel-coder may be used to improve people communication through a multi-path channel. This is done by observing that to communicate **p** to its destination at 2 pm (as implied by $\mathit{\Lambda}_{\mathbf{f}} = 2$ hrs/**p** and a starting time of 12:00 noon) the mover must depart to Marco Island at least ½ hr earlier in time, i.e. at 11:30 am. Thus the life-time channel-coder directs the mover to depart at 11:30 am along the channel path yielding the least life-time use. At the receiving end the life-time channel-decoder has the people exiting the mover. Clearly the improved performance achieved with the CMI-coder yields an added cost. This cost is the need to design and implement a life-time channel-decoder as well as the need to increase the life-time rate from 2 hrs/**p** to 2 ½ hrs/**p.**

The life-time physical methodology that seeks to achieve the channel-stay of (9) is called life-time channel-coding. Furthermore using our duality perspective we have that life-time channel-coding is called 'the physical theory of communication', and is a special case of P-LT that may cover other certainty life-time topics.

5. Physical Information Theory (P-IT)

The principal task of P-IT is to guide the design of life-space compressors using the performance-bounds of retainer-entropy N and sensor-scope $\boldsymbol{I}$ for a noisy life-space sensor. The units of life-space are physical surface area units, thus in this kind of setting information theory offers a physical guidance methodology for life-space compressor designs. The physical surface area of life-space accompanies energy retention across a time interval (or time-dislocation) and is recognized across a space surface area via an uncertainty, or noisy life-space sensor of given scope. Moreover, the physical surface area unit of life-space has a passing of time uncertainty nature since its origin is traced to an uncertainty signal-retainer for the time-dislocation of physical signals. Furthermore, the expected life-space for any required time-dislocation of a physical signal can never be zero due to an upper bound on energy retention. This upper bound is the 'uncertainty' pace of dark in a black-hole (the uncertainty space-dual of the speed of light in a vacuum) which tells us that the maximum time-dislocation for energy per cubic meter is of 6.1123×10^{63} seconds [7]. The uncertainty nature of P-IT is modeled by assigning probability distributions to the uncertainty signal-retainer life-space that is driven by the passing of time, e.g. of uniform distributions for the thermal bottle microstates of tea molecules that are being time-dislocated.

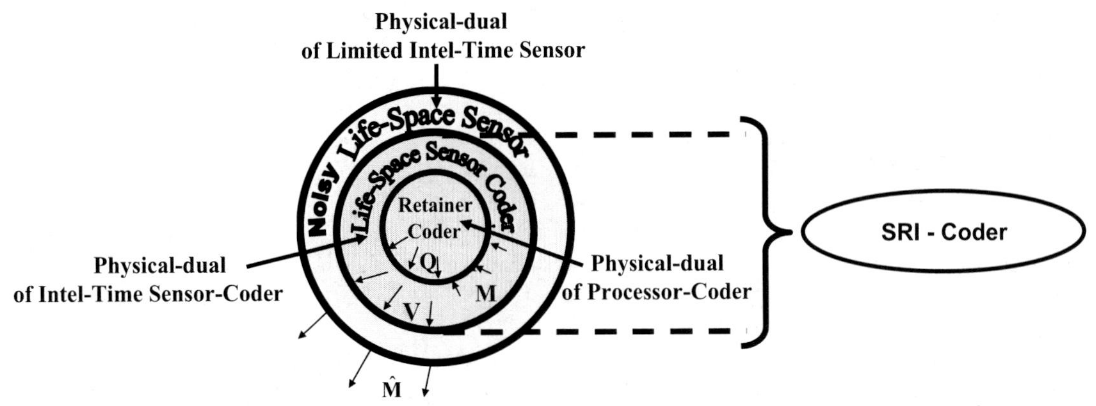

Fig. 7 Physical Information Theory's Recognition System For Life-Space Compression

In Fig. 7 the P-IT's recognition system for life-space compression is shown. It consists of a given physical signal **M** to be recognized and a given noisy life-space sensor plus a retainer-coder and a life-space sensor-coder to be designed. Furthermore, when designed together, the two coders form a sensor and retainer integrated (SRI) coder. In particular, the following three mathematical-physical dualities between the P-IT and the M-LT subsystems are highlighted: 1) the life-space sensor-coder is the physical dual of the intel-time sensor-coder; 2) the retainer-coder is the physical dual of the processor-coder; and 3) the noisy life-space sensor is the physical dual of the limited intel-time sensor. Also a special case of the retainer-coder is any starting retainer-coder which is denoted as the signal-retainer and is the physical-dual of a signal-processor. The signal-retainer retains a random vector physical signal $\mathbf{M} \in \{\mathbf{m}_1, \mathbf{m}_2, .., \mathbf{m}_\Omega\}$ with Ω vector outcomes $\{\mathbf{m}_i\}$ or microstates where each vector outcome $\mathbf{m}_i$ is of dimension U where this dimension is equal to the number of fundamental physical entities, e.g. molecules, present in the signal-retainer. Thus, each vector outcome of **M** represents an identical physical mass for the retained energy.

The expected amount of surface area of the signal-retainer is called the retainer-rate R_R. For instance, consider U tea molecules that are being time-dislocated in the cylindrical thermos bottle of Fig. 8a. It is noted that it takes the thermos 70 cm^2 of life-space surface area for its storage. Since it is assumed that every possible microstate of the tea has the same expected life-space of 70 cm^2, the retainer-rate for this case is then given by R_R=70 cm^2/**M**. The uncertainty in the microstate set $\{\mathbf{m}_i\}$ is modeled with the retainer-probability set $\{P_R[\mathbf{m}_i]\}$. These probabilities are then used to derive the expected life-space or retainer-information $I_R[\mathbf{m}_i]$ in surface area units, of each member of the set $\{\mathbf{m}_i\}$. For instance, for the substance retainer system of Fig. 8a, it is noted that the expected retainer-information is of 70 cm^2 for each tea microstate when the retainer-probability set $\{P_R[\mathbf{m}_i]\}$ is uniformly distributed. More specifically, we have the following expected retainer-information for all the microstates

$$I_R[\mathbf{m}_i] = f_R(P_R[\mathbf{m}_i]) = 70 \text{ cm}^2 \text{ for all i.} \tag{10}$$

A general expression for $I_R[\mathbf{m}_i] = f_R(P_R[\mathbf{m}_i])$ in surface area units that may be viewed as the life-space recognition dual of the intel-space communication source-information expression $I_S(x_i) = \log_2(1/P_S[x_i])$ in bit units has already been obtained in [7] and is given

$$I_R[\mathbf{m}_i] = \eta \frac{T_i}{V_i} \log_2(1/P_R[\mathbf{m}_i]) \quad \text{in } m^2 \text{ units} \tag{11}$$

where: a) T_i is the retention-time of the microstate $\mathbf{m}_i$; b) V_i is the uncertainty volume where the microstate $\mathbf{m}_i$ resides; and c) η is the retainer-information constant of 2.473 x 10^{-133} m^5/sec. The value of the retention constant η is a normalizing constant that is derived in [7] using black-hole conditions as will be noted shortly. More specifically, the expression for η from which its value is derived is

$$\eta = \frac{1920}{c\text{X}^2 \ln 2} = \frac{h^2 G^2}{480\pi^2 c^5 \ln 2} \tag{12}$$

where: a) c is the speed of light in a vacuum; b) h is Plank's constant of 6.6206896 x 10^{-34} Joule.sec; c) G is the gravitational constant of 6.673 x 10^{-11} m^3/kg.sec^2; d) X is the pace of dark in a black hole of 6.1123 x 10^{63} sec/m^3 whose value is derived from the pace of dark expression [7]

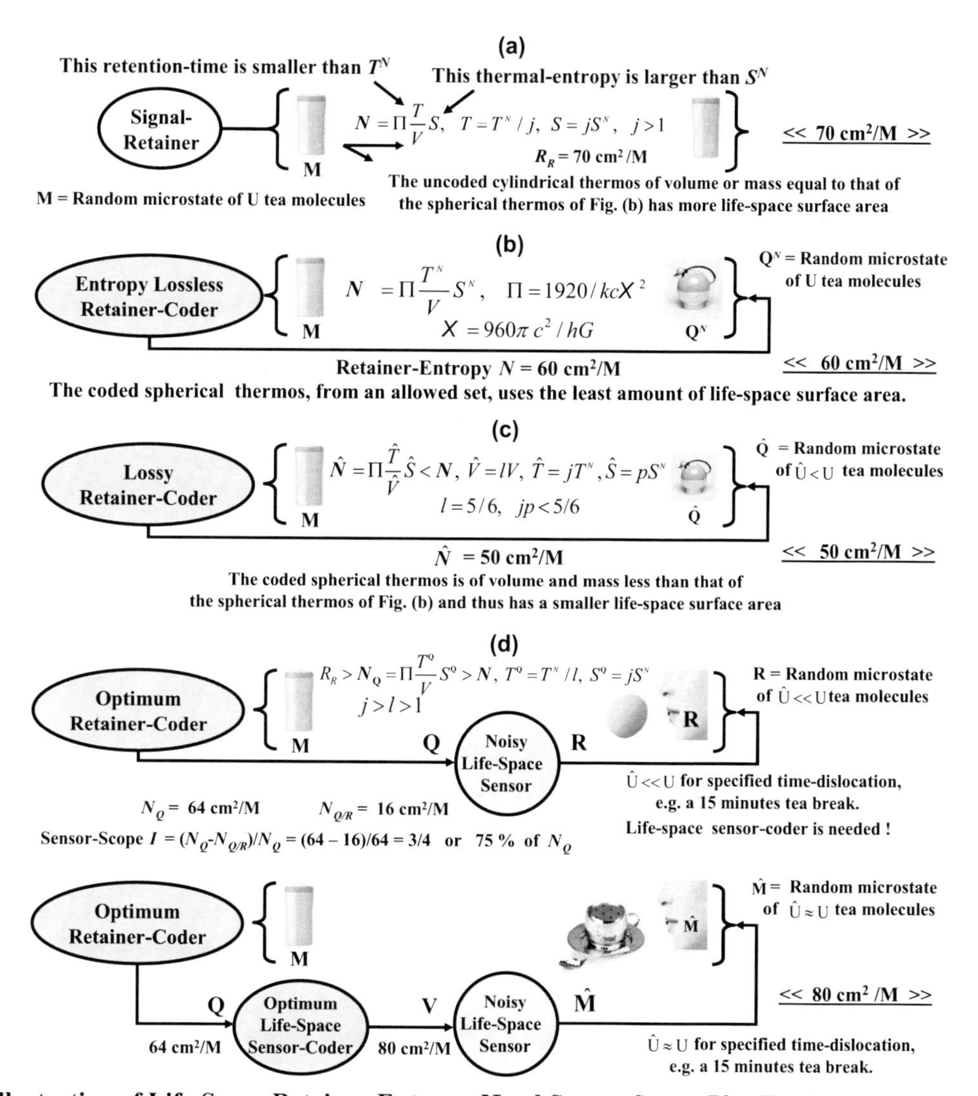

Fig. 8. Illustration of Life-Space Retainer-Entropy N and Sensor-Scope I for Tea Time-Dislocation System

$$X = 960\pi \frac{c^2}{hG} \tag{13}$$

and its derivation is once again given in Appendix A for the benefit of the reader. The derivation of (13) in Appendix A is essentially the same as that of Appendix A in [7] except that some variable symbols have been altered to conform with the notation of the current paper. The expected value of the retainer-information (10), called the retainer-entropy and expressed by

$$N = \sum_{i=1}^{\Omega} P_R[\mathbf{m}_i] I_R(\mathbf{m}_i) \tag{14}$$

then guides as a lower performance bound retainer-coder designs. For instance, for the retainer system of Fig. 8a the retainer-entropy N is of 60 cm^2 as seen from Fig. 8b where a spherical thermos is used to retain the tea**.** Notice that this spherical thermos is the thermos of smallest life-space surface area for the assumed **M** and thus cannot be compressed any further.

An important special case for which an analytical expression can be found for the retainer-entropy is when the the microstates of **M** are equally likely and the ratio of T_i/V_i is the same for all microstates in **M**, i.e. $T_i/V_i = T/V$ for i=1,...,Ω. Then (11) can be used in (14) together with the microstate thermodynamic entropy $S = k \log_2 \Omega / \ln 2$, with k being the Boltzmann's constant 1.3806504 x 10^{-23} Joules/Kelvin, to yield for N the following expression

$$N = \sum_{i=1}^{\Omega} P_R[\mathbf{m}_i]I_R(\mathbf{m}_i) = \eta \sum_{i=1}^{\Omega} \frac{T_i}{V_i} P_R[\mathbf{m}_i]\log_2(1/P_R(\mathbf{m}_i)) = \eta \frac{T}{V}\log_2 \Omega = \eta \frac{T}{V}\frac{\ln 2 \bullet S}{k} = \Pi \frac{T}{V} S \quad (15)$$

$$\Pi = \frac{1920}{kc X^2} = \frac{\ln 2}{k}\eta \cdot \quad (16)$$

where Π is the retainer-entropy constant of 1.2416 x 10^{-110} m^5Kelvin/(sec.Joules). It should be noticed that the analytical as well as quantum-gravity [11] and holographic [12] retainer-entropy expression of (15), i.e.

$$N = \eta \frac{T}{V}\log_2 \Omega = \eta \frac{T}{V}\frac{\ln 2 \bullet S}{k} = \Pi \frac{T}{V} S, \quad (17)$$

was first derived in [7] and is given there by expressions (3.7) and (3.17). Next it is noted that (17) has the desirable property that when the retainer-coder is a black-hole it reduces to Hawking's thermal-entropy $S=\pi Akc^3/2hG$ [11]-[12] where A is the surface area of the black-hole. In Appendix A of this paper as well as that of [7] it is shown that the ratio of retention-time T to retention-volume V of any substance in a black-hole is constant. More specifically this ratio is the pace of dark $T/V = X$ (derived from (A.10) by setting the initial time of mass retention equal to zero, i.e. t_i=0) and thus $N_{BH} = \Pi X\ S$. Secondly and last using the pace of dark expression (13), the retainer-entropy constant expression (16) and replacing N_{BH} with A in $N_{BH} = \Pi X\ S$ the Hawking's thermal-entropy expression $S=\pi Akc^3/2hG$ is derived. For the illustrative tea example of Fig. 8 it is assumed that expression (17) holds where **U** tea molecules with a random microstate **M** of Ω possible vector outcomes with a constant volume each are being retained or time-dislocated in a thermos. For the retainer system of Fig. 8a the retainer-entropy N is of 60 cm^2/**M** as seen from Fig. 8b. Four observations can be made about these two cases. They are: 1) both thermos have identical volume but the surface area of the spherical thermos is the smallest that can be attained for any fixed volume and thus results in the greatest surface area savings; 2) it is noted that the larger thermal-entropy and smaller retention-time of the cylindrical thermos is consistent with its more extensive surface area; 3) it is possible for both retainers to have the same retainer-entropy because the product of retention-time and thermal-entropy remains the same for each case; and 4) it is noted that the retention-time of any thermos readily follows if its thermal-entropy can be calculated.

The retainer-coder of Fig. 7 is characterized by the retainer-coder rate R_{RC}. In particular, when the life-space sensor is not present and R_{RC} is greater than or equal to the retainer-entropy N a lossless retainer-coder can always be designed that yields a random microstate **Q** of U tea molecules. In Fig. 8b the retainer-coder is shown for the case where it is lossless and its rate is the same as that of the retainer-entropy, i.e. $R_{RC} = N$. On the other hand, the retainer-coder is said to be lossy when its rate is less than the retainer-entropy. For instance, in Fig. 8c the retainer-coder is lossy since only 5/6 of the mass (or 5/6 of the original U molecules) of the physical signal **M** is retained. The rate for this lossy coder is of 50 cm^2 which is 83.3 % of the retainer-entropy N= 60 cm^2. Two observations can be made about this case. They are: 1) the retainer-coder rate for this lossy but spherically shaped retainer-coder is the same as its retainer-entropy $\hat{N}$; 2) the retainer-entropy $\hat{N}$ is less than that for the two thermos of Figs. 8a and 8b as expected.

Yet, when the life-space sensor of Fig. 7 is present an undesirable degradation of the recognition life-space may occur. Thus when this occurs it becomes necessary to use a life-space sensor-coder that advances overhead knowledge to satisfy the life-space scope needs introduced by the noisy life-space sensor. More specifically, for our illustrative case it becomes necessary to use a life-space source-coder that introduces sufficient overhead life-space, such as the surface area of a tea cup to elicit satisfactory 'space corrections' when the retaining space (surface area) so require it. This overhead life-space is an unavoidable sensor-induced life-space penalty. To guide the design of a SRI-coder, which includes the life-space sensor-coder, an upper performance bound is then defined which tell us about the maximum possible percentage of the sensed life-space that is not overhead life-space. This upper performance bound is the sensor-scope I, that for a noisy life-space sensor with a random vector input $\mathbf{Q} \in \{\mathbf{q}, \mathbf{q}_2, .., \mathbf{q}_\Omega\}$ and a random vector output $\mathbf{R} \in \{\mathbf{r}_1, \mathbf{r}_2, .., \mathbf{r}_\Omega\}$ with the U dimension of their vector outcomes $\{q_i\}$ and $\{r_i\}$ being the same as the number of occurring physical entities, e.g. tea molecules, is defined by

$$I = (N_{\mathbf{Q}} - N_{\mathbf{Q/R}})/N_{\mathbf{Q}} = \max_{\{P_R[\mathbf{q}_i]\}} (N_{\mathcal{Q}} - N_{\mathcal{Q}\mathcal{R}})/N_{\mathcal{Q}} \quad (18)$$

where **Q** and **R** are the $\mathcal{Q}$ and $\mathcal{R}$ cases with a probability distribution $\{P_R[\mathbf{q}_i]\}$ for **Q** that maximizes the mutual retainer-information ratio $(N_{\mathcal{Q}} - N_{\mathcal{Q}\mathcal{R}})/N_{\mathcal{Q}}$ where $N_{\mathcal{Q}\mathcal{R}}$ is the *sensor-induced life-space penalty*.

In Fig. 8d an illustration is given of the sensor-induced life-space penalty associated with a noisy life-space sensor, i.e. the mouth of a generally unknown tea drinker, which results in a noisy tea recognition over some specified time-dislocation, e.g. a 15 minutes tea break, when a life-space sensor-coder is not used. The noisy life-space sensor may use as overhead knowledge the relevant properties of the physical signal that is being recognized. The sensor-induced life-space penalty is assumed to be of 20 $cm^2/\mathbf{M}$ for this case with the sensor-scope I of 0.75 covering 75% of the optimum retainer-entropy $\mathcal{N}_Q$ = 64 $cm^2/\mathbf{M}$. Notice that $\mathcal{N}_Q$ is also assumed to be larger by 4 $cm^2/\mathbf{M}$ than the retainer-entropy case displayed in Fig. 8b of 60 $cm^2/\mathbf{M}$ for the case where a noisy life-space sensor is not used. In the same figure it is also shown how an optimum life-space sensor-coder in the form of a tea cup may be used to improve tea recognitions across a noisy life-space sensor, i.e. the tea drinker mouth, over some specified time-dislocation. Clearly the improved performance of the SRI-coder is achieved with an added cost. This cost is the need to design and implement a life-space sensor-coder as well as the need to increase the life-space rate from 64 $cm^2/\mathbf{M}$ to 80 $cm^2/\mathbf{M}$.

The life-space physical methodology that seeks to achieve the sensor-scope of (18) is called life-space sensor-coding. Furthermore using our duality perspective we have that life-space sensor-coding is called 'the physical theory of recognition', and is a special case of P-IT that may cover other uncertainty life-space topics.

6. The Latency-Information Theory Revolution

The LIT revolution of Fig. 9 conveniently displays in four quadrants the principal dualities that exist between four methodologies, two mathematical and two physical, for the study of systems. Together they advance a guidance theory for intelligence and life system designs. Quadrant 'I' pertains to certainty life-time compressor designs for channel-mediated communication systems that are guided by physical latency theory or P-LT. The physical time interval of life-time accompanies energy motion through space, and is communicated through a time interval via a certainty, or multi-path life-time channel of given stay. More specifically, channel and mover integrated coders or CMI-coders are derived for the compression of a physical signal's motion life-time in units of time. These designs are enabled by the laws of motion in physics, of a configuration of space certainty nature, such as the speed of light $\boldsymbol{c}$ in a vacuum limit. Quadrant II pertains to uncertainty intel-space compressor designs for channel-mediated communication systems that are guided by mathematical information theory or M-IT. The bits of intel-space are sourced through time, and are communicated through space via an uncertainty, or noisy intel-space channel of given capacity. More specifically, channel and source integrated coders or CSI-coders are derived for the compression of a mathematical signal's sourced intel-space in units of bit. These designs are enabled by universal source compression schemes, of a passing of time uncertainty nature, such as MMSE predictive-transform source-coding [8]. Quadrant III pertains to uncertainty life-space compressor designs for sensor-mediated recognition systems that are guided by physical information theory or P-IT. The physical surface area of life-space accompanies energy retention across a time interval, and is recognized across a space surface area via an uncertainty, or noisy life-space sensor of given scope. More specifically, sensor and retainer integrated coders or SRI-coders are derived for the compression of a physical signal's retention life-space in units of surface area. These designs are enabled by the laws of retention in physics, of a passing of time uncertainty nature, such as the pace of dark $\boldsymbol{X}$ in a black-hole limit and the retainer-entropy $\mathcal{N}$. Quadrant IV pertains to certainty intel-time compressor designs for sensor-mediated recognition systems that are guided by mathematical latency theory or M-LT. The bors of intel-time process across a space surface area, and are recognized across a time interval via a certainty, or limited intel-time sensor of given consciousness. More specifically, sensor and processor integrated coders or SPI-coders are derived for the compression of a mathematical signal's processing intel-time in units of bor. These designs are enabled by universal processor compression schemes, of a configuration of space certainty nature, such as the time-dual of transform source-coding treated in [1] and [13].

In particular the LIT revolution highlights several types of dualities guiding compressor designs. They are:

1. The uncertainty–information-space/certainty-latency-time duality exhibited by the quadrant pairs (II,III)/(I,IV). This duality tell us that the uncertainty mathematical intel-space and uncertainty physical life-space methods of quadrants (II,III) have as duals the certainty physical life-time and certainty mathematical intel-time methods of quadrants (I,IV) and visa-versa. Examples are M-IT/M-LT, P-LT/P-IT dualities, etc.

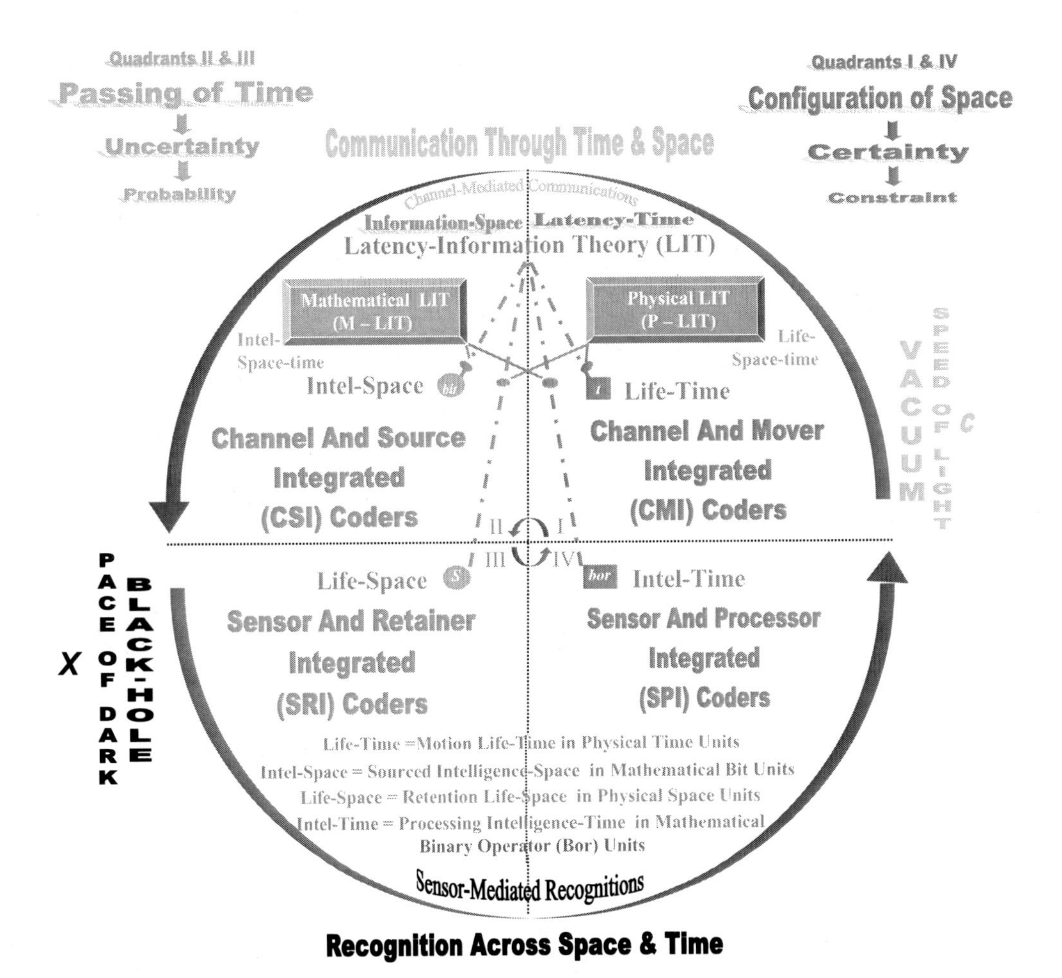

Fig. 9 The Latency-Information Theory Revolution

2. The channel-communication/sensor-recognition duality exhibited by the quadrant pairs (I,II)/(III,IV). This duality tell us that the certainty physical life-time and uncertainty mathematical intel-space methods of quadrants (I,II) enabling communications through time and space have as duals uncertainty physical life-space and certainty mathematical intel-time methods of quadrants (III,IV) enabling recognitions across space and time and visa-versa. Examples are M-IT/P-IT, P-LT/M-LT dualities, etc.

3. The mathematical-intelligence/physical-life duality exhibited by the quadrant pairs (II,IV)/(I,III). This duality tell us that uncertainty mathematical intel-space and certainty mathematical intel-time methods of quadrants (II,IV) have as duals uncertainty physical life-space and certainty physical life-time methods of quadrants (I,III) and visa-versa. Examples are source-entropy/retainer-entropy, mover-ectropy/processor-ectropy dualities, etc.

4. The uncertainty-intel-space/certainty-intel-time duality exhibited by quadrants II/IV. This duality tell us that sourcing-methods of quadrant II have as duals processing-methods of quadrant IV and visa-versa. Examples are transform-source-coding/transform-processing-coding dualities [1], [13], etc.

5. The certainty-life-time/uncertainty-life-space duality exhibited by quadrants I/III. This duality tell us that motion-methods of quadrant 'I' have as duals retention-methods of quadrant III and visa-versa. Examples are laws-of-motion/laws-of-retention dualities such as speed-of-light-in-a-vacuum/pace-of-dark-in-a-black-hole limits, etc.

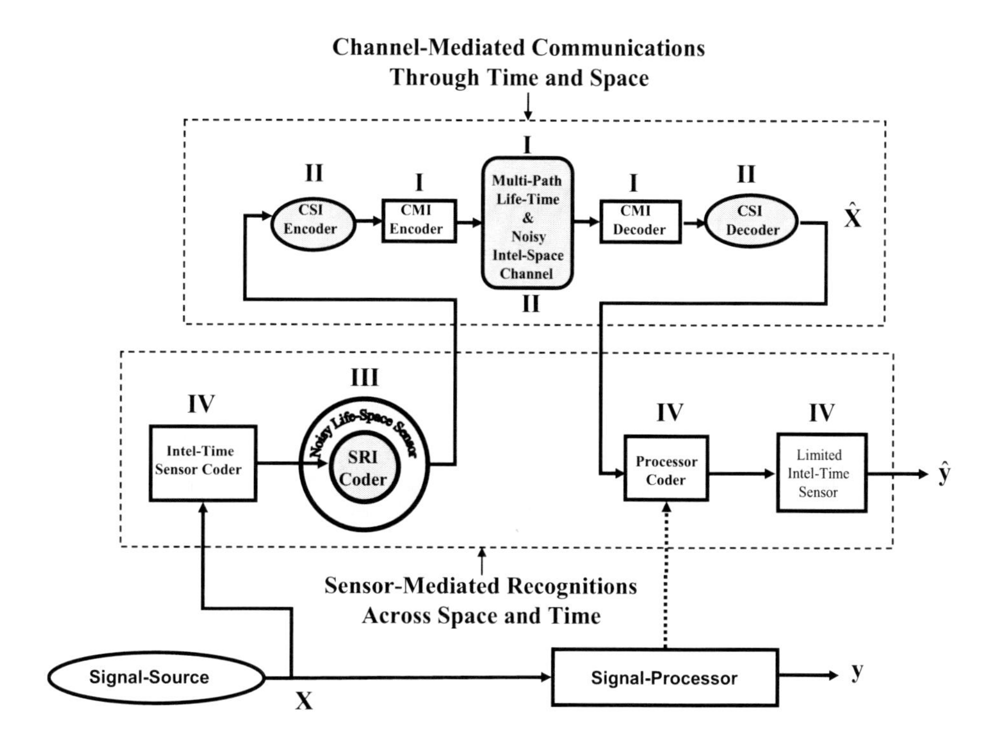

Fig. 10. The Communication Embedded Recognition (CER) System

6. The certainty-life-time/certainty-intel-time duality exhibited by quadrants I/IV. This duality tell us that motion-methods of quadrant 'I' have as duals processing-methods of quadrant IV and visa-versa. Examples are mover-ectropy/processor-ectropy duality, etc.

7. The uncertainty-intel-space/uncertainty-life-space duality exhibited by quadrants II/III. This duality tell us that sourcing-methods of quadrant II have as duals retention-methods of quadrant III and visa-versa. Examples are source-entropy/retainer-entropy duality, etc.

Next in Fig. 10 it can be seen how the four signal compressors of the LIT revolution can be integrated to form a communication embedded recognition (CER) system. The operation of the CER system starts with the intel-time sensor-coder deriving prior-knowledge from a signal-source, e.g. in the form of synthetic aperture radar (SAR) imagery [13]. This prior-knowledge is then stored in the SRI-coder where it is assumed that both the intel-time sensor-coder and SRI-coder reside in the same location at some distance away from the processor-coder. In this same remote location there is also a CSI-encoder followed by a CMI-encoder that communicate the SRI-coder content, available via a noisy life-space sensor, to the processor-coder location. However, before the processor-coder receives the SRI-coder content a CMI-decoder operates on the output of the twofold communication channel, i.e. a noisy intel-space and multi-path life-time channel, and routes it to the input of the CSI-decoder. The output of the CSI-decoder is then fed to the processor-coder with its output sensed by a limited intel-time sensor. This sensing action by the limited intel-time sensor completes the operation of the CER system.

As a concluding remark it should be noted that when one views the unifying LIT revolution via the CER system prism, it becomes apparent that the use of the advanced guidance theory for intelligence and life system designs applies not only to the study of artificial systems but also to living systems. For instance, both intel-space and intel-time compressors can be assumed to exist in the neuron networks of our brains, and both life-time and life-space compressors can be assumed to exist in the heart, arteries and veins of our circulatory system. Clearly, living systems, whose signal compressor evolutionary designs efficiently interact through and across space and time to achieve their own survival in a resources limited environment, should be investigated and used as role models to emulate by the designers of signal compressors of any type. Finally in Appendix B both visual and written reminders are offered whose goal is to genially recall the many symbols, terminology, methods and unifying ideas of the LIT revolution.

APPENDIX A
Derivation of the Pace of Dark X

The derivation begins with the power expression

$$P(t\ ) = -\frac{dE(t\ )}{dt} = -c^2\frac{dm(t)}{dt} \tag{A.1}$$

where $P(t)$ is the rate of change of the energy of an uncharged and nonrotating black hole (UNBH) [11] at the instant of time t and $m(t)$ is the UNBH mass that is related to E(t) via the energy-mass equation E(t)=c^2m(t). $P(t)$ is then noted to be equal to the black body luminance $L(t)$ resulting in the expression

$$P(t) = L(t) = (\pi^2 k^4 / 60\hbar^3 c^2) A \Gamma^4(t) \tag{A.2}$$

where k is Boltzmann's constant, $\hbar = h/2\pi$ is Plank's reduced constant, c is the speed of light and $\Gamma(t)$ is the temperature of the UNBH: the radiation frequency of the black body, or equivalently Hawking's radiation frequency $f(t)$ for a black hole, is related to $\Gamma(t)$ via the expression $f(t) = k\Gamma(t)/2\hbar$. In addition, A is the surface area of the spherically shaped UNBH. Next it is noted that $\Gamma(t)$ is given by the reciprocal of the rate of change of the UNBH thermodynamical entropy $S(t)$ with respect to $E(t)$ where the $S(t)$ is given by the Hawking entropy [11]. Thus

$$\Gamma(t) = \left(\partial S(t)/\partial E(t)\right)^{-1} \tag{A.3}$$

$$S(t) = \frac{kc^3}{4\hbar G}A. \tag{A.4}$$

Next using Schwarzschild's radius in the expression for A in (A.4) and then replacing $m(t)$ with its energy equivalence one obtains the following expression for $S(t)$ as a function of $E(t)$:

$$S(t) = \frac{kc^3}{4\hbar G}A = \frac{k\pi c^3}{\hbar G}\left(\frac{2Gm(t)}{c^2}\right)^2 = \frac{k4\pi G}{\hbar c^5}E^2(t) \tag{A.5}$$

Next using (A.5) in the evaluation of (A.3) one finds

$$\Gamma(t) = \frac{\hbar c^3}{8\pi k G m(t)} \tag{A.6}$$

Using (A.6) in (A.2) and equating the result with (A.1) the following nonlinear differential equation is derived

$$\frac{dm(t)}{dt} + \frac{\hbar c^4}{15360\pi G^2 m^2(t)} = 0 \tag{A.7}$$

The solution of this differential equation is analytical and yields

$$m^3(t) = m^3(t_i) - (\hbar c^4 / 5120\pi G^2)t \tag{A.8}$$

where $m(t_i)$ is the initial UNBH mass. One then sets expression (A.8) to zero to derive the final time t_f when the black hole ends its existence, i.e.,

$$t_f = \frac{5120\pi G^2}{\hbar c^4}m^3(t_i) = \frac{5120\pi G^2}{\hbar c^4}\left(\frac{c^2 r}{2G}\right)^3 = \frac{640\pi c^2}{\hbar G}(r)^3 = \frac{480c^2}{\hbar G}V \tag{A.9}$$

where r and V are the mass retention radius and volume, respectively, of the UNBH at the initial time of t_i. The retention-time T for the UNBH is then given by the expressions

$$T = t_f - t_i = (480c^2/\hbar G)V - t_i \tag{A.10}$$

$$V = 4\pi r^3/3 = 4\pi\left(2Gm(t_i)/c^2\right)^3/3 = 4\pi\left(2GE(t_i)/c^4\right)^3/3 \tag{A.11}$$

The rate of change of the retention-time T with respect to the retention-space V (or 'pace of energy retention' which is the 'uncertainty' space-dual of the certainty 'speed of energy motion', i.e. the rate of change of motion-space s with respect to motion-time t) is then derived from (A.10) to give us the sought after pace of dark X for a black-hole, i.e.,

$$X = \frac{dT}{dV} = \frac{480c^2}{\hbar G} = \frac{960\pi c^2}{hG}. \tag{A.12}$$

Appendix B

Latency-Information Theory Genially Remembered

Latency information theory is genially remembered with Fig. B and the following eight lucky reminders:

1) **Like a friendly Cassper ghost** the pace of dark equation $\boldsymbol{hG = 960\pi c^2/X}$ radiating the **LIT** figure permeates information-space of either an intel-space or life-space nature as well as latency-time of either a life-time or intel-time nature. The supreme '2' of this revealing expression further tell us that 'dualities' span vertically, horizontally and diagonally the fabric of information-space and latency-time. The vertical duality has an uncertainty/certainty nature that is exhibited by the information-space of quadrants II and III and the latency-time of quadrants I and IV. The horizontal duality has a channel-communication/sensor-recognition nature that is exhibited by the 'through' integration of motion life-time and sourced intel-space of quadrants I and II and the 'across' integration of retention life-space and processing intel-time of quadrants III and IV. The last diagonal duality has a mathematical-intelligence/physical-life nature that is exhibited by the integration of sourced intel-space and processing intel-time of quadrants II and IV where M-LIT resides and is symbolized by an intelligent brain, and the integration of motion life-time and retention life-space of quadrants I and III where P-LIT resides and is symbolized by a living heart.

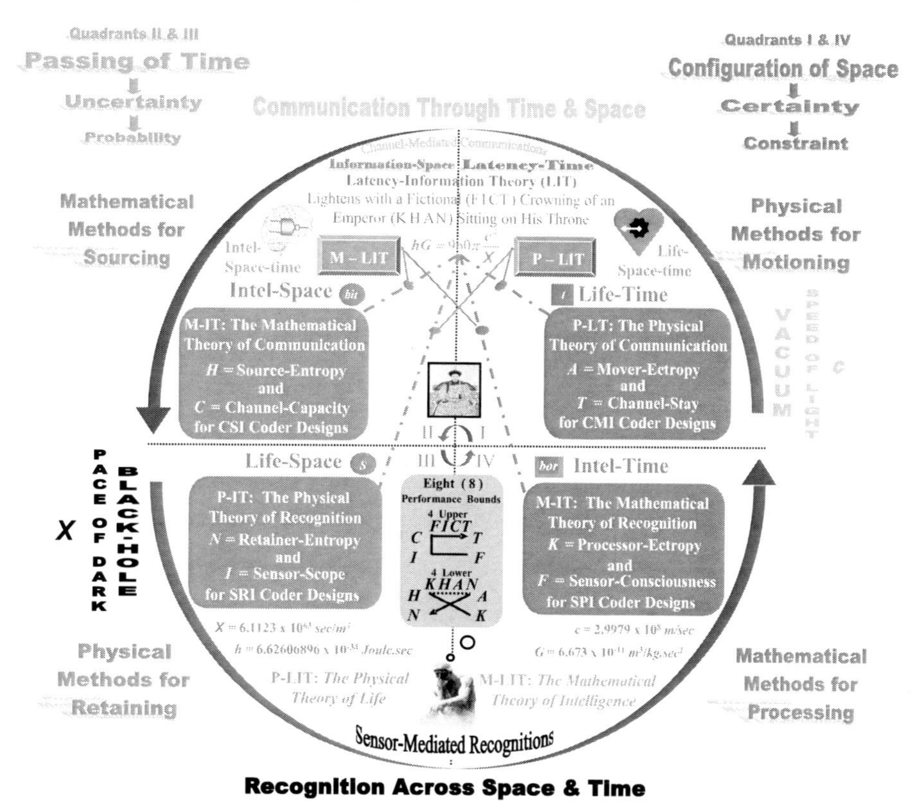

Fig. B. A Genial Reminder for Latency-Information Theory

2) The fictional (**FICT.**) crowning of an emperor (**KHAN**) also permeates information-space and latency-time. The four letters of KHAN denote the 'lower' performance-bounds *K*=processor-ectropy, *H*=source-entropy, *A*=mover-ectropy and *N*=retainer-entropy guiding respectively, processor, source, mover and retainer coder designs. On the other hand, the four letters of FICT denote the 'upper' performance-bounds *F*=sensor-consciousness of a limited intel-time sensor, *I*=sensor-scope of a noisy life-space sensor, *C*=channel-capacity of a noisy intel-space channel and *T*=channel-stay of a multi-path life-time channel guiding respectively, sensor and processor integrated (**SPI**), sensor and retainer integrated (**SRI**), channel and source integrated (**CSI**) and channel and mover integrated (**CMI**) coder designs.

3) Following the letter positions in the LIT figure the FICT crown is drawn starting first with the recognition case from the bottom right *F* to the bottom left *I*, and then moving on to the communication case from the top left *C* to the top right *T* with the caveat that one must be very patient when moving from *I* to *C* since this movement is only possible via Hawking radiation as one emanates from a black hole in the third LIT quadrant. The FICT crown, however, has the good fortune of forming the letter ***c*** which stands for the *speed of light* and should help in speeding up things a '***bor***'. Next the KHAN emperor chair is drawn starting with the intelligence case from the bottom right *K* to the top left *H* which is then followed by the life case from the top right *A* to the bottom left *N*, where it is also noticed that the communication move from the top left *H* to the top right *A* has been ignored because it is done so fast, i.e. at the speed of light! The KHAN chair also has the good fortune of forming the letter ***X*** which stands for both the *pace of dark* and the unknown nature of a black-hole.

4) Returning to $hG=960\pi c^2/X$ the Plank constant *h* is to the left of *hG* just like the uncertainty in information, of a passing of time origin, is to the left of the figure. On the other hand, the Gravitational constant *G* is to the right of *hG* just like the certainty in latency, of a configuration of space origin, is to the right of the figure.

5) The ratio c/X in $hG = 960\pi c^2/X$ is easily remembered from the *c* shaped FICT crown sitting on top of the *X* shaped KHAN chair.

6) The constant **960π** in $hG = 960\pi c^2/X$ is equal to the product of **480** by **2π**. The '**4**' in the 480 tell us that there are four LIT quadrants as one moves by 2π on the figure, i.e. by a single **LIT revolution** that starts with the **physical latency theory** located in the first LIT quadrant, then the **mathematical information theory** of the second quadrant, then the **physical information theory** of the third and finally the **mathematical latency theory** of the fourth. Furthermore the '**8**' in the 480 tell us that there are eight performance-bounds, i.e. *F, I, C, T, K, H, A, N* to be found in the LIT figure where it should be noted that the number eight is a lucky number for the Chinese KHAN emperor.

7) The *I, N* letters of '**the physical theory of recognition**' stand for retaining energy **IN** a black-hole, and the *A, T* letters of '**the physical theory of communication**' stand for moving energy **AT** some distance in a vacuum.

8) Finally, what the letters *C, H* of '**the mathematical theory of communication**' and the letters *F, K* of '**the mathematical theory of recognition**' stand for I will leave for the readers to figure out on their own. Nevertheless, if asked, I would suggest that these letters may stand for the names of artificial and/or former living systems acting as friendly ghosts, where each ghost should of course exemplify a constructive as well as revolutionary intelligence

DEDICATION

This manuscript is dedicated to the memory of Claude E. Shannon 1916-2001

REFERENCES

[1] Feria, E. H., "Latency-information theory: A novel latency theory revealed as time-dual of information theory", *Proceedings of IEEE Signal Processing Society: DSP Workshop and SPE Workshop*, Marco Island, Florida, Jan. 2009

[2 Shannon, C. E., "A mathematical theory of communication", *Bell System Technical Journal*, vol. 27, pp. 379-423, 623-656, July, October, 1948

[3] Ramsey, D. and Weber, M.,"Claude Shannon father of the information age," *University of California Television*, San Diego, 2002.

[4] Wozencraft, J. M. and Jacobs, I. M., "Principles of communication engineering," *Waveland Press,* Inc. 1965

[5] Feria, E. H., "Matched processors for optimum control", *City University Of New York (CUNY)*, Ph.D. in E.E., Aug. 1981 (the research leading to this dissertation started in 1977 under the guidance of Prof. Frederick E. Thau, however, prior to this time from 1973 to 1977 two other Ph.D. dissertation topics were explored, the first from 1973 to 1975 on the design of optimum waveforms for communication systems with the support of Prof. Donald L. Schilling and the second on the brain's motor control systems for bio-feedback applications sponsored by Dr. Julius Korein of the New York University Medical Center and also supervised by Prof. Ralph Mekel). Feria, E. H., "Matched processors for quantized control: A practical parallel processing approach," *International Journal of Controls*, Vol. 42, Issue 3, pp. 695-713, Sept. 1985.

[6] Kalman, R.,"A new approach to linear filtering and prediction problems," ASME Transactions Journal of Basic Engineering, 82(1):35-45, 1960

[7] Feria, E. H., "Latency-information theory and applications, Part III: On the discovery of the space-dual for the laws of motion in physics," *Proceedings of SPIE Defense and Security Symposium,* 17-21, March 2008

[8] Feria, E. H., "Predictive-transform source coding with subbands", *Proceedings of 2006 IEEE System, Man & Cybernetics Conference,* Oct. 8-11, Taipei, Taiwan, 2006

[9] A. Moffat, A. Turpin, *Compression and Coding Algorithms*, Kluwer, 2002

[10] Mano, M. M and Ciletti, M. D., Digital Design, Prentice Hall, 2007

[11] Lloyd, S., "Ultimate physical limits to computation", 406, 1047, *Nature,* Aug. 31, 2000

[12] Bekenstein, J.D., "Information in the Holographic Universe", pp. 66-73, *Scientific American Reports*, Spring 2007.

[13] Feria, E. H., "On a nascent mathematical-physical latency-information theory, Part I: The revelation of powerful and fast knowledge-unaided power-centroid radar," *Proceedings of SPIE Defense, Security and Sensing 2009,* Orlando, Florida

Hardware Based Segmentation in Iris Recognition and Authentication Systems

Bradley J. Ulis[a], Randy P. Broussard[b], Ryan N. Rakvic[a], Robert W. Ives[a], Neil Steiner[c], Hau Ngo[a]
[a]Electrical and Computer Engineering Department, U.S. Naval Academy, Annapolis, MD
[b]Weapons and Systems Engineering Department, U.S. Naval Academy, Annapolis, MD
[c]Information Sciences Institute East, University of Southern California, Arlington, VA

ABSTRACT

Iris recognition algorithms depend on image processing techniques for proper segmentation of the iris. In the Ridge Energy Direction (RED) iris recognition algorithm, the initial step in the segmentation process searches for the pupil by thresholding and using binary morphology functions to rectify artifacts obfuscating the pupil. These functions take substantial processing time in software on the order of a few hundred million operations. Alternatively, a hardware version of the binary morphology functions is implemented to assist in the segmentation process. The hardware binary morphology functions have negligible hardware footprint and power consumption while achieving speed up of 200 times compared to the original software functions.

Keywords: Iris Recognition, Hardware Implementation, Binary Morphology, Dilation and Erosion, Segmentation, Computer Vision

1. INTRODUCTION

Within the last decade, governments and organizations around the world have invested heavily in biometric authentication for increased security at critical access points. Biometrics differentiate individual people by their physical characteristics such as their face or fingerprints. These methods of identification depend on relatively unchangeable features and thus biometric identification methods are more accurately defined as authentication. By using biometrics to positively identify a person seeking to gain access, the authenticating system has more confidence that the person being identified is in fact who they claim to be. They can more easily detect fraudulent access. For example, the United Arab Emirates employs biometric systems to regulate the people traffic across their borders. Subsequently, several biometrics systems have attracted much attention, such as facial recognition and iris recognition.

While facial recognition is something humans are exceptionally familiar with on a daily basis when recognizing friends and family members, iris recognition is more abstract. Iris recognition analyzes the fine details of the colored part of the eye as seen in Fig. (1) to extract information unique to the individual person. Thus, the iris of a person is like a fingerprint or an indicator of identity and conveys information about the uniqueness and identity of the person. However, a human lacks the capacity to differentiate one iris from another beyond simply the color of the iris and thus we must rely on computers to perform iris recognition.

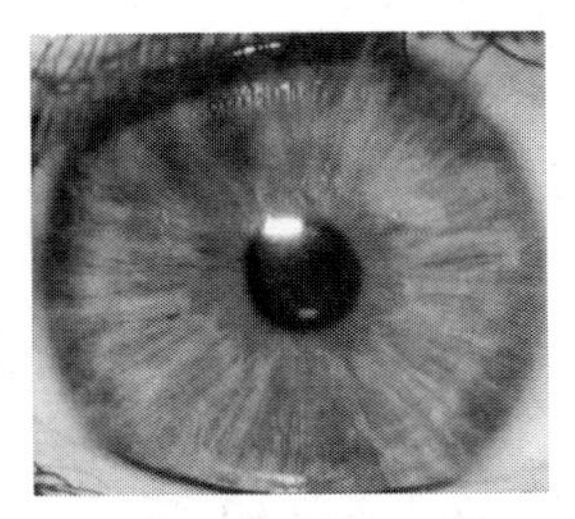

Fig. 1. Typical human eye with colorful iris

Computer systems which this paper will refer to as general purpose systems are electronic machines that execute instructions known as software. Many computer systems have operating systems that run in the background along with many other concurrent processes or programs that can slow the general purpose system down. Iris recognition algorithms

Mobile Multimedia/Image Processing, Security, and Applications 2009, edited by Sos S. Agaian, Sabah A. Jassim,
Proc. of SPIE Vol. 7351, 73510W · © 2009 SPIE · CCC code: 0277-786X/09/$18 · doi: 10.1117/12.820239

are usually written in C code and compiled into an application that the operating system runs among other these programs. For the purpose of optimization, the iris recognition software would ideally be isolated from the parent system and concurrent processes. This could allow a processor to focus its full throughput on the execution of the algorithm software. Furthermore, general purpose systems are not especially mobile (with the exception of very lightweight laptop computers) and thus may not serve every venue where iris recognition systems could be deployed. Thus, a hardware implementation of an iris recognition system is especially interesting as it could be exceptionally faster than its general purpose counterpart while also being small enough to be part of a digital camera or camera phone.

1.1 Ridge Energy Direction (RED) Algorithm

The iris recognition algorithm discussed in this paper is the Ridge Energy Direction (RED) algorithm which is currently being developed at the U.S. Naval Academy by Dr. Robert Ives et. al. This algorithm has three key steps in order to go from an image of a person's eye to positively identifying that person; segmentation, template generation and template matching. The first step, segmentation, extracts the iris from the image. Like any algorithm that depends on computer vision to properly execute, this part of the algorithm requires the most accuracy for there to be any chance of success in the later steps. Next, template generation is where this algorithm gets its name, using finite impulse response (FIR) filtering techniques to quantify the existence of ridges on an energy map of the iris and assigning information bits based on the ridge orientations. A template is a bit vector that uniquely represents the information contained in the iris. In biometric algorithms, it is necessary to record the *information* that the pixels of an image convey rather than the pixels themselves. This is because pixels will differ from one image to another but the information should be consistent if the iris within the image is in fact the same. Finally, template matching computes the hamming distance of the iris information under test with previously stored iris information. The RED algorithm is comparable to other algorithms but for the most part is not especially relevant to this paper.

Each step of the algorithm requires varying times of execution. Depending on the number of templates that need to be matched, template matching can require the most time. Matching the information of a few people such as those in a household or a small business would take a negligible amount of time while a database of an entire nation's people would require significant time to process. The other major part of the algorithm to consume processing resources is the segmentation process. To accurately detect the location of an iris within an image, the algorithm uses extensive image processing techniques. Image processing without the assistance of a co-processor such as a video card in a general purpose system can be noticeably slow. This processing time can even be worsened when migrating the iris recognition system from a general purpose system with a high efficiency and high speed processor to a slower "mobile" solution. Template generation execution time is essentially negligible.

1.2 Iris Recognition Segmentation

Segmentation searches for the iris within an image of an eye. For template generation to be successful, the segmentation process must properly extract the iris from the image. As humans, we have an exceptionally powerful image processor working for us as a result of billions of years of development that allows our brains to effortlessly segment out objects in our field of view. For a computer, segmenting objects from an image is similar to trying to inspect a photograph by looking through a coffee straw. Computers lack the ability to see the "big picture" and must use lengthy and complex image processing functions to break down the image.

The iris is bounded by the pupil at the inner radius and the limbic boundary at the outer radius. Finding these boundaries begins with finding the pupil since it is assumed to be the easiest part of the image to segment with the assumption that it is simply a large black circle somewhere in the image. To identify the pupil, the image undergoes contrast limited histogram equalization (CLAHE) to adjust contrast and better differentiate the difference between minimum and maximum values across the image before thresholding all the values. All values above the threshold (typically the mean plus one standard deviation) are assigned a value of '0' and all values below are assigned a '1.' This leaves only the pupil and other very dark objects such as some eyelashes and perhaps some of the eyebrow. However, due to the nature of the cornea being at its apex right above the pupil, there are often glare artifacts directly over the pupil and cause the pupil to be obfuscated and irregular. To correct for artifacts and find the existence of objects in the image, the segmentation process uses a series of binary morphology functions to dilate, erode and assign values to the objects based on area and perimeter.

Once the pupil is detected by comparing the different objects in the image, the pupil's center is used as a starting point for finding the limbic boundary. Local statistics compute the kurtosis of 3 by 3 windows all around the image. The algorithm then picks values radially from around the area of the limbic boundary and accumulates the values. Segmentation then steps left and right of the pupil center recalculating the sum of the local kurtosis. The smallest value is most likely the true location of the limbic boundary and the iris can be located and unwrapped for template generation.

1.3 Binary Morphology

Binary morphology functions operate on the image after it has been thresholded in order to find the existence of relevant objects such as the pupil. When an image is thresholded, the pixel values in the image are reduced to a single bit for logical operation. Approximately half of the algorithm's segmentation execution time is consumed by binary morphology functions. The most difficult of these functions are the dilation and erosion functions which are slow even when pre-processing the image to find objects for dilation and erosion to operate on. Fig. (2) illustrates a single step dilation and erosion and how the resulting object is roughly the same shape but without the artifact in the center.

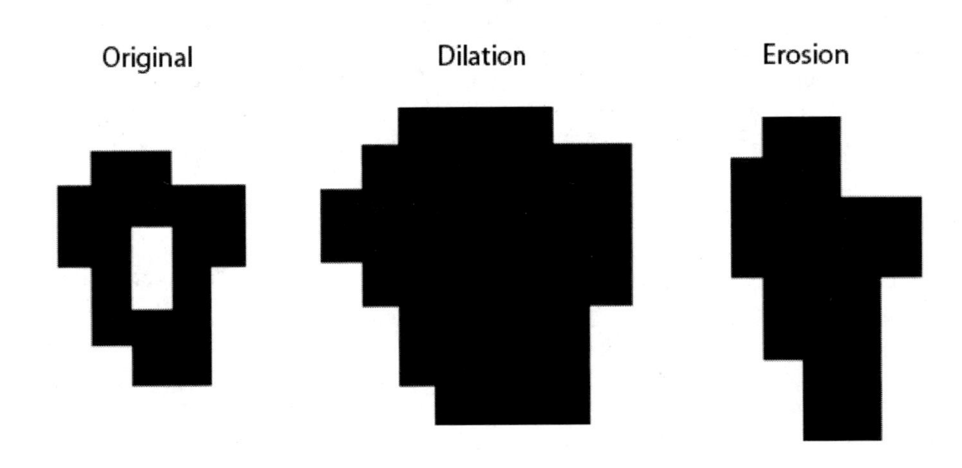

Fig. 2. Left: image before dilation and erosion. Middle: one step of dilation. Right: one step of erosion. Note that the artifact (hole) was filled in the final image but the general shape is retained.

Logically, dilation and erosion are the same function. Dilation looks at all the '1' values in the image and places a '1' in every immediately adjacent pixel above, below, left and right. Similarly, erosion places a '0' value pixel adjacent to every '0' pixel in the image. These functions must scan all the relevant pixels and make non-linear adjustments around the image. Therefore, binary morphology even when ideally optimized is a $\Theta(n^3)$ growth function. However, with many objects and many pixels in the image, binary morphology can become increasingly messy and complex, consuming much of the general purpose processor's attention.

1.4 Hardware Implementation in Field Programmable Gate Arrays (FPGAs)

Because the binary morphology functions are software implementing binary logic, these functions are ideal for translating into a hardware equivalent. However, hardware design requires finite and defined input and output whereas software can execute functions of many varying inputs. Thus, software can implement any hardware function but hardware can only execute hard-bounded software functions. For example, software multiplication such as squaring could make use of an 'add' and a 'count' variable while hardware could use a look-up table to find the answer as shown in Fig. (3). Software multiplication is more portable since it can make use of the same basic binary logical add as long as the registers that hold the variables are large enough to prevent overflow. Hardware on the other hand requires the input to be hard bounded to the size of the input of the look up table and would produce a hard bounded output with twice as many bits as the input. Therefore, software can operate on variable inputs and outputs such as packets while hardware implementation would be impractical. Yet, hardware is capable of tasks software can have difficulty executing. In the multiplication example, the software requires a loop to repeatedly make use of an add function while the hardware equivalent could have the answer in a single step. In systems where the hardware design can be changed to fit the application (such as an application specific integrated circuit), having a hardware squaring function compute the square could be advantageous for speed and power if the software often squares large numbers.

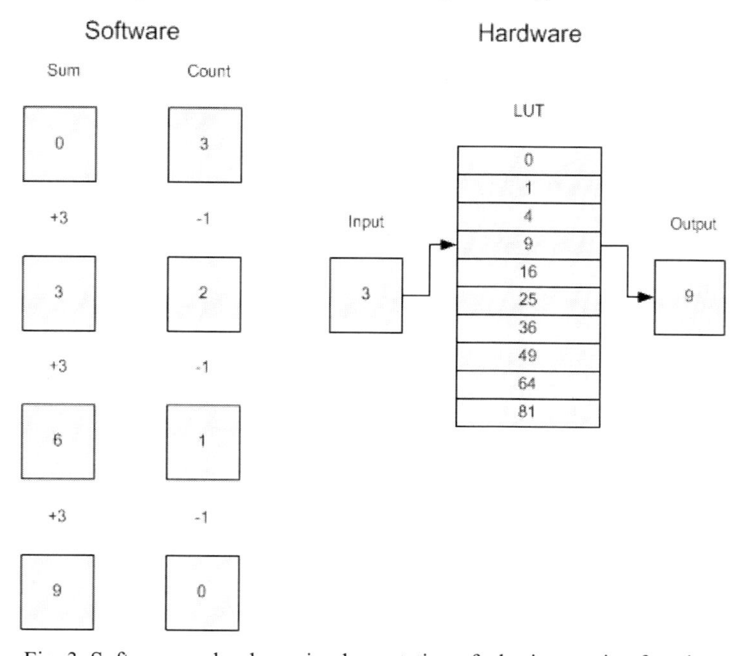

Fig. 3. Software vs. hardware implementation of a basic squaring function.

Field programmable gate arrays are increasingly popular devices that allow designers to develop both the hardware and software implementation. FPGAs are a kind of "blank slate" integrated circuit that can be programmed to execute any hardware function from simple binary logic to full scale microprocessors. These can be especially useful when prototyping new systems to evaluate hardware designs for integrated circuits. For the purpose of this paper, an FPGA is used to evaluate an equivalent software and hardware implementation of the binary morphology dilation and erosion functions.

2. METHODOLOGY

This paper will take one part of the RED algorithm and compare a software and hardware implementation of the same function in order to demonstrate the effectiveness of a hardware equivalent function. Hardware based segmentation could be faster, slower or more difficult depending on the function being implemented. These functions are directly comparable to other standard software based image processing functions. Also, there are many opportunities for alternate implementation in the algorithm and implementation of the binary morphology dilation and erosion functions are just a couple of many possible functions that can be implemented. In order to properly implement the hardware equivalent, we investigate the software implementation of dilation and erosion in order to understand how the logical function should behave in hardware.

To properly evaluate the conversion from software to hardware, a test system is developed to execute the entire algorithm in software on the target device. The FPGA is programmed with a test system that can execute C code and the details of this implementation are beyond the scope of this paper. The FPGA system is different than a general purpose system because the processor on the FPGA runs only the algorithm code and no other processes. Additionally, the speed of the FPGA system is generally slower than a general purpose system. This FPGA system is fed an image of an eye where the segmentation, template generation and template matching processes begin just as they would on a general purpose system. Note that the programming of an FPGA is different from the programming of software into a computer. This programming defines the interconnection of basic logic elements inside the FPGA so that the desired logic (processor, memory controller and test hardware device) is realized. However, unlike the general purpose system, the FPGA system has the ability to call on a specially designed peripheral hardware component inside the FPGA to process data and can read the result back almost seamlessly.

Upon converting the function from software to hardware, the software based function call is followed by a new function that calls on the hardware device for data processing. The functions under test are isolated in the algorithm by placing reference points before and after the function call that records start and end times. The output of the software function can be used to assert the valid output of the hardware based function and the execution times can be compared.

2.1 Software implementation

The software version of the algorithm preprocesses the image by scanning the image for objects that have an area that is at least the minimum area for the pupil. These objects are individually passed into the dilate and erosion functions for processing in an effort to avoid processing the entire image in a brute force function. The software function calls appear in the algorithm as shown in Tables (1) and (2).

Table. 1. Binary morphology function calls from FindPupil function.

```
...

label=ThresholdUINT16Image(im,
    (UINT16)threshval);

areas=label->BWLabel(MIN_PUPIL_AREA);

bwobjs=ConvertLabelImageToObjects(label,
    areas);

for (i=0;i<bwobjs->Nobj;i++) {
    BWDilateInPlace(bwobjs->obj[i]->
        data,se1); // Dilate
    BWErodeInPlace(bwobjs->obj[i]->
        data,se2); // Erode

...

}
...
```

Table. 2. BWDilateInPlace function definition. Note that the return of the image is inverted

```
void BWDilateInPlace(Image *in, Image *se)
{

        ...

    for (i=1; i<in->ActualRows-1; i++)
        for (j=1;j<in->ActualCols-1;j++)
            if (in->matrix[i][j])
            {
                in1->
                matrix[i+offset][j+offset]=1;
                if((in->matrix[i-1][j]==0)||
                 (in->matrix[i+1][j]==0)||
                 (in->matrix[i][j-1]==0)||
                 (in->matrix[i][j+1]==0))
                {
                    ioffset=i+offset;
                    joffset=j+offset;
                    for(ii=0;ii<numones;ii++){
                    in1->matrix[ioffset+
                    rowindex[ii]]
                    [joffset+colindex[ii]]=1;

                };
            };

        ...
}
```

The software implementation uses three FOR loops to step through the objects in the image and perform dilate/erode processing. Each function call steps through all relevant pixels and assigns the appropriate bit based on adjacent black and white values. Additionally, each function must be called 15 times for a total of 15 dilates and 15 erodes to fully process each object in the image. With a large image that has many objects covering enough area to possibly be the pupil, the functions can take a lot of time to process. Next, we will investigate a hardware alternative to these functions.

The business portion of the software implementation is the looping within each binary morphology functions. These loops use basic logical 'or' of the neighboring pixel above, below, left and right to determine each bit assignment. These loops are also the most time consuming since they are the highest order nest in the function calls. A hardware implementation could reduce these loops to a $\Theta(n)$ growth function.

The software implementation takes into account edge effects of by padding the outside edges of the image with zeros. The borders of the image complicate the dilation and erosion functions since pixels on the far outside of image will reference pixels not defined within the rows and columns of the image. Instead, these areas are filled with null values so that they do not contribute to the dilation or erosion of objects very close to the borders.

Another issue with hardware is that a hardware implementation takes advantage of replication to speed up processing power. The preprocessing of the image to find objects before dilation and erosion is not necessary to increase speed. Instead, the initial object detection of the algorithm will be moved to after dilating and eroding all objects in the image simultaneously. Theoretically, this reordering of operations should not affect the output.

2.2 Hardware Implementation (VHDL)

Hardware design is *inferred* similar to the way software is coded using special programming languages such as Verilog Hardware Description Language (Verilog HDL) or Very high speed integrated circuit Hardware Description Language (VHDL). Designs are compiled using a synthesizer and fitter that build a netlist describing how the fundamental logic elements within the FPGA are wired together.

The primary function to be implemented is the logical 'or' function that we identified from the software function loops. These 'or' functions are pixel oriented and directly translate to an elementary unit in hardware that we'll call a BM_Element which represents the pixel being operated on. The value stored within this BM_Element is driven by similar hardware elements holding values for pixels adjacent to the pixel stored in this BM_Element. Thus, these BM_Elements are easiest to imagine as though they were arranged in the 'shape' of an *m* x *n* image. The logic for these BM_Elements also takes into account both the dilation and erosion since they are complimentary binary logic functions.

The coding procedure for this hardware is to instantiate a full dilation and erosion hardware entity called BM_Dilate_Erode that will receive the pixel data from the processor one pixel at a time (with a width of only a single bit for '1' or '0') and transmit the resulting pixel data back to the processor one pixel at a time. BM_Dilate_Erode will act as a three-state state machine that receives the pixel data, processes the data and then sends the data back. Within the overarching BM_Dilate_Erode instantiation are the individual BM_Element blocks and an interconnect fabric that allows for pixel data to be loaded into each BM_Element and for each BM_Element exchange relevant pixel data among each other. Data enters the BM_Dilate_Erode device as a shift register as shown in Fig. (4).

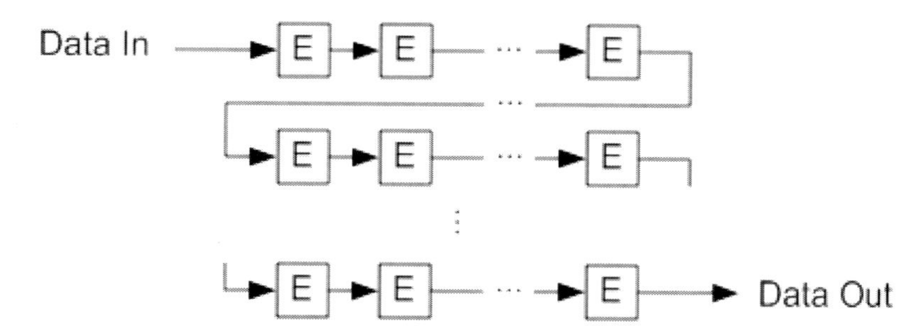

Fig. 4. The input shifts the data into the elements starting at the first and ending at the last BM_Element represented by the labeled "E" blocks. The processor shifts the last pixel of the thresholded image in first and when processing is complete the data is shifted out last pixel first.

The interconnect fabric within the BM_Dilate_Erode machine is the most challenging part of the device. Unlike the software implementation, this fabric cannot account for edge effects using the technique of padding which places null values around the edges. Instead, it circularly connects the top and bottom, left and right elements together so that these connections are not undefined. The effects of these circular connections are negligible since a segmentable pupil will be located somewhere within the image borders and not partially cut off at the edge of the image. The interconnect fabric is actually a long string of "nodes" numbered from zero to the total number of BM_Elements within the device. These nodes represent the connect points between different BM_Elements and each BM_Element references adjacent elements by their numbered connection in this fabric. This is illustrated in Fig. (5).

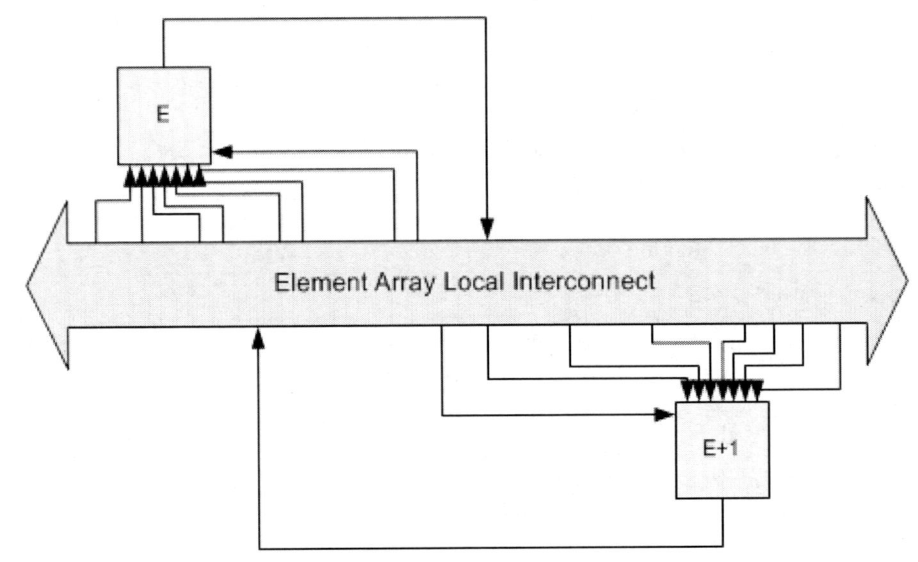

Fig. 5. The local interconnect fabric is series of nodes labeled from zero to the total number of BM_Elements in the BM_Dilate_Erode device. Each BM_Element inputs the relevant adjacent pixel value by referencing various nodes in the fabric and outputs its own held pixel value to the interconnect fabric node associated with that specific BM_Element.

Each BM_Element computes the dilation and erosion using only a few lines of code as shown in Table (3) and produces the logic blocks seen in Fig. (6). Like a function call in software, the BM_Element encapsulates recurring code into a single logic block with finite inputs and outputs. These inputs reference the neighboring four nodes and necessary control signals. It outputs its value back into the local interconnect fabric node allotted for that BM_Element. This is accomplished using the code shown in Table (4).

Table. 3. BM_Element functional logic

```
process(aload, adatain, clk)
begin
...
        if (aload = '1') then
                adataout <= pixel;
                pixel_step <= adatain;
        else
                if (rising_edge(clk)) then
                        if(clk_enable = '1') then
                                if(operation = '0') then      -- Dilate
                                        pixel_step <=topmid or midleft or midright or
                                        bottommid;
                                else
                                        pixel_step <= not(topmid or midleft or midright or
                                        bottommid);
                                end if;
                ...
process(clk)
begin
        if(rising_edge(clk)) then
                if(operation = '0') then
                        output <= pixel;
                else
                        output <= not(pixel);
                end if;
        end if;
end process;
```

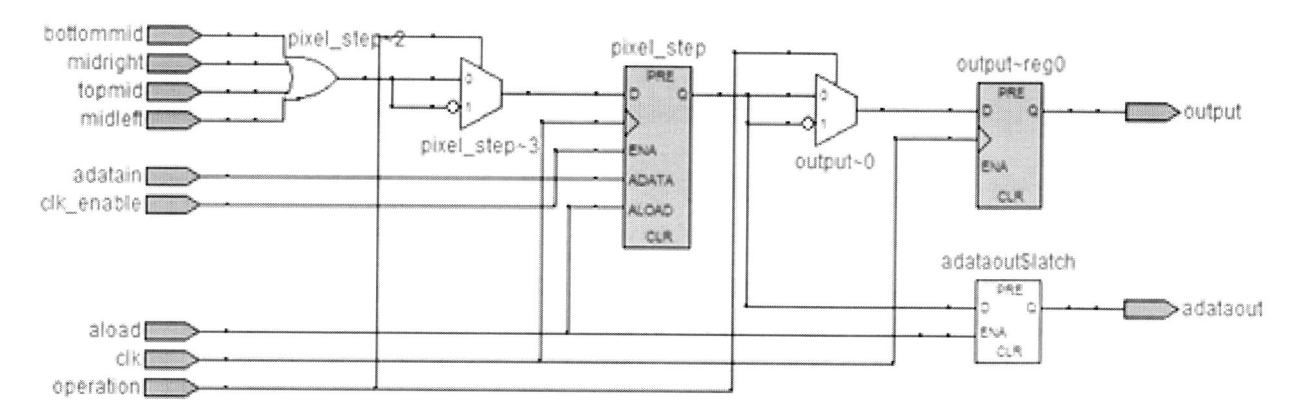

Fig. 6. Internal logic of one BM_Element block inferred from the code seen in Table (3). The device takes in 4 inputs from the local interconnect and 'or's the logic together. This value is then latched depending on the operation (dilate or erode). There is also control logic to shift the value in this BM_Element on to the next BM_Element and input the value from the previous BM_Element in the chain as demonstrated in Fig. (4).

Table. 4. BM_Element instantiation and interconnect logic

```
signal local_interconnect : std_logic_vector(ROWS*COLS downto 0);
...

BM_rows: for j in 0 to ROWS - 1 generate
begin
        BM_cols: for i in 0 to COLS - 1 generate
        begin
                U1: BM_Element
                port map(
                        clk => avs_s1_clk,
                        clk_enable => clk_e,
                        operation => operation,
                        aload => shift_load,
                        adatain => local_data(j*COLS + i),
                        adataout => local_data(j*COLS + i + 1),
                        topmid => local_interconnect(((j + ROWS - 1) mod ROWS)*COLS + ((i
                        + COLS) mod COLS) ),
                        midleft => local_interconnect((j*COLS)+((i+(COLS-1)) mod COLS)),
                        midright => local_interconnect((j*COLS)+((i+(COLS+1)) mod COLS)),
                        bottommid => local_interconnect(((j + ROWS + 1) mod ROWS)*COLS +
                        ((i + COLS) mod COLS) ),
                        output => local_interconnect( j*COLS + i)
                );
        end generate;

end generate;
```

Once the entire design is put together, it must be incorporated into the overall FPGA system so that the performance of the hardware design can be evaluated relative to the original software design. In this system, the design is included as a peripheral component to the microprocessor in the system and the compiler provides the microprocessor with a pointer to the input of the BM_Dilate_Erode device. The algorithm software code is rewritten to send the entire image for dilation and erosion and the result is read back for object detection. This peripheral totally eliminates the dilation and erosion binary morphology functions from the algorithm in exchange for simple copy functions that move data from SDRAM into the peripheral device and back out.

3. DATA

3.1 Computing the Execution Times

The functions for computing the binary dilation and erosion are isolated and the start and end times are recorded to compute the time in milliseconds the functions took to execute. The data is tabulated in Table (5). The images tested are standard 640 by 480 VGA images from the BATH database.

Table. 5. Timing comparisons of software and hardware implementation. 1000 iterations of the functions comprise these timing characteristics.

Software Functions		Hardware Functions	
Mean	1.341s	Mean	6.144ms
Dilation	611.7ms		
Erosion	729.3ms	Standard Deviation	313μs
Standard Deviation	72.476ms	Range	1.943ms to 2.846ms
Range	1.090s to 1.470s		

4. RESULTS

4.1 Comparison of Software and Hardware Dilation and Erosion Functions

The hardware function outperforms the software in speed of execution. The software timing shows that the binary morphology dilation and erosion functions together take approximately 218 times longer to execute. In large processing applications that must process many images at a time this can be a significant increase. Furthermore, the increase makes the segmentation of the pupil from the image nearly negligible and increases total segmentation time by about 37%.

There are tradeoffs, however, and the hardware takes up additional space within the FPGA which also includes the microprocessor, memory controller and other peripherals. With today's synthesizing software that effectively optimizes the design for fit and power, the logic consumption takes up less than 1% of the logic elements and registers in the Stratix III EP3SL150F1152C3ES which has approximately 150,000 logic elements and 1 Mbit of on-chip memory. Subsequently, the power consumption in the overall system is reduced since the hardware implementation uses substantially less logic than the software loops used.

5. CONCLUSIONS

Application specific hardware design can be a complex process that may or may not be worth the cost of design for just small enhancements over software. Theoretically, an algorithm's execution could be nearly totally implemented with hardware functions and the processor would merely need to move data from one peripheral hardware device to another, assuming sufficient space on the programmable logic device or integrated circuit. This design is certainly not the only part of the RED iris recognition algorithm that can be implemented using hardware functions. However, some things are simply easier to do in software even though they may be slower.

REFERENCES

[1] Zhang, D. D., [Automated Biometrics: Technologies and Systems], Kluwer Academic Publishers, 1-18 (2000).

[2] Sims, D., "Biometrics Recognition: Our Hands, Eyes and Faces Give Us Away," IEEE Computer graphics and Applications, 0272-17-16/94 (1994).

[3] Williams, G. O., "Iris Recognition Technology," IEEE, 0-7803-3537-6-9/96 (1996).

[4] Boles, W. W., "A Security System Based on Human Iris Identification Using Wavelet Transform," First International Conference on Knowledge-Based Intelligent Electronic Systems, Adelaide, 21-23 (1997).

[5] Kennell, L. R., Ives, R. W., Gaunt, R. M., "Binary morphology and local statistics applied to iris segmentation for recognition," Proc. of the 13th Annual International Conference on Image Processing, (2006).

[6] Daugman, J., "Statistical richness of visual phase information." International Journal of Computer Vision 45(1), 25-38 (2001).

[7] Ives, R. W., Broussard, R. P., Kennell, L. R., Rakvic, R. N., Etter, D. M., "Recognition using the Ridge Energy Direction Algorithm" presented at the 42nd Asilomar Conference on Signals, Systems and Computers, Pacific Grove, California, (2008).

[8] Daugman, J., "Probing the uniqueness and randomness of IrisCodes: Results from 200 billion iris pair comparisons." Proc. of the IEEE 94(11), 1927-1935.

[9] Daugman, J., "High Confidence Visual Recognition of a Persons By a Test of Statistical Independence," IEEE Trans. on PAMI 15, (1993).

[10] Altera Inc., "Stratix III Datasheet." [Online datasheet], Available at http://altera.com/literature/hb/stx3/stx3_siii52001.pdf, (2008).

Image Steganography in Fractal Compression

Mei-Ching Chen*[a], Sos S. Agaian[a], C. L. Philip Chen[a], Benjamin M. Rodriguez[b]
[a]Dept. of Electrical and Computer Engineering, The University of Texas at San Antonio,
One UTSA Circle, San Antonio, TX, USA 78249-1644;
[b]Space Department, Johns Hopkins University Applied Physics Laboratory
11100 Johns Hopkins Road, Laurel, MD, USA 20723-6099

ABSTRACT

This paper proposes a steganographic scheme utilizing within and/or after fractal encoding procedures on images for data security. Fractal generation exploits the concepts of iterated function systems (IFS) consisting of a collection of contractive transformations. Fractal images make use of partitioned iterated function systems (PIFS) to determine self similarity within images along with approximating the original uncompressed image. The transformed coefficients are stored in a fractal code table in order to decode the image as an alternative to storing or transmitting image pixel values directly. The proposed steganographic algorithm conceals secret information in the contrast/scaling and brightness/shifting coefficients in the code table, resulting in a stego fractal code table. Using fractal transform as a means of steganography provides a new embedding domain other than the existing steganography tools. The advantages of using fractal compression for securing information within images are: no current fractal detection methods exist, the hidden information is disseminated throughout the image in the spatial domain, the capacity of the image can be increased, and the decoding of the stego table results in a visually undistorted image.

Keywords: Fractal steganography, fractal encoding, fractal decoding, fractal compression, image steganography

1. INTRODUCTION

Backbone Security [2] has identified over 725 available steganography tools with a large number of these methods applied to digital images. The majority of image steganalysis tools up to date tend to detect hidden information in uncompressed images, such as bitmaps, and compressed images, for instance, JPEGs, GIFs and PNGs [13]. In many instances, steganography is used to store and transmit pertinent information from a malicious third party observer. New steganographic algorithms on different image domains are needed in order to have secure communication for vital information.

Fractal generation exploits the concepts of iterated function systems (IFS). This consists of a collection of contractive transformations in an iterative process until convergence, which is known as an attractor [3,4]. Fractal compressed images employ partitioned iterated function systems (PIFS) to find self similarity within images along with approximating the corresponding uncompressed images [10,11]. Fractal compression utilizes the theory of signal compression in order to preserve memory space. In order to generate a fractal image, the transformed coefficients are stored in a fractal code table for decoding the image instead of storing or transmitting image pixel values directly. Companies such as Microsoft and IBM have investigated and use fractals as a means to store the large file size images in practice [4].

In the paper, fractal compression containing both encoding and decoding algorithms is applied to create a new image steganography domain. The steganographic process is applied during and/or after the encoding procedure. In the presented stego fractal method, the scaling factors applied to the transformation are used to embed a series of binary bits converted from confidential digital files. In order to increase the number of embeddable bits in the transformed coefficients, fractal tables generated after encoding the image are also used for embedding.

Only a few papers have discussed the steganographic process using fractal images, both artificial and natural images. For artificial fractal images, Khadivi integrates the idea of cryptography into fractal steganography by defining the number of stego message units and the number of alphabets used in the message [12]. The stego message is hidden within a set of affine transformation elements before generating an artificial fractal image. Agaian and Susmilch [1] develop a fractal steganographic process through the use of color modification when generating artificial fractal images. For natural

Mobile Multimedia/Image Processing, Security, and Applications 2009, edited by Sos S. Agaian, Sabah A. Jassim,
Proc. of SPIE Vol. 7351, 73510Y · © 2009 SPIE · CCC code: 0277-786X/09/$18 · doi: 10.1117/12.818976

images, Davern and Scott incorporate the stego message within the fractal image compression procedure by selecting alternative domain pools [7]. The presented method takes advantage of embedding during the fractal encoding along with using the transformed coefficients in order to increase the capacity of embedding.

The remainder of this paper is structured as follows. In this paper, a steganographic process within and/or after fractal encoding techniques is applied to generate steganographic fractal images. In Section 2, an overview of the fractal concepts is presented. The fractal compression is described in Section 3 along with the steganographic scheme. Section 4 demonstrates the analyses followed by the conclusion and future work in Section 5.

2. MATHEMATICAL FOUNDATION

This section describes the theoretical background referred to in fractals. To generate fractals considers an IFS consisting of a collection of contractive affine transformations. The iteration procedure of applying a number of transforms is terminated when convergence is met, i.e., an attractor [4,15,17]. The criterion of convergence is dependent on a minimum allowable error.

2.1 Iterated function systems

Iterated function systems [4,6,8] consist of a collection of contractive affine transformations as defined in Equation (1).

$$A_{n+1} = \bigcup_{j=1}^{J} w_j(A_n) \tag{1}$$

For a given shape A_n, where n indicates the iteration, a set of J transformations has been defined and employed on A_n, denoted as w_j, where $j = 1, 2, \ldots, J$. The shapes after each transformation are united, resulting in the shape of the next iteration A_{n+1}.

2.2 Contraction mapping

In fractals, contraction mappings/transformations are used within an iterated function system in order for the system to converge. In other words, an attractor will be generated after repetition. A simple definition of mapping is described in Equation (2). Given a domain X and a range Y,

$$y = f(x) \tag{2}$$

where y is a mapping function of x, $\forall\ x \in \mathrm{X}$ and $\forall\ y \in \mathrm{Y}$.

The concept of contraction mapping is the distance between the two elements, $y_1, y_2 \in \mathrm{Y}$, Y being less than or equal to the corresponding elements, $x_1, x_2 \in \mathrm{X}$.

$$d(y_1, y_2) \leq c \cdot d(y_1, y_2)$$

where $0 \leq c < 1$, $y_1 = f(x_1)$, and $y_2 = f(x_2)$. d is any distance measurement specified by the user. In other words, a map is contractive if it brings points closer together defined by a metric d. If a shape is given, the output of its contraction mapping is the same shape of smaller size.

2.3 Affine Transformations

The affine transformations consist of three basic mappings: scaling, rotation, and translation/shifting. The mappings can be used individually, as a subset or a combination of all three. As seen in Equation (3), for a given point in a homogeneous coordinate system, affine transformations are vector/matrix operations. In the transformation matrix, six parameters, a, b, c, d, e, and f, are defined, in which a, b, c, and d are regarding to scaling and rotation, while e and f are translation factors.

$$\begin{bmatrix} x_{n+1} \\ y_{n+1} \\ 1 \end{bmatrix} = \begin{bmatrix} a & b & e \\ c & d & f \\ 0 & 0 & 1 \end{bmatrix} \begin{bmatrix} x_n \\ y_n \\ 1 \end{bmatrix} \tag{3}$$

In Fig. 1, an illustration of three examples of artificial fractal images are shown.

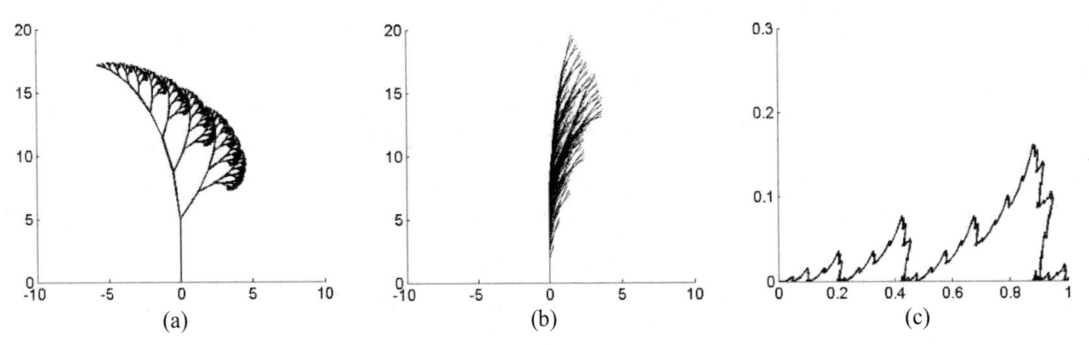

Fig. 1. Fractal examples (a) a fern (b) a reed in the wind (c) blowing fire.

3. FRACTAL IMAGE STEGANOGRAPHY

Based on the mathematical foundation of fractals described in Section 2, the concept has been extended and utilized in image compression, including both encoding and decoding algorithms. Originated from Jacquin [10,11], a number of fractal compression techniques have been developed [15,17]. A fractal image employs a PIFS to find self similarity within the image for compression. This section describes a simple and straightforward fractal compression technique in [18] along with a steganographic process.

3.1 Fractal image encoding and decoding

Fractal image compression techniques utilize the concept of fractal generation, specifically a set of contractive affine transformation mappings within an image. In other words, a block within an image may be represented by another block within the image by applying a set of transformations. The transformation coefficients are stored in a fractal code table in order to decode the image instead of storing image pixel values. The fractal compression scheme used in this paper is shown in Fig. 2.

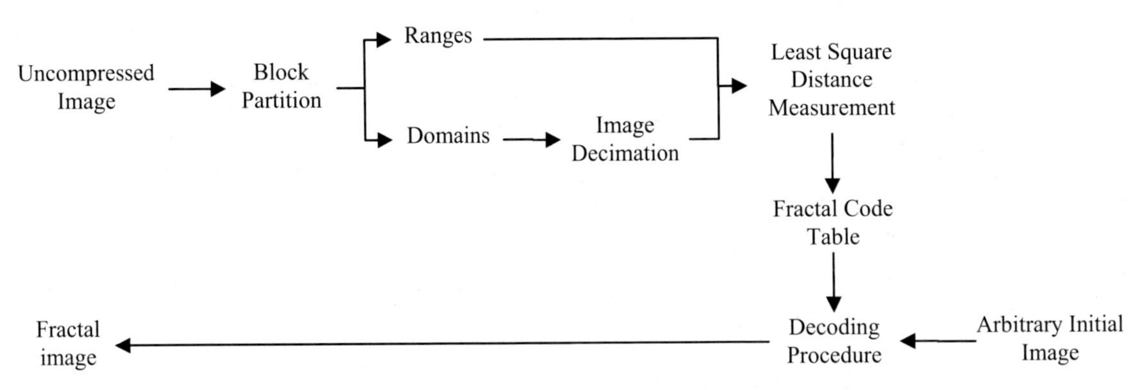

Fig. 2. A schematic view of fractal compression.

The image to be compressed is denoted as $\mathbf{M}$. Two sets of block functions, $R(\mathbf{M})$ and $D(\mathbf{M})$, are defined resulting in a set of range blocks and a set of domain blocks, respectively. The block partitioning technique used for computer simulation in this paper is non-overlapped fixed block partition (uniform partition) [17]. The image is divided into k non-overlapped range blocks denoted as R_k, where the blocks are of size $n \times n$. Domain blocks are derived by partitioning the image into l non-overlapped blocks, D_l, of size $2n \times 2n$. Note that $l = k/4$. The goal of fractal image compression is for each range block to find a best match from the set of domain blocks, represented by four fractal coefficients including row and column indices of the selected domain block, a scaling factor α, and a shifting factor β. A fractal code table is then constructed.

Since the domain blocks are twice the size of the range blocks, a block reduction method is needed for shrinking larger domain blocks to the size of $n \times n$, resulting in a set of shrunken domain blocks $\tilde{D}_l$. Averaging every four pixels for size reduction is used for the analysis in Section 4. Other sampling methods may be applied. For each range block $\mathbf{R}_i$, where $1 \le i \le k$, a shrunken domain block $\tilde{\mathbf{D}}_j$ from all available l blocks is sought after, i.e., $1 \le j \le l$, that best approximates the range block after a geometric transformation. The procedure tends to find the minimum error with the associated range blocks, domain blocks, and a set of predefined scaling factors α. The error for selecting the best match is defined as a function of $\mathbf{r}_i$, $\tilde{\mathbf{d}}_j$, and α, referred to in Equations (4)-(6). The function computes the error between the shrunken domain block $\tilde{\mathbf{D}}_j$ and the range block $\mathbf{R}_i$ by mapping both from matrices into one dimensional vectors, represented as a shrunken domain vector $\tilde{\mathbf{d}}_j$ and a range vector $\mathbf{r}_i$. The error and the corresponding shifting factor β are returned. For all $\tilde{\mathbf{d}}$ and α, the error is computed using the range block and the affine transformed domain block as in Equation (4) and (5).

$$\varepsilon = \tilde{\boldsymbol{\varepsilon}}' \cdot \tilde{\boldsymbol{\varepsilon}} \tag{4}$$

$$\begin{aligned}\tilde{\boldsymbol{\varepsilon}} &= \left(\mathbf{r}_i - \overline{\mathbf{r}}_i\right) - \alpha\left(\tilde{\mathbf{d}}_j - \overline{\tilde{\mathbf{d}}}_j\right) \\ &= \mathbf{r}_i - \alpha\tilde{\mathbf{d}}_j - \frac{1}{k}\sum_i \mathbf{r}_i + \alpha\frac{1}{l}\sum_j \tilde{\mathbf{d}}_j \\ &= \mathbf{r}_i - \alpha\tilde{\mathbf{d}}_j - \left(\frac{1}{k}\sum_i \mathbf{r}_i - \alpha\frac{1}{l}\sum_j \tilde{\mathbf{d}}_j\right) \\ &= \mathbf{r}_i - \left(\alpha\tilde{\mathbf{d}}_j + \beta\right)\end{aligned}, \tag{5}$$

In Equation (5), $0 \le \alpha \le 1$ and can be determined recursively by comparing $\mathbf{r}_i$ with each $\tilde{\mathbf{d}}_j$, and

$$\beta = \frac{1}{k}\sum_i \mathbf{r}_i - \alpha\frac{1}{l}\sum_j \tilde{\mathbf{d}}_j \tag{6}$$

such that α and β are identified subject to minimizing the error $\varepsilon = E\left(\mathbf{r}_i, \tilde{\mathbf{d}}_j, \alpha\right)$. Therefore, the row index and column index of the selected $\mathbf{D}_j$, α, and β, are assigned in regards to $\mathbf{R}_i$.

For the decoding procedure, the function takes the fractal code table created by the encoder and generates the fractal image. This is done by taking an initial image $\hat{\mathbf{M}}_0$ with the condition that $\hat{\mathbf{M}}_0$ and $\mathbf{M}$ are the same size. The image $\hat{\mathbf{M}}_0$ is divided into k non-overlapped partitions generating from a set of domain blocks, $\hat{D}_l$, where $l = k/4$. For each partitioned block, Equation (7) is applied in an iterative manner applying the transformation using the scaling factor α and the shifting factor β found during encoding.

$$\hat{\mathbf{r}} = \alpha\hat{\tilde{\mathbf{d}}} + \beta \tag{7}$$

where $\hat{\tilde{\mathbf{d}}}$ is the one dimensional mapping of the shrunken domain block of $\hat{\tilde{\mathbf{D}}}$. A new image $\hat{\mathbf{M}}_t$ is created by inserting reconstructed range blocks $\hat{\mathbf{R}}$ based on the stored row index, column index, α and β. The range block $\hat{\mathbf{R}}$ is created by reshaping the vector $\hat{\mathbf{r}}$ in Equation (8) to blocks of size $n \times n$. This continues in an iterative procedure until the root mean square error (RMSE) between images $\hat{\mathbf{M}}_t$ and $\hat{\mathbf{M}}_{t-1}$ becomes less than a desired error, e.g., rmse($\hat{\mathbf{M}}_t$, $\hat{\mathbf{M}}_{t-1}$) < 1. This indicates $\hat{\mathbf{M}}_t \approx \hat{\mathbf{M}}_{t-1}$, in which t is the number of iteration. A fractal image is created for approximating the uncompressed image. Fig. 3 shows an example of an original uncompressed Lena image and a fractal Lena image. The histogram of difference image between the original and the fractal image has the characteristics of presenting a Gaussian-like distribution. In other words, the fractal image can be considered as the original additive with Gaussian-like noises. Fig. 3(f) is a zoom-in histogram having the same scale as Fig. 3(c).

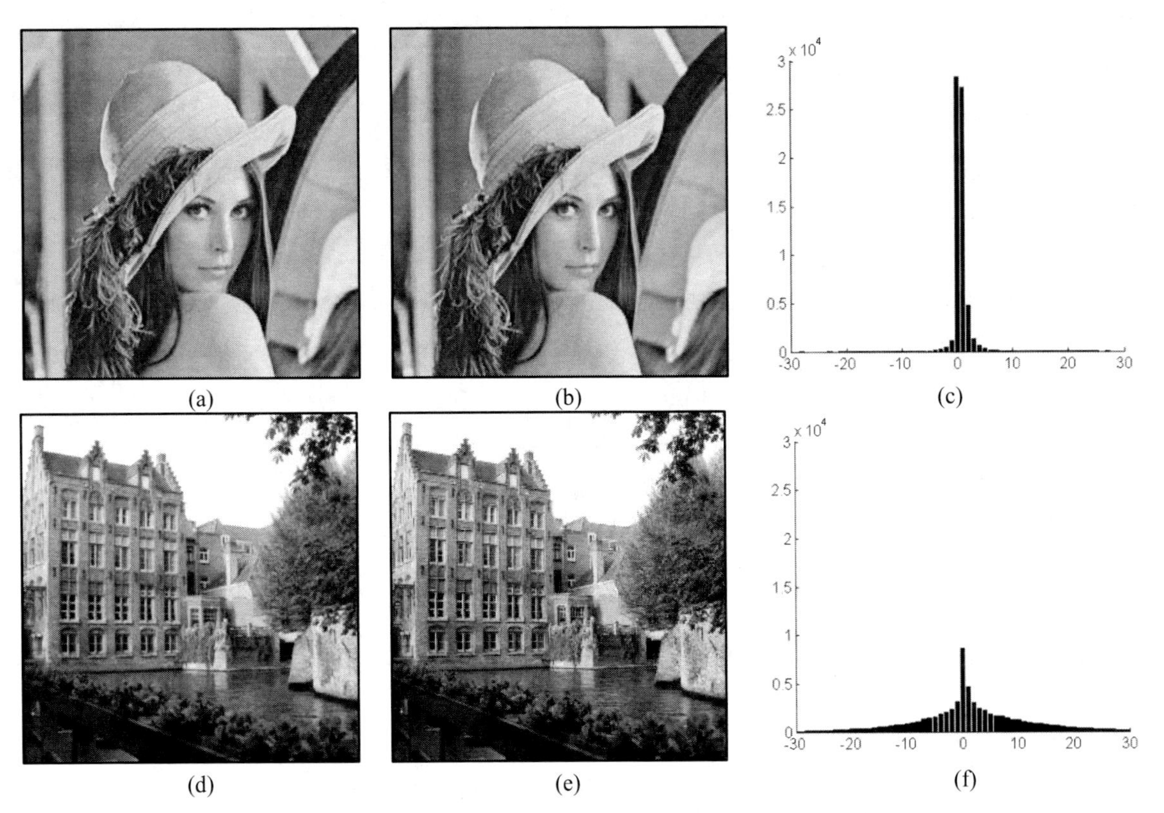

Fig. 3. (a) Uncompressed Lena image (b) Fractal Lena image (c) The histogram of pixel differences between (a) and (b) (d) Uncompressed Building image (e) Fractal Building image (c) The histogram of pixel differences between (d) and (e).

3.2 Embedding Algorithm

In a fractal code table, four coefficients, row index, column index, α, and β, are stored for each range block. The proposed steganographic process embeds the stego message within and/or after the encoding procedure, where the stego message is represented as a set of binary bit, b, called stego binary bits. Let scaling factor α predefined contains {0, 0.1, 0.2, 0.3, . . . , 0.9}. If the embedding bit is 0, let α_e = {0, 0.2, 0.4, 0.6, 0.8}. The error term and shifting factor β are calculated as

$$\tilde{\boldsymbol{\varepsilon}}_e = \mathbf{r}_i - \left(\alpha_e \tilde{\mathbf{d}}_j + \beta_e\right)$$

where

$$\beta_e = \frac{1}{k}\sum_i \mathbf{r}_i - \alpha_e \frac{1}{l}\sum_j \tilde{\mathbf{d}}_j , \quad \beta = \beta_e$$

Hence,

$$\varepsilon = \tilde{\boldsymbol{\varepsilon}}_e' \cdot \tilde{\boldsymbol{\varepsilon}}_e$$

If the embedding is 1, let $\alpha_o = \{0.1, 0.3, 0.5, 0.7, 0.9\}$. The error term and shifting factor β are calculated as

$$\tilde{\boldsymbol{\varepsilon}}_o = \mathbf{r}_i - \left(\alpha_o \tilde{\mathbf{d}}_j + \beta_o\right)$$

where

$$\beta_o = \frac{1}{k}\sum_i \mathbf{r}_i - \alpha_o \frac{1}{l}\sum_j \tilde{\mathbf{d}}_j, \quad \beta = \beta_o$$

Therefore,

$$\varepsilon = \tilde{\boldsymbol{\varepsilon}}_o' \cdot \tilde{\boldsymbol{\varepsilon}}_o$$

The set of scaling factors can be denoted as $\alpha = \{\alpha_e, \alpha_o\}$. When identifying the best estimate of $\mathbf{R}_i$, use the appropriate α ensuring the bit is properly embedded. This implies the embedding in α requires the steganographic process to be executed during the encoding procedure. A key is incorporated for randomness embedding, i.e., to select which range block for embedding. The highest embedding capacity apparently then depends on the size of image and range blocks. The larger the number of range blocks, the higher the embedding capacity.

For hiding in the shifting coefficients β, the embedding of bits is performed after encoding, i.e., after generating the fractal code table. The β coefficients are $\in \mathbb{R}$, which allow the embedding to occur in each digit location respectively, for example, tens, ones, tenths, hundredths, thousandths, etc. Note that the higher digit location is used for embedding results in lower image quality. Hence, to increase the capacity, the floating points can be used for embedding as well but lower compression capability. A simple least significant bit embedding method is used for demonstration in Section 4. Other embedding methods, such as matrix embedding, may be utilized. This embedding method rounds the randomly selected shifting coefficient $\hat{\beta}$ to be $\in \mathbb{Z}$. The random sequence of the $\hat{\beta}$'s index is denoted as x, where x is the same length as $\hat{\beta}$. Let $\tilde{x}$ represent the index of the number of bits, b, to be embedded such that $\tilde{x}$ and b are the same length. Equation (8) describes the embedding of the stego binary bits, b, into the carrier, shifting coefficients $\hat{\beta}$.

$$\hat{\beta}_{\tilde{x}} = f\left(\hat{\beta}_{\tilde{x}}, b\right) \tag{8}$$

where $f\left(\hat{\beta}_{\tilde{x}}, b\right)$ is defined as

$$f\left(\hat{\beta}_{\tilde{x}}, b\right) = \begin{cases} \hat{\beta}_{\tilde{x}}, & \text{if } \hat{\beta}_{\tilde{x}} \text{ is even and } b = 0 \\ \hat{\beta}_{\tilde{x}} +/-1, & \text{if } \hat{\beta}_{\tilde{x}} \text{ is even and } b = 1 \\ \hat{\beta}_{\tilde{x}} +/-1, & \text{if } \hat{\beta}_{\tilde{x}} \text{ is odd and } b = 0 \\ \hat{\beta}_{\tilde{x}}, & \text{if } \hat{\beta}_{\tilde{x}} \text{ is odd and } b = 1 \end{cases}$$

3.3 Retrieving Algorithm

As the fractal compression functions, the steganographic process is incorporated. A stego fractal code table is generated. As seen in Fig. 2, the fractal code table is stored or transmitted through a public channel in order for decoding. Hence, a receiver takes delivery of the fractal code table and is able to extract the stego message with specified keys during encoding for randomness. By knowing the order of the embedding the bits are retrieved and the message is reconstructed using the collected bits. The stego key in the random sequence seed for the index $\tilde{x}$ of $\hat{\beta}$. Equation (9) represents the bit extraction for α and Equation (10) represents the bit extraction for $\hat{\beta}$

$$\hat{b}_\alpha = \tilde{f}_\alpha(\alpha) \tag{9}$$

where $\tilde{f}_\alpha(\alpha)$ is defined as

$$\tilde{f}_{\alpha}(\alpha)=\begin{cases}0 & \text{if } \alpha=\{0.0,0.2,0.4,0.6,0.8\}\\ 1 & \text{if } \alpha=\{0.1,0.3,0.5,0.7,0.9\}\end{cases}$$

and

$$\hat{b}_{\beta}=\tilde{f}_{\beta}\left(\hat{\beta}_{\tilde{x}}\right) \tag{10}$$

where $\tilde{f}_{\beta}\left(\hat{\beta}_{\tilde{x}}\right)$ is defined as

$$\tilde{f}_{\beta}\left(\hat{\beta}_{\tilde{x}}\right)=\begin{cases}0 & \text{if } \hat{\beta}_{\tilde{x}} \text{ is even}\\ 1 & \text{if } \hat{\beta}_{\tilde{x}} \text{ is odd}\end{cases}$$

4. COMPUTER SIMULATION

The image sets used for the demonstration are from BOWS2 [5] and USC-SIPI [16] image database. The stego message is a portion of prologue from the book "The Da Vinci Code", which describes a crime scene. A total of 100 cover grayscale images of size 256 × 256 are used. The amount of embedding is denoted in percentage p%, indicating that there is p% of the fractal coefficients containing the secret message in bits. For example, p = 25% for α embedding indicates there is 25% in the scaling factor which contain the stego message. A key used by a pseudo-random number generator is incorporated within the steganographic process. For visualized demonstration, the Lena image and the Building image are presented.

Fig. 4(a)-(f) show several stego fractal Lena images with different percentage of embedding in the scaling factor α and the shifting factor β, as well as in Fig. 5(a)-(f) with stego fractal Building images. As can be seen, the decoded fractal images in the spatial domain contain a secret message and are perceptually unnoticeable. Fractal Lena and Building images with 0% embedding are shown in Fig. 3(b) and 3(e) considered as clean images. In Fig. 4(g)-(l) and Fig. 5(g)-(l) show the difference images between each fractal image in Fig. 4(a)-(f) and Fig. 5(a)-(f) and their corresponding clean images. As can be seen, the α embedding mainly results in an influence on edge pixels. Since the neighboring pixels around edges have significant contrast comparing to smooth regions, the steganographic content is unobservable when embedding on edges. The β embedding is more distributive in spatial domain after decoding. The advantage is the flexibility that the capacity can be increased by considering the floating point digits. Note that the β embedding demonstrated here only incorporates the least significant bit of the integer values. If tenths and/or thousands floating points are used during the steganographic process, the resulting stego image after decoding will not be apparently different.

Fig. 6 and Fig. 7 are histograms of pixel values and pixel value differences corresponding to the images shown in Fig. 4 and Fig. 5 along with the uncompressed and fractal Lena and Building images without embedding. Note that the distributions of fractal stego images keep the shape when compared to both the uncompressed and fractal image with 0% embedding. Table 1 shows an average of RMSE between 100 fractal stego images and their corresponding clean images. The larger differences occur in 25% α embedding due to the embedding method described in Section 3. If different embedding methods are incorporated for β embedding, then the characteristics of the results will be different. However, the difference images and pixel value histograms will not have significant differences since the stego message has been disseminated throughout spatial domain.

Table 1. Average error measurement after embedding 100 images.

Fractal Stego Images RMSE		β embedding				
		0%	25%	50%	75%	100%
α embedding	0%	0.000	1.901	2.102	2.314	2.478
	25%	4.292	4.415	4.517	4.603	4.667
	50%	3.791	3.857	3.929	3.985	4.041
	75%	3.464	3.540	3.583	3.623	3.680
	100%	3.742	3.798	3.840	3.898	3.927

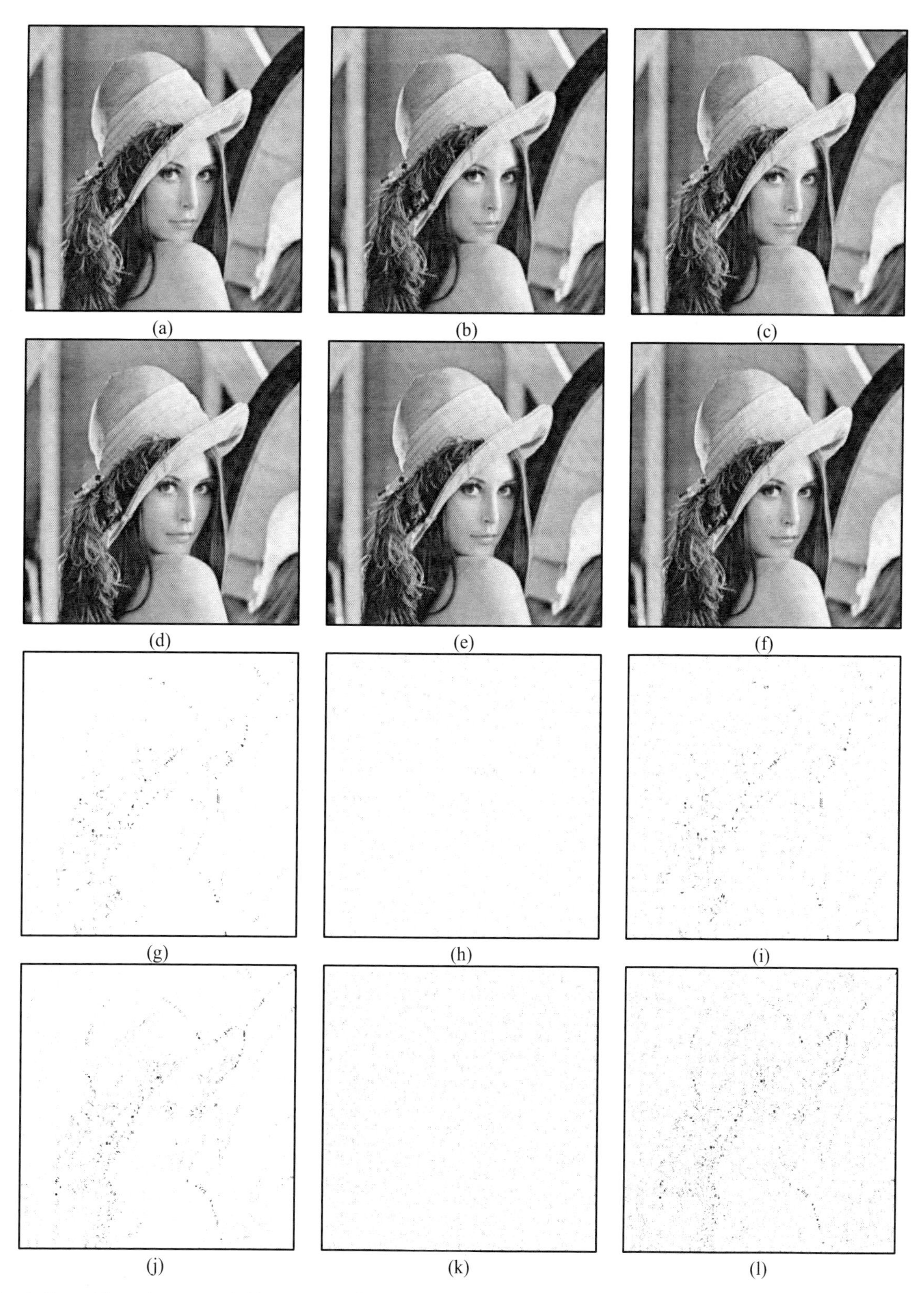

Fig. 4. Fractal Lena images and the corresponding difference images (a)(g) 50% α embedding (b)(h) 50% β embedding (c)(i) 50% α and 50% β embedding (d)(j) 100% α embedding (e)(k) 100% β embedding (f)(l) 100% α and 100% β embedding.

Fig. 5. Fractal Building images and the corresponding difference images (a)(g) 25% α embedding (b)(h) 100% β embedding (c)(i) 25% α and 100% β embedding (d)(j) 50% α embedding (e)(k) 75% β embedding (f)(l) 50% α and 75% β embedding.

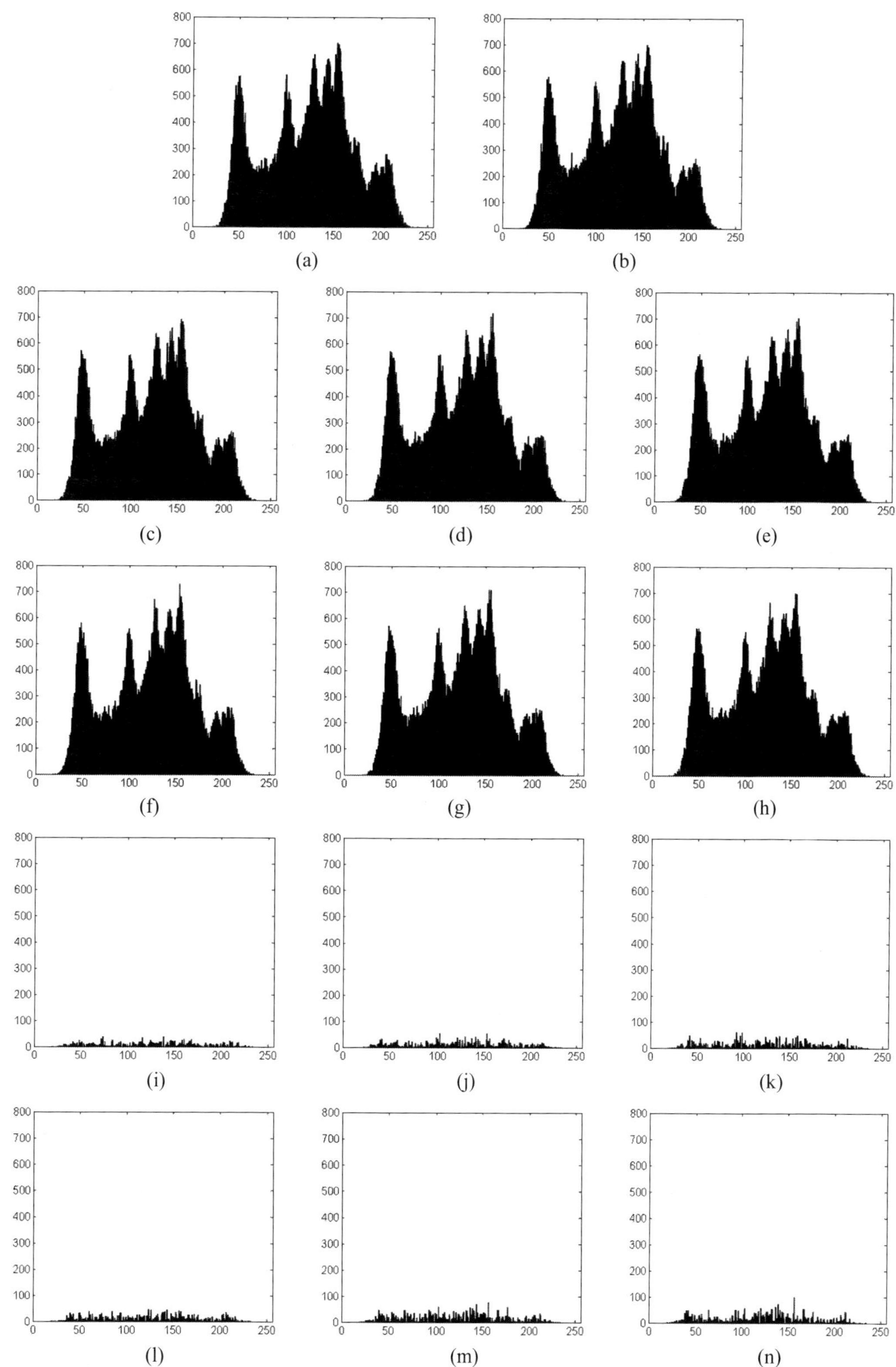

Fig. 6. The histogram of pixel values and pixel value differences between the clean and the stego fractal Lena image (a) Uncompressed Lena image (b) Fractal Lena image (c)(i) 50% α embedding (d)(j) 50% β embedding (e)(k) 50% α and 50% β embedding (f)(l) 100% α embedding (g)(m) 100% β embedding (h)(n) 100% α and 100% β embedding.

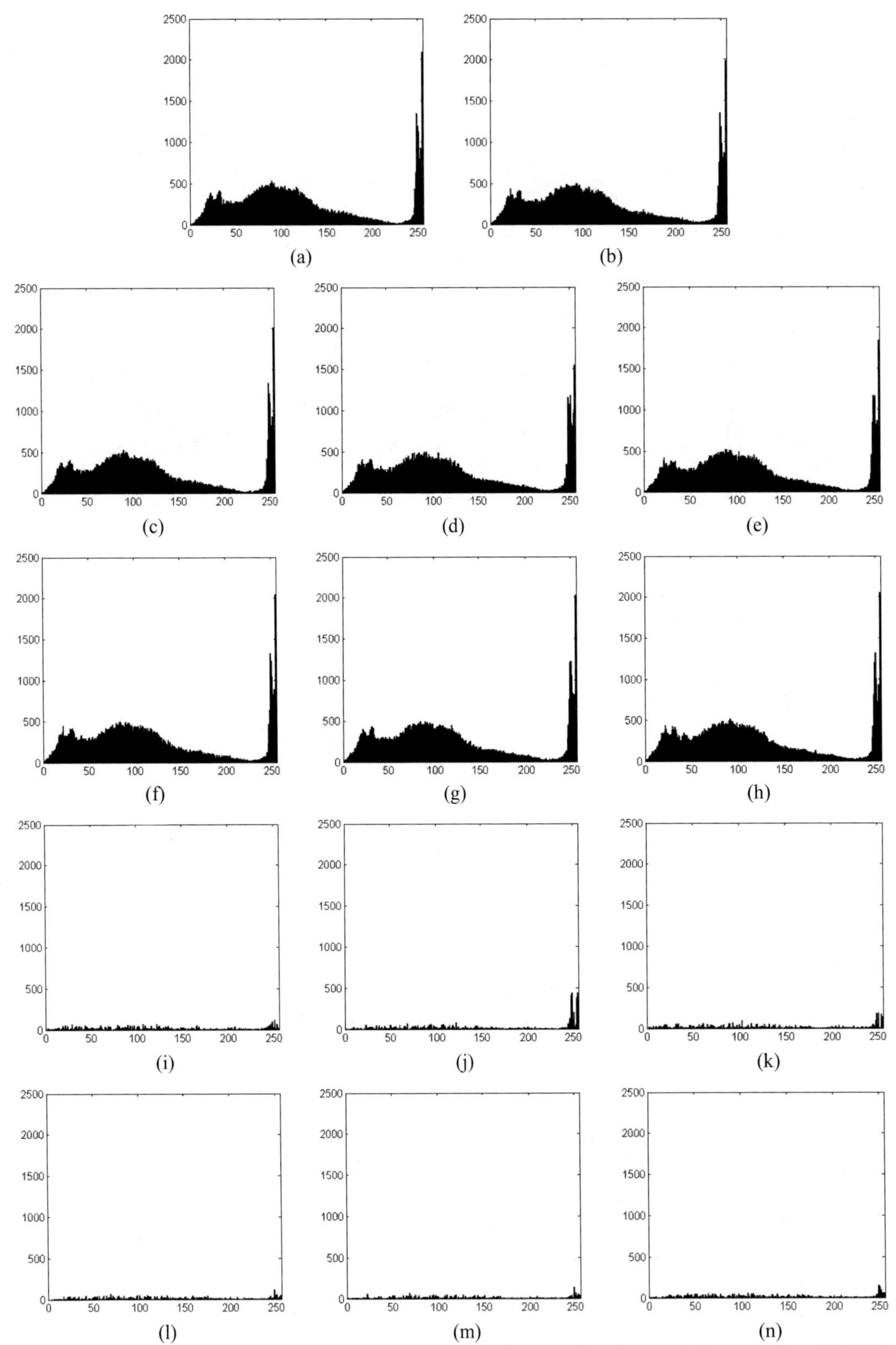

Fig. 7. The histogram of pixel values and pixel value differences between the clean and the stego fractal Building image (a) Uncompressed Building image (b) Fractal Building image (c)(i) 25% α embedding (d)(j) 100% β embedding (e)(k) 25% α and 100% β embedding (f)(l) 50% α embedding (g)(m) 75% β embedding (h)(n) 50% α and 75% β embedding.

5. CONCLUSION

This paper presented a new fractal steganography method for securing pertinent information. The proposed steganographic algorithm hides the secret information in both contrast/scaling and brightness/shifting coefficients in the fractal code table. The steganographic algorithm may be used in different fractal compression algorithms, for instance, various image partitioning techniques and different metrics. The contributions of the work in this paper include: (1) an embedding method (1) considers a new image compression domain, (2) uses a simple fractal encoding method [18] as a basis for other fractal encoders, (3) allows the capacity to be increased based on the message to be hidden and (4) disseminates the hidden information throughout the image in the spatial domain resulting in visually undistorted stego fractal images.

ACKNOWLEDGEMENT

A special thank goes to the Department of Electrical and Computer Engineering, The University of Texas at San Antonio for partially funding this research.

REFERENCES

[1] Agaian, S. S. and Susmilch, J. M., "Fractal steganography using artificially generated images," Proc. of SPIE 6982, 69820B-1-69820B-9 (2008).

[2] Backbone Security, http://www.sarc-wv.com/.

[3] Barnsley, M. F., Fractals Everywhere, Academic Press, Boston (1988).

[4] Barnsley, M. F., "Fractal image compression," Notices of the AMS 43(6), 657-662 (1996).

[5] Break Our Watermarking System, 2nd Ed., Retrieved from http://bows2.gipsa-lab.inpg.fr/, (2008).

[6] Fisher, Y., Fractal Image Compression, Springer Verlag, New York (1997).

[7] Davern, P. and Scott, M., "Fractal based image steganography," Proc. of the First Intl. Workshop on Information Hiding, Lecture Notes in Computer Science 1174, 279-294 (1996).

[8] Hutchinson, J., "Fractals and self-similarity," Indiana Univ. J. Math 30, 713-747 (1981).

[9] Hussain, S. J., Siddiqui, M. S., Boghani, A. A. and Daniyal, A., "IFS-based image coding," IEEE Multitopic Conference, 152-156 (2006).

[10] Jacquin, A. E., "Image coding based on a fractal theory of iterated contractive image transformations," IEEE Trans. on Image Processing 1(1), 18-30 (1992).

[11] Jacquin, A. E., "Fractal image coding: a review," Proc. of the IEEE 81(10), 1451-1465 (1993).

[12] Khadivi, M. R., "IFS and its use in cryptography and steganography," Department of Mathematics, Jackson State University.

[13] Kharrazi, M., Sencar, H. T., and Memon, N, "Performance study of common image steganography and steganalysis techniques," Journal of Electronic Imaging, 15 (4), 041104-1-041104-16 (2006).

[14] Puate, J. and Jordan, F. D., "Using fractal compression scheme to embed a digital signature into an image," Proc. of SPIE 2915, 108-118 (1997)

[15] Saupe, D. and Hamzaoui, R., "A review of the fractal image compression literature," Computer Graphics 28(4), 268-276 (1994).

[16] The USC-SIPI Image Database, Retrieved from http://sipi.usc.edu/database/, (2007).

[17] Wohlberg, B. and De Jager, G., "A review of the fractal image coding literature," IEEE Trans. on Image Processing 8(12), 1716-1729 (1999).

[18] Xu, C. M. and Zhang Z. Y., "A fast fractal image compression coding method," Journal of Shanghai University, 5(1), 57-59 (2001).

*axf710@my.utsa.edu; phone 210-458-5594; fax 210-458-5947

Author Index

Numbers in the index correspond to the last two digits of the six-digit citation identifier (CID) article numbering system used in Proceedings of SPIE. The first four digits reflect the volume number. Base 36 numbering is employed for the last two digits and indicates the order of articles within the volume. Numbers start with 00, 01, 02, 03, 04, 05, 06, 07, 08, 09, 0A, 0B ... 0Z, followed by 10-1Z, 20-2Z, etc.